Lecture Notes in Applied and Computational Mechanics

Volume 86

Series editors

This series aims to report new developments in applied and computational mechanics—quickly, informally and at a high level. This includes the fields of fluid, solid and structural mechanics, dynamics and control, and related disciplines. The applied methods can be of analytical, numerical and computational nature.

More information about this series at http://www.springer.com/series/4623

Jurica Sorić · Peter Wriggers
Olivier Allix
Editors

Multiscale Modeling of Heterogeneous Structures

Springer

Editors
Jurica Sorić
University of Zagreb
Zagreb
Croatia

Olivier Allix
LMT Cachan
Cachan
France

Peter Wriggers
Institute of Continuum Mechanics
Leibniz Universität Hannover
Hannover
Germany

ISSN 1613-7736 ISSN 1860-0816 (electronic)
Lecture Notes in Applied and Computational Mechanics
ISBN 978-3-319-65462-1 ISBN 978-3-319-65463-8 (eBook)
https://doi.org/10.1007/978-3-319-65463-8

Library of Congress Control Number: 2017949492

Printed on acid-free paper

This Springer imprint is published by Springer Nature
The registered company is Springer International Publishing AG
The registered company address is: Gewerbestrasse 11, 6330 Cham, Switzerland

Preface

Many engineering materials have a heterogeneous structure, especially at the microscopic scale. These are often referred to as multiphase materials, composite or heterogeneous materials. From an engineering point of view, multiphase materials are desirable because they can be tailor-made to take advantage of particular properties of each constituent. The size, shape, spatial distribution, volume fraction, and properties of the constituents at microstructural level have a significant impact on the behavior of material properties observed at the macroscale. Additionally, the external loading applied at macroscale might cause changes in the microstructural morphology, e.g., void formation, damage as well as cracking, which can put structural integrity at risk. In order to assess structural integrity and to predict structural lifetime, an analysis of the evolving microstructure is necessary. An efficient computational strategy enabling more realistic material description as well as deformation response is still a challenge in computational mechanics. Various multiscale techniques have been developed that model materials at multiple levels. Moreover, various modern experimental techniques provide access to a detailed characterization of the internal structure and processes taking place in materials at small scales, paving the way to new routes for model validation.

This book contains 18 papers that resulted from selected presentations at the workshop "Multiscale Modeling of Heterogeneous Structures" held September 21–23, 2016 in Dubrovnik, Croatia. The workshop focused on multiscale approaches and homogenization procedures as well as damage evaluation and crack initiation. Recent advances in the analysis and discretization of heterogeneous materials were addressed. The state of the art in this research area was highlighted with respect to different computational methods, software development, and applications to engineering structures.

The papers were allotted to four topics: Composites, Computational Solution Approaches, Gradient Enhanced Modeling, and Multiphysics and associated experimental techniques. The topic Composites covers defects in composite materials including their numerical and experimental investigations. Elastic as well as elastoplastic constitutive models are considered, where the modeling has been performed at macro- and microlevels. The second group of the papers is more focused on novel computational schemes applied at the different scales. The validation of numerical results has been discussed. The quasi-brittle and the ductile damage using the gradient enhanced approach are considered in the frame of the topic Gradient Enhanced Modeling. Finally, the thermoplasticity, the solid-liquid mixture as well as the ferroelectric models are discussed in the fourth topic.

The workshop was held under the auspices of the German Association for Computational Mechanics (GACM), the Central European Association for Computational Mechanics (CEACM), the ENS Cachan, the Leibniz Universität Hannover, the Faculty of Mechanical Engineering, and Naval Architecture of the University of Zagreb. It was supported by the Alexander von Humboldt Foundation, the Deutsch-Französische Hochschule, and the Deutsche Forschungsgemeinschaft. The editors express their deep gratitude to all sponsoring institutions. Furthermore, the editors would like to thank Ms. Schulte and Dr. Lesičar for their engagement in the organization of the workshop as well as Dr. Weißenfels for his valuable assistance in preparing the book.

Zagreb, Croatia — Jurica Sorić
Hannover, Germany — Peter Wriggers
Cachan, France — Olivier Allix

Contents

Part I
Composites

Evolution of Failure Mechanisms in Composite Shell Structures Using Different Models

Werner Wagner and Friedrich Gruttmann

Abstract Modelling of structures on different scales has been a popular subject in the past. Within such a strategy the structural behaviour is modeled on a macro-level, describing the structure itself, whereas the material behaviour is modeled on a micro-level. Here typically RVEs are used. The proper choice of boundary conditions for the RVE is a difficult task in case of shell structures. Here, results have been presented for homogeneous and layered structures for composite materials in (Gruttmann & Wagner, Int J Num Meth Eng 94:1233–1254, 2013) [10]. In the present paper we discuss the influence of material nonlinear behaviour, especially the damage behaviour of fiber reinforced polymers, within the above described setting in comparison to other modeling techniques.

1 Introduction

Finite shell elements which are based on the first–order shear deformation theory are in general able to describe the global deformation behaviour of thin plate and shell structures. In [20] we presented results that show remarkable robustness of mixed formulations in nonlinear applications. Modifications of such mixed formulations to layered structures are developed in [9]. With respect to damage behaviour a layer-wise numerical integration has to be used. A similar approach can be achieved in case of solid shell elements, e.g. [14, 15]. Choosing one element in thickness direction

W. Wagner (✉)
Institut für Baustatik, Karlsruher Institut für Technologie,
Kaiserstr. 12, 76131 Karlsruhe, Germany
e-mail: w.wagner@kit.edu

F. Gruttmann
Fachgebiet Festkörpermechanik, Technische Universität Darmstadt,
Franziska-Braun-Str. 7, 64287 Darmstadt, Germany
e-mail: gruttmann@mechanik.tu-darmstadt.de

J. Sorić et al. (eds.), *Multiscale Modeling of Heterogeneous Structures*,
Lecture Notes in Applied and Computational Mechanics 86,
https://doi.org/10.1007/978-3-319-65463-8_1

again a layerwise approach has to be added. However for some stress components only an average shape through the thickness can be obtained. Various methods have been developed to obtain the complicated local stress state in inhomogeneous thin structures. So-called multi-director shell formulations with an appropriate number of global degrees of freedom at the nodes yield approximate solutions of the three-dimensional boundary value problem, e.g. [8, 18]. The application of brick elements or solid shell elements provides likewise a computationally expensive approach, e.g. [14, 15], but allow the description of warping or other effects in the cross section of the shell. For laminates each layer must be discretised with several elements in thickness direction to obtain satisfactory results. The enhancement of the displacement field by layer-wise linear (zig-zag) functions through the thickness, see e.g. [1], could be another option, which leads to a more precise deformation behaviour. New actual promising results for locally extended shell formulations can be found in [11] for the elastic case. A further alternative is the treatment of shells as a homogeneous continuum in a 2D shell environment with effective properties obtained through a homogenisation procedure to avoid large-scale computations. A large number of papers exists on computational homogenisation methods for general heterogeneous materials, see e.g. [5, 23] for a survey and new developments. Based on the formulation in [20] we have derived a two-scale model with a variational formulation and an associated linearisation for the coupled global–local boundary value problem in [10] and an adaptive application to local elasto-plastic material in [22]. In this paper we present the applicability to damaged composite shell structures. For that we compare different discretisation models

- layered solid shell models [14, 15]
- layerwise solid shell models [14, 15]
- layerwise shell models [9, 20]
- shell models with an internal FE^2-approach [2, 6, 10].

To do this we describe in the next two sections briefly the main equations of a two-scale shell model and the main equations of damage models of Hashin, e.g. [7, 12], Puck, e.g. [16] and Cuntze, e.g. [3, 4].

2 Two-Scale Shell Model

2.1 *Theoretical Background*

At first the basic equations of a Reissner-Mindlin shell model are summarised. Based on a reference surface the thickness coordinate $\xi^3 = z$ is defined, where h^- and h^+ are the z-coordinates of the outer surfaces. The shell is loaded statically by loads $\bar{\mathbf{p}}$ in Ω and by boundary forces $\bar{\mathbf{t}}$ on Γ_σ. The displacement field of the Reissner-Mindlin theory is obtained with

$$\bar{\mathbf{u}} = \bar{\mathbf{u}}_0 + z\,(\bar{\mathbf{d}} - \bar{\mathbf{D}}) \quad \bar{\mathbf{u}}_0 = \mathbf{x} - \mathbf{X}\,, \tag{1}$$

where $\mathbf{x}$, $\mathbf{X}$ denote the position vectors of the initial and current reference surface, respectively. The unit director vectors are denoted by $\bar{\mathbf{D}}$, $\bar{\mathbf{d}}$, where $\bar{\mathbf{d}}$ is a function of the rotational parameters $\bar{\omega}$.

The shell strains are derived from the Green-Lagrangian strain tensor using kinematic assumption (1) and are arranged in a vector as

$$\varepsilon(\bar{\mathbf{u}}_0, \bar{\omega}) = [\varepsilon_{11}, \varepsilon_{22}, 2\,\varepsilon_{12}, \kappa_{11}, \kappa_{22}, 2\,\kappa_{12}, \gamma_1, \gamma_2]^T\,. \tag{2}$$

Furthermore the vector of stress resultants σ with membrane forces $n^{\alpha\beta}$, bending moments $m^{\alpha\beta}$ and shear forces q^{α} is introduced via

$$\sigma = [n^{11}, n^{22}, n^{12}, m^{11}, m^{22}, m^{12}, q^1, q^2]^T\,. \tag{3}$$

Further details on the remarkable robust mixed formulation are described in [20] and [9]. According to Fig. 1 a representative volume element (RVE) at an integration point i of a particular finite shell element is introduced. The domain $\mathscr{B}_i$ extends through the total thickness h of the shell and has the size $l_x \times l_y \times h$. The displacement field is split in an averaged part $\bar{\mathbf{u}}$ and a fluctuation part $\tilde{\mathbf{u}}$.

$$\mathbf{u} = \bar{\mathbf{u}} + \tilde{\mathbf{u}}\,. \tag{4}$$

The averaged displacements $\bar{\mathbf{u}}$ according to (1) exhibit a linear shape of the thickness coordinate, whereas $\tilde{\mathbf{u}}$ describes warping and thickness change. The weak form of equilibrium of the coupled problem can now be written with $\mathbf{v} = [\bar{\mathbf{u}}_0, \bar{\omega}, \mathbf{u}]^T$ and associated admissible variations

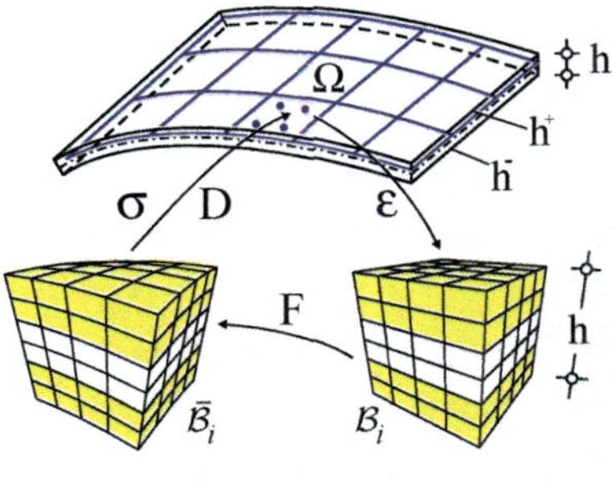

Fig. 1 Computational homogenisation of a layered shell

$$
\begin{aligned}
g(\mathbf{v},\delta\mathbf{v}) = & \int_{\Omega_1} (\sigma\cdot\delta\varepsilon - \bar{\mathbf{p}}\cdot\delta\bar{\mathbf{u}}_0)\,dA_1 + \int_{\Omega_2} (\sigma\cdot\delta\varepsilon - \bar{\mathbf{p}}\cdot\delta\bar{\mathbf{u}}_0)\,dA_2 \\
& + \sum_{e=1}^{ne2}\sum_{i=1}^{NGP} \frac{1}{A_i}\int_{\Omega_i}\int_{h^-}^{h^+} \mathbf{S}\cdot\delta\mathbf{E}\,\bar{\mu}\,dz\,dA - \int_{\Gamma_\sigma} \bar{\mathbf{t}}\cdot\delta\bar{\mathbf{u}}_0\,ds = 0\,.
\end{aligned} \tag{5}
$$

The structure is divided in parts Ω_1 without and Ω_2 with a two-scale model, respectively. Furthermore $ne1$ and $ne2$ denote the associated number of shell elements within a discretisation. NGP is the number of Gauss points for each element and $A_i = l_x\,l_y$ is the reference area of the RVE. On the RVE $\mathbf{S}$ denotes the Second Piola-Kirchhoff stress tensor with $\mathbf{P} = \mathbf{F}\,\mathbf{S}$ and the virtual Green-Lagrangian strain tensor is introduced via $\delta\mathbf{E} = \frac{1}{2}(\delta\mathbf{F}^T\mathbf{F} + \mathbf{F}^T\delta\mathbf{F})$. For the finite element formulation of the next section we need to derive the linearisation of Eq. (5). With conservative loads $\bar{\mathbf{p}}$ and $\bar{\mathbf{t}}$ one obtains

$$
L\,[g(\mathbf{v},\delta\mathbf{v}),\Delta\mathbf{v}] := g(\mathbf{v},\delta\mathbf{v}) + Dg\cdot\Delta\mathbf{v} \tag{6}
$$

where $g(\mathbf{v},\delta\mathbf{v})$ is given in (5) and

$$
\begin{aligned}
Dg\cdot\Delta\mathbf{v} = & \int_{\Omega_1} (\Delta\sigma\cdot\delta\varepsilon + \sigma\cdot\Delta\delta\varepsilon)\,dA_1 + \int_{\Omega_2} (\Delta\sigma\cdot\delta\varepsilon + \sigma\cdot\Delta\delta\varepsilon)\,dA_2 \\
& + \sum_{e=1}^{ne2}\sum_{i=1}^{NGP} \frac{1}{A_i}\int_{\Omega_i}\int_{h^-}^{h^+} (\Delta\mathbf{S}:\delta\mathbf{E} + \mathbf{S}:\Delta\delta\mathbf{E})\,dz\,dA
\end{aligned} \tag{7}
$$

with $\Delta\sigma = \mathbf{D}\Delta\varepsilon$, $\Delta\mathbf{S} = \mathbf{C}\Delta\mathbf{E}$ and $\Delta\delta\mathbf{E} = \frac{1}{2}(\delta\mathbf{F}^T\Delta\mathbf{F} + \Delta\mathbf{F}^T\delta\mathbf{F})$. The material matrix $\mathbf{C}$ is a standard output of a library of constitutive laws in a material description. The linearised virtual shell strains $\Delta\delta\varepsilon$ are derived for finite rotations in [20]. The stress resultant vector σ and the matrix of linearised stress resultants $\mathbf{D}$ are specified within the finite element formulation in the next section.

2.2 *Finite Element Formulation*

The reference surface of the shell is discretised with $ne = ne1 + ne2$ quadrilateral isoparametric shell elements, using bilinear shape functions $N_I(\xi,\eta)$ which are arranged in the matrix $\mathbf{N}$. The nodal degrees of freedom are three displacements and two or three rotations. Inserting these interpolation functions into the linearised weak form (6) considering (5) and (7) yields

$$L\,[g(\mathbf{v}^h,\delta\mathbf{v}^h),\Delta\mathbf{v}^h] = \sum_{e=1}^{ne1} \delta\mathbf{v}_e^G \mathbf{k}_e^G \Delta\mathbf{v}_e^G + \mathbf{f}_e^G + \sum_{e=1}^{ne2}$$

$$\begin{bmatrix} \delta\mathbf{v}^G \\ \delta\mathbf{V}_1 \\ \vdots \\ \delta\mathbf{V}_i \\ \vdots \\ \delta\mathbf{V}_{NGP} \end{bmatrix}_e^T \left\{ \begin{bmatrix} \mathbf{k}^G & \mathbf{0} & \vdots & \mathbf{0} & \vdots & \mathbf{0} \\ \mathbf{0} & \mathbf{K}_1^L & \vdots & \mathbf{0} & \vdots & \mathbf{0} \\ \dots & \dots & \ddots & \mathbf{0} & \dots & \dots \\ \mathbf{0} & \mathbf{0} & \mathbf{0} & \mathbf{K}_i^L & \mathbf{0} & \mathbf{0} \\ \dots & \dots & \dots & \mathbf{0} & \ddots & \dots \\ \mathbf{0} & \mathbf{0} & \dots & \mathbf{0} & \dots & \mathbf{K}_{NGP}^L \end{bmatrix}_e \begin{bmatrix} \Delta\mathbf{v}^G \\ \Delta\mathbf{V}_1 \\ \vdots \\ \Delta\mathbf{V}_i \\ \vdots \\ \Delta\mathbf{V}_{NGP} \end{bmatrix}_e + \begin{bmatrix} \mathbf{f}^G(\sigma_i) \\ \mathbf{F}_1^L \\ \vdots \\ \mathbf{F}_i^L \\ \vdots \\ \mathbf{F}_{NGP}^L \end{bmatrix}_e \right\} \tag{8}$$

The indices G and L refer to the global and local boundary value problems, respectively. The matrices of the first row in (8) follow from the global part of the linearised weak form. The element residual vector and the tangential element stiffness matrix read

$$\begin{aligned} \mathbf{f}^G(\sigma_i) &= \int_{\Omega_e} (\mathbf{B}^T\sigma - \mathbf{N}^T\bar{\mathbf{p}})\,dA - \int_{\Gamma_{\sigma e}} \mathbf{N}^T\bar{\mathbf{t}}\,ds\mathbf{k}^G(\mathbf{D}_i) \\ &= \int_{\Omega_e} (\mathbf{B}^T\mathbf{D}\,\mathbf{B} + \mathbf{G})\,dA \end{aligned} \tag{9}$$

where the matrices $\mathbf{B}$ and $\mathbf{G}$ are derived in [20]. The vector of stress resultants σ_i and linearised stress resultants $\mathbf{D}_i$ are specified below. The matrices of the second to the last row in (8) are associated with the local boundary value problems at Gauss points $1 \le i \le NGP$ of shell element e and occur only, if a two-scale model is used. A local boundary value problem can be defined at Gauss point i

$$\delta\mathbf{V}_i^T\,(\mathbf{K}_i^L\,\Delta\mathbf{V}_i + \mathbf{F}_i^L) = \frac{1}{A_i}\sum_{e=1}^{Ne} \delta\mathbf{v}_e^T\,(\mathbf{k}_e^L\,\Delta\mathbf{v}_e + \mathbf{f}_e^L)\,. \tag{10}$$

Here, the total number of elements used for the discretisation of the RVE is denoted by Ne. The element residual vector $\mathbf{f}_e^L$ and the tangential element stiffness matrix $\mathbf{k}_e^L$ read

$$\mathbf{f}_e^L = \int_{(V_e)} \tilde{\mathbf{B}}^T\mathbf{S}\,dV \qquad \mathbf{k}_e^L = \int_{(V_e)} (\tilde{\mathbf{B}}^T\mathbf{C}\,\tilde{\mathbf{B}} + \tilde{\mathbf{G}})\,dV \tag{11}$$

where $\tilde{\mathbf{B}}$ and $\tilde{\mathbf{G}}$ are the virtual strain displacement matrix and the geometrical matrix of 8-noded solid shell elements, respectively. The element displacement vector $\mathbf{v}_e$ is now split in a part $\mathbf{v}_\Omega$ with internal displacements and a part $\mathbf{v}_\Gamma$ which contains displacements on the boundary Γ_u of the RVE

$$\mathbf{v}_e = \begin{bmatrix} \mathbf{v}_\Omega \\ \mathbf{v}_\Gamma \end{bmatrix} = \begin{bmatrix} \mathbf{a}_e\,\mathbf{V}_i \\ \mathbf{A}_e\,\varepsilon_i \end{bmatrix}. \tag{12}$$

In Eq. (12) $\mathbf{a}_e$ is a standard assembly matrix. $\mathbf{A}_e$ is defined for *nel* nodes on the element with

$$\mathbf{A}_e = [\delta_1 \, \mathbf{A}_1, \ldots, \delta_I \, \mathbf{A}_I, \ldots, \delta_{nel} \, \mathbf{A}_{nel}]^T \tag{13}$$

with $\delta_I = 0$ for internal nodes and $\delta_I = 1$ for boundary nodes. Assuming small strains the relation of the boundary displacements to the shell strains ε can be written as

$$\begin{bmatrix} v_x \\ v_y \end{bmatrix}_I = \mathbf{A}_I(x, y, z) \, \varepsilon_i = \begin{bmatrix} x & 0 & \frac{1}{2} y & xz & 0 & \frac{1}{2} yz & z & 0 \\ 0 & y & \frac{1}{2} x & 0 & yz & \frac{1}{2} xz & 0 & z \end{bmatrix}_I \begin{bmatrix} \varepsilon_{11} \\ \varepsilon_{22} \\ 2\varepsilon_{12} \\ \kappa_{11} \\ \kappa_{22} \\ 2\kappa_{12} \\ \gamma_1 \\ \gamma_2 \end{bmatrix}_i \tag{14}$$

Based on the element displacement split (12) one can introduce submatrices of $\mathbf{k}_e^L$ and $\mathbf{f}_e^L$ in (10)

$$\begin{aligned}
&\delta \mathbf{V}_i^T \, (\mathbf{K}_i^L \, \Delta \mathbf{V}_i + \mathbf{F}_i^L) \\
&= \frac{1}{A_i} \sum_{e=1}^{Ne} \begin{bmatrix} \delta \mathbf{V}_i \\ \delta \varepsilon_i \end{bmatrix}^T \left\{ \begin{bmatrix} \mathbf{a}_e^T \, \mathbf{k}_{\Omega\Omega} \, \mathbf{a}_e & \mathbf{a}_e^T \, \mathbf{k}_{\Omega\Gamma} \, \mathbf{A}_e \\ \mathbf{A}_e^T \, \mathbf{k}_{\Gamma\Omega} \, \mathbf{a}_e & \mathbf{A}_e^T \, \mathbf{k}_{\Gamma\Gamma} \, \mathbf{A}_e \end{bmatrix}_e \begin{bmatrix} \Delta \mathbf{V}_i \\ \Delta \varepsilon_i \end{bmatrix} + \begin{bmatrix} \mathbf{a}_e^T \, \mathbf{f}_\Omega \\ \mathbf{A}_e^T \, \mathbf{f}_\Gamma \end{bmatrix}_e \right\} \\
&= \frac{1}{A_i} \begin{bmatrix} \delta \mathbf{V}_i \\ \delta \varepsilon_i \end{bmatrix}^T \left\{ \begin{bmatrix} \mathbf{K} & \mathbf{L} \\ \mathbf{L}^T & \mathbf{M} \end{bmatrix} \begin{bmatrix} \Delta \mathbf{V}_i \\ \Delta \varepsilon_i \end{bmatrix} + \begin{bmatrix} \mathbf{F}_\Omega \\ \mathbf{F}_\Gamma \end{bmatrix} \right\}.
\end{aligned} \tag{15}$$

The internal degrees of freedom $\Delta \mathbf{V}_i$ can now be eliminated from the set of equations which yields the final form of Eq. (15)

$$\delta \mathbf{V}_i^T \, (\mathbf{K}_i^L \, \Delta \mathbf{V}_i + \mathbf{F}_i^L) = \delta \varepsilon_i^T \, (\mathbf{D}_i \, \Delta \varepsilon_i + \sigma_i) \tag{16}$$

where the stress resultants and linearised stress resultants of Gauss point i are defined using $\mathbf{KX} = \mathbf{L}$ and $\mathbf{KY} = \mathbf{F}_\Omega$

$$\sigma_i = \frac{1}{A_i} (\mathbf{F}_\Gamma - \mathbf{L}^T \mathbf{Y}) \quad \mathbf{D}_i = \frac{1}{A_i} (\mathbf{M} - \mathbf{L}^T \mathbf{X}) \, . \tag{17}$$

Finally (16) is inserted into the linearised coupled global-local boundary value problem (8)

$$L\left[g(\mathbf{v}^h,\delta\mathbf{v}^h),\Delta\mathbf{v}^h\right]=\sum_{e=1}^{ne1}\delta\mathbf{v}_e^G\mathbf{k}_e^G\Delta\mathbf{v}_e^G+\mathbf{f}_e^G+\sum_{e=1}^{ne2}$$
$$\begin{bmatrix}\delta\mathbf{v}^G\\ \delta\varepsilon_1\\ \vdots\\ \delta\varepsilon_i\\ \vdots\\ \delta\varepsilon_{NGP}\end{bmatrix}_e^T\left\{\begin{bmatrix}\mathbf{k}^G(\mathbf{D}_i) & \mathbf{0} & \vdots & \mathbf{0} & \vdots & \mathbf{0}\\ \mathbf{0} & \mathbf{D}_1 & \vdots & \mathbf{0} & \vdots & \mathbf{0}\\ \dots & \dots & \ddots & \mathbf{0} & \dots & \dots\\ \mathbf{0} & \mathbf{0} & \mathbf{0} & \mathbf{D}_i & \mathbf{0} & \mathbf{0}\\ \dots & \dots & \dots & \mathbf{0} & \ddots & \dots\\ \mathbf{0} & \mathbf{0} & \dots & \mathbf{0} & \dots & \mathbf{D}_{NGP}\end{bmatrix}_e\begin{bmatrix}\Delta\mathbf{v}^G\\ \Delta\varepsilon_1\\ \vdots\\ \Delta\varepsilon_i\\ \vdots\\ \Delta\varepsilon_{NGP}\end{bmatrix}_e+\begin{bmatrix}\mathbf{f}^G(\sigma_i)\\ \sigma_1\\ \vdots\\ \sigma_i\\ \vdots\\ \sigma_{NGP}\end{bmatrix}_e\right\} \tag{18}$$

It has been shown, see e.g. [21], that the equivalence of macroscopic and microscopic stress power for shell structures, the so called Hill condition, holds

$$\frac{1}{h}\sigma\dot{\epsilon}=\frac{1}{V}\int_V \mathbf{S}:\dot{\mathbf{E}}\,dV=\frac{1}{V}\int_A \mathbf{t}\cdot\dot{\bar{\mathbf{u}}}\,dA \quad \text{with} \quad \dot{\bar{\mathbf{u}}}_I=\begin{bmatrix}\dot{\bar{u}}_x\\ \dot{\bar{u}}_y\end{bmatrix}_I=\mathbf{A}_I\,\dot{\epsilon} \tag{19}$$

3 Failure Models of Hashin, Puck and Cuntze

For fiber reinforced composite structures a range of failure mechanisms as fiber fracture (FF), inter fiber failure (IFF) and delamination may occur. Several failure models are proposed to describe FF (Fig. 2a, b) and IFF, usually each distinct in a tensile(t) and compressive(c) mode, (Fig. 2c, d). Improved models subdivide the IFF compressive mode in a shear dominant (Fig. 2d) and purely compression dominant mode (Fig. 2e). FF is the most severe failure and generally leads to structural collapse.

Based on a large number of models we choose the modified Hashin-model [7] as one of the most commercial ones. Furthermore we employ the Puck-model [16] and the Cuntze-model [3], which provide approaches to the physical aspects of the fracture. We describe the main equations. For a further discussion we refer to the original references.

3.1 *The Modified Hashin-Model*

One of the used failure models is the modified Hashin-model [7], based on the original version in [12]. The model consists of five failure criteria where the equations separate tensile and compressive matrix(M)- and fiber(F)-failure (M^t is used if $S_{22} > 0$, M^c if $S_{22} < 0$).

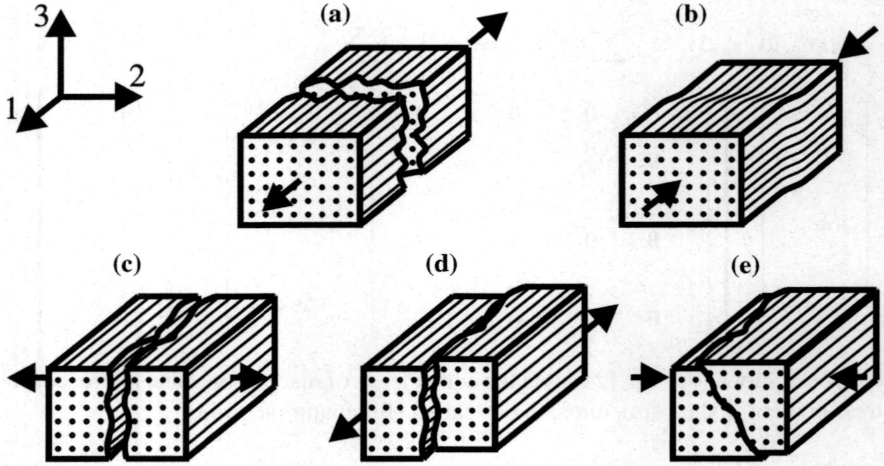

Fig. 2 Possible failure modes of a single ply of composite laminates: **a** tensile FF **b** compressive FF **c** tensile IFF **d** shear-dominant IFF **e** compressive IFF

$$
\begin{aligned}
&\mathrm{F}^t: && \left(\frac{S_{11}}{R^t_{\parallel}}\right)^2 = 1, \quad S_{11} > 0\\
&\mathrm{F}^c: && \left(\frac{S_{11}}{R^c_{\parallel}}\right)^2 = 1, \quad S_{11} < 0\\
&\mathrm{M}^t: && \left(\frac{S_{22}}{R^t_{\perp}}\right)^2 + \left(\frac{S_{12}}{R_{\perp\parallel}}\right)^2 + \left(\frac{S_{13}}{R_{\perp\parallel}}\right)^2 + \left(\frac{S_{23}}{R_{\perp\perp}}\right)^2 = 1\\
&\mathrm{M}^c: && \left(\frac{S_{22}}{2R_{\perp\perp}}\right)^2 + \left[\left(\frac{R^c_{\perp}}{2R_{\perp\perp}}\right)^2 - 1\right]\frac{S_{22}}{R^c_{\perp}} + \left(\frac{S_{12}}{R_{\perp\parallel}}\right)^2 + \left(\frac{S_{13}}{R_{\perp\parallel}}\right)^2 + \left(\frac{S_{23}}{R_{\perp\perp}}\right)^2 = 1\\
&\mathrm{FMS}: && \left(\frac{\langle -S_{11}\rangle}{R^c_{\parallel}}\right)^2 + \left(\frac{S_{12}}{R_{\perp\parallel}}\right)^2 + \left(\frac{S_{13}}{R_{\perp\parallel}}\right)^2 = 1
\end{aligned}
\tag{20}
$$

The most important modification is the fiber-matrix-shear (FMS)-condition. This cut-off considers the shear load additional via

$$
\langle -S_{11}\rangle = \begin{cases} 0, & S_{11} \geq 0 \\ S_{11}, & S_{11} < 0 \end{cases} . \tag{21}
$$

The failure model is used ply-by-ply, thus every single layer is treated exclusively. S_{ij} are stresses referring to a local coordinate system, where the 1-direction specifies the fiber-direction, the 2-direction the in-plane direction normal to the fibers and the 3-direction is the through-thickness direction. Associated material strength values are defined typically with $R^t_{\|}$, $R^c_{\|}$, $R^t_{\perp}$, $R^c_{\perp}$, $R_{\perp\|}$, $R_{\perp\perp}$ where subscripts $\|$ and $\perp$ denote the directions parallel and transverse to the fiber direction.

3.2 The Puck-Model

The Puck-model is used as a second model to predict failure of composite laminates. All equations and contents of this subsection refer to [16]. To describe the individual failure mode Puck introduces four different failure conditions. Each layer can fail in different modes, a FF and three IFF-conditions with tensile (IFFA), compressive (IFFC) and shear dominant inter fiber failure (IFFB).

$$
\begin{aligned}
&\text{FF:} && \varepsilon_{FF} = \left(\frac{\langle S_{11}\rangle}{R^t_{\|}}\right)^2 + \left(\frac{\langle -S_{11}\rangle}{R^c_{\|}}\right)^2 = 1 \\
&\text{IFF A:} && \varepsilon_{IFF\,A} = \sqrt{\left(\frac{S_{12}}{R_{\perp\|}}\right)^2 + \left(1 - p^t_{\perp\|}\frac{R^t_{\perp}}{R_{\perp\|}}\right)^2\left(\frac{S_{22}}{R^t_{\perp}}\right)^2} + p^t_{\perp\|}\frac{S_{22}}{R_{\perp\|}} = 1, \quad S_{22} > 0 \\
&\text{IFF B:} && \varepsilon_{IFF\,B} = \frac{1}{R_{\perp\|}}\left(\sqrt{S_{12}^2 + \left(p^c_{\perp\|}S_{22}\right)^2} + p^c_{\perp\|}S_{22}\right) = 1 \\
&\text{IFF C:} && \varepsilon_{IFF\,C} = \left[\left(\frac{S_{12}}{2\left(1 + p^c_{\perp\perp}\right)R_{\perp\|}}\right)^2 + \left(\frac{S_{22}}{R^c_{\perp}}\right)^2\right]\frac{R^c_{\perp}}{(-S_{22})} = 1
\end{aligned}
\tag{22}
$$

Equation (22) defines efforts $Eff_{(mode)}$. Failure occur for $Eff_i > 1$. Curve fitting parameters appear which can be used to fit the failure model to a single material or an experiment. For carbon fiber reinforced plastics Puck [17] proposes $p^t_{\perp\|} = 0.35$, $p^c_{\perp\|} = 0.30$, $p^t_{\perp\perp} = 0.25 - 0.30$, $p^c_{\perp\perp} = 0.25 - 0.30$.

3.3 The Cuntze-Model

As a third model for composite laminates the Cuntze failure model is used. All equations and contents of this subsection refer to [3, 4]. To describe an individual failure mode Cuntze [3] introduces five failure conditions. Each layer can fail in different modes, two FF and three IFF-conditions. The FF-conditions distinguish tensile (FF1) and compressive fiber failure (FF2), and the IFF-conditions can be divided into tensile (IFF1), compressive (IFF3) and shear dominant inter fiber failure (IFF2).

$$
\begin{aligned}
&\text{FF1:} \quad \frac{I_1}{R^t_{\parallel}} = 1\\
&\text{FF2:} \quad \frac{-I_1}{R^c_{\parallel}} = 1\\
&\text{IFF1:} \quad \frac{I_2 + \sqrt{I_4}}{R^t_{\perp}} = 1\\
&\text{IFF2:} \quad \frac{\sqrt{I_3^3 + b_{\perp\parallel}(I_2 I_3 - I_5)}}{(R_{\perp\parallel})^3} = 1\\
&\text{IFF3:} \quad \frac{(b^\tau_{\perp} - 1) I_2 + b^\tau_{\perp}\sqrt{I_4}}{R^t_{\perp}} = 1
\end{aligned}
\tag{23}
$$

with

$$
\begin{aligned}
I_1 &= S_{11} & I_2 &= S_{22}\\
I_3 &= S_{22}^2 + S_{13}^2 & I_4 &= S_{22}^2 + 4S_{23}^2\\
I_5 &= S_{22}(S_{13}^2 - S_{12}^2) - 4S_{12}S_{13}S_{23}
\end{aligned}
\tag{24}
$$

In Eq. (24) two curve fitting parameters appear which can be used to fit the failure model to a single material or an experiment. For carbon fiber reinforced plastics Cuntze [4] propose $0.05 < b_{\perp\parallel} < 0.15$, and $1.0 < b^\tau_{\perp} < 1.15$). Equation (23) defines stress efforts $Eff_{(mode)}$ for each failure mode which depends only on one material strength. They can be combined to a more realistic numerical model via

$$
Eff^{\dot{m}}_{(res)} = \sum_{i=1}^{5} Eff^{\dot{m}}_{(i)}. \tag{25}
$$

Here, a third curve fitting parameter $\dot{m}$ ('rounding-off-parameter') is used, which considers an interaction of the failure modes. A value of $\dot{m} \approx 3.0$ is recommended, see [4].

4 Example-Four Point Bending Test

4.1 Problem Description

The developed algorithms and elements are implemented in an extended version of the general finite element program FEAP [19]. The investigated example is a four point bending test, depicted in Fig. 3. The geometrical data are L = 400 mm, B = 12.5 mm, D = 20 mm and a layer thickness of t = 2.5 mm together with a stacking sequence $[0°/90°]_{2s}$. With respect to symmetry only one half of the structure is

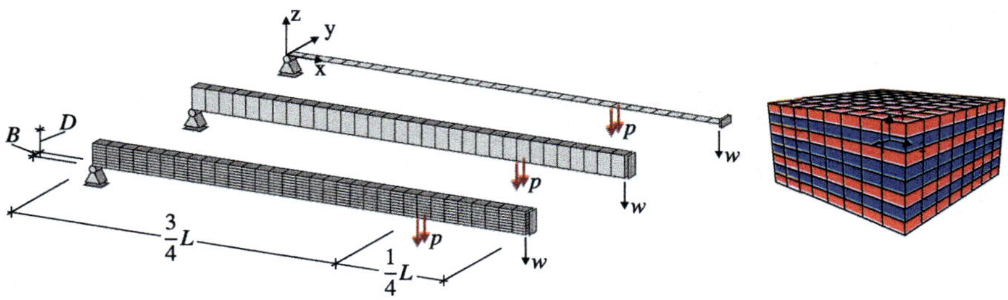

Fig. 3 Four point bending test: geometry, loading and discretisation options

discretised. A discretisation with solid shell elements [15] is chosen with 32 × 1 elements in x-y-plane. Furthermore the 1-Element-8-Layer model ('Solid Shell 1(8)') has one element in thickness direction and 8 layers, whereas the 8-Element-1-Layer model ('Solid Shell 8(1)') has for each layer one element in thickness direction. The discretisation with shell elements [20] is chosen with 32 × 1 elements in x-y-plane ('Shell (8)'). The thickness direction is modeled within a layerwise approach. With respect to the multi-scale approach the same shell model is used. Here, on RVE-level, discretisations using solid shell elements are applied. Again an option with 1-Element-8-Layer (8 × 8 × 1(8) named 'Shell FE2 1(8)') and an option with 8-Element-1-Layer (8 × 8 × 8(1) named 'Shell FE2 8(1)') in thickness direction could be used. The global boundary conditions are chosen as $u_z = 0$ at $x = 0$ and symmetry conditions at $x = L$. Furthermore plain strain conditions are assumed with $u_y = 0$ at $y = 0$ and $y = B$. It is well known, that in case of local stress based failure models a mesh dependency of solutions may occur. This is not topic of present paper. As stated above, we compare different discretisation options but introduce comparable meshes. The underlying material data for A-S Epoxy1 are depicted in Tables 1 and 2. Results are presented for the different shell and failure models in the following. The analysis is performed geometrically and material nonlinear on global as well as on local level. An arc-length scheme with displacement control is adopted. Load-deflection curves and damage distributions are depicted for the different cases. Different failure modes are shown with values between 0 (no damage) and 1 (fully

Table 1 A-S Epoxy1 stiffness values

$E_\parallel$ [MPa]	$E_\perp$ [MPa]	$\nu_{\parallel\perp}$ [-]	$G_{\parallel\perp}$ [MPa]	$G_{\perp\perp}$ [MPa]
140000	10000	0.3	6000	3335

Table 2 A-S Epoxy1 strength values

$R^t_\parallel$ [MPa]	$R^c_\parallel$ [MPa]	$R^t_\perp$ [MPa]	$R^c_\perp$ [MPa]	$R_{\perp\parallel}$ [MPa]
1990	1500	38	150	70

damaged) for solid shell and multi-scale models. The failure behaviour is plotted layerwise since no averaging in thickness direction is allowed.

The general deformation behaviour, see Fig. 5, is as follows. A nearly linear load-deflection behaviour occurs until a value of approximately w = 70 mm. The failure modes start with a matrix-failure in the 90°-layer at bottom in the loading region, when reaching $R^t_\perp$. The drastic reduction of the load is then based mainly on the fiber failure in the 0°-layer at top in the loading region, when reaching $R^c_\parallel$. Note $R^c_\parallel < R^t_\parallel$. Further loading leads to another matrix-failure in the 90°-layer at top in the loading region, with respect to $R^c_\perp$. Further minor mixed shear failure modes occur.

4.2 Hashin-Model

Figure 4 presents the deformed meshes at the final deformation of w = 100 mm for the different models. Figure 5 depicts load-deflection curves for different discretisation models for the case of the Hashin failure model. Depicted are results for the external load p [N/mm^2] with respect to the center deflection w [mm], see Fig. 3. Relatively similar results are produced for a discretisation 'Solid Shell 1(8) and a standard shell formulation 'Shell (8)'. Results for 'Solid Shell 1(8)' and 'Solid Shell 8(1)' differ in the post-failure region. 'Solid Shell 8(1)' and 'Shell (8)' lead to nearly similar results, even in the post-failure region. The multi-scale solutions 'Shell FE2 1(8)' and 'Shell FE2 8(1)' deviate from the 'Solid Shell 8(1)' solutions. A lower failure load and different post-failure paths are reached. This will be discussed in more detail in the following. Figure 6 presents the failure behaviour at w = 100 mm in the solid model 'Solid Shell 8(1)', which is dominated by failures $M^c(0°)$ and $M^t(90°)$, $M^c(90°)$. A much more detailed analysis is possible for the multi-scale model, which is presented in Fig. 7. Here different failure modes are presented on the RVE, chosen here for

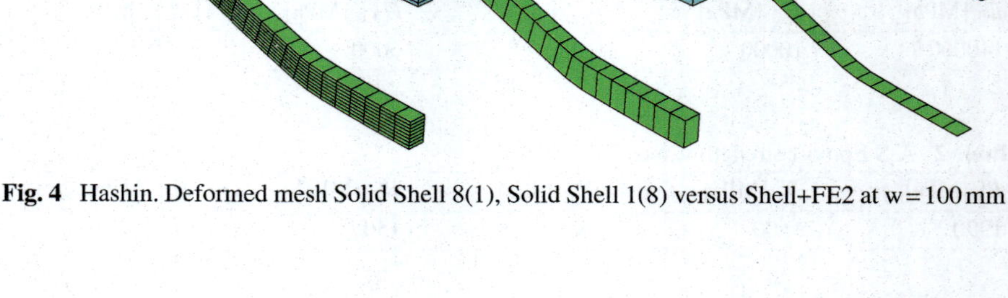

Fig. 4 Hashin. Deformed mesh Solid Shell 8(1), Solid Shell 1(8) versus Shell+FE2 at w = 100 mm

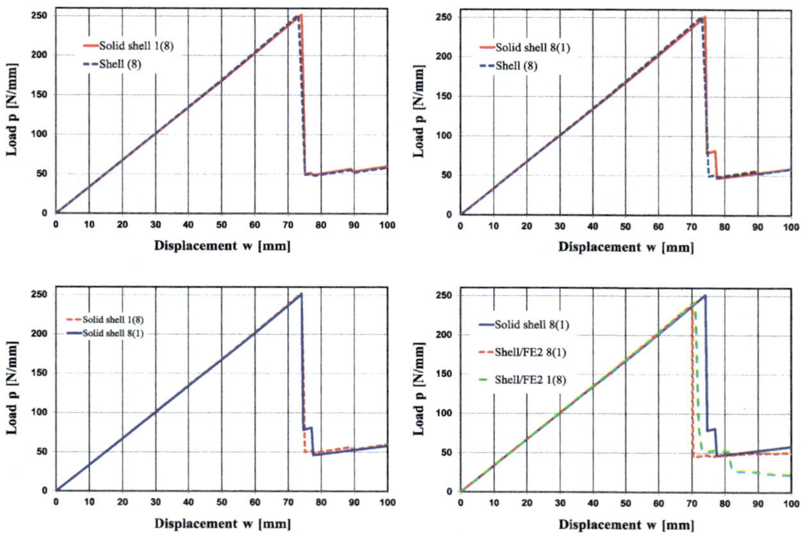

Fig. 5 Hashin-Failure model. Load-deflection curves for different discretisation models

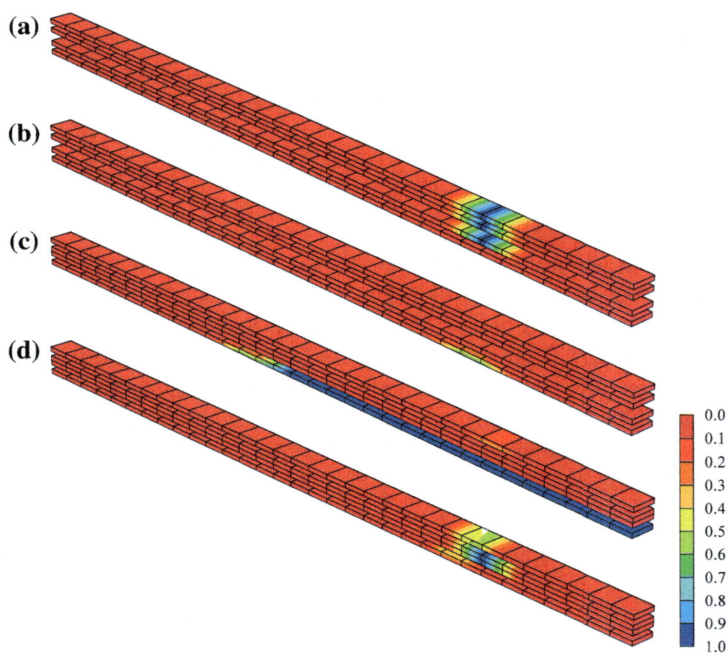

Fig. 6 Hashin-Failure model in Solid Shell at w = 100 mm **a** $FF(0°)$, **b** $M^t(0°)$, **c** $M^t(90°)$, **d** $M^c(90°)$

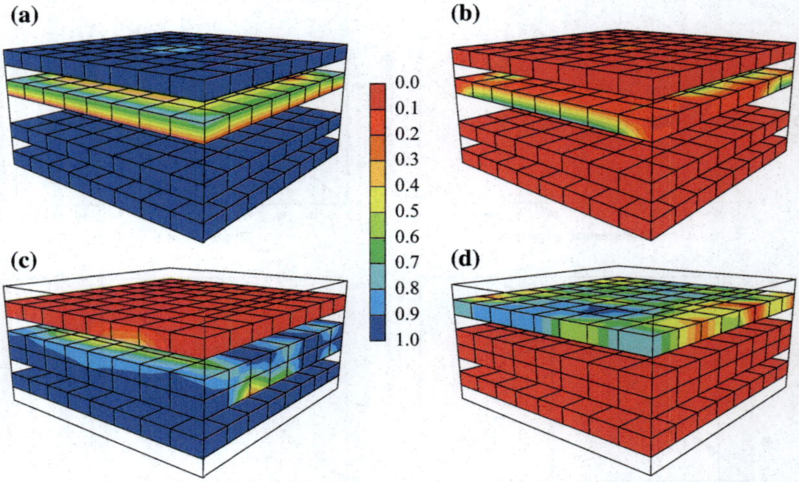

Fig. 7 Hashin-Failure model in Shell-FE2 at w = 100 mm **a** $FF(0°)$, **b** $FMS(0°)$, **c** $M^t(90°)$, **d** $M^c(90°)$

element 24, Gauss-point 1, which is near the loading area. As can be seen from Fig. 7, the failure behaviour in the 'Shell-FE2 8(1)' model, is dominated by failure $FF(0°)$ at top (c) and bottom (t), $FMS(0°)$ and $M^t(90°)$, $M^c(90°)$.

4.3 *Puck-Model*

Again load-deflection curves for different discretisation models, now in case of the Puck failure model, are depicted in Fig. 8. As can be seen from the diagram similar results are produced for a discretisation 'Solid Shell 1(8) and a standard shell formulation 'Shell (8)'. These results deviate from the model 'Solid Shell 8(1)', where a more complex post-failure region occurs. The multi-scale solutions 'Shell FE2 1(8)' and 'Shell FE2 8(1)' have lower failure loads than 'Solid Shell 8(1)' but a similar post-critical behaviour. The distribution of damaged areas can be seen again in Figs. 9 and 10. The detailed failure behaviour is similar to the Hashin-model. Failure modes occur for $FF(0°)$ at top (c) and bottom (t), $IFF\ B(0°)$ and $IFF\ A = M^t(90°)$, $IFF\ C = M^c(90°)$.

4.4 *Cuntze-Model*

The Cuntze-model is based on similar concepts as the Puck-model. Thus it could be expected, that the load-deflection behaviour for this model lies in the same range, see Fig. 11. Again similar results are produced for discretisation 'Solid Shell 1(8)' and 'Shell (8)'. Also these results do not show the more complex post failure behaviour, which occur for model 'Solid Shell 8(1)'. The multi-scale solutions 'Shell

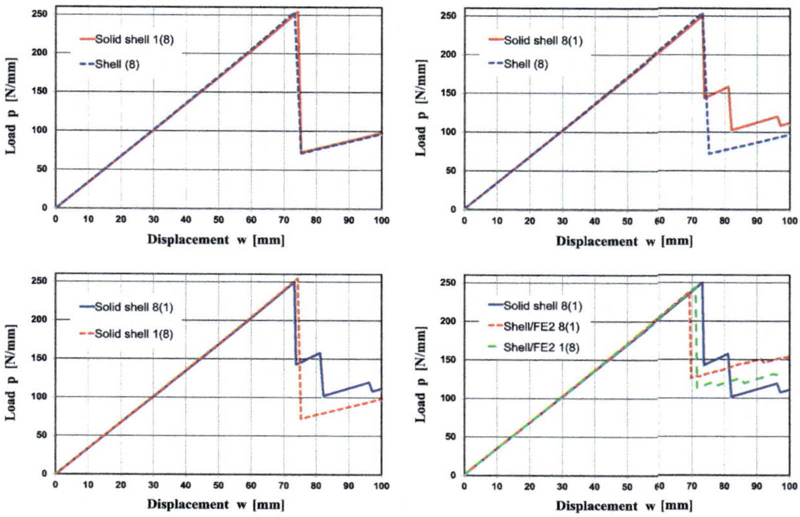

Fig. 8 Puck-Failure model. Load-deflection curves for different discretisation models

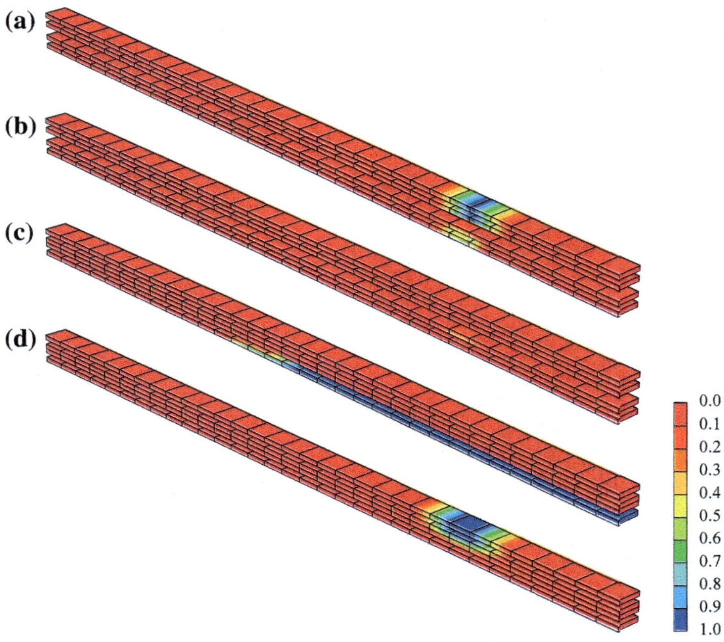

Fig. 9 Puck-Failure model in Solid Shell at w = 100 mm **a** $FF(0°)$, **b** $IFF\ A(0°)$, **c** $IFF\ A = M^t(90°)$, **d** $IFF\ C = M^c(90°)$

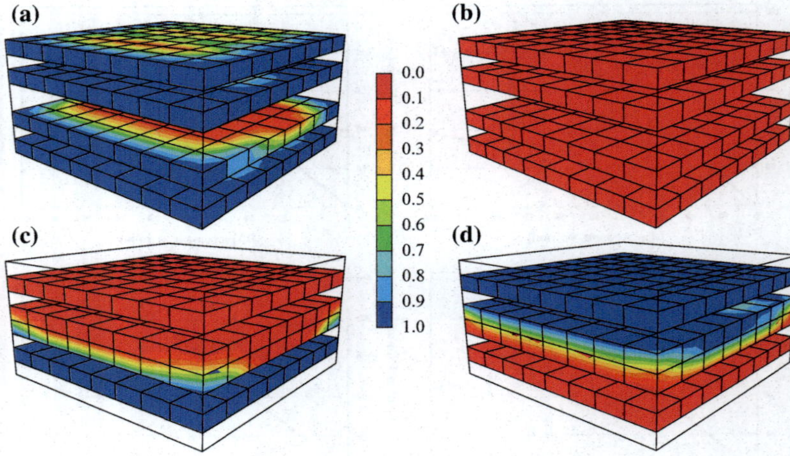

Fig. 10 Puck-Failure model in Shell-FE2 at w = 100 mm **a** $FF(0°)$, **b** $IFF\ B(0°)$, **c** $IFF\ A = M^t(90°)$, **d** $IFF\ C = M^c(90°)$

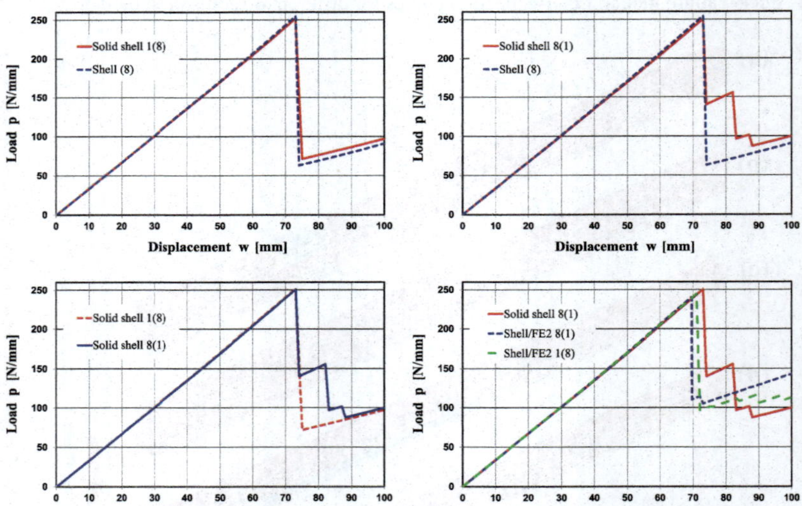

Fig. 11 Cuntze-Failure model. Load-deflection curves for different discretisation models

FE2 1(8)' and 'Shell FE2 8(1)' have lower failure loads than 'Solid Shell 8(1)' but describe the post-critical behaviour in a similar way. The distribution of damaged areas can be seen again in Figs. 12 and 13. The detailed failure behaviour for the Cuntze-model is governed by $FF(0°)$ at top (c) and bottom (t), $IFF1(0°)$ and $IFF1 = M^t(90°)$, $IFF3 = M^c(90°)$ and is close to the results of the Puck-model.

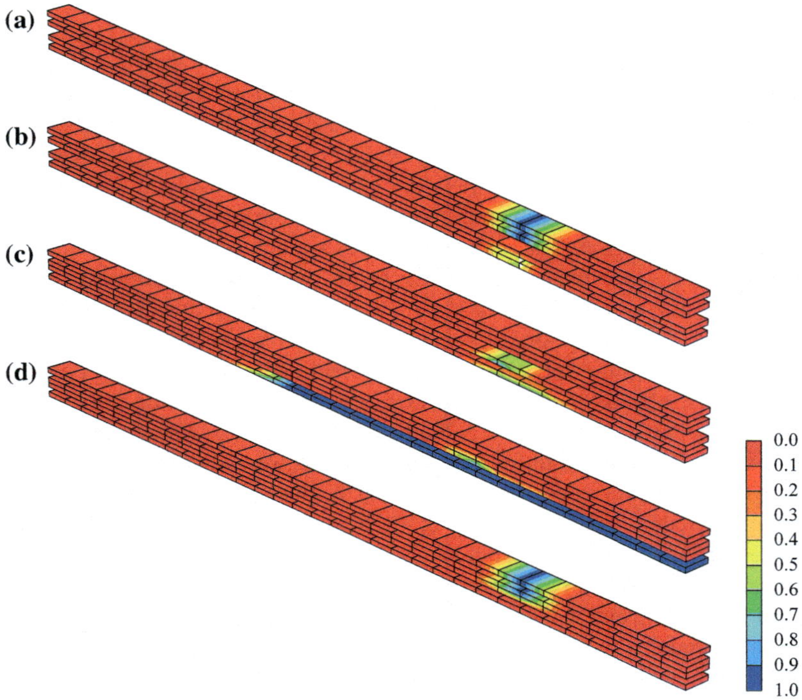

Fig. 12 Cuntze-Failure model in Solid Shell at w = 100 mm **a** $FF(0°)$, **b** $IFF1 = M^t(0°)$, **c** $IFF1 = M^t(90°)$, **d** $IFF3 = M^c(90°)$

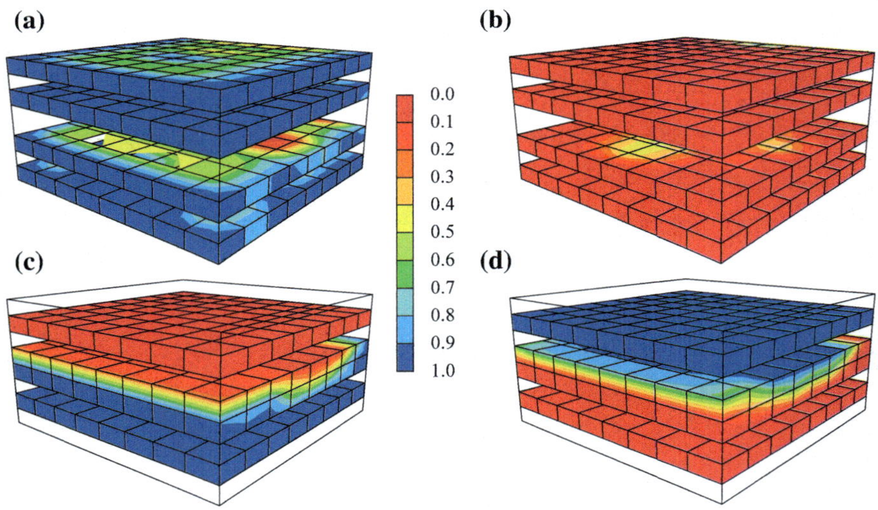

Fig. 13 Cuntze-Failure model in Shell-FE2 at w = 100 mm **a** $FF(0°)$, **b** $IFF1 = M^t(0°)$, **c** $IFF1 = M^t(90°)$, **d** $IFF3 = M^c(90°)$

5 Conclusions

Different discretisation concepts for thin shell structures with respect to damage behaviour have been discussed in this paper. These concepts are layered solid shell models, layerwise solid shell models, layerwise shell models and shell models with a FE^2-approach. Layerwise solid shell and shell models possess a similar number of unknowns. The layered solid shell models allows the studying of possible warping in thickness direction. The multi-scale models are chosen with associated RVE discretisations. Results for the pre-damage behaviour are very close together for all models. No influences of different kinematic models occur for the chosen example, see Figs. 5, 8 and 11. Damage behaviour is described via three different failure models. Obviously these models lead to different estimations for the post-failure behaviour. Nevertheless the general failure behaviour could be described with all models, when matrix-failure in the 90°-layer at bottom, fiber failure in the 0°-layer at top and another matrix-failure in the 90°-layer at top occur. However, in detail, different load deflection curves are predicted in the post-failure region. The influence of the failure models are presented in Fig. 14. Results using layerwise solid shell elements for the Puck- and Cuntze-model are comparable, whereas the Hashin-model leads to a more conservative interpretation in the post-failure region. This is also reflected when choosing the shell-FE^2 approach. With respect to Figs. 7, 10 and 13 it could be stated that the shell-FE^2-approach gives much more insight into the local behaviour at integration points. Finally it should be stated that such detailed results could be reached only after time-consuming calculations. Improvements are adaptive schemes, see e.g. [22] as well as parallelisation techniques, see e.g. [13], for which multi-scale models are well suited ([2, 6]).

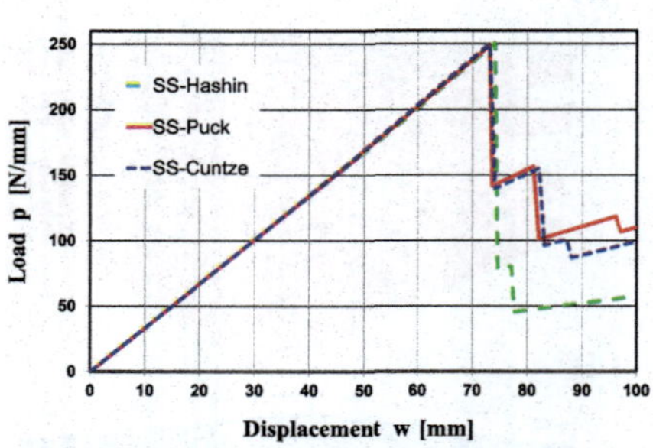

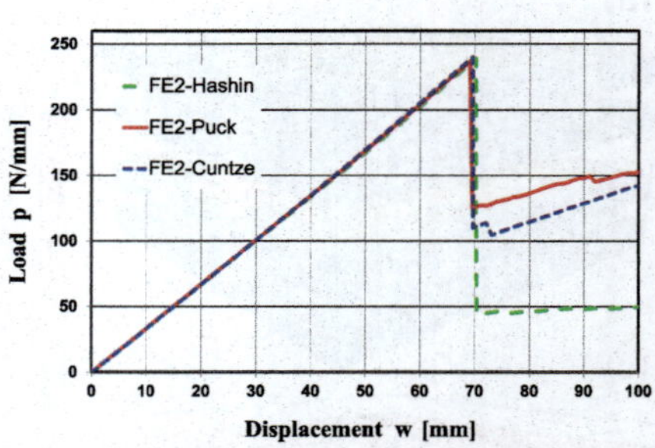

Fig. 14 Load-deflection curves for different discretisation and failure models

References

1. Carrera, E.: Historical review of zig-zag theories for multilayered plates and shells. Appl. Mech. Rev. **56**, 237–308 (2003)
2. Coenen, E., Kouznetsova, V., Geers, M.: Computational homogenization for heterogeneous thin sheets. Int. J. Num. Method Eng. **83**, 1180–1205 (2010)
3. Cuntze, R.G., Freund, A.: The predictive capability of failure mode concept-based strength criteria for multi-directional laminates. Compos. Sci. Technol. **64**, 343–377 (2004)
4. Cuntze, R.G.: The predictive capability of failure mode concept-based strength criteria for multi-directional laminates—part b. Compos. Sci. Technol. **64**, 487–516 (2004)
5. De Borst, R., Ramm, E.: Multiscale methods in computational mechanics: progress and accomplishments. In: Springer Series: Lecture Notes in Applied and Computational Mechanics, vol. 55. Springer, Berlin, Heidelberg (2011)
6. Geers, M., Coenen, E., Kouznetsova, V.: Multi-scale computational homogenization of structured thin sheets. Model. Simul. Mater. Sci. Eng. **15**, 393–404 (2007)
7. Goyal, V.K., Jaunky, N., Johnson, E.R., Ambur, D.R.: Intralaminar and interlaminar progressive failure analyses of composite panels with circular cutouts. Compos. Sci. Technol. **64**, 91–105 (2004)
8. Gruttmann, F., Wagner, W.: Coupling of 2D- and 3D-composite shell elements in linear and nonlinear applications. Comp. Meth. Appl. Mech. Eng. **129**, 271–287 (1996)
9. Gruttmann, F., Wagner, W.: Structural analysis of composite laminates using a mixed hybrid shell element. Comput. Mech. **37**, 479–497 (2006)
10. Gruttmann, F., Wagner, W.: A coupled two-scale shell model with applications to layered structures. Int. J. Num. Method Eng. **94**, 1233–1254 (2013)
11. Gruttmann, F., Wagner, W., Knust, G.: A coupled global-local shell model with continuous interlaminar shear stresses. Comput. Mech. **57**, 237255 (2016)
12. Hashin, Z.: Failure criteria for unidirectional fiber composites. J. Appl. Mech. **47**, 329–334 (1980)
13. Jarzebski, P., Wisniewski, K., Taylor, R.L.: On parallelization of the loop over elements in FEAP. Comput. Mech. **56**, 77–86 (2015)
14. Klinkel, S., Gruttmann, F., Wagner, W.: A continuum based 3d-shell element for laminated structures. Comput. Struct. **71**, 43–62 (1999)
15. Klinkel, S., Gruttmann, F., Wagner, W.: A robust non-linear solid shell element based on a mixed variational formulation. Comp. Meth. Appl. Mech. Eng. **195**, 179–201 (2006)
16. Puck, A., Schürmann, H.: Failure analysis of FRP laminates by means of physically based phenomenological models. Compos. Sci. Technol. **58**, 1045–1067 (1998)
17. Puck, A., Kopp, J., Knops, M.: Guidelines for the determination of the parameters in Pucks action plane strength criterion. Compos. Sci. Technol. **62**, 371378 (2002)
18. Reddy, J.N.: Mechanics of Laminated Composite Plates and Shells: Theory and Analysis. CRC Press, London (2004)
19. Taylor, R.: FEAP. http://www.ce.berkeley.edu/projects/feap/
20. Wagner, W., Gruttmann, F.: A robust nonlinear mixed hybrid quadrilateral shell element. Int. J. Num. Meth. Eng. **64**, 635–666 (2005)
21. Wagner, W., Gruttmann, F.: A consistently linearized multi-scale model for shell structures. In: Pietraszkiewicz, W., Górski, J. (eds.) Shell Structures: Theory and Applications, vol(3), pp. 453–456. Taylor & Francis, London (2013)
22. Wagner, W., Gruttmann, F.: An adaptive strategy for the multi-scale analysis of plate and shell structures with elasto-plastic material behaviour. Technische Mechanik **36**(1–2), 132–144 (2016)
23. Zohdi, T., Wriggers, P.: Introduction to Computational Micromechanics. Springer Series: Lecture Notes in Applied and Computational Mechanics, vol. 20. Springer, Berlin, Heidelberg (2005)

Micro-Macro Modelling of Metallic Composites

Rex Bedzra, Stefanie Reese and Jaan-Willem Simon

Abstract This contribution describes a scale bridging approach for modelling pressure independent elastoplastic unidirectional metallic composite materials by making use of an anisotropic elastoplastic constitutive model. The material under investigation is tungsten fiber reinforced copper (W/Cu) composite. To identify the yield surface of the composite, a finite element model of a repeating unit cell (RUC) is set-up (micro-model). Through virtual experiments, the yield surface of the composite is identified. An anisotropic elastoplastic constitutive model based on the identified yield surface, which makes use of the concept of structural tensors, is developed. This material model serves as the material model for macro computations. To ensure a good agreement between constitutive model and RUC during plastic evolution, multiple hardening functions are employed. The parameters of the constitutive model are identified and the constitutive model is validated against the response of the RUC.

1 Introduction

Metal matrix composite (MMC) such as tungsten fiber reinforced copper (W/Cu) composites, are being considered for use in high temperature propulsion environments due to their ability to withstand high thermal loads and their relatively low density [1, 2].

R. Bedzra (✉)
Aachen Institute for Advanced Studies in Computational
Engineering Science (AICES), RWTH Aachen University, Schinkelstraße 2,
52062 Aachen, Germany
e-mail: rex.bedzra@rwth-aachen.de

S. Reese · J.-W. Simon
Institute of Applied Mechanics, RWTH Aachen University,
Mies-van-der-Rohe-Straße 1, 52074 Aachen, Germany
e-mail: stefanie.reese@rwth-aachen.de

J.-W. Simon
e-mail: jaan.simon@rwth-aachen.de

J. Sorić et al. (eds.), *Multiscale Modeling of Heterogeneous Structures*,
Lecture Notes in Applied and Computational Mechanics 86,
https://doi.org/10.1007/978-3-319-65463-8_2

To model such a material, different length scales have to be taken into account in order to capture the influence of the intrinsic microstructure of the composite material on the macro-scale. A method to accomplish this is to employ a computational homogenization scheme [3–5]. This method is based on the construction of a micro-scale boundary value problem which is used to determine the governing behaviour of the material on the macro-scale. If the boundary value problem is solved simultaneously, a fully nested solution of two boundary value problems is obtained, often referred to as the FE^2 method. Although this method allows one to asses the macroscopic influence on the microscopic parameters, it is computationally expensive and time consuming. An alternative approach, which is less computationally expensive, is to develop a constitutive model which is able to predict the response of the micro-scale boundary value problem. This approach has been exploited in [6] for modelling the elastoplastic behaviour of pneumatic membranes and also in [7, 8] for modelling layered fiber reinforced composites. In the current contribution, it is used to model the elastoplastic response of a unidirectional fiber reinforced metal matrix composite.

The developed anisotropic elastoplastic model employs the concept of structural tensors to describe the initial yield surface of the composite. The structural tensors are second order tensors which are characterized by unit vectors pointing in privileged directions of the material. The structural tensors may be fixed [9–11] or may evolve with plastic flow [12–14]. In the present case, they are considered to be fixed.

The paper is organized as follows: In Sect. 2, a finite element model of the unidirectional composite model is set-up, and the yield surface of the composite is identified by performing virtual experiments. In Sect. 3, the derivation of the anisotropic constitutive model is briefly described and validated.

2 Unidirectional Composite (Micro-scale)

To identify the yield surface of the composite as well as the material parameters of the continuum anisotropic constitutive model and for later validation, a repeating unit cell (RUC) of a unidirectional composite with random fiber distribution is generated.

2.1 Geometry Generation

The fiber distribution in the RUC is generated through a uniform random point generator function in matlab. The length of the RUC in longitudinal fiber direction as well as in transverse fiber direction is 33.98 μm. The diameter of the fibers is 7 μm, the fiber volume ratio is 50% and the total number of fibers is 10. The RUC is assumed to be part of a much larger material specimen. Therefore, periodic boundary conditions are applied through equation constraints in the finite element package

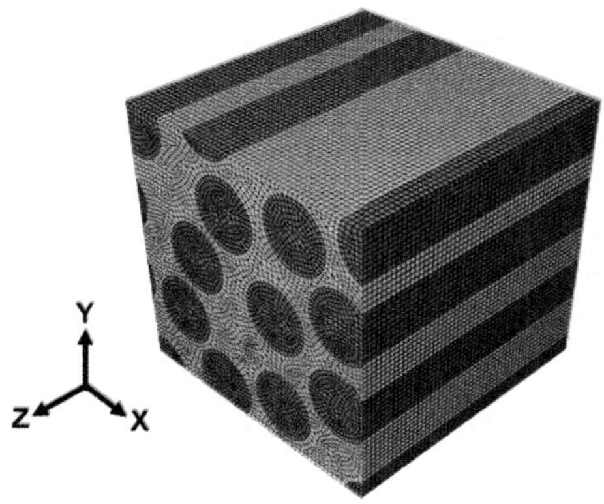

Fig. 1 Geometry of unidirectional composite RUC with random fiber distribution

ABAQUS/standard [15]. For details on the application of periodic boundary condition in ABAQUS/standard, see [16].

For efficiency reasons, reduced integration eight node linear solid elements (C3D8R) are used to descritize the RUC. In total, the model is made up of 104, 192 number of elements. The meshed geometry of the unidirectional composite RUC is shown in Fig. 1.

2.2 *Fiber and Matrix Material*

The fibers are modelled as General Electric 218CS (GE218CS) continuous tungsten wire and the matrix is modelled as a cold worked oxygen free high conductivity copper (OFHC) material. The constitutive model for both GE218CS and OFHC is summarized in Table 1 where σ^D represents the devatoric part of the Cauchy stress tensor σ which in turn is related to the elastic part of the strain tensor ε_e through the fourth-order isotropic elastic stiffness tensor $\mathbb{C}$, which is a function of the Poisson's ratio ν and the Young's modulus E, ε is the total strain which is additively decomposed into an elastic part ε_e and a plastic part ε_p. The yield function Φ is assumed to be a standard von Mises yield function. The isotropic hardening stress R is chosen as a Voce's exponential hardening function, where κ is the isotropic hardening variable. Furthermore, β and Q_0 are material parameters which describe isotropic hardening. Also, the Kuhn-Tucker conditions for loading and unloading as well as the consistency condition are listed.

The fit of the constitutive model to experimental uniaxial stress-strain curves, obtained from [2] for GE218CS and OFHC, is shown in Figs. 2 and 3, respectively. Also, the material parameters identified through fitting the constitutive model to the experimental results is given in Table 2.

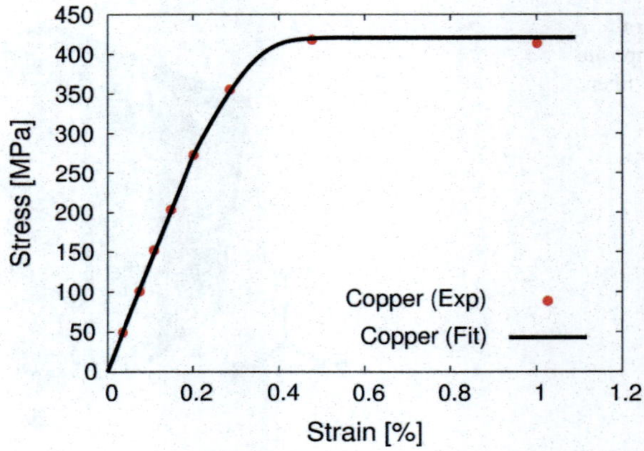

Fig. 2 Fit to experimentally obtained uniaxial stress-strain curve of copper, from [Sanfeliz]

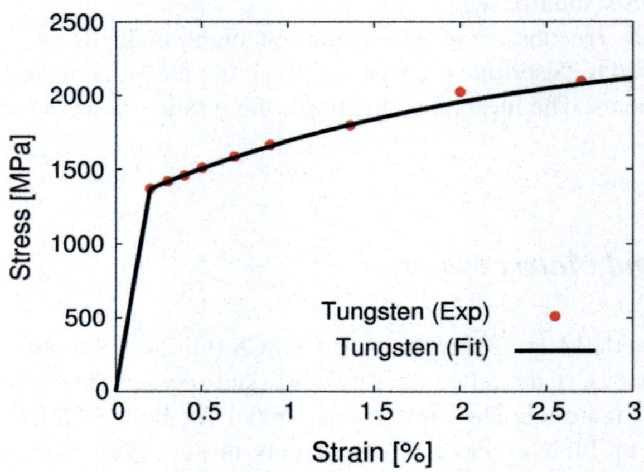

Fig. 3 Fit to experimentally obtained stress-strain curve of tungsten wire, from [Sanfeliz]

Table 1 Fiber and matrix material model

Elastic law	$\sigma = \mathbb{C}\,[\varepsilon_e]$	Hardening law	$R = Q_0(1 - e^{-\beta\kappa})$
Strain	$\varepsilon = \varepsilon_e + \varepsilon_p$	Flowrule	$\dot{\varepsilon_p} = \dot{\lambda}\frac{\partial \Phi}{\partial \sigma}$, $\dot{\kappa} = \dot{\lambda}\frac{\partial \Phi}{\partial R}$
Yield function	$\Phi = \sqrt{\sigma^D \cdot \sigma^D} - \sqrt{\frac{2}{3}}(\sigma_Y - R)$	K.T conditions	$\dot{\lambda} \geq 0$, $\Phi \leq 0$, $\dot{\lambda}\Phi = 0$

Table 2 Tow and matrix material constants

Fit	E [MPa]	ν [−]	σ_y [MPa]	Q_0 [MPa]	β [−]
Tungsten	677887	0.28	1368.87	987.20	52.60
Copper	135248	0.34	272.76	148.10	2994

2.3 *Determination of Unidirectional Composite Yield Surface*

To obtain a profile of the composite yield surface, the unidirectional fiber reinforced composite RUC is subjected to a series of biaxial tensile tests as well as combined shear tension tests. The load ratios, σ_{xx}/σ_{yy}, σ_{xy}/σ_{xx}, σ_{xx}/σ_{zz}, σ_{xz}/σ_{xx}, σ_{xy}/σ_{yy}, and σ_{xz}/σ_{zz}, employed to identify the yield surface of the composite in 11/22- and 11/33-principal stress space, are summarized in Table 3. Due to geometrical and material symmetry, there is no need to identify the yield surface in 22/33-principal stress space since it will result in the same yield surface as identified in 11/33-principal stress space. The obtained yield surface in 11/22- and 22/33-principal stress space respectively, are displayed in Figs. 4 and 5.

Table 3 Load ratios for biaxial and combined shear-tension virtual experiments

σ_{xx}/σ_{yy}	σ_{xx}/σ_{zz}	1/0	4/1	2/1	4/3	1/1
σ_{xx}/σ_{yy}	σ_{xx}/σ_{zz}	0/1	1/4	1/2	3/4	1/1
σ_{xy}/σ_{xx}	σ_{xz}/σ_{zz}	1/0	4/1	2/1	4/3	0/1
σ_{xy}/σ_{yy}	σ_{xz}/σ_{xx}	1/0	4/1	2/1	4/3	0/1

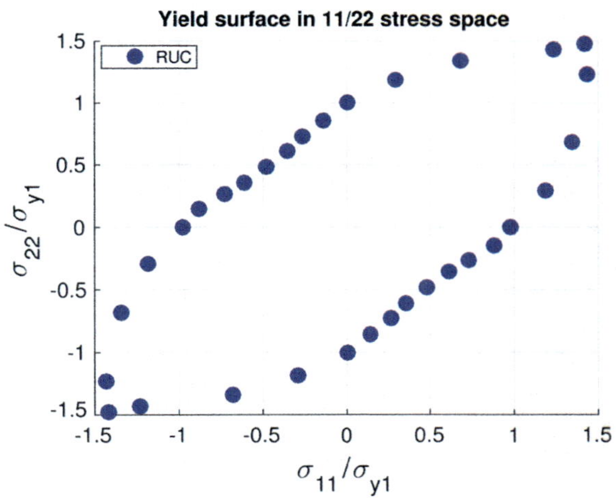

Fig. 4 Yield surface in 11/22-principal stress space normalized by uniaxial yield stress in x-direction σ_{y1}

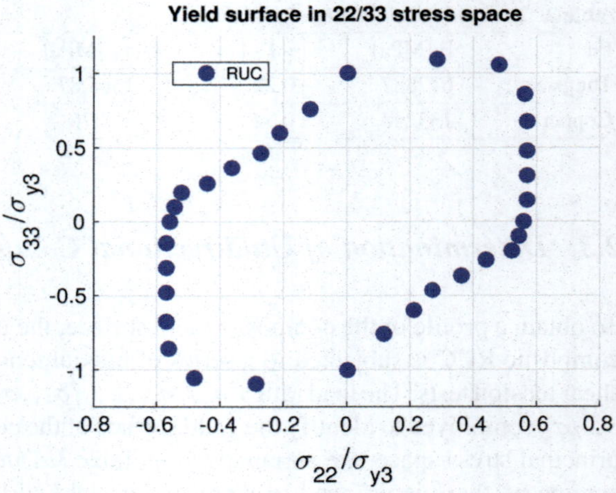

Fig. 5 Yield surface in 22/33-principal stress space normalized by uniaxial yield stress in z-direction σ_{y3}

3 Macro-model

To capture the identified yield surface of the unidirectional composite RUC, a constitutive model is developed.

3.1 Constitutive Model of Unidirectional Composite

The Helmholtz free energy per unit volume is given in the form

$$\psi = \psi_e \left(\varepsilon_e, M_1, M_2\right) + \sum_{i=1}^{6} \psi_i^{iso}\left(\kappa\right) \tag{1}$$

The first part ψ_e of the energy terms describes the macroscopic anisotropic elastic properties of the material. ε_e represents the elastic strain which is obtained from the additive decomposition of the total strain tensor ε into an elastic part and a plastic part ε_p

$$\varepsilon = \varepsilon_e + \varepsilon_p \tag{2}$$

and M_i $(i = 1, 2)$ are structural tensors, more details to follow. The terms ψ_i^{iso} represent the additional amount of stored energy due to isotropic hardening and they are a function of the accumulated plastic strain κ.

Utilizing the Clausius-Duhem second law of thermodynamics, $-\dot{\psi} + \sigma \cdot \dot{\varepsilon} \geqslant 0$, a relation for the Cauchy stress tensor σ and the isotropic hardening stresses R_i are obtained as:

$$\sigma = \frac{\partial \psi_e}{\partial \varepsilon}, \quad R_i = -\frac{\partial \psi_i^{iso}}{\partial \kappa} = -Q_i(1 - e^{-\beta_i \kappa}) \tag{3}$$

The expression for the isotropic hardening stresses corresponds to Voce's exponential hardening as given in [17], where Q_i and β_i are material parameters used to describe nonlinear isotropic hardening. Furthermore, to ensure thermodynamic consistency, the following evolution equations for associative plastic flow and accumulated plastic strain are specified.

$$\dot{\varepsilon}_p = \dot{\lambda} \frac{K}{\|K\|} \quad \text{where} \quad K = \frac{\partial \Phi}{\partial \sigma}, \quad \dot{\kappa} = \|\dot{\varepsilon}_p\| = \dot{\lambda} \tag{4}$$

where Φ and $\dot{\lambda}$ are the yield function and the plastic multiplier, respectively.

To capture the yield surface obtained from the virtual experiments carried out on the unidirectional composite RUC, a yield surface corresponding to Hill [18] is employed and is given here as:

$$\Phi = \frac{1}{2} \sigma^D \cdot \mathbb{A} \left[\sigma^D \right] - 1 \tag{5}$$

The superscript D denotes the deviator of a second-order tensor, which produce $\mathsf{B}^D = \mathsf{B} - \frac{1}{3} (\mathrm{tr}\mathsf{B})\, 1$. The fourth order anisotropy tensor $\mathbb{A}$ written in terms of structural tensors has the form:

$$\mathbb{A} = 3 \left(a_1 \mathbb{I} + a_2 \mathsf{M}_1 \otimes \mathsf{M}_1 + a_3 M_2 \otimes \mathsf{M}_2 + a_4 \left(\mathsf{M}_1 \otimes \mathsf{M}_2 + \mathsf{M}_2 \otimes \mathsf{M}_1 \right) + a_5 \mathbb{D}_1 + a_6 \mathbb{D}_2 \right) \tag{6}$$

A second-order structural tensor as developed by [19–21] is defined as:

$$\mathsf{M}_i = \mathbf{n}_i \otimes \mathbf{n}_i, \quad i = 1, 2, 3 \tag{7}$$

where $\mathbf{n}_i$ are unit vectors pointing in privileged directions of the material. Additionally, these unit vectors $\mathbf{n}_i$ are orthogonal to each other, and hence the following relations hold

$$\sum_{i=1}^{3} \mathsf{M}_i = \mathbf{n}_1 \otimes \mathbf{n}_1 + \mathbf{n}_2 \otimes \mathbf{n}_2 + \mathbf{n}_3 \otimes \mathbf{n}_3 = 1, \quad \mathbf{n}_i \cdot \mathbf{n}_j = \delta_{ij}, \quad i, j = 1, 2, 3 \tag{8}$$

δ_{ij} refers to the Kronecker symbol and 1 is the second order identity tensor. Furthermore, the fourth order tensors $\mathbb{D}_1$ and $\mathbb{D}_2$ are defined as:

$$\mathbb{D}^{\alpha}_{ijkl} = M^{\alpha}_{ik} \delta_{jl} + M^{\alpha}_{jk} \delta_{il}, \quad \alpha = 1, 2 \tag{9}$$

The tensor $\mathbb{I}$ represents the fourth order identity tensor.

The coefficients a_i ($i = 1, ..., 6$), which describe the initial state of anisotropy, are obtained by setting them in relation to the classical Hill coefficients F, G, H, L, M, and N, see [22] for details:

$$
\begin{aligned}
a_1 &= \frac{2}{3}(L + M - N) \\
a_2 &= \frac{2}{3}(F + 4G + H - 2M) \\
a_3 &= \frac{2}{3}(4F + G + H - 2L) \\
a_4 &= \frac{2}{3}(2F + 2G - H - L - M + N) \\
a_5 &= \frac{2}{3}(N - L) \\
a_6 &= \frac{2}{3}(N - M)
\end{aligned}
\tag{10}
$$

To take into account anisotropic plastic evolution, the classical Hill coefficients are redefined as:

$$
\begin{aligned}
F &= \frac{1}{2}\left[\frac{1}{(\sigma_{y2} - R_2)^2} + \frac{1}{(\sigma_{y3} - R_3)^2} - \frac{1}{(\sigma_{y1} - R_1)^2}\right] \\
G &= \frac{1}{2}\left[\frac{1}{(\sigma_{y3} - R_3)^2} + \frac{1}{(\sigma_{y1} - R_1)^2} - \frac{1}{(\sigma_{y2} - R_2)^2}\right] \\
H &= \frac{1}{2}\left[\frac{1}{(\sigma_{y1} - R_1)^2} + \frac{1}{(\sigma_{y2} - R_2)^2} - \frac{1}{(\sigma_{y3} - R_3)^2}\right] \\
L &= \frac{1}{2}\left[\frac{1}{(\sigma_{y23} - R_{23})^2}\right] \\
M &= \frac{1}{2}\left[\frac{1}{(\sigma_{y13} - R_{13})^2}\right] \\
N &= \frac{1}{2}\left[\frac{1}{(\sigma_{y12} - R_{12})^2}\right]
\end{aligned}
\tag{11}
$$

where $\sigma_{y1}, \sigma_{y2}, \sigma_{y3}, \sigma_{y23}, \sigma_{y13}$ and σ_{y12} are the initial yield stresses. It should be noted that the isotropic hardening stresses R_4, R_5, and R_6 correspond to R_{23}, R_{13}, and R_{12}, respectively.

The final form of the constitutive equations of the model is summarized below:

- Stress tensor

$$
\sigma = \frac{\partial \psi_e}{\partial \varepsilon}
$$

- Yield function

$$\Phi = \frac{1}{2}\sigma^D \cdot \mathbb{A}\left[\sigma^D\right] - 1$$

- Evolution equations

$$\dot{\varepsilon}_p = \dot{\lambda}\frac{\mathbb{P} : \mathbb{A}\left[\sigma^D\right]}{\left\|\mathbb{P} : \mathbb{A}\left[\sigma^D\right]\right\|} \quad \text{where} \quad \mathbb{P} = \mathbb{I} - \frac{1}{3}\left(1 \otimes 1\right) \qquad \dot{\kappa} = \dot{\lambda}$$

- Kuhn-Tucker conditions

$$\dot{\lambda} \geqslant 0, \quad \Phi \leqslant 0, \quad \dot{\lambda}\Phi = 0$$

The energy term ψ_e is specified as a linearized version of the strain energy function proposed in [23] for modelling bidirectional composite materials and is given as:

$$\begin{aligned} \psi_e\left(\varepsilon_e, M_1, M_2\right) = {} & 4K_1^{iso}\left(I_1\right)^2 + 4K_2^{iso}\left(I_2 + I_1\right) + 4K_1^{ani1}\left(I_4\right)^2 + 4K_2^{ani1}\left(I_5 + I_4\right) \\ & + 4K_1^{ani2}\left(I_6\right)^2 + 4K_2^{ani2}\left(I_7 + I_6\right) + 4K^{kop1}\left(I_1\right)\left(I_4\right) + 4K^{kop2}\left(I_1\right)\left(I_6\right) \\ & + 4K^{kop12}\left(I_4\right)\left(I_6\right) \end{aligned} \tag{12}$$

where I_1 and I_2 represent the first and second invariant of the elastic strain tensor. Additionally, I_4, I_5, I_6, and I_7 denote the first invariant of $\varepsilon_e M_1$, $\varepsilon_e^2 M_1$, $\varepsilon_e M_2$, and $\varepsilon_e^2 M_2$, respectively.

$$\begin{aligned} & I_1 := \mathrm{tr}\varepsilon_e \quad I_2 := \frac{1}{2}\left[\left(\mathrm{tr}\varepsilon_e\right)^2 - \mathrm{tr}\left(\varepsilon_e^2\right)\right] \quad I_4 := \mathrm{tr}\left(\varepsilon_e M_1\right) \quad I_5 := \mathrm{tr}\left(\varepsilon_e^2 M_1\right) \\ & I_6 := \mathrm{tr}\left(\varepsilon_e M_2\right) \quad I_7 := \mathrm{tr}\left(\varepsilon_e^2 M_2\right) \end{aligned} \tag{13}$$

Furthermore, the coefficients K_1^{iso}, K_2^{iso}, K_1^{ani1}, K_2^{ani1}, K_1^{ani2}, K_2^{ani2}, K^{kop1}, K^{kop2}, and K^{kop12} are related to the Young's moduli E_1, E_2, E_3, the Poisson's ratios ν_{12}, ν_{13}, ν_{23}, and the shear moduli G_{12}, G_{13}, G_{23}. Also, the energy terms ψ_i^{iso} are specified as:

$$\psi_i^{iso}\left(\kappa\right) = Q_i\left(\kappa + \frac{e^{-\beta_i \kappa}}{\beta_i}\right) \tag{14}$$

In total, the constitutive model has 27 material parameters which are E_1, E_2, E_3, ν_{12}, ν_{13}, ν_{23}, G_{12}, G_{13}, G_{23}, σ_{y1}, σ_{y2}, σ_{y3}, σ_{y23}, σ_{y13}, σ_{y12}, Q_i $(i = 1, \ldots, 6)$, and β_i $(i = 1, \ldots, 6)$. The material parameters can be obtained by performing virtual experiments on the unidirectional composite RUC.

The numerical integration of the evolution equations is performed by means of the backward Euler integration scheme at the Gauss point level. The integrated plastic flow rule and the yield function in a residuum format reads

$$\mathbf{r} = \varepsilon^p - \varepsilon^p_n - \bar{\lambda}\frac{\mathbb{P} : \mathbb{A}\left[\sigma^D\right]}{\left\|\mathbb{P} : \mathbb{A}\left[\sigma^D\right]\right\|}$$

$$r = \Phi \tag{15}$$

where $\bar{\lambda} := \dot{\lambda}\Delta t$. The system of equations consists of 7 nonlinear scalar equations which are solved iteratively by means of the Newton method. The integration of the evolution equation for isotropic hardening is also descritized by means of the backward Euler integration scheme as:

$$\kappa = \kappa_n + \bar{\lambda} \tag{16}$$

The accumulated plastic strain κ is computed at the end of each local Newton iteration via the converged value of the plastic multiplier $\bar{\lambda}$.

3.2 *Parameter Identification and Validation*

To identify the parameters of the constitutive model, the fibers are assumed to coincide with the z-direction of the cartesian coordinate system. The unidirectional composite RUC is subjected to uniaxial tension in the direction of the fibers (z-direction) and in the transverse direction of the fibers (x- or y-direction). Furthermore, the RUC is subjected to shearing in the plane of the fibers (xz- or yz-plane) and shearing in the plane transverse to the direction of the fibers (xy-plane). From the obtained stress-strain curves, the elasticity constants E_1, E_2, E_3, ν_{12}, ν_{13}, ν_{23}, G_{12}, G_{13}, and G_{23} are identified. Also the uniaxial yield stresses σ_{y1}, σ_{y2}, σ_{y3} as well as the shear yield stresses σ_{y23}, σ_{y13}, σ_{y12} are identified.

Next, the constants of the six exponential hardening functions are identified by manually fitting the isotropic hardening stresses R_i to the flow curves obtained from the six virtual experiments. From the fits, the constants Q_i $(i = 1, \ldots, 6)$ and β_i $(i = 1, \ldots, 6)$ are identified. To confirm the fit, simulations are performed using the identified material parameters and are compared to the stress-strain curves obtained from virtual experiment. The plot of the fit is shown in Figs. 6 and 7 for uniaxial tensile test in the principal material directions and shear directions, respectively. Also summarized in Tables 4 and 5 are the identified anisotropic elasticity constants and plasticity constants respectively.

To validate the constitutive model, a geometry similar to the unidirectional composite RUC with the same mesh is generated. The geometry is assigned the constitutive model and it is subjected to periodic boundary conditions as well as the load ratios given in Table 3. The obtained yield surfaces resulting from the constitutive model are compared to the yield surfaces obtained from the unidirectional composite RUC. Displayed in Figs. 8 and 9 are plots comparing the yield surfaces obtained from the constitutive model to those obtained from the unidirectional composite RUC in 11/22- and 22/33-principal stress space, respectively.

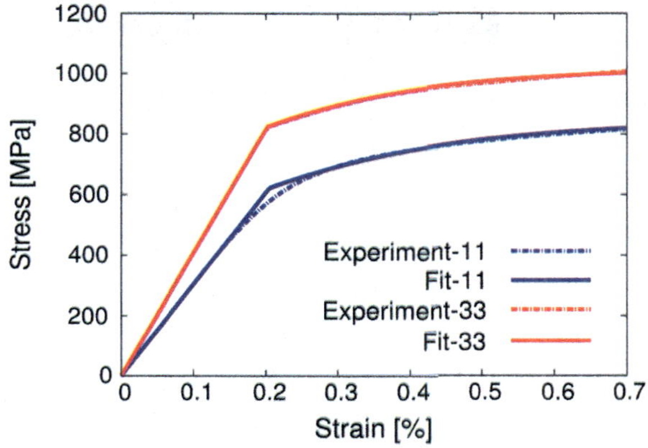

Fig. 6 Fit of the constitutive model to uniaxial stress-strain curves

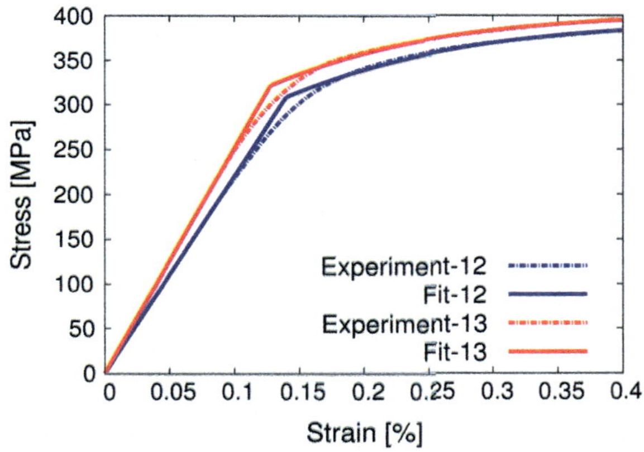

Fig. 7 Fit of the constitutive model to shear yield stresses

Table 4 Anisotropic elasticity constants

	Youngs modulus [MPa]		Shear modulus [MPa]		Poisson's ratio [−]
E_1	304319.78	G_{12}	220220.03	ν_{12}	0.3525
E_2	304319.78	G_{13}	251178.75	ν_{13}	0.2297
E_3	407914.16	G_{23}	251178.75	ν_{23}	0.2297

From Figs. 8 and 9, it can be observed that although the yield surface profile obtained from the constitutive model is in a good agreement with that obtained from the unidirectional composite RUC in the region corresponding to combined shear tensile test, some discrepancies are observed in the region corresponding to biaxial tensile test. This can be attributed to the fact that the yield function does not allow control of biaxial yielding. An extension of the yield function to take into account

Table 5 Plasticity constants

	σ_{yi} [MPa]	Q_i [MPa]	β_i [−]
1	619.99	220.60	420.50
2	619.99	220.60	420.50
3	822.55	189.60	469.60
32	320.71	80.00	1435.00
13	320.71	80.00	1435.00
12	307.73	84.04	1380.00

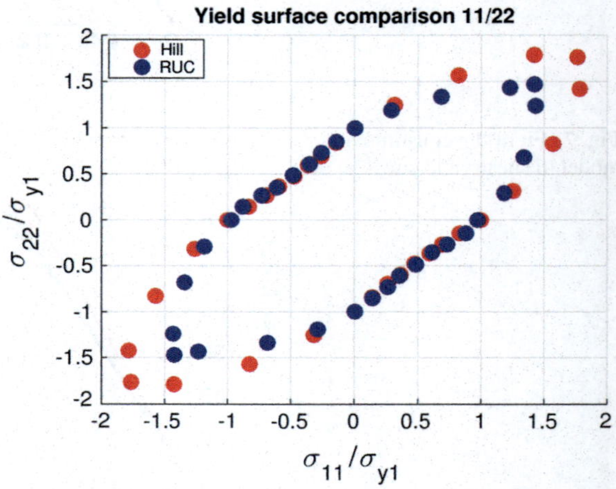

Fig. 8 Comparison between yield surface of constitutive model and RUC in 11/22-principal stress space normalized by unidirectional yield stress in x-direction σ_{y1}

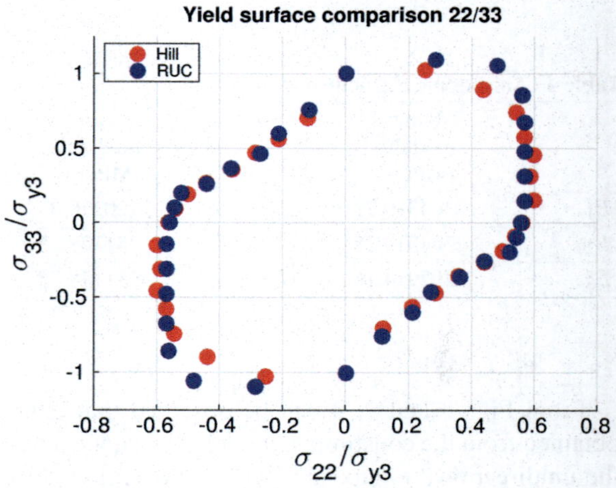

Fig. 9 Comparison between yieldsurface of constitutive model and RUC in 22/33-principal stress space normalized by unidirectional yield stress in z-direction σ_{y3}

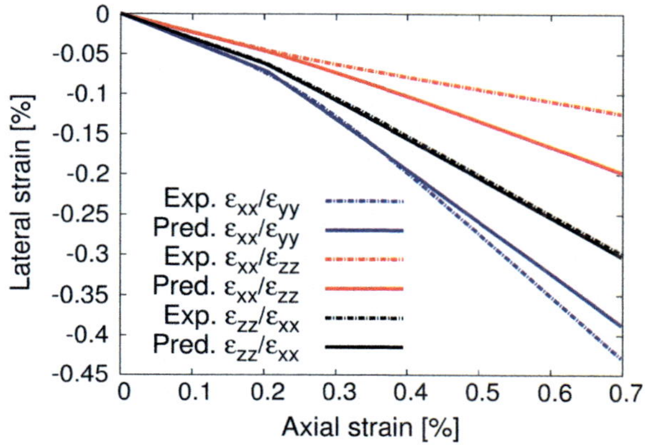

Fig. 10 Comparison between predicted lateral strain versus axial strain for uniaxial tension in x- and z-direction to that generated from the unidirectional composite RUC

biaxial yielding could result in a better agreement between the constitutive model and the unidirectional composite RUC.

Furthermore to study the ability of the constitutive model to accurately predict lateral contraction during uniaxial tension in the principal material directions, the comparison between predicted lateral contraction versus axial strain to that generated from the unidirectional composite RUC is given in Fig. 10.

From the plot, it can be observed that the constitutive model is able to predict the response of the unidirectional composite RUC for lateral contraction in x-direction versus axial strain in z-direction and also to some extent for lateral contraction in y-direction versus axial strain in x-direction. However, for lateral contraction in z-direction versus axial strain in x-direction, the curve obtained from the constitutive model differs from that obtained from the unidirectional composite RUC in the plastic flow regime. This discrepancy can be resolved by employing a non-associative flow rule which allows the plastic Poisson's ratios to be controlled.

4 Conclusion

The goal of this contribution, was to develop a general anisotropic elastoplastic material model which is able to accurately describe the pressure independent elastoplastic response of a unidirectional metallic composite RUC.

For this purpose, a finite element model of the RUC was setup. It was subjected to various biaxial tensile test as well as combine shear tensile tests. From these virtual experiments, the initial yield surface of the RUC was identified in principal stress-space. To capture the yield surface, a Hill type yield function was employed. Furthermore, to capture the anisotropic hardening behaviour of the yield surface, multiple isotropic hardening functions were employed. After parameter identification through fitting the constitutive model to stress-strain curves generated from the RUC, it was validated by comparing the predicted yield surface to that obtained from the RUC. Although the yield surface predicted by the constitutive model was in good agreement with the yield surface of the RUC in the region corresponding to combined shear tensile test, the same could not be said for the region corresponding to biaxial tensile test. This was attributed to the fact that the yield function could not be controlled in biaxial yielding. An extension of the yield function to take into account biaxial yielding may rectify this problem. Also, the ability of the constitutive model to capture lateral strain versus axial strain response obtained from the RUC for uniaxial tension in x- and z-direction was investigated. Here, the response of the constitutive model was in an excellent agreement with the RUC response for lateral contraction in x-direction versus axial strain in z-direction and in a good agreement for lateral contraction in y-direction versus axial strain in x-direction. However, for lateral contraction in z-direction versus axial strain in x-direction, the response of the constitutive model differed from the RUC response in the plastic regime. This was attributed to lack of control of the plastic Poisson's ratios. By employing a non-associative flow rule, this problem can be rectified.

Acknowledgements The authors gratefully acknowledge the financial support from the Deutsche Forschungsgemeinschaft through Grant GSC 111. The last author is grateful for the financial support provided by the Ministry of Innovation, Science and Research of the State of North Rhine-Westphalia. The authors would also like to acknowledge Dipl.-Ing. Bertram Stier for providing the finite element model of the unidirectional composite repeating unit cell (RUC) used in this work.

References

1. Doychak, J.: Metal- and intermettallic-matrix composites for aerospace propulsion and power systems. J. Min. Met. Mat. S. **44**, 46–51 (1992)
2. Sanfeliz, J.G.: Elastic-plastic finite element analyses of a unidirectional, 9 vol % tungsten fiber reinforced copper matrix composite. Nasa Technical Memorandum 106304. NASA (1993)
3. Miehe, C., Schotte, J., Schröder, J.: Computational micro-macro transitions and overall moduli in the analysis of polycrystals at large strains. Comput. Mater. Sci. **16**, 372–382 (1999)
4. Kanouté, P., Boso, B.P., Chaboche, J.L., Schrefler, B.A.: Multiscale methods for composites: a review. Arch. Comput. Meth. Eng. **16**, 31–37 (2009)
5. Geers, M.G.D., Kouznetsova, V.G., Brekelmans, W.A.M.: Multi-scale computational homogenization: trends and challenges. J. Comput. Appl. Math **234**, 2175–2182 (2010)
6. Reese, S.: Meso-macro modelling of fiber-reinforced rubber-like composites exhibiting large elastoplastic deformation. Int. J. Solids Struct. **40**, 951–980 (2003)
7. Stier, B., Simon, J.W., Reese, S.: Finite element analysis of layered fiber composite structures accounting for the materials microstructure and delamination. Appl. Compos. Mater. **21**, 1–17 (2014)
8. Simon, J.W., Höwer, D., Stier, B., Reese, S.: Meso-mechanically motivated modeling of layered fiber reinforced composites accounting for delamination. Compos. Struct. **122**, 477–487 (2015)
9. Gaser, T.C., Holzapfel, G.A.: A rate-independent elastoplastic constitutive model for biological fiber-reinforced composites at finite strains: continuum basis, algorithmic formulation and finite element implementation. Comput. Mech. **29**, 340–360 (2002)
10. Vladimirov, I.N., Pietryga, M.P., Reese, S.: Anisotropic finite elastoplasticity with nonlinear kinematic and isotropic hardening and application to sheet metal forming. Int. J. Plast. **26**, 659–687 (2010)

11. Vogler, V., Rolfes, R., Camanho, P.P.: Modelling the inelastic deformation and fracture of polymer composites—part I: plasticity model. Adv. Appl. Mech. **59**, 50–64 (2013)
12. Svendsen, B., Levkovitch, V., Wang, J., Reusch, F., Reese, S.: Application of the concept of evolving structural tensors to the modelling of initial and induced anisotropy at large deformation. Comput. Struct. **84**, 1077–1085 (2006)
13. Noman, M., Clausmeyer, T., Barthel, C., Svendsen, B., Huétink, J., van Riel, M.: Experimental characterization and modeling of the hardening behavior of the sheet steel LH800. Mater. Sci. Eng. A **527**, 2515–2526 (2010)
14. Pietryga, M.P., Vladimirov, I.N., Reese, S.: A finite deformation model for evolving flow anisotropy with distorsional hardening including experimental validation. Mech. Mater. **44**, 163–173 (2012)
15. Abaqus finite element software: Dessault systémes. Providence, RI (2015)
16. Barbero, E.J.: Finite Element Analysis of Composite Materials using Abaqus. CRC Press (2008)
17. Chaboche, J.L.: A review of some plasticity and viscoplasticity constitutive theories. Int. J. Plast. **24**, 1642–1693 (2008)
18. Hill, R.: A theory of the yielding and plastic flow of anisotropic metals. Proc. R. Soc. London, Ser. A **193**, 95–102 (1948)
19. Spencer, A.J.M.: Theory of invariants. In: Continuum Physics. Academic Press, New York (1971)
20. Boehler, J.P.: A simple derivation of representations for non-polynomial constitutive equations in some cases of anisotropy. Z. Angew. Math. Mech. **59**, 157–167 (1979)
21. Zhang, J.M., Rychlewski, J.: Structural tensors for anisotropic solids. Arch. Mech. **42**, 267–277 (1990)
22. Vladimirov, I.N., Pietryga, M.P., Reese, S.: On the influence of kinematic hardening on plastic anisotropy in the context of finite strain plasticity. Int. J. Mater. Form. **4**, 255–267 (2011)
23. Reese, S., Raible, T., Wriggers, P.: Finite element modelling of orthotropic material behaviour in pneumatic membranes. Int. J. Solids Struct. **38**, 9525–9544 (2001)

Comparison of Mechanical Tests for the Identification of Composite Defects Using Full-Field Measurements and the Modified Constitutive Relation Error

E. Barbarella, O. Allix, F. Daghia, E. Jansen and R. Rolfes

Abstract Composite parts manufactured in large batches always present defects. These may not influence the behavior of the structure or might on the contrary be seriously detrimental to the performance of the component. In the first case, their presence is negligible, in the other case it is fundamental to be aware of their presence to foresee countermeasures. In this framework, being able to localize and estimate the intensities of flaws is extremely interesting. In this article, we present an approach based on the Modified Constitutive Relation Error to characterize defects, employing as input the displacements field measured from simple static and dynamic tests. The identification capabilities from tensile, bending, vibration and compression tests are compared using pseudo-experimental results as input data; then the identification is shown on a real case for buckling experiments to show the potential of the method.

1 Introduction

Inverse approaches [1] can be exploited in the framework of defect characterization. Historically, a large number of works exist, which address the issue of material identification exploiting experimental results. The material identification techniques

E. Barbarella (✉) · O. Allix · F. Daghia
LMT Cachan (ENS Cachan, CNRS, Univ Paris-Saclay), 61 avenue du Président Wilson, 94230 Cachan, France
e-mail: elena.barbarella@gmail.com

O. Allix
e-mail: olivier.allix@ens-paris-saclay.fr

F. Daghia
e-mail: federica.daghia@ens-paris-saclay.fr

E. Jansen · R. Rolfes
ISD, Leibniz Universität Hanover, Appelstraße 9A, 30167 Hanover, Germany
e-mail: e.jansen@isd.uni-hannover.de

R. Rolfes
e-mail: r.rolfes@isd.uni-hannover.de

J. Sorić et al. (eds.), *Multiscale Modeling of Heterogeneous Structures*, Lecture Notes in Applied and Computational Mechanics 86,
https://doi.org/10.1007/978-3-319-65463-8_3

[2] can be successfully employed for determining the intensity of a distributed flaw affecting the material properties.

However, when dealing with localized macro-defects it is also required to determine the position of the flaws: macro-defect characterization requires, in addition to the intensity, to determine the position of each defect. For the purpose of characterizing the flaw position, another family of inverse problems can be used: the model updating techniques [3]. These are originally exploited to identify modeling errors, by evaluating the discrepancy between measurements and the model behavior.

Those two methods, material identification and model updating, shall be used in synergy to guarantee to determine both position and intensity of the defects. A technique that couples the two, providing, in addition to the correction (intensity), also its geometric support (position and dimensions), is the Modified Constitutive Relation Error (MCRE).

Born as a model updating technique [4], the Constitutive Relation Error (CRE) has been extended to the field of identification problems [5–7]. In its modified form it shows increased robustness and it is less influenced by measurement errors [8–11].

The defects we will focus on in this work are macro-defects, that is localized flaws that affect the macro or meso-scales, for instance macro-pores or fibre waviness. In particular the focus is set on material flaws: flaws which can be modeled as local modifications in material properties. Geometric effect, as for instance an initial crookedness, even if not characterized in this work, will be considered, being unavoidable.

The first section of the article deals with the description of the technique used to categorize the effects based on the Modified Constitutive Relation Error. A special focus is set on the methodology and on the formulation of the MCRE. The general technique is explained and specific details on the formulation are given for each type of test: tensile and bending static tests, vibration tests and compressive (buckling) tests.

Then the second section is devoted to present the identification results employing pseudo-experimental measurements, simulated using the finite element software Cast3M. A comparison of the identification results and of their quality is proposed for two cases: for a specimen affected by a small geometric defect and for one presenting a bigger geometric imperfection (specifically an initial crookedness of amplitude 5% and 50% of the thickness respectively).

The third section presents the identification results on a real specimen for the case of compressive tests. This type of experiment is the most tricky and the one presenting the greatest difficulties, due to the necessity to deal with instability and therefore large out-of-plane displacements. In this section the entire process is explained. Specimens manufactured with an induced fibre waviness are tested. The out of plane displacement is obtained employing StereoDIC [12, 13]: this technique turns out to be extremely powerful and accurate in reconstructing the full field measurement even for a specimen buckling considerably out of plane. The complete post-processing of the experimental data is presented, in order to obtain the inputs for the identification. To conclude, the identification results are presented and compared to the flaws visible on the surface of the specimen.

Remarks on the method and the results conclude the paper.

2 MCRE Method and Formulation

The Modified Constitutive Relation Error measures the discrepancy between the response of a real body (in our case a specimen affected by material defects) and the model of the body (defect free). The specificity of this technique, which enables us to localize the position of the flaws, is the use of an energetic functional—the modified constitutive relation error functional. Due to its formulation, the areas of the body in reality containing defects—which have not been included in the "defect free" model—correspond to higher values of the MCRE functional.

The identification procedure is an iterative procedure. Each iteration consists in two steps:

1. Localization step: through the computation of the error functional the geometric support, the position of the defects is obtained
2. Regularization and parameter updating step: the material parameters, corresponding to the areas detected at the preceding step, are updated, in this way the intensity of the defect is determined.

The high regularization property is first given by the fact that the correction of the parameter is limited to the zones that are localized as defective and second that each of these zones, thus each detected flaw, is updated as a unique entity; these two factors reduce the number of unknowns and therefore the ill-posedness of the problem.

2.1 The MCRE Functional

The first principle of the Modified Constitutive Relation Error is to divide the problem equations and available experimental information into two sets: the reliable ones, which are prescribed exactly, and the non reliable ones. In particular, the compatibility and equilibrium equations are considered reliable. Among the non reliable information one distinguish the one associated with the model itself:

- displacement and stress fields, parameters of the model, which will be iteratively modified according to the minimization of a specific functional;
- the one associated with the experiments.

For the latter two cases have to be distinguished. In the first case the measurement provides a valuable information even though possibly affected by some noise or

Table 1 Fundamentals of the constitutive relation error formulation

	Reliable	Non reliable
Theoretical	• Compatibility • Equilibrium	Parameters of the Hooke's tensor $(\boldsymbol{\sigma}, \boldsymbol{u})$ stress and displacement fields
Experimental	Initial shape Ω_0, prescribed forces and displacements	Measured displacements $\tilde{\boldsymbol{u}}$

uncertainty. In that case those values are not taken as possible unknowns and are not prone to possible modification. In the second case important measurements error can affect the data and a specific treatment of the uncertain experimental information can be included in the formulation as proposed in [9, 10]. In this paper the material is supposed to be elastic and the associated model error is defined by a strain energy measure term. The experimental measurement of displacement although not strongly prescribed will not be modified at all. Table 1 summarizes these choices.

Starting from this concept, an energy functional can be defined and the MCRE is formulated as follows:

Find the displacement field $\boldsymbol{u}$, the stress field $\boldsymbol{\sigma}$ and the parameters $\boldsymbol{p}$ that minimize :

$$E_m^2(\boldsymbol{u}, \boldsymbol{\sigma}, \boldsymbol{p}) = \int_\Omega (\boldsymbol{\sigma} - \mathcal{K}(\boldsymbol{p}) : \boldsymbol{\varepsilon}(\boldsymbol{u})) : \mathcal{K}(\boldsymbol{p})^{-1} : (\boldsymbol{\sigma} - \mathcal{K}(\boldsymbol{p}) : \boldsymbol{\varepsilon}(\boldsymbol{u}))\, d\Omega + \frac{r}{1-r} \parallel \Pi \boldsymbol{u} - \tilde{\boldsymbol{u}} \parallel^2_{\mathcal{K}} \tag{1}$$

where $\boldsymbol{\varepsilon}$ is the strain tensor, $\boldsymbol{p}$ are the parameters of the Hooke's tensor $\mathcal{K}$, $\tilde{\boldsymbol{u}}$ are the experimental displacement measurements and r is a weighting factor. Here, the two terms of the MCRE functional measure the constitutive relation error and the discrepancy between the measurements and the model response, respectively.

To solve the problem, a constraint condition is imposed. For the four loadings studied in this paper, the equilibrium equation is employed for linear static tests (tensile and bending), the eigenvalue equation for vibrations and the linear buckling equation for compressive tests (buckling).

In the following section, we specify the formulation of the modified constitutive relation error for buckling for which specific simplifications have to be made and will be discussed. For the other load cases the equilibrium equations are the usual ones applied in the infinitesimal regime and will therefore not be detailed.

2.2 The MCRE Formulation for Linearized Buckling

For the sake of simplification it has been decided in the first instance to analyze the buckling experiments in the framework of the linearized theory of buckling. This of course has also some drawbacks which will be discussed along the paper. The detailed theory behind the buckling formulation and the formulation itself can be found in [14, 15]. In particular, it should be noted that two types of experimental information can be obtained from a compressive test: in addition to the deformed shape at different instants during the test, the critical buckling load of the geometrically perfect specimen can be extracted from the experimental measurements using the Southwell plot. Of these two information, the deformed shape is considered non reliable in the following: indeed, the deformed shape is the result of a full geometrically non-linear problem, and it does not necessarily correspond to the buckling mode shape as required by the linear buckling model used in the present formulation. The experimental critical load, on the other hand, can be obtained with great precision from the Southwell plot: for this reason, it is considered to be reliable and it is enforced directly in the constraint condition.

The MCRE problem for linearized buckling can be written as follows:

Find the kinematically admissible field $\boldsymbol{u}(x) \in \mathcal{U}^{KA}(\tilde{\boldsymbol{u}})$ and the statically admissible field $\boldsymbol{\sigma}(x) \in \mathcal{S}^{SA}(\tilde{\boldsymbol{f}}, P, \boldsymbol{\Sigma_0})$ that minimize:

$$E_m^2(\boldsymbol{u}, \boldsymbol{\sigma}, \boldsymbol{p}) = \int_\Omega (\boldsymbol{\sigma} - \mathcal{K}(\boldsymbol{p}) : \boldsymbol{\varepsilon}(\boldsymbol{u})) : \mathcal{K}(\boldsymbol{p})^{-1} : (\boldsymbol{\sigma} - \mathcal{K}(\boldsymbol{p}) : \boldsymbol{\varepsilon}(\boldsymbol{u}))\, d\Omega + \frac{r}{1-r} \parallel \boldsymbol{\Pi u} - \tilde{\boldsymbol{u}} \parallel_{\mathcal{K}}^2 \tag{2}$$

under the constraint: $\forall \boldsymbol{u}^* \in \mathcal{U}^{KA}(0)$

$$\int_{\Omega_0} Tr\left(\boldsymbol{\sigma}\boldsymbol{\epsilon}(\boldsymbol{u}^*)\right) d\Omega_0 = \tilde{P}_{cr} \int_{\Omega_0} Tr\left(\boldsymbol{\Sigma}_0 \frac{{}^t\partial \boldsymbol{u}}{\partial \boldsymbol{M_0}} \frac{\partial \boldsymbol{u}^*}{\partial \boldsymbol{M_0}}\right) d\Omega_0 \tag{3}$$

where the constraint condition Eq. (3) is the the linearized buckling equation, with $\tilde{P}_{cr}$ the experimental critical load and $\boldsymbol{\Sigma}_0$ the pre-stress operator.

To be able to solve the inverse problem, the mixed formulation has to be transformed into a pure displacement one, suited for the finite element framework. The MCRE problem can therefore be written by dualization, after discretization over a finite element subspace, as follows:

Find the kinematically admissible nodal displacements **U** and ***V*** that minimize:

$$\mathscr{E}_m^2\,(\mathbf{U},\mathbf{V},\boldsymbol{p}) = \frac{1}{2}\{\mathbf{U}-\mathbf{V}\}^T[K\,(\boldsymbol{p})]\{\mathbf{U}-\mathbf{V}\} + \frac{1}{2}\frac{r}{1-r}\{\mathbf{U}-\tilde{\mathbf{U}}\}^T[G_{\tilde{u}}\,(\boldsymbol{p})]\{\mathbf{U}-\tilde{\mathbf{U}}\} \tag{4}$$

under the constraint:

$$[K\,(\boldsymbol{p})]\{\mathbf{V}\} = \tilde{P}_{cr}[K_\sigma\,(\boldsymbol{p})]\{\mathbf{U}\} \tag{5}$$

2.2.1 The Localization Step

The purpose of this step is to determine the geometric support for the updating of the parameters, namely the position of the flaws. The defective areas correspond to high values of the MCRE functional of Eq. (4), it is thus necessary to compute the error distribution, i.e. the local values of $\mathscr{E}_m^2(\mathbf{U},\mathbf{V},\boldsymbol{p})$ over the finite elements.

In order to compute the error values, taken an initial set of parameter $\boldsymbol{p}$ supposed known, it is necessary to compute the nodal displacement **U** and **V**.

The admissible displacements fields of the finite element problem **U** and **V** are solutions of the constrained minimization problem (Eqs. (4), (5)). This can be solved by introducing a Lagrange multiplier $\boldsymbol{\Lambda}$, as an additional vector of unknowns. The corresponding Lagrangian writes:

$$\begin{aligned} L(\mathbf{U},\mathbf{V},\boldsymbol{\Lambda}) =& \frac{1}{2}\{\mathbf{U}-\mathbf{V}\}^T[K]\{\mathbf{U}-\mathbf{V}\} + \frac{1}{2}\frac{r}{1-r}\{\mathbf{U}-\tilde{\mathbf{U}}\}^T[G_{\tilde{u}}]\{\mathbf{U}-\tilde{\mathbf{U}}\}+ \\ & \{\boldsymbol{\Lambda}\}^T([K]\{\mathbf{V}\} - \tilde{P}_{cr}[K_\sigma]\{\mathbf{U}\}) \end{aligned} \tag{6}$$

where the dependence on the sought-after parameters $\boldsymbol{p}$ is implied. The three unknowns can be obtained by imposing the stationarity of L:

$$\begin{aligned} \delta L =& \{\delta\mathbf{U}-\delta\mathbf{V}\}^T[K]\{\mathbf{U}-\mathbf{V}\} + \frac{r}{1-r}\{\delta\mathbf{U}\}^T[G_{\tilde{u}}]\{\mathbf{U}-\tilde{\mathbf{U}}\}+ \\ & \{\boldsymbol{\Lambda}\}^T([K]\{\delta\mathbf{V}\} - \tilde{P}_{cr}[K_\sigma]\{\delta\mathbf{U}\}) + \{\delta\boldsymbol{\Lambda}\}^T([K]\{\mathbf{V}\} - \tilde{P}_{cr}[K_\sigma]\{\mathbf{U}\}) = 0 \\ & \left(\forall\ \delta\mathbf{U},\delta\mathbf{V} \in \mathscr{U}^{KA}\right)(\forall\ \delta\boldsymbol{\Lambda}) \end{aligned} \tag{7}$$

Since the value of the critical load can be obtained from the experiment and is considered reliable, the system obtained to solve Eq. (7) with respect to the three unknowns $(\mathbf{U},\mathbf{V},\boldsymbol{\Lambda})$ is linear,

$$\begin{cases} \delta\mathbf{U} : [K]\{\mathbf{U}-\mathbf{V}\} + \frac{r}{1-r}[G_{\tilde{u}}]\{\mathbf{U}-\tilde{\mathbf{U}}\} - \tilde{P}_{cr}[K_\sigma]\{\mathbf{\Lambda}\} = 0 \\ \delta\mathbf{V} : -[K]\{\mathbf{U}-\mathbf{V}\} + [K]\{\mathbf{\Lambda}\} = 0 \\ \delta\mathbf{\Lambda} : [K]\{\mathbf{V}\} - \tilde{P}_{cr}[K_\sigma]\{\mathbf{U}\} = 0 \end{cases} \tag{8}$$

The elimination of the Lagrange multipliers $\mathbf{\Lambda}$, using the second equation, yields:

$$\begin{bmatrix} [K] + \frac{r}{1-r}[G_{\tilde{u}}] - \tilde{P}_{cr}[K_\sigma] & -[K] + \tilde{P}_{cr}[K_\sigma] \\ -\tilde{P}_{cr}[K_\sigma] & [K] \end{bmatrix} \begin{Bmatrix} \mathbf{U} \\ \mathbf{V} \end{Bmatrix} = \begin{Bmatrix} \frac{r}{1-r}[G_{\tilde{u}}]\tilde{\mathbf{U}} \\ 0 \end{Bmatrix} \tag{9}$$

which leads to the computation of $\mathbf{U}$ and $\mathbf{V}$ and thus of the error distribution.

When different loading conditions are employed, what changes is the constraint condition in Eq. (5), which is replaced

- for static tests by

$$[K(\boldsymbol{p})]\{\mathbf{V}\} = \{\tilde{\mathbf{F}}\} \tag{10}$$

  where $\tilde{\mathbf{F}}$ is the vector of imposed loads (either tensile or bending).
- for vibrations by

$$[K(\boldsymbol{p})]\{\mathbf{V}\} = \omega^2[M]\{\mathbf{U}\} \tag{11}$$

  where the stiffness matrix K is considered non-reliable, and the mass matrix M is considered reliable.

Therefore also the system (9) changes but the method of obtaining $\mathbf{U}$ and $\mathbf{V}$ does not.

To switch from the error distribution to the location of defects it was decided to use a threshold e_{th}. This is defined as the ratio between the local error value and the maximum value of the local error. All the elements whose normalized error value exceeds the threshold e_{th} contribute to compose the defective area. If a low e_{th} is chosen, the algorithm tends to over-localize the defect amplitude, while if the value is too high the identification tends to be long and more computational expensive. Commonly $e_{th} = 0.5$ or $e_{th} = 0.7$ are a good compromise.

2.2.2 Regularization and Parameter Updating Step

The regularization and parameter updating step introduces a strong regularization into the problem by reducing the number of unknowns and therefore the ill-posednesss of the problem. As output of the preceding step, the high error elements are selected to be corrected. The parameters tune the Young‘s moduli: $p_1 = E_{11}/E_{11}^0$ define the Young’s modulus in fiber direction normalized with respect to the nominal value and $p_2 = E_{22}/E_{22}^0$ the normalized one in the perpendicular direction. A possibility would be to update the parameters element by element. This choice, although giving high flexibility, increases the number of unknowns (two parameters for each element) and drastically increase the ill-posedness of the inverse problem.

It has instead been decided to gather together adjacent elements and consider them a flaw. All the elements belonging to a flaw are corrected as a unique entity. This choice has the advantage of simplicity but presents a main drawback: when the defective area is over-localized in terms of dimensions it is not possible to determine the proper intensity. This is why the choice of the threshold plays here an important role in terms of quality of the identification.

The parameter updating step can now be detailed. A gradient steepest descent methods is employed to correct the parameters $\boldsymbol{p}$. The set of good parameters is the one minimizing the cost function. With respect to the parameters at the step k, the updated values are computed:

$$p^{(k+1)} = p^{(k)} + \alpha^{(k)} g^{(k)} \tag{12}$$

where $\alpha^{(k)}$ is the step length, an objective manner to estimate its value is given by the Armijo rule, $g^{(k)} = \nabla J(\boldsymbol{p}^{(k)})$ is the search direction, with $J(\boldsymbol{p})$ being the cost function. The particularity is in the choice of the cost function J to minimize. This is taken equal to the error functional $\mathscr{E}_m^2$, choice which guarantees robustness of the correction and which eases the computation thanks to the stationarity properties of $(\mathbf{U}, \boldsymbol{V}, \boldsymbol{\Lambda})$ leading to:

$$\partial J = d\mathscr{E}_m^2 = \cancel{\frac{\partial \mathscr{E}_m^2}{\partial \mathbf{U}}} d\mathbf{U} + \cancel{\frac{\partial \mathscr{E}_m^2}{\partial \mathbf{V}}} d\mathbf{V} + \cancel{\frac{\partial \mathscr{E}_m^2}{\partial \boldsymbol{\Lambda}}} d\boldsymbol{\Lambda} + \frac{\partial \mathscr{E}_m^2}{\partial p_i} dp_i \tag{13}$$

Even if an analytic estimation of the cost function would be possible when considering as defect a drop in the Young's modulus (linearly proportional to the parameters of the Hooke's tensor), it is here decided for the sake of generality to compute numerically the gradient of the cost function (in general this is necessary when dealing with a more complex description of the defects or when the Young's modulus is not directly updated).

Once the new set of parameters computed, the accuracy of the correction is evaluated by subjecting the updated model to a successive identification step. For the purpose of evaluating the defectiveness of the model, a global error indicator is employed. It has been decided to use for the purpose the modified constitutive relation error functional normalized by its initial value:

$$GEI = \sqrt{\frac{\mathscr{E}_m}{\mathscr{E}_m^0}} \tag{14}$$

3 Comparison of the Identification Results for Different Loading Conditions

In this section we deal with the results of the identification for different loading conditions. In order to test the potential of the different loading schemes, the MCRE

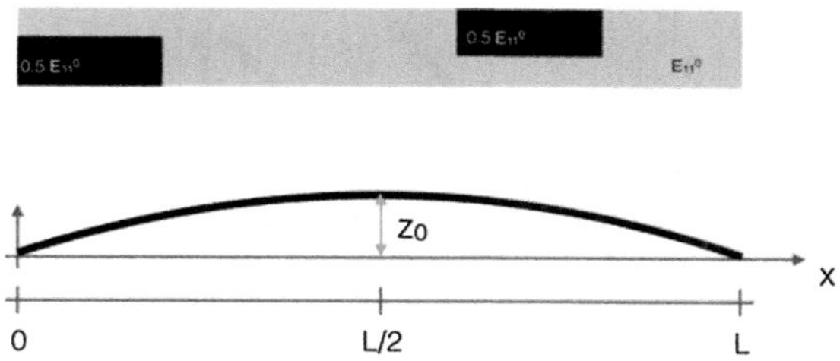

Fig. 1 Simulated specimen with two material defects (a 50% drop in Young's Modulus) and a geometric defect of initial crookedness

approach is carried out using, as pseudo-experimental measurements, the responses of a specimen with known defect as simulated using the finite element software Cast3M. To enable the comparison of the defect characterization, both in terms of position and intensity of the material defects, the specimen employed is the same for every loading conditions.

3.1 The Simulated Specimen

The pseudo-experimental specimen has the shape of an elongated plate of dimensions $200 \times 20 \times 2$ (in mm). It is affected by two material defects, the black areas in Fig. 1, that simulate a 10° fiber misalignment: $E_{11} = 0.5E_{11}^0$ and $E_{22} \simeq E_{22}^0$.

To perform non-linear analysis and to increase the verisimilitude of the simulation, a geometric defect is introduced. It consists in an initial crookedness in the form of an half sine wave. Two cases are considered, with different amplitude of the initial imperfection, respectively an almost negligible geometric defect of maximum amplitude z_0 of 5% of the thickness, and one of relatively big size, of z_0/t of 50%.

3.2 Pseudo-experimental Tests: Simulated Responses of the Specimen

In this paragraph, we present the response of the two configurations of the pseudo-experimental specimen (same material defects and different amplitude of the geometric defect) under the different loading cases and we treat the post-processing of the experimental data in order to obtain the inputs required for the identification procedure.

Vibration tests are, both theoretically and experimentally, an eigenvalue problem. For this reason, the direct output of the experiments are couples frequency-mode, which can be directly introduced in the identification procedure. In our case, no synthetic noise—noise error adedd to the simulated measurements- was introduced (even if noise may affect real measurements) and we make use of only one couple

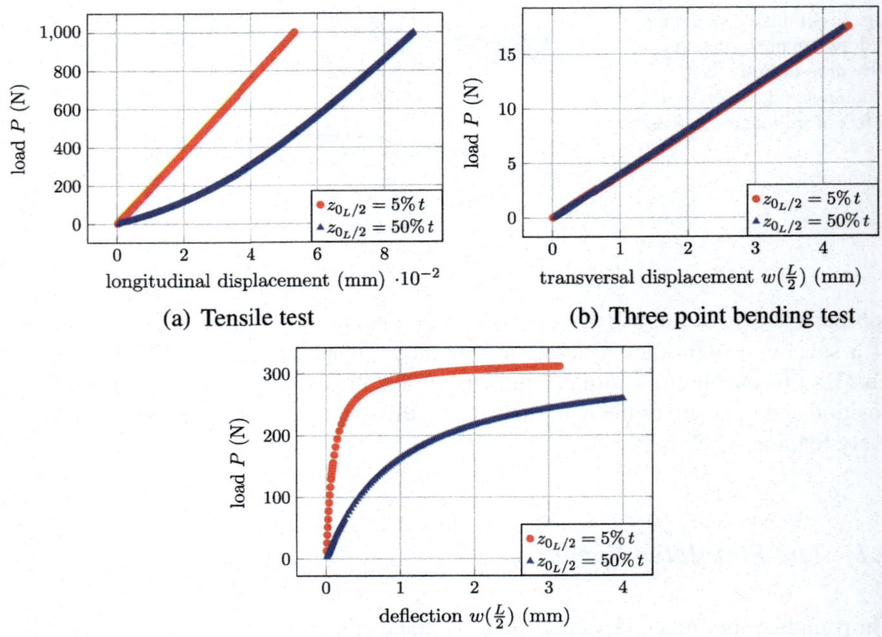

(a) Tensile test (b) Three point bending test (c) Compression test

Fig. 2 Non-linear results of the pseudo-experiments for the static loadings

frequency-mode for the identification. Commonly, for identification purposes, up to a dozen couples can be employed.

For all of the other tests considered here, on the other hand, the data that can be retrieved from experiments is the global load-displacement response (see Fig. 2), as well as the deformed shapes corresponding to different load levels, if a full-field measurement technique is used. The pseudo-experimental data for tension, bending and compressive tests are obtained here using geometrically non-linear simulations, in order to best take into account the initial geometric defect. As it can be seen in Fig. 2, the non-linearity of the response increases with the amplitude of the geometric defect; particularly exemplary is the case of compressive tests, where, in presence of a big geometric defect, the specimen's performances drop considerably (cfr. the blue line in Fig. 2c).

For tensile and bending tests, the constraint condition considered within the MCRE formulation is the linear static equilibrium equation. For this reason, it is enough to extract from the pseudo-experimental response the load vector $\tilde{\mathbf{F}}$, as well as the deformed shape $\tilde{\mathbf{U}}$ for the same load level. Here, a load of 20 N and the corresponding deformed shape are chosen.

For compressive tests, the MCRE formulation considers as a constraint condition the linearized buckling, which is an eigenvalue problem. The required information is thus the critical load $\tilde{P}_{cr}$ of the structure with material defects and the corresponding eigenmode. This data cannot be directly obtained from experiments, since for imperfect structures the eigenvalue solution is never obtained and the behaviour in compression is non-linear [16].

When dealing with beam like structures, a tool of the linearized buckling theory can be employed: the Southwell Plot [17]. This enables to obtain the first critical load of the geometrically perfect structure by post-treating appropriately the experimental data (load and out of plane displacement). In addition it also provides the amplitude of the initial imperfection. If the load P—deflection w data are plotted on a $(w\text{-}w/P)$ plane, they tend to follow a linear trend. The data diverge from the linear trend at the extremities of the curve: for small values of P/P_{cr} and for big values of the out-of plane displacement w, which is outside the range of validity of the linearized theory. The best fit curve of the data carries the information of P_{cr} and of z_0; the slope is proportional to the eigenvalue, i.e. the first critical load, and the intercept is proportional to the amplitude of the initial imperfection. It is possible to find all the theory and the assumption behind the Southwell plot in [16].

The P_{cr} and z_0 evaluated this way show an high accuracy, in [14] the value obtained from the Southwell plot are compared to the output of an eigenvalue calculation, showing for the P_{cr} an error below 1%. Therefore the P_{cr} evaluated thanks to the post-processing of the experimental data is used here as a reliable input of the MCRE identification procedure based on linearized buckling.

An issue which remains open concerns the reconstruction of the modal shape from experiments: a way of obtaining it from non-linear experimental data has not yet been found. For this reason, the non-linear deformed shape taken from the experiment is here used as mode, i.e. the second input of the identification procedure. Every deflected shape inside the range of validity of the linearized theory can be used as modal shape. For coherence with the choice made for the other types of test, the one relative to 20 N load is employed.

Now that the post-processing of the experimental results has been explained and that the inputs are available, it is possible to present the identification results.

3.3 *Identification Results: The Case of a Small Geometric Defect*

This section treats the identification of a specimen affected by an unknown geometric defect with a comparatively small size. The procedure proves its ability to identify defects even in presence of an imperfection.

In presence of a small unknown geometric defect, the procedure works well for all the loading cases analyzed.

The first identification step for all methods is summarized in Table 2. The initial values of the global error indicator and the first error density maps are shown and the resulting localization and updating are reported. Considering the error density

Table 2 Identification results for the specimen of Fig. 1 affected by a small geometric defect. Comparison of the four loading conditions, initial step

	Initial Error Map	First localization
Tension	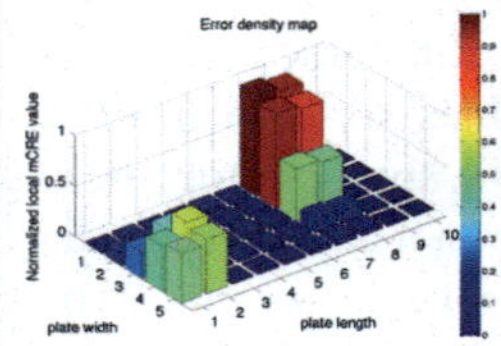	 1 0.4375
Bending	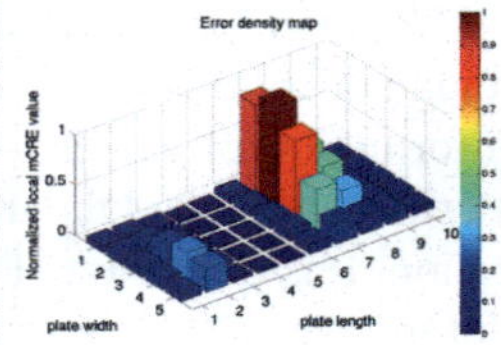	 1 0.547
Buckling	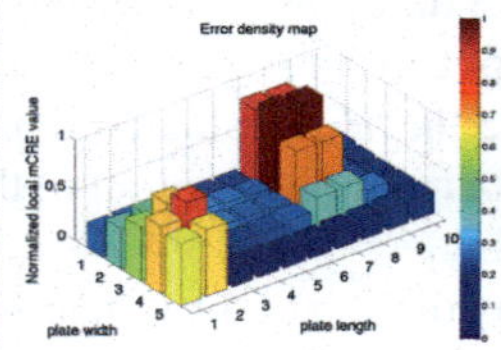	 0.592 0.375
Vibration	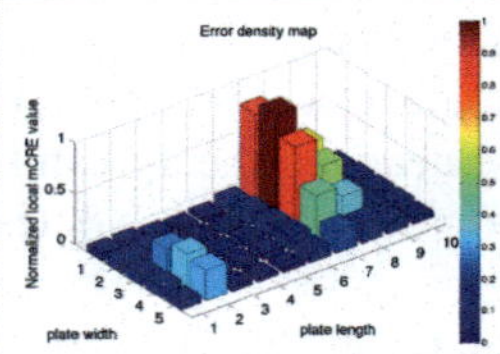	 1 0.531

map, the local error appear to be negligible everywhere except for the defects areas, the peak corresponding to the central flaw being particularly enhanced. For tension, bending and vibrations, the choice of fixing the threshold e_{th} at 0.7 brings to the localization of elements of the central flaw. For compression, on the other hand, the error map displays higher value in the second defective zone. Fixing the threshold at 0.7 entails the localization of elements on both areas: the entire *defect 2* and one element of *defect 1*.

Table 3 displays the final results of the identification. The fastest technique is the one based on buckling, which also gives the best localization results. Instead, the parameters defining the intensity of the defects are overestimated in the correction procedure. This is supposed to be due to the strong geometrical non-linearity of the

Table 3 Identification results for the specimen of Fig. 1 affected by a small geometric defect. Comparison of the four loading conditions, final step

	Final localization	Updated $p_1^{def_1}$	Error map after correction
Tension	Step 3  0.521 0.501	0.326	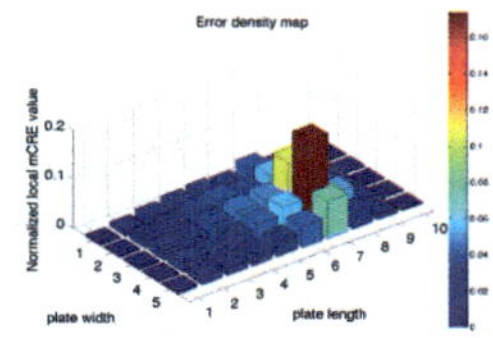
Bending	Step 3 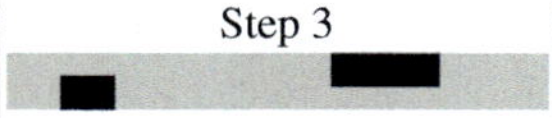 0.5 0.498	0.579	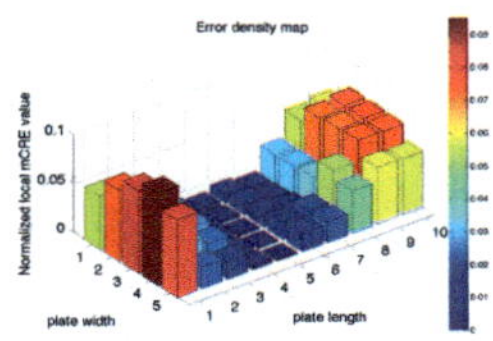
Buckling	Step 2  0.307 0.375	0.768	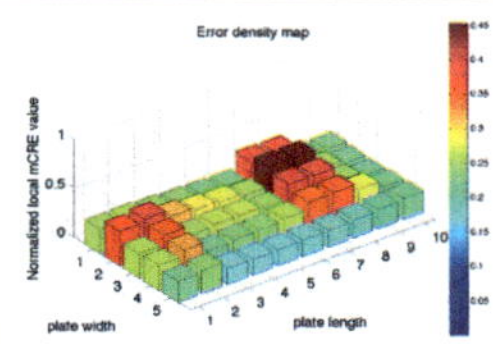
Vibration	Step 5 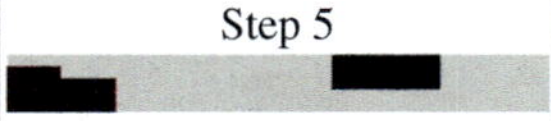 0.519 0.488	0.430	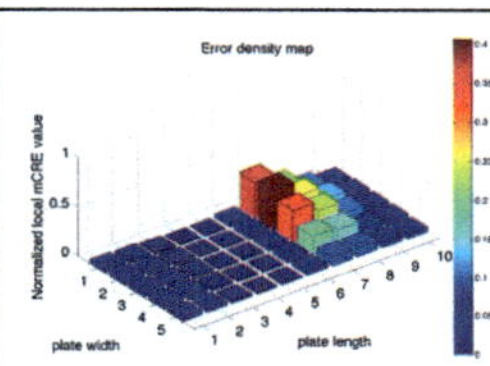

compressive test, and to their use as input of a technique formulated employing the linearized buckling theory.

The other three methods need a higher number of iterations: one more for tension and bending and three more for vibrations. For all these techniques the localization is acceptable and the updating is more accurate, the best results being obtained for bending.

Table 4 Identification results for the specimen of Fig. 1 affected by a big geometric defect. Comparison of the four loading conditions, initial step

	Initial Error Map	First localization
Tension	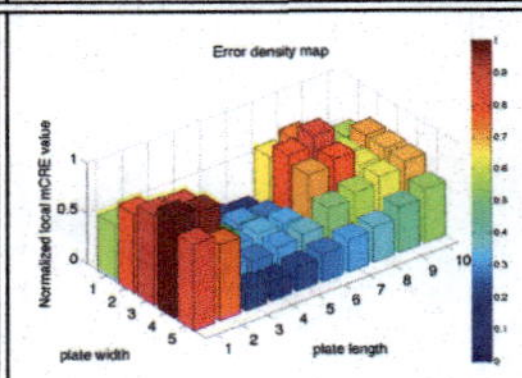	 0.433 0.437
Bending	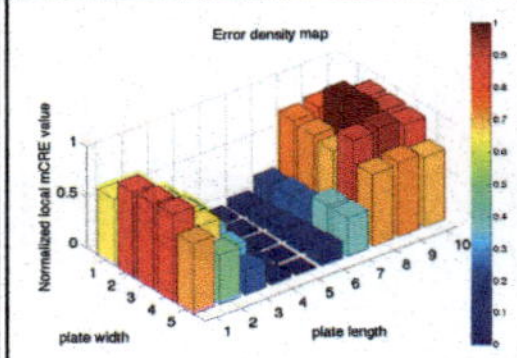	 0.441 0.375
Buckling	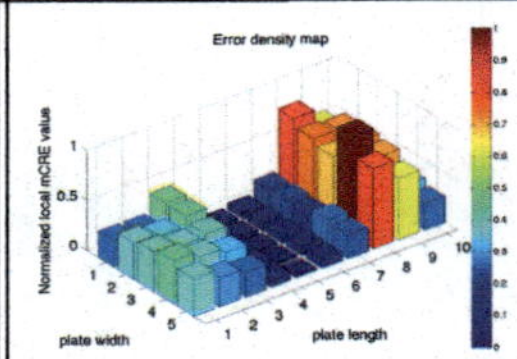	 0.787 0.4375
Vibration	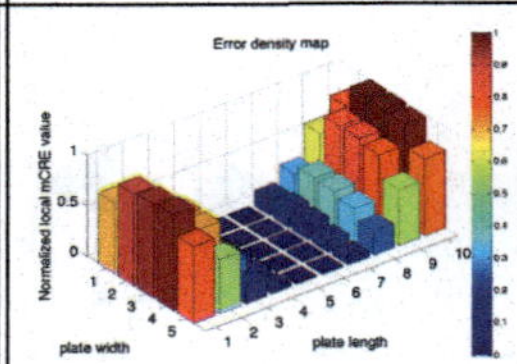	 0.708 0.379

3.4 *Identification Results: The Case of a Big Geometric Defect*

Compared to the others, this example is particularly extreme, as a geometric defect of this intensity strongly changes the behavior. This can be observed from the load-displacement curves of Fig. 2. A presentation of the different identification results is proposed in Table 4.

A comparison of the initial step for the different tests is proposed in Table 4. After inspection of the first error density maps, one may notice that in all cases the defective areas are not detected. Instead two wider zones, adjacent to the boundaries, are the ones showing a higher local error. For all loading cases, the subsequent steps of the procedure identify and correct larger and larger portions of the specimen, until almost all of the specimen is involved in the correction. This behavior of the algorithm denoted an overall error in the initial model, which corresponds to not accounting for the large initial geometric defect.

Referring to the complete identification procedure, buckling is the only technique that warns the user of the wrong updating: through the iterations the global error indicator (Eq. (14)) increases to signal that the correction carried on is incorrect.

To conclude, the identification procedure is not able to handle the presence of a big geometric defect. A possibility would be either to reconstruct the initial geometry thanks to the digital image correlation or to introduce in the procedure a step of geometry correction. In this eventuality, buckling would bring an advantage as the Southwell plot computed z_0 can serve as aid for geometry reconstruction.

3.5 Conclusions on the Comparison

In this section the effectiveness of the identification procedure is inspected for different experimental tests. In this simple case of unidirectional composite plies with the defect affecting the whole thickness, all the methods show, for the case of small unknown geometric defect, the capability to accurately identify and correct the flaws. The best results in terms of parameter estimation are obtained for bending and in terms of localization are retrieved for buckling. The fact of employing as inputs the results of strongly non-linear experiments is responsible of the poorer buckling updating results. It is also important to state that, in presence of intense geometric defect, none of these techniques is suited, without introducing a step of geometry correction.

As perspectives, two points should be assessed. First, one may want to inquire how the four techniques reacts when a more complex description of the defect is proposed, namely in presence of a laminate with a general stacking sequence and a flaw affecting only some plies through the thickness. Then, another aspect to take into account is also the simplicity of the test. In that case tension and bending are particularly appealing.

One of the advantages of buckling is given by the possibility to estimate z_0, made possible by the use of the Southwell plot. This value can be used to have an *a priori* knowledge of the identifiability, i.e. to be able even before the procedure to have a feedback on the quality of the identification. This is one of the reason for choosing buckling as the type of experiment to tests in this work, even if the intensity is in general overestimated.

4 Real Tests: Identification Employing Buckling Experimental Results

At last, we tackle the issue of identification of material defects on real specimens using buckling tests. Two types of specimen have been manufactured and tested: nominally perfect specimens, presenting no visible material defect, and defective specimens, which present a visible defect in the form of a fiber waviness induced during manufacturing.

The inputs required to perform the identification procedure are: the critical load and the modal shape. As it has been discussed in the previous chapter this translate into requiring:

- load and out-of plane displacement data to evaluate the P_{cr} using the Southwell plot
- full field measurements of the deformed shape shape, used as modal shape

The first two, load and deflection, can be obtained as direct output of the test machine and of a linear variable differential transducer (LVDT) that follows the out-of-plane displacement of a measurement point. Reconstructing the deformed shape means to be able to reconstruct the displacement through all the test of a considerable amount of measurement points on the surface: the StereoDIC is a powerful tool for this purpose.

Therefore in addition to the standard tension/compression testing machine, an LVDT and an optical system composed of lights and two cameras have been added to the rig.

The results of the tests are shown in Fig. 3 for four specimens (two nominally perfect and two defective).

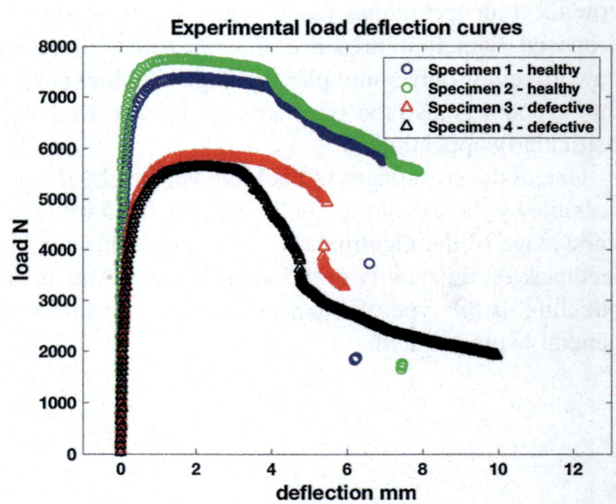

Fig. 3 Experimental results of the compression tests

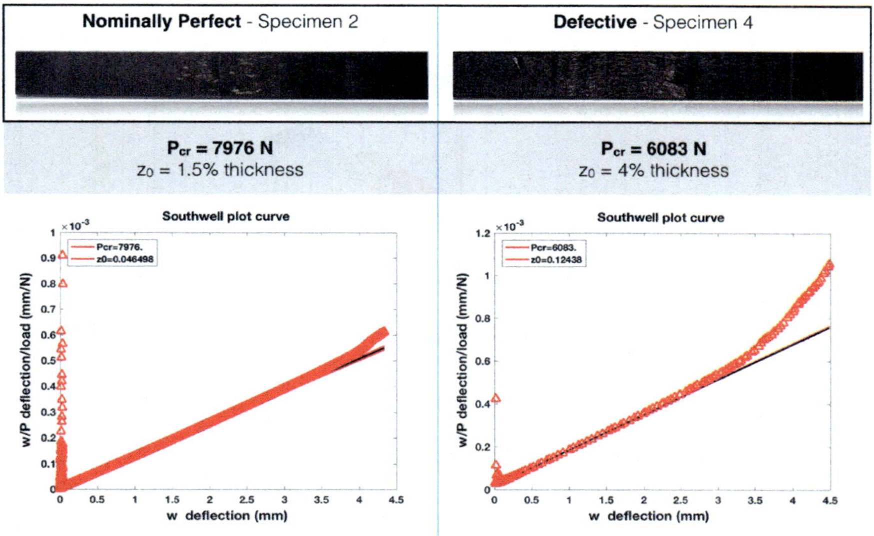

Fig. 4 Post-processing via Southwell plot

One specimen of each type, in particular specimens 2 and 4 are selected. These two will further undergo the procedure for defect identification. Before that, it is necessary to present the post-processing of the experimental results.

Let us start with the critical load. Figure 4 shows the value of the P_{cr} obtained from the Southwell plot. In addition to the critical load, the Southwell plot provides also the amplitude of the initial imperfection z_0, which can give a hint on the quality of the identification results. In both cases $z_0 < 5\%$ of the thickness, guaranteeing the identifiability.

Concerning the modal shape, Fig. 5 shows one of the deformed shape of the defective specimen during the test reconstructed using the StereoDIC. In addition to the deflected shape obtained from the full field displacements during the entire test, this technique also provides a measure of the noise. The principle is simple: by performing the StereoDIC between multiple couples of pictures taken for the unloaded specimens, the deformation obtained (supposed to be zero) is instead the noise. What is obtained for our specimens is a noise in the range $[-4, 4] \times 10^{-3}$ mm. This value, in addition to showing the low noise, is employed in the choice of the deflected shape to use as input of the identification procedure: it is decided to use a deformed shape at the initial stages of buckling but whose amplitude overpasses the noise value, to be sure that the influence of the noise is limited and that the identification results are not misrepresented.

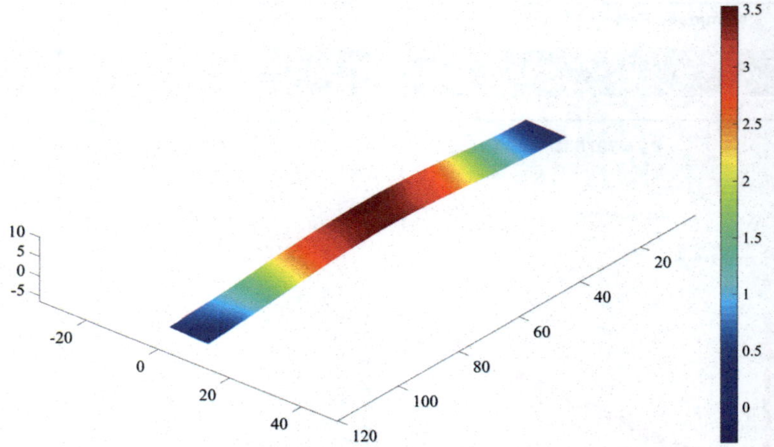

Fig. 5 Deformed shape of the deflected specimen reconstructed using StereoDIC (in mm)

Fig. 6 Defective specimen

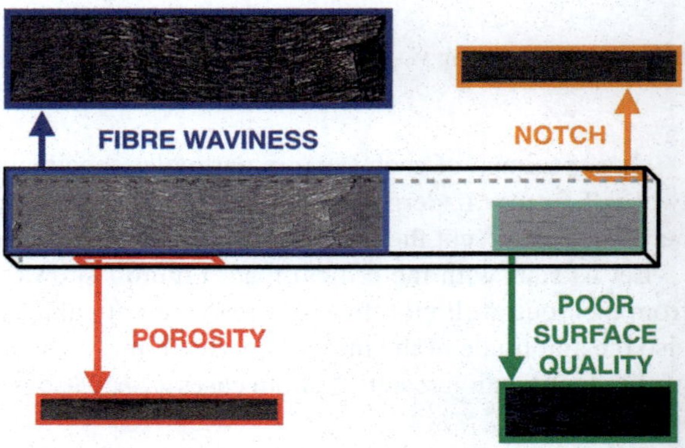

4.1 Identification Results for a Defective Specimen (Specimen 4)

Let us treat first the specimen where a defect of fiber waviness has been induced. The defective specimen 4 is shown in Fig. 6, where particular emphasis is given to the position of the flaws on the surface. In particular, the most important is the fibre waviness that occupies almost half of the surface. In the area where the undulation shows the bigger extent, an angular misalignment of the fibers of 14° is measured. This corresponds to a drop in Young's Modulus E_{11} of roughly 50–60%. In addition to this major flaw, other defects are visible: an higher porosity through the thickness in the area where the fibre undulation is present, a poor surface quality zone and a notch-like defect. To verify the latter, visual control is not sufficient and a technique as tomography would be required.

Fig. 7 Identification results for a defective specimen

The identification results are shown in Fig. 7, together with and superposed to the surface of the specimen affected by the flaw. This juxtaposition of the results enables to verify the quality of the localization: a good agreement between the position of the wavy fibers area and the zone detected by the MCRE algorithm can be found. In addition, the intensity is well estimated: a value of p_1 of 0.4 is obtained, a 60% drop in the Young's modulus, which corresponds to a 10°–20° fiber misalignment, fairly close to the 14° measured on the surface.

To conclude, bot the localization and the intensity are in good accordance with what can be visually verified. The MCRE is able to satisfactorily identify flaws when employing experimental results obtained from buckling tests.

4.2 Identification Results for a Nominally Perfect Specimen (Specimen 2)

Let us now deal with the nominally perfect specimen. The manufacturing of this batch of specimen is a particularly controlled one and, as a result, no flaws are visible on the surface. Nevertheless the identification procedure on this specimen is performed.

Figure 8 shows the identification results both at steps 3 and 4 of the identification procedure, again in comparison with the specimen surface.

At Step 3, the algorithm clearly detects that the actual, realistic boundary conditions are not represented by the boundary conditions of the defect-free model (see Fig. 8), this is plausible due to the decision of simulating perfect fixed ends boundary conditions for the identification procedure. At Step 4, the specimen is completely localized but with a very feeble correction of the parameter $p_1 = E_{11}/E^0_{11}$.

These two successive identifications highlight a unique thing: no defects are detected. This finding are in good accordance with the observation of the surface which do not show any flaw.

Both for the nominally perfect and defective specimen, the identification results correspond well with the knowledge of the defect we have. The results are therefore absolutely satisfactory.

Fig. 8 Identification results for the nominally perfect specimen: surface of the specimen, identification at Step 3 and identification at Step 4

5 Conclusions

At first, the formulation of the MCRE in the field of identification has been proposed for different loading conditions. The different tests: tension, bending, buckling and vibrations, are simulated and the identification performed. All behave in a comparable way: the position and the intensity are well estimated. Only buckling presents an overestimation of the flaw intensity, due to the fact of employing measures obtained form a strongly non-linear experiment in the framework of the linearized buckling theory.

It has therefore been intentionally decided to use buckling as loading condition for real tests, this to show the big potential of the technique. Buckling is indeed one of the trickiest test type, it requires post-processing of the results and due to the strong non-linearity leads to an overestimation of the defect intensity. Nonetheless, even with these evident difficulties, the identifications are in good accordance with the presence/absence of the flaw.

To conclude, it stems from the analysis that the MCRE is a powerful tool to localize and identify defects from simple non-destructive experimental results.

References

1. Bui, H.D.: Inverse Problems in The Mechanics of Materials: An introduction. CRC Press (1994)
2. Bonnet, M., Constantinescu, A.: Inverse problems in elasticity. IOP Publishing, Inverse problems 21(2): R1 (2005)
3. Mottershead, J.E., Friswell, M.I.: Model updating in structural dynamics: a survey. J. Sound Vib. **167**(2), 347–375 (1993)
4. Ladeveze, P.: Comparison of models of continuum media. Université Pierre et Marie Curie, Paris, These de doctorat d'État (1975)
5. Rota, L.: An inverse approach for identification of dynamic constitutive equations. ISIP **94**, 251–256 (1994)
6. Bonnet, M.: Exploiting partial or complete geometrical symmetry in 3D symmetric Galerkin indirect BEM formulations. Int. J. Numer. Methods Eng. **57**(8), 1053–1083 (2003)
7. Constantinescu, A.: An inverse approach for identification of dynamic constitutive equations. Taylor Francis Inverse Probl. **1**(4), 293–313 (1995)

8. Ladevèze, P., Leguillon, D.: Error estimate procedure in the finite element method and applications. SIAM J. Numer. Anal **20**(3), 485–509 (1983)
9. Allix, O., Feissel, P., Nguyen, H.M.: Identification strategy in the presence of corrupted measurements. Eng. Comput. **22**(5/6), 487–504 (2005)
10. Feissel, P., Allix, O.: Modified constitutive relation error identification strategy for transient dynamics with corrupted data: the elastic case. Comput. Methods Appl Mech. Eng. **196**(13), 1968–1983 (2007)
11. Feissel, P.: Vers une stratégie d'identification en dynamique rapide pour des données incertaines. Ph.D. thesis (2003). École normale supérieure de Cachan-ENS Cachan
12. Beaubier, B., Dufour, J.E., Hild, F., Roux, S., Sylvain, Lavernhe, Lavernhe-Taillard, K.: CAD-based calibration and shape measurement with stereoDIC. Exp. Mech. **54**(3), 329–341 (2014)
13. Dufour, J.E., Beaubier, B., Hild, F., Roux, S.: CAD-based displacement measurements with stereo-DIC. Exp. Mech. **55**(9), 1657–1668 (2015)
14. Barbarella, E., Allix, O., Daghia, F., Lamon, J., Jollivet, T.: A new inverse approach for the localization and characterization of defects based on compressive experiments. Comput. Mech. **57**(6), 1061–1074 (2016)
15. Barbarella, E.: Towards the localization and characterization of defects based on the modified error in constitutive relation: focus on the buckling test and comparison with other type of experiments. Ph.D. thesis (2016). École normale supérieure de Cachan-ENS Cachan
16. Bažant, Z., Cedolin, L.: Stability of structures. In: Elastic Inelastic Fracture and Damage Theories. World Scientific (2010)
17. Southwell, E.V.: On the analysis of experimental observations in problems of elastic stability. In: Problems of Elastic Stability Proceedings of Royal Society of London, vol. 135, pp. 601–616 (1932)

Snap-Through of Bistable Configurations Generated from Variable Stiffness Composites

Ayan Haldar, José Reinoso, Eelco Jansen and Raimund Rolfes

Abstract Structures made of variable stiffness (VS) composites possess a rich design space for bistable configurations that demonstrate different values of curvatures and out-of-plane displacements. In this study, several VS composites are investigated which can yield cylindrical bistable shapes similar to those generated from unsymmetrical cross-plies. Such configurations have been found favorable as a component for certain morphing applications. A semi-analytical model based on the Rayleigh-Ritz approach is presented to calculate the thermally induced multistable shapes as well as the snap-through forces particularly taking into account the curvilinear paths of VS composites. A nonlinear finite element analysis is performed to check the accuracy of the semi-analytical method. Cylindrical shapes generated from VS laminates and the corresponding straight fiber cross-ply laminates are analyzed and compared. The snap-through forces are subsequently calculated and compared for different VS laminates and the straight fiber cross-ply. It is observed that certain VS composites have a significant reduction of snap-through forces but with a marginal difference in out-of-plane displacement as compared to the corresponding straight fiber cross-ply laminate.

1 Introduction

In the field of aerospace and wind industry, multistable or bistable structures have shown a great potential in morphing applications [1–5]. This is especially due to the existence of multiple stable shapes and their ability to remain in these stable states without any external forces. Several concepts have been investigated to include

A. Haldar (✉) · E. Jansen · R. Rolfes
Institute of Structural Analysis, Leibniz Universität Hannover,
Appelstrasse 9A, 30167 Hannover, Germany
e-mail: a.haldar@isd.uni-hannover.de

J. Reinoso
Elasticity and Strength of Materials Group, School of Engineering,
Universidad de Sevilla, Camino de Los Descubrimientos s/n, 41092 Seville, Spain
e-mail: jreinoso@us.es

J. Sorić et al. (eds.), *Multiscale Modeling of Heterogeneous Structures*,
Lecture Notes in Applied and Computational Mechanics 86,
https://doi.org/10.1007/978-3-319-65463-8_4

multistable components in morphing aerofoils. Diaconu et al. [6] explored three concepts using bistable elements in a morphing aerofoil by changing the camber, the chord length of the section and also by using it as an adaptive flap. Arrieta et al. [7] studied several composites with different layups to obtain stiffness variability to achieve a distributed compliance for passive load alleviation in a morphing airfoil. Conventionally, such bistable shapes are constructed using unsymmetrical cross-plies.

It is known that an ideal morphing system asks for a structure with highly anisotropic internal architecture to fulfill the contradictory requirements of compliance and stiffness [8]. With the advancement of fiber placement technology, it is possible to manufacture fibers even with curvilinear paths or so-called variable stiffness (VS) composites. This ability to locally tailor the fiber orientation provides a much wider design space and allows composite structures with desired optimal properties to be developed.

Previous studies regarding the mechanical performance of VS composites demonstrated their higher critical buckling loads in comparison to those complying with conventional straight fibers. This capability directly stems from their ability to redistribute the pre-buckling stresses [9]. The enhanced postbuckling behavior of VS composites was also demonstrated by Wu et al. [10] and Coburn et al. [11] for plates with different boundary conditions under uniform edge-compressive displacements and in-plane shear loading [12], among others.

VS laminates similar to those of unsymmetric straight fiber laminates exhibit multistable shapes when cured from high temperature to room temperature [13, 14]. Panesar et al. [15] used bistable tow steered blended laminates to study the behavior of the stable states in the trailing edge flap, and also found the optimum fiber direction for maximum out-of-plane displacement and maximum angle of attack. The optimized results showed varying values of fiber orientation at every discrete part of the flap. Sousa et al. [13] modified the structural example presented by Mattioni et al. [16] that contained a rectangular plate consisting of partly symmetric and unsymmetric layup, with VS composites. The multistable shapes of the VS composite were analyzed using a commercial FE package and were compared with straight fibers. The curvilinear fiber path used by Sousa et al. ensured much smoother fiber continuity between the two adjacent regions, thus avoiding stress concentrations at the delimiting interface. This aspect is quite advantageous when a multistable part is integrated into a larger structure for morphing applications. Further, the FE calculations predicted that the VS laminates required lower snap-through forces for changing from one stable shape to another, which could be beneficial for morphing structures [13, 17]. Recently, Haldar et al. [14] conducted a comprehensive parametric study to determine the relation between different angle parameters and the resulting bistable shapes of a VS composite plate. From this study, it was seen that certain VS laminates that satisfy the condition $\phi = 45°$ and $T_0 + T_1 = 90°$ generated cylindrical bistable shapes with low twisting curvatures similar to the bistable shapes produced from unsymmetric cross-plies (Fig. 1).

It has been seen from previous works [2, 3, 6] that the stable shapes generated from cross-plies have attracted attention in applications to morphing in the aerospace

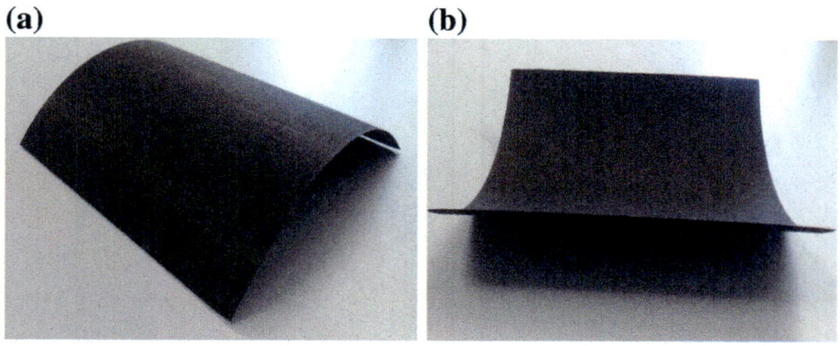

Fig. 1 Cured shapes of unsymmetrical cross-ply laminate, $[0°/90°]_T$

industry. This is particularly because of its excellent bistable characteristics. Therefore, in this work equivalent VS laminates are investigated that yields similar cylindrical bistable shapes. The bistable shapes are predicted and compared using the semi-analytical techniques and the FE approach. Further, the snap-through forces of the investigated VS and the cross-ply laminates are also calculated using the semi-analytical and the FE approach. The primary objective this work is to check the effect on the out-of-plane displacements and the snap-through forces of the bistable shapes by changing the fiber orientation of the VS laminates.

The manuscript is organized as follows. Section 2 briefly describes the curvilinear fiber path definition and defines various angle parameters. The extended CLT for VS composites is outlined in Sect. 3. Key aspects concerning the FE computations of VS composites like the cool-down and the snap-through process are given in Sect. 4. Section 5 presents the semi-analytical results and their correlation with the corresponding FE simulations. The calculated snap-through from semi-analytical and FE calculations are presented in Sect. 6. Finally, the concluding remarks of this investigation are given in Sect. 7.

2 Variable Stiffness Model

In this study instead of straight fiber, a curvilinear fiber path definition for the composite laminate is employed. Although there exist different ways to vary the fiber path in a curvilinear pattern, the linear variation of the fiber orientation angle proposed by Gürdal et al. [9] is considered in this work. Such linear variation is more practical due to its adaptability to manufacturing constraints, and also gives the possibility to build simpler analytical models.

The fiber orientation angle θ is defined as follows:

$$\theta(x') = \phi + \frac{(T_1 - T_0)\,|x'|}{d} + T_0 \tag{1}$$

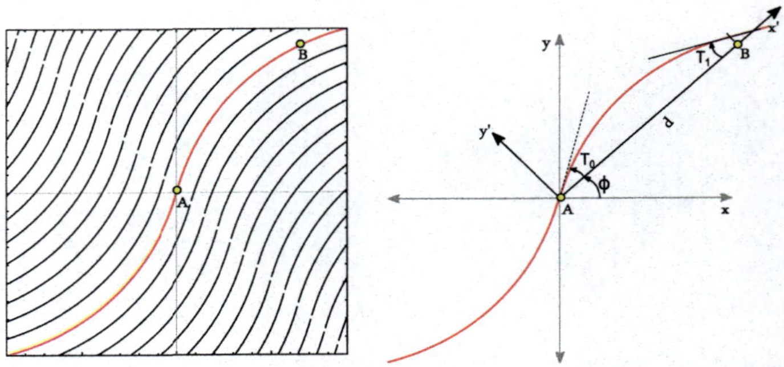

Fig. 2 Parameters defining the curvilinear fiber path [9]

where

$$x' = x \cos \phi + y \sin \phi \quad (2)$$

Here, the fiber orientation linearly varies from angle T_0 at Point A to angle T_1 at Point B (Eq. 2). The points A and B are separated by a distance d on the x' axis, referred as the characteristic length (Fig. 2). The angle parameters T_0 and T_1 are defined with respect to the x' and y' axes, which are rotated by angle ϕ with respect to the reference Cartesian coordinate axes (Fig. 2). The value d can be determined from the coordinates of point $A(X_0, Y_0)$ and $B(X_1, Y_1)$ where the angles T_0 and T_1 are defined.

In Fig. 2, the fiber marked in red is the reference fiber path. All the other fiber paths can be generated by shifting the reference fiber path in the y' direction (known as the shifted fiber method [18]. Although the fiber is changing its orientation along x' direction, if viewed from the perspective of Cartesian coordinate axes, the fiber orientation is a function of x and y: $\theta = \theta(x, y)$. The standard notation to define a particular VS laminate with the above-mentioned three parameters is as follows: $\phi\langle T_0|T_1\rangle$. Throughout this work, T_0 is defined as the angle at the center of the plate, whereas T_1 is defined as the angle at the characteristic length of the plate.

3 Semi-analytical Approach

3.1 Governing Equations

In the present study, the Dano and Hyer model [19] is extended, whereby the fiber orientation is defined according to Eq. (1). An unsymmetric square plate of side length L and thickness t is considered as reference structural definition, where neither external mechanical forces nor hygroscopic effects are accounted for in the analysis. It should also be noted that as the fiber orientation is a function of x and y, the

ABD matrix also varies along the coordinates of a plate. This flexibility to change the stiffness terms of the plate as a function of the coordinates of the composite gives the designer a wide range of tailoring possibilities. The following equations systematically derive the total potential energy with an assumed polynomial for the out-of-plane displacement and strain field and follow the Rayleigh-Ritz approach to determine the coefficients of the displacement functions.

A material point in the deformed configuration can be expressed as $\mathbf{x} = \mathbf{X} + \mathbf{u}$, where $\mathbf{u}(u, v, w)$ denotes the displacement vector in the x, y and z direction, whereas **x**, **X** identify the position vectors in the undeformed and in the reference configuration, respectively. The components of the displacement vector are defined accordingly as:

$$u(x, y, z) = u_0(x, y) - z\frac{\partial w_0}{\partial x}, \quad v(x, y, z) = v_0(x, y) - z\frac{\partial w_0}{\partial y}, \quad w(x, y, z) = w_0(x, y) \tag{3}$$

where the subscript 0 identifies the mid-plane displacements.

The strain components include non-linear von Kármán strains under the assumption of small strains and moderate rotations are given by:

$$\epsilon_{xx} = \frac{\partial u}{\partial x} + \frac{1}{2}\left(\frac{\partial w}{\partial x}\right)^2, \quad \epsilon_{yy} = \frac{\partial v}{\partial y} + \frac{1}{2}\left(\frac{\partial w}{\partial y}\right)^2, \quad \gamma_{xy} = \frac{\partial u}{\partial y} + \frac{\partial v}{\partial x} + \frac{\partial w}{\partial x}\frac{\partial w}{\partial y} \tag{4}$$

which shows the non-linear strain displacement relationships. By inserting Eq. (3) into Eq. (4), the strain relations can be rearranged as:

$$\epsilon = \begin{bmatrix} \epsilon_{xx} \\ \epsilon_{yy} \\ \gamma_{xy} \end{bmatrix} = \begin{bmatrix} \varepsilon_{xx} \\ \varepsilon_{yy} \\ \varepsilon_{xy} \end{bmatrix} + z \begin{bmatrix} \kappa_{xx} \\ \kappa_{yy} \\ \kappa_{xy} \end{bmatrix} = \begin{bmatrix} \frac{\partial u_0}{\partial x} + \frac{1}{2}\left(\frac{\partial w_0}{\partial x}\right)^2 \\ \frac{\partial v_0}{\partial y} + \frac{1}{2}\left(\frac{\partial w_0}{\partial y}\right)^2 \\ \frac{\partial u_0}{\partial y} + \frac{\partial v_0}{\partial x} + \frac{\partial w_0}{\partial x}\frac{\partial w_0}{\partial y} \end{bmatrix} + z \begin{bmatrix} -\frac{\partial^2 w_0}{\partial x^2} \\ -\frac{\partial^2 w_0}{\partial y^2} \\ -2\frac{\partial^2 w_0}{\partial x \partial y} \end{bmatrix} = \boldsymbol{\varepsilon} + z\boldsymbol{\kappa}, \tag{5}$$

where $\boldsymbol{\varepsilon}$ and $\boldsymbol{\kappa}$ represent the mid-plane strain and curvature vectors, respectively.

The kinematic field proposed by Dano and Hyer [20] is considered in this study which accounts for both the bending and the twisting curvature in its kinematic description. Particularly, the mid-plane strains are approximated as the following set of complete polynomials:

$$\begin{aligned} \varepsilon_x^0 &= c_1 + c_2 x^2 + c_3 y^2 + c_4 xy \\ \varepsilon_y^0 &= c_5 + c_6 x^2 + c_7 y^2 + c_8 xy \\ w_0 &= \frac{1}{2}(c_9 x^2 + c_{10} y^2 + c_{11} xy) \end{aligned} \tag{6}$$

The following function leads to the development of constant curvatures through the plate. The curvatures can therefore be written as:

$$\kappa_{xx} = \frac{\partial^2 w_0}{\partial x^2} = -c_9, \ \kappa_{yy} = \frac{\partial^2 w_0}{\partial y^2} = -c_{10}, \ \kappa_{xy} = 2\frac{\partial^2 w_0}{\partial x \partial y} = -c_{11} \quad (7)$$

It is important to note that the assumed strain and displacement fields satisfy the compatibility equation of a curved surface, as given in [21]:

$$\det \kappa = \kappa_{xx}\kappa_{yy} - \kappa_{xy}^2 = \frac{\partial^2 \varepsilon_y}{\partial x^2} + \frac{\partial^2 \varepsilon_x}{\partial y^2} - 2\frac{\partial^2 \varepsilon_{xy}}{\partial x \partial y} \quad (8)$$

Using Eq. (6) in the expression for extensional strains ε_x^0, ε_y^0, the in-plane displacements can be computed as:

$$\begin{aligned} u_0(x, y) &= \int \left[\varepsilon_x^0 - \frac{1}{2}\left(\frac{\partial w^0}{\partial x}\right)^2 \right] dx + c_{12}y + c_{13}y^3 \\ v_0(x, y) &= \int \left[\varepsilon_y^0 - \frac{1}{2}\left(\frac{\partial w^0}{\partial y}\right)^2 \right] dy + c_{15}x + c_{14}x^3 \end{aligned} \quad (9)$$

Solving Eq. (9) the in-plane displacement field can be expressed as:

$$\begin{aligned} u_0(x, y) = c_1 x + c_{12} y + \frac{1}{2}\left(c_4 - \frac{c_9 c_{11}}{2}\right) x^2 y + \left(c_3 - \frac{c_{11}^2}{2}\right) x y^2 + \\ \frac{1}{3}\left(c_2 - \frac{c_9^2}{2}\right) x^3 + \frac{1}{3} c_{13} y^3 \\ v_0(x, y) = c_{15} x + c_5 y + \left(c_6 - \frac{c_{11}^2}{8}\right) x^2 y + \frac{1}{2}\left(c_8 - \frac{c_{10} c_{11}}{2}\right) x y^2 + \\ \frac{1}{3}\left(c_7 - \frac{c_{10}^2}{2}\right) y^3 + \frac{1}{3} c_{14} y^3 \end{aligned} \quad (10)$$

In order to remove rigid body motion from the assumed displacement field, the first order terms of the variable x and y are needed to be equated, which results in $c_{15} = c_{12}$. The shear strain can simply be calculated from the expression:

$$\gamma^0 = \frac{\partial u^0}{\partial y} + \frac{\partial v^0}{\partial x} + \frac{\partial w^0}{\partial y}\frac{\partial w^0}{\partial x} \quad (11)$$

Based on the above derivations, it can be seen that the coefficients from c_1 to c_{14} correspond to the set of unknowns, which are determined using Rayleigh–Ritz method.

Integrating the stress-strain relation over the thickness, one obtains force and moments resultant vectors as follows

$$\begin{bmatrix} \mathbf{N} \\ \mathbf{M} \end{bmatrix} = \begin{bmatrix} \mathbf{A}(x,y) & \mathbf{B}(x,y) \\ \mathbf{B}(x,y) & \mathbf{D}(x,y) \end{bmatrix} \begin{bmatrix} \boldsymbol{\varepsilon} \\ \boldsymbol{\kappa} \end{bmatrix} - \begin{bmatrix} \mathbf{N}^{th} \\ \mathbf{M}^{th} \end{bmatrix}, \tag{12}$$

where **N** and **M** are the vectors containing the resultant forces and moments respectively and **A**, **B** and **D** represent the in-plane, coupling and bending matrices respectively. The main difference between the formulation used in a typical straight fiber laminate and the one corresponding to VS composites are the terms present in the **A**, **B** and **D** matrices, which are constants for straight fiber while they vary as a function of spatial coordinates in case of VS composites as shown in Eq. (12). The resultant quantities with the superscript *th* in Eq. (12) denote the thermal contributions. Therefore, one can express

$$\begin{aligned} \left[A_{ij}(x,y), B_{ij}(x,y), D_{ij}(x,y)\right] &= \int_{-t/2}^{t/2} \overline{Q}_{ij} \left[1, z, z^2\right] \mathrm{d}z \\ &= \sum_{k=1}^{N_{ply}} \overline{Q}_{ij}^{(k)}(x,y) \left[z_{k+1} - z_k, z_{k+1}^2 - z_k^2, z_{k+1}^3 - z_k^3\right] \end{aligned} \tag{13}$$

where N_{ply} identifies the total number of plies for each laminate configuration.

The value of $\overline{Q}_{ij}^{(k)}$ can also be written as a function of invariants of orthotropic material U_i [9, 22].

$$\begin{aligned} \bar{Q}_{11} &= U_1 + U_2 \cos\left[2\theta(x,y)\right] + U_3 \cos\left[4\theta(x,y)\right] \\ \bar{Q}_{12} &= U_4 - U_3 \cos\left[4\theta(x,y)\right] \\ \bar{Q}_{22} &= U_1 - U_2 \cos\left[2\theta(x,y)\right] + U_3 \cos\left[4\theta(x,y)\right] \\ \bar{Q}_{66} &= U_5 - U_3 \cos\left[4\theta(x,y)\right] \\ \bar{Q}_{16} &= \frac{1}{2} U_2 \sin\left[2\theta(x,y)\right] + U_3 \sin\left[4\theta(x,y)\right] \\ \bar{Q}_{26} &= \frac{1}{2} U_2 \sin\left[2\theta(x,y)\right] - U_3 \sin\left[4\theta(x,y)\right] \end{aligned} \tag{14}$$

The **ABD** matrix can be written in terms of the material invariants, whose details can be found in Haldar et al. [14]. This splitting of stiffness coefficients $\overline{Q}_{ij}$ in terms of invariants makes the computations much faster and avoids higher powers of trigonometric terms in the formulation. This is especially suitable for this work where the fiber orientation angle depends on the spatial coordinates of the plate.

The response of the laminated structure is determined through the Minimum Potential Energy Theorem, where the potential energy of the structure in the absence of external mechanical actions is given by:

$$\Pi = \int_{-L/2}^{L/2} \int_{-L/2}^{L/2} \left(\frac{1}{2} \begin{bmatrix} \varepsilon \\ \kappa \end{bmatrix}^T \begin{bmatrix} \mathbf{A}(x, y) & \mathbf{B}(x, y) \\ \mathbf{B}(x, y) & \mathbf{D}(x, y) \end{bmatrix} \begin{bmatrix} \varepsilon \\ \kappa \end{bmatrix} - \begin{bmatrix} \mathbf{N}^{th}(x, y) \\ \mathbf{M}^{th}(x, y) \end{bmatrix}^T \begin{bmatrix} \varepsilon \\ \kappa \end{bmatrix} \right) \mathrm{d}x\, \mathrm{d}y \tag{15}$$

For square plates, the Rayleigh-Ritz method can be applied using Eq. (15) as starting point, minimizing the potential energy of the structure ($\delta\Pi = 0$). The displacements depend on a certain number of unknowns denoted as c_i (i = 1, nn being the total number of unknowns) that need to be determined. For a particular ΔT, the total potential energy can be expressed in terms of unknowns c_i as shown in Eq. (16).

$$\Pi \approx \Pi_N(\mathbf{c}), \mathbf{c} = \{c_i\}, i = 1, \ldots, nn \tag{16}$$

Using the Principle of Minimum Potential Energy, the final deformed shapes of the multistable unsymmetric laminates are calculated, as it is cooled down from its cure temperature.

3.2 *Snap-Through Force*

The cool-down from the curing temperature to working or room temperature leads to the generation of one of the stable shapes. Snap-through from one stable shape to another can be achieved using actuators, smart memory alloys or simply a mechanical force. In this work, a concentrated force is applied at each corner of the plate to facilitate snap-through. Therefore, the external force contribution due to these concentrated forces to the virtual energy equation can be written as:

$$\delta V = F_x.\delta u + F_y.\delta v + F_z.\delta w \tag{17}$$

As the force is applied at the z-direction, the components $F_x = F_y = 0$. The values of the displacement fields can be directly computed using the kinematic relation described in Sect. 3.1. Substituting Eq. (6) in Eq. (17), one obtains:

$$\delta V = \frac{L^2}{8} F_z \left(\delta c_9 + \delta c_{10} + \delta c_{11}\right) \tag{18}$$

Accordingly, the modified virtual work principle with the external force contribution V and internal strain energy Π can be written as follows:

$$\delta W_T = \delta\Pi - \delta V = 0 \tag{19}$$

$$\frac{\partial W_T(\mathbf{c})}{\partial c_i} = 0 \tag{20}$$

$$\left(\frac{\partial \Pi_N}{\partial c_9} - \frac{L^2}{8}F_z\right)\delta c_9 + \left(\frac{\partial \Pi_N}{\partial c_{10}} - \frac{L^2}{8}F_z\right)\delta c_{10} + \left(\frac{\partial \Pi_N}{\partial c_{11}} - \frac{L^2}{8}F_z\right)\delta c_{11} + \cdots \frac{\partial \Pi_N}{\partial c_i}\delta c_i = 0 \tag{21}$$

This results in a set of highly non-linear system of equations (Eq. 20) which are solved using the Newton-Raphson technique. Finally, the stability of the solution is evaluated by means of the construction of the Jacobian matrix **J**, that reads:

$$\mathbf{J} = \frac{\partial^2 W_T}{\partial c_i \partial c_j}, \text{ i, j=1, ..., nn} \tag{22}$$

An equilibrium configuration is stable, if and only if the corresponding Jacobian matrix Eq. (22) is positive definite. All the symbolic computation described were accomplished using the routines written in *Mathematica.*

The bistable shapes obtained after the cool down process can be determined by setting the value of the applied force to zero. The snap-through force can subsequently be found by gradually increasing the value of the applied force until a single solution is found. It should be noted that Dano and Hyer [19] and Diaconu et al. [23] describes this approach for straight fiber laminates.

4 Finite Element Analysis

In this section, the procedure to model multistable variable stiffness composites using FEM is described. As outlined in the previous section, it is clear that with the semi-analytical method can be used to analyze simple geometries. Through the use of finite elements, there is no restriction on the complexity of the laminate geometry. Although being accurate, finite elements are computationally much more expensive and therefore unsuitable for carrying out parametric or optimization studies.

In this study, a nonlinear FE analysis is performed in ABAQUS using 2304 four-node quadrilateral shell elements (S4R) for all the results presented in the next sections. The curing process is simulated by using a predefined uniform temperature field over the entire domain under analysis. The composite plate is allowed to cool down from curing temperature to room temperature, resulting in a curved bistable configuration. The plate is considered to be fixed at the center node during the cool-down process to preclude rigid body motions.

The curvilinear fiber path in the variable stiffness composite is approximated by considering a piecewise function, where each element assumes a straight fiber orientation. The corresponding fiber angle at each element is computed at its centroid from Eq. (1). Figure 3 illustrates the adopted approach, where the fiber orientation within an individual element, remains constant but varies from one element to the other in a linearly piecewise manner. It is quite obvious that with finer meshes the current approximation of the curved fiber paths is improved. Mesh convergence

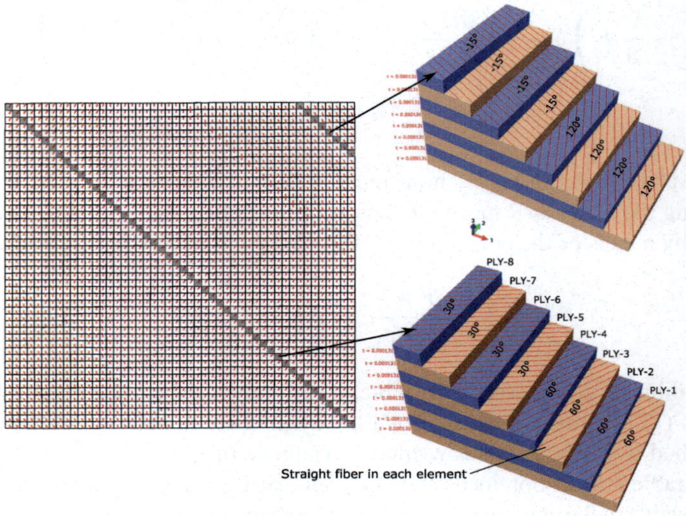

Fig. 3 FE modeling of variable stiffness composites

studies were carried out, and this particular mesh size (2304 elements) was chosen to achieve a good balance between the computational efforts and accuracy of the analysis.

To model the fibers using shifted method [18] a reference fiber path passing through the origin is defined. The other fiber paths are generated by directly translating the reference fiber path in the direction of the y' axis. However, for $\phi = 0$, the set of elements that have the same fiber orientation is parallel to the rectangular Cartesian y axis. However, for $\phi = 45°$, the new coordinate axes (x' and y') are rotated by 45° with the Cartesian coordinate axes. Therefore, all the diagonal elements in the FE element model have the same fiber orientation (as shown in Fig. 3).

4.1 Cool-Down Process

The cool down process is simulated for the given composite plate using a commercial FE package from the curing temperature to room temperature. No temperature dependence on the material properties is considered. It is assumed that the temperature is distributed uniformly on the plate surface. The following steps obtain the cooling down process for VS composites:

1. The curvilinear fiber path is defined by ϕ, T_0 and T_1 values. The fiber orientation at the centroid of each element is assigned using Eq. (1). The coordinates of the centroid can be transformed in terms of Cartesian coordinate axes using Eq. (2).

2. A linear eigenvalue buckling problem is solved under the uniform prescribed thermal loading over the plate domain.
3. The resulting eigenmodes are subsequently applied as an initial geometrical imperfection to the real composite laminates resulting in one of the stable states. Using a different eigenmode as initial geometrical imperfection may lead to the other stable shape.
4. The cool down process is simulated by using static finite element method considering geometrical nonlinear effects. The analysis is carried out using stabilization with artificial damping or viscous forces in order to facilitate converged equilibrium solutions along the loading path.

During the cool down process, the plate is considered to be initially in a stress-free state. Environmental effects like moisture absorption and shrinkage are not considered in this analysis. However, to obtain the stable shapes, it is important to check that the tangent stiffness matrix does not have negative eigenvalues. If negative eigenvalues are found, the numerical scheme jumps over the bifurcation point, and an unstable solution is found. The problem is assumed as quasi-static, and an automatic stabilization with constant damping factor is applied [24] to achieve convergence in the FE simulation. In particular, viscous forces are added in the form of:

$$F_v = c\mathbf{M}^* v \tag{23}$$

where c is the damping factor, $\mathbf{M}$ is the artificial mass matrix, v is the vector for nodal velocities. The viscous force vector is added to the global equilibrium equations as follows:

$$F_{ext} - F_{int} - F_v = 0 \tag{24}$$

where F_{ext} and F_{int} are the external and the internal force vectors. To replicate the results of [23] for straight fibers, a certain damping factor was used during the analysis. The damping factor is high enough to reduce the local instabilities when convergence is not achieved. However, it should not be chosen too high as this results in inaccurate solutions. The artificial damping factor equal to 10^{-7} is found to be appropriate upon some preliminary FE computations and lead to solutions with positive eigenvalues, without deteriorating the accuracy of the solution.

4.2 *Snap-Through Process*

In order to characterize a multistable laminate, it is important to study its snap-through behavior. To use it in the context of morphing applications, it is important that snap-through forces are low enough so that it can be realistically achieved. The transition from one stable solution to the other stable configuration can also be achieved through mechanical actions, as is the case of the present investigation. In particular, all the described steps are well illustrated in Fig. 4, where the transition

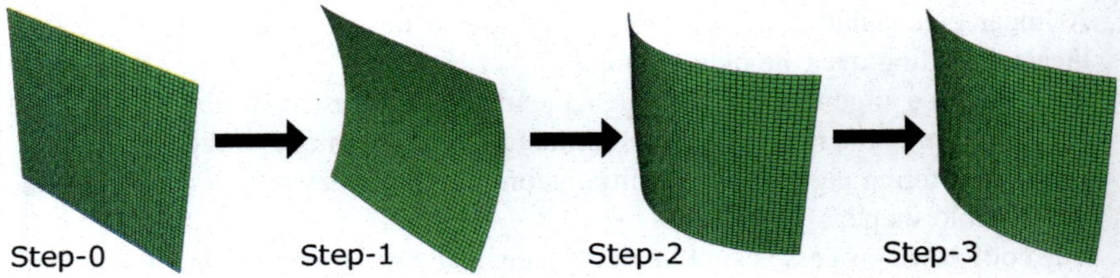

Fig. 4 Performed steps in the FE analysis of the snap-through process

between stable shapes is depicted. This procedure consists of different steps, which are outlined as follows:

1. After identifying the first stable configuration resulting from the cool-down process (Step-1 in Fig. 4), the snap-through phenomenon is investigated by changing the boundary conditions of the FE simulation from the cool-down step. In particular, the displacements along the 'z' direction are restrained at the corner nodes. Additionally, the center node is restrained from translational motion in 'x' and 'y' direction and rotational motion in the 'z' direction.
2. The second step concerns on the application of the external load on the obtained cured shape. This force should be greater than the required snap-through force so that the analysis crosses the limit point (Step-2 in Fig. 4).
3. Subsequently, at the final stage, the applied load is removed from the center node, so that the plate comes to equilibrium and returns to the second stable shape (Step-3 in Fig. 4).

5 Results and Discussion

The multistable shapes of VS composite subjected to thermal loading with temperature difference $\Delta T = 180\,^\circ\text{C}$, are calculated using the semi–analytical approach as described in Sect. 3. The corresponding results are compared with the FE simulations following the modeling guidelines outlined in Sect. 4. The laminates studied are square and have a length L equal to 200 mm, with eight layers each of 0.131 mm-thick plies of graphite-epoxy prepreg. The material properties at ply-level are given as:

$$\begin{aligned} &E_1 = 164\,\text{GPa},\ E_2 = 12\,\text{GPa},\ G_{12} = 4.6\,\text{GPa} \\ &\nu_{12} = 0.3,\ \alpha_1 = -1.8 \times 10^{-8}/^\circ\text{C},\ \alpha_2 = 3 \times 10^{-5}/^\circ\text{C} \end{aligned} \tag{25}$$

The layup data and the values of angle parameters ϕ, T_0 and T_1 of the investigated VS laminates are given in Table 1. As reported in Haldar et al. [14], all the VS composites satisfying $\phi = 45^\circ$ and $T_0 + T_1 = 90^\circ$ leads to cylindrical bistable shapes

Table 1 Fiber orientation and layup data for the investigated straight cross-ply and various VS composites

Type	ϕ	T_0	T_1	Layup
Straight	45	± 45	± 45	$[0_4/90_4]_T$
VS-1	45	± 15	± 75	$[45\langle 15\|75\rangle_4/45\langle -15\| -75\rangle_4]_T$
VS-2	45	± 30	± 60	$[45\langle 30\|60\rangle_4/45\langle -30\| -60\rangle_4]_T$
VS-3	45	± 60	± 30	$[45\langle 60\|30\rangle_4/45\langle -60\| -30\rangle_4]_T$
VS-4	45	± 75	± 15	$[45\langle 75\|15\rangle_4/45\langle -75\| -15\rangle_4]_T$

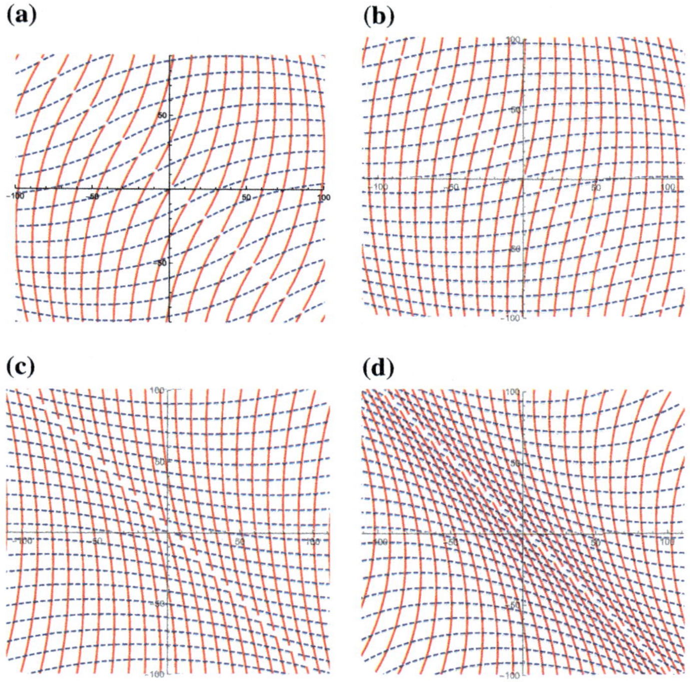

Fig. 5 Investigated VS composites **a** VS-1 **b** VS-2 **c** VS-3 **d** VS-4. All of them yield cylindrical bistable shapes similar to unsymmetric cross-ply laminate

with low twisting curvatures, similar to those obtained from cross-ply laminates. The VS laminates considered in this study satisfy this particular condition. The investigated VS composites are then compared with $[0_4/90_4]$ straight fiber laminate, which generates a similar shape as seen in [23].

The fiber paths of all the investigated VS laminates are illustrated in Fig. 5. For VS-1 and VS-2, the value of angle parameter $T_0 < T_1$ and thus the fiber concentration is more localized at the edges and decreases at the center. On the other hand, for VS-3

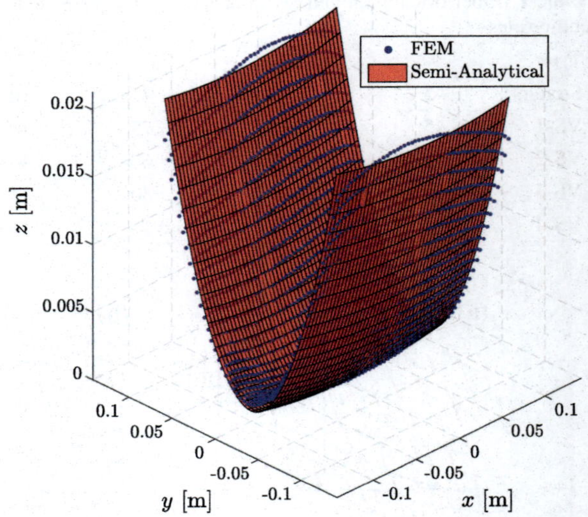

Fig. 6 Comparison between finite element and semi-analytical results for VS laminate with $T_0 = 60°$, $T_1 = 30°$, $\phi = 60°$

and VS-4, the angle parameters satisfy $T_0 > T_1$ where the largest fiber concentration is localized at the center of the plate and is thus, reduced towards the edges.

As was previously stated, all the investigated VS composite laminates generate two stable shapes. The proposed semi-analytical method is assessed by comparing the stable configurations at the room temperature with the FE results. Figure 6 shows the out-of-plane displacement of one of the stable shapes at room temperature for the laminate VS-3 ($\phi = 45°$, $T_0 = 60°$ and $T_1 = 30°$). The FE results (dotted line) are superimposed over the semi–analytical solutions (solid line) to depict the difference. Analyzing this graph, it can be observed that the FE results do not a show a constant curvature throughout the composite plate and manifest a reverse curvature near the edges. Similar observations are made by Gidding et al. [25] for the straight fiber composites and Haldar et al. [14] for VS laminates. All the investigated VS laminates show good agreement between semi-analytical and FE results.

Details of the semi-analytical and FE comparison at the corner and middle edge point can be found in Table 2, which reports the corner displacement and the edge displacements, measured in mm, (at $x = L_x/2$, $y = 0$ and $x = 0$, $y = L_y/2$) of all the two stables shapes of $[0_4/90_4]$ cross-ply and the VS laminates. In this table, w_1 and w_2 are the out-of-plane displacements of the two stable shapes. The unstable shape is not investigated in this study, and therefore its corresponding displacements are not calculated. The difference between the FE and semi-analytical results at the corners in case of the straight fiber cross-ply is 13.9, and 21.8% for VS-1, 18.7% for VS-2, 11.9% for VS-3 and 10.5% for VS-4. At the edge point $x = L_x/2$, $y = 0$, laminates exhibit discrepancies of 1.4% for the straight cross-ply, and 8.8% for VS-1, 8.6% for VS-2, 1.3% for VS-3 and 2.8% for VS-4. Therefore, it can be observed that the differences are lower at the edges of the plate than at the corner points. According

Table 2 Out-of-plane displacement of cross-ply and VS plates–in (mm)

Laminate	$[0_4/90_4]_T$	VS-1	VS-2	VS-3	VS-4
Displ. at $(L_x/2, L_y/2)$					
Analytical					
w_1	−21.07	−16.15	−18.23	−21.65	−19.73
w_2	21.07	16.15	18.23	21.65	19.73
FEM					
w_1	−18.5	−13.26	−15.36	−19.34	−17.86
w_2	18.5	13.26	15.36	19.34	17.86
Displ. at $(0, L_y/2)$; $(L_x/2, 0)$					
Analytical					
w_1	−21.20	−16.75	−19.09	−20.54	−18.48
	0.13	0.54	0.55	−0.76	−1.17
w_2	−0.13	−0.54	−0.55	0.76	1.17
	21.20	16.75	19.09	20.54	18.48
FEM					
w_1	−21.90	−15.39	−17.58	−20.82	−19.02
	−0.08	−2.84	−1.88	1.18	1.53
w_2	0.08	2.84	1.88	−1.18	−1.53
	21.90	15.39	17.58	20.82	19.02

to the previous results, it can be concluded that the edge effects are more prominent near the corners points than the center of the edges. This observation can also be verified from Fig. 6.

The analysis is complemented through performing a comparison between different stable configurations for several VS laminates. Figure 7a compares the FE result of the two stable shapes of the straight fiber cross-ply laminate Fig. 7. Figure 7b depicts the two stable shapes of VS-1 where the value of $T_0 < T_1$, whereas Fig. 7c depicts the two stable shapes of VS-4 where the value of $T_0 > T_1$ computed using the FE analysis. Based on these results, it is observed that the straight fiber cross-ply exhibits a uniform distribution of out-of-plane displacement at the center of the plate. However, VS-1 and VS-4 shows non-uniform curvature at the center region on the plate. In order of further analyze the distribution of out-of-plane displacements at the center region of the plate, a graph is constructed which depicts the value of out-of-plane displacement of the first stable shape against the x-axis (Fig. 8). For the straight fiber cross-ply, the value of out-of-plane displacement does not vary much along the x-axis. On the other hand, for VS-1 the value of out-of-plane displacement varies from zero at the center of the plate to negative values at the edge of the plate.

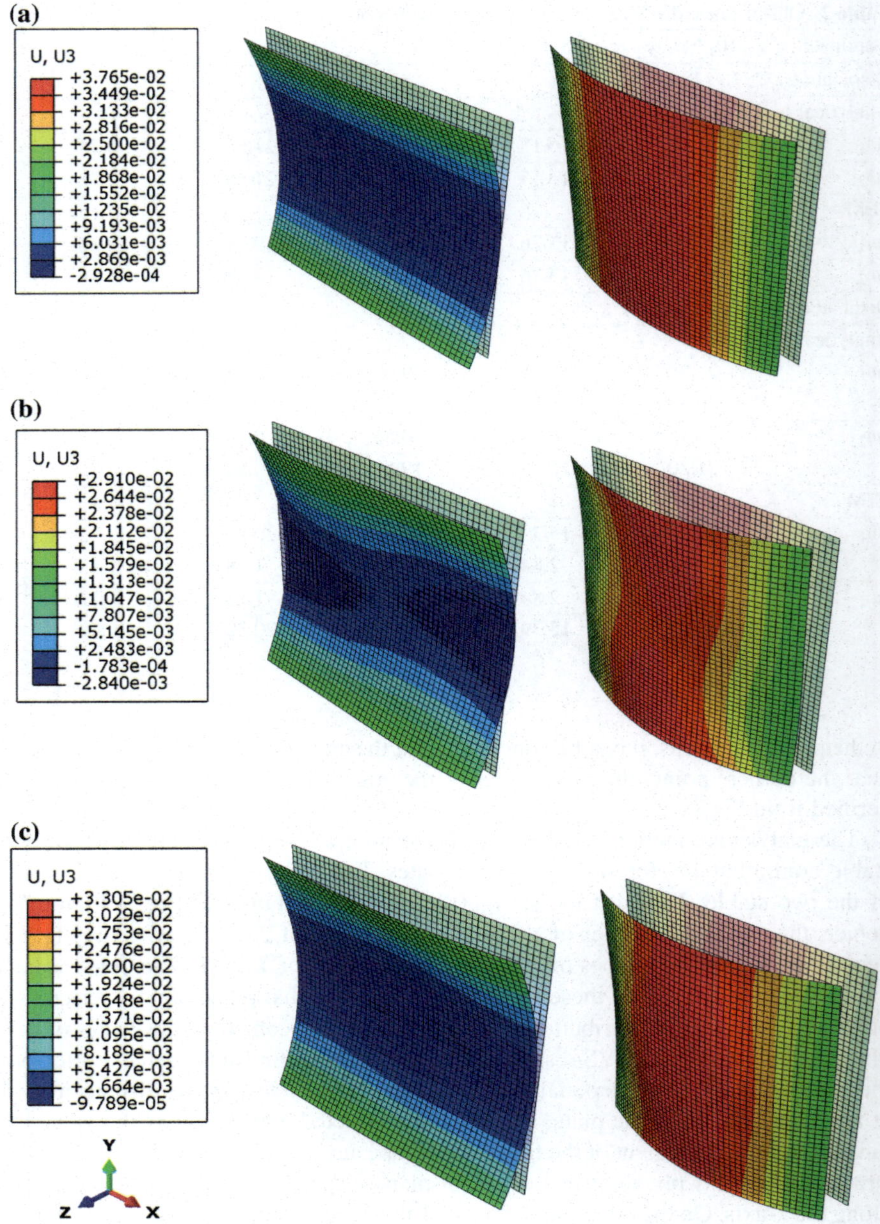

Fig. 7 Two stable shapes with out-of-plane displacement plots–in (mm) of **a** straight fiber cross-ply and **b** VS laminate with $T_0 < T_1$ (VS-1) **c** VS laminate with $T_0 > T_1$ (VS-4) calculated using FEM

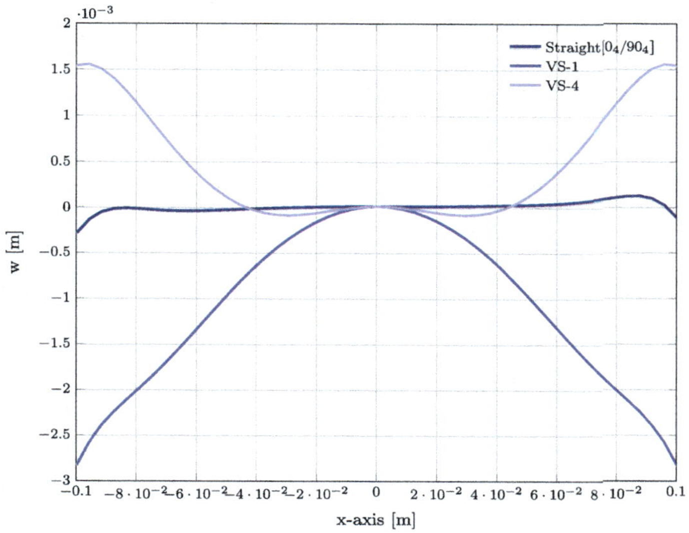

Fig. 8 Plot of out-of-plane displacement with x-axis. The graph shows the formation of local curvatures in VS composites

This results in a dip at the center of the plate. Such local deformation makes the plate conducive to snap-through to the second stable shape. An opposite effect is observed for the laminate VS-4, where values of out-of-plane displacement changes from zero at the center to positive values of out-of-plane displacement at the edges, making it unfavorable for the snap-through process.

6 Calculation of Snap-Through Forces

The first stable shape is generated when the laminate is cooled from curing to room temperature. In order to snap from this stable shape to another, a certain load is applied at the center of the plate.As described in Sect. 4.2 a geometrically nonlinear FE calculation was carried out in ABAQUS to simulate the snap-through process. The snap-through is a dynamic process and requires a stabilization technique to damp the local instabilities. Therefore, numerical stabilization is introduced in the form of viscous forces when instabilities are detected at the stiffness matrix of the system. This method is equivalent to the load-controlled test performed in experiments.

The out-of-plane displacement at the center of the plate is measured at each load increment of all the VS laminates and the straight fiber cross-ply. The resultant load-displacement curve is shown in Fig. 9. It can be seen that with the applied load, the structure deforms elastically featuring linear load-displacement curves. Once the critical point is detected, the structure snaps from one stable shape to another.

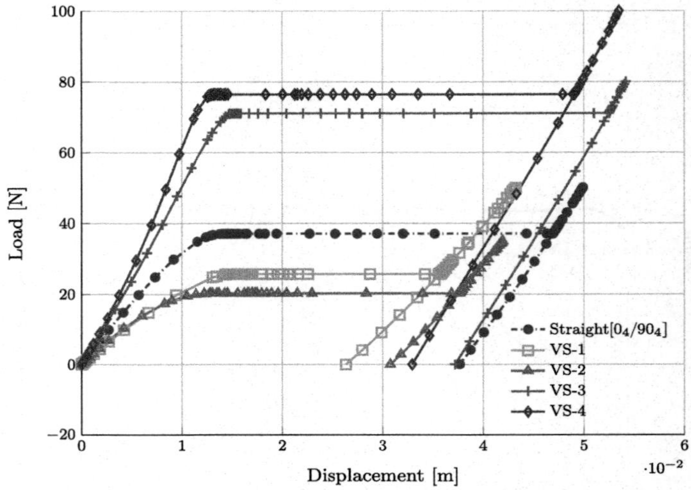

Fig. 9 Stabilised load-displacement diagram depicting snap-through phenomenon

The snap-through force can be directly determined from the load-displacement curve. It can be observed from the figure that among all the laminates investigated, VS-4 requires the maximum snap-through force (76.5 N) followed by VS-3 (71.0 N), VS-1 (25.7 N) and VS-2 (20.2 N). The straight fiber cross-ply $[0_4/90_4]$ requires a snap-through force of 36.8N, the value of whose are verified from the calculation carried out by Diaconu et al. [23]. It can be observed that laminate VS-2 has 45% lower snap-through forces than the straight cross-ply laminate, with a just 14% lower out-of-plane displacement. Figure 10 depicts the variation of the reaction forces at the corners of the plate and the out-of-plane displacement. Through this curve, the unstable equilibrium path after the limit point can also be observed. The difference between the curve depicted in Figs. 9 and 10 shows the effect of viscous damping in the nonlinear FE analysis. The intermediate unstable shape described previously by Hyer [26] in the performed semi-analytical can be seen in this unstable equilibrium path. Furthermore, as previously described by Potter and Weaver [27], such snap-through behavior is not a single event, but it corresponds to a multiple event phenomenon. Similar phenomena can also be observed in the case of VS composites.

The snap-through forces calculated from the semi-analytical approach is compared with FE results in Table 3. Analyzing these data, some deviation between the FE and the semi-analytical predictions can be observed. These differences are attributed to the development of complex intermediate unstable shapes during the snap-through process. However, such complex, unstable geometries cannot be captured using the current semi-analytical procedure since this model restricts the plate to deform only within the cylindrical modes, which is incorrect as observed in the FE analysis.

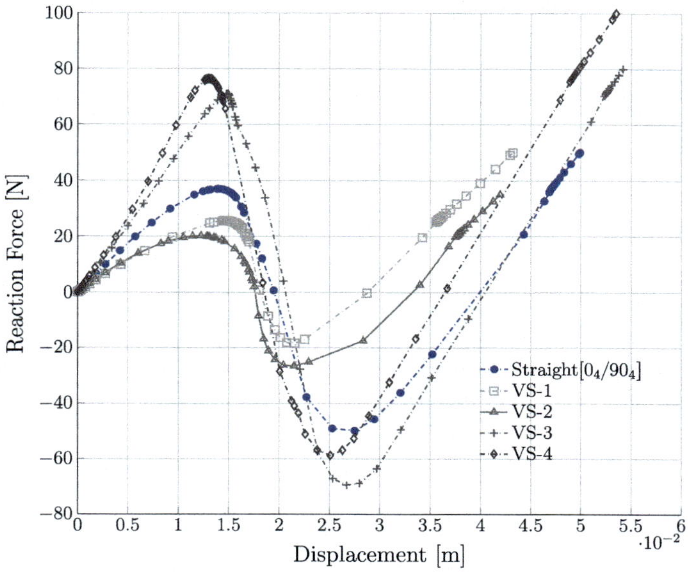

Fig. 10 Reaction force-displacement diagram showing the intermediate unstable path

Table 3 Comparison of snap-through loads for various VS laminates and straight fiber laminate

Laminate	$[0_4/90_4]_T$	VS-1	VS-2	VS-3	VS-4
Analytical (N)	54.4	33.6	38.0	92.0	106.0
FEM (N)	36.8	25.7	20.2	71.0	76.4

Figure 11 shows an exaggerated picture of the snap-through process for VS-1 ($T_0 < T_1$) and VS-4 ($T_1 < T_0$) involving complex intermediate shapes prior to snap-through. This graph also illustrates the formation of a dip at the center-line of the VS-1 laminate as described in Sect. 5 which favors the snap-through process, and requires a lower external force for the corresponding transition between stable shapes. On the other hand, VS-4 laminate shows a bulge at the center of the plate making the snap-through process unfavorable. This leads to higher snap-through forces for VS-4. A similar argument can be made for VS-2 and VS-3 laminates.

From this study, an important aspect of using VS composite as a multistable structure can yet be noticed. Due to a large design space owing to huge tailoring options, VS composites possess the capability to allow large out-of-plane displacement with lower snap-through loads than the straight fiber laminates. Multistable VS composites can thus be used in morphing applications enabling large deflections with lower snap-through forces than what is achieved using conventional straight fiber laminates.

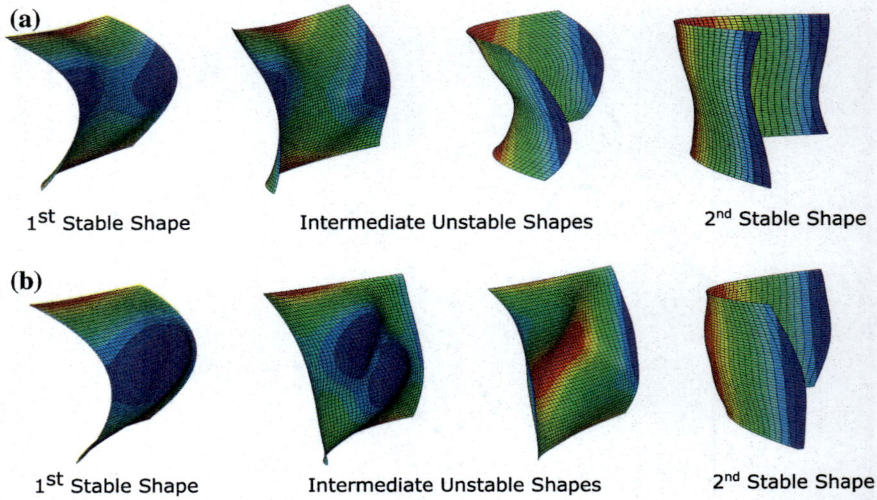

Fig. 11 Transition from one stable shape to another when subjected to concentrated force at the center of the VS plates with **a** $T_0 < T_1$ **b** $T_0 > T_1$

7 Conclusion

In this work, VS composites that yield bistable cylindrical shapes with low twisting curvatures were thoroughly investigated. A semi-analytical tool based on Rayleigh-Ritz approach is used for calculating the thermally induced multistable behavior of unsymmetric laminates with curvilinear fiber paths. The framework of the current approach builds on the Dano-Hyer model, which assumes in-plane strains and out-of-plane displacement fields with polynomial functions. The fiber angle of the VS composite used in this work is considered to vary linearly from the center of the plate to the edge.

The developed technique proves to be an efficient and a relatively simple tool, providing accurate estimations of all the stable shapes for the unsymmetric composite plates under investigation. Four different VS laminates were considered satisfying the condition: $\phi = 45°$ and $T_0 + T_1 = 90°$. All of them generated a bistable cylindrical shape similar to that produced from a cross-ply laminate. VS-1 and VS-2 had angle parameter $T_0 < T_1$ whereas laminates VS-3 and VS-4 satisfied $T_1 < T_0$. Unlike the cross-ply laminate, the VS laminate exhibited local deformations at the center region of the plate. The stable shape of VS-1 and VS-2 have a dip in a direction favorable for the snap-through phenomenon, whereas VS-3 and VS-4 have a bulged out local deformation at the center counteracting the snap-through process.

Local deformations occurring in VS composites at the center region of the plate and deviation at the edges may not be accurately captured by the Rayleigh-Ritz method as it assumes constant curvatures. However, finite element computations can handle such complexities and can accurately predict the local deformations. The snap-through forces were also calculated using the semi-analytical method. However, due to the development of complex intermediate unstable shapes during the snap-through process, the current semi-analytical method yield some deviations in comparison to the FE predictions of the snap-through forces. Such differences are attributed to the fact that the proposed semi-analytical model restricts the plate to deform only within the cylindrical modes, which is incorrect as observed in the FE analysis.

From the non-linear FE analysis, it was observed that the snap through force varies with different VS composites. As expected the snap-through forces were much higher for the laminates VS-3 and VS-4, where the fibers are more concentrated at the center of the plate. The laminates VS-1 and VS-2, on the other hand, required lower snap-through forces than the straight fiber cross-ply with a marginal reduction of corner displacements. Such VS composites can advantageously be embedded as a component in a larger structure to achieve morphing, with low snap-through forces.

Acknowledgements The authors acknowledge funding by the German Research Foundation (DFG) within the International Research Training Group (IRTG) 1627—"Virtual Materials and Structures and their Validation". Ayan Haldar is grateful to Prof. Olivier Allix for simulating discussions and Federica Daghia for helping with the experiments. José Reinoso is grateful for the support of the Spanish Ministry of Economy and Competitiveness (Projects MAT2015-71036-P and MAT2015-71309-P) and the Andalusian Government (Projects of Excellence P11-TEP-7093 and P12-TEP-1050).

References

1. Schultz, M.R.: A concept for airfoil-like active bistable twisting structures. J. Intell. Mater. Syst. Struct. **19**(2), 157–169 (2007)
2. Mattioni, F., Weaver, P.M., Potter, K.D., Friswell, M.I.: The application of thermally induced multistable composites to morphing aircraft structures. In: The 15th International Symposium on: Smart Structures and Materials & Nondestructive Evaluation and Health Monitoring, International Society for Optics and Photonics, pp. 693012–693012 (2008)
3. Arrieta, A.F., Kuder, I.K., Rist, M., Waeber, T., Ermanni, P.: Passive load alleviation aerofoil concept with variable stiffness multi-stable composites. Compos. Struct. **116**, 235–242 (2014)
4. Daynes, S., Nall, S. J., Weaver, P.M., Potter, K.D., Margaris, P., Mellor, P.H.: On a bistable flap for an airfoil. In: Collection of Technical Papers—AIAA/ASME/ASCE/AHS/ASC Structures, Structural Dynamics and Materials Conference (2009)
5. Kuder, I.K., Arrieta, A.F., Rist, M., Ermanni, P.: Aeroelastic response of a selectively compliant morphing aerofoil featuring integrated variable stiffness bi-stable laminates. J. Intell. Mater. Syst. Struct. **27**(14), 1949–1966 (2016)
6. Diaconu, C.G., Weaver, P.M., Mattioni, F.: Concepts for morphing airfoil sections using bi-stable laminated composite structures. Thin-Walled Struct. **46**(6), 689–701 (2008)
7. Arrieta, A.F., Kuder, I.K., Waeber, T., Ermanni, P.: Variable stiffness characteristics of embeddable multi-stable composites. Compos. Sci. Technol. **97**, 12–18 (2014)
8. Kuder, I.K., Arrieta, A.F., Ermanni, P.: Design space of embeddable variable stiffness bi-stable elements for morphing applications. Compos. Struct. **122**, 445–455 (2015)
9. Gürdal, Z., Tatting, B., Wu, C.: Variable stiffness composite panels: Effects of stiffness variation on the in-plane and buckling response. Compos. A Appl. Sci. Manuf. **39**(5), 911–922 (2008)
10. Wu, Z., Raju, G., Weaver, P.M.: Postbuckling analysis of variable angle tow composite plates. Int. J. Solids Struct. **50**(10), 1770–1780 (2013)

11. Coburn, B.H., Wu, Z., Weaver, P.M.: Buckling analysis of stiffened variable angle tow panels. Compos. Struct. **111**, 259–270 (2014)
12. Raju, G., Wu, Z., Weaver, P.M.: Buckling and postbuckling of variable angle tow composite plates under in-plane shear loading. Int. J. Solids Struct. **58**, 270–287 (2015)
13. Sousa, C., Camanho, P., Suleman, A.: Analysis of multistable variable stiffness composite plates. Compos. Struct. **98**, 34–46 (2013)
14. Haldar, A., Reinoso, J., Jansen, E., Rolfes, R.: Thermally induced multistable configurations of variable stiffness composite plates: Semi-analytical and finite element investigation, in press (2017), https://doi.org/10.1016/j.compstruct.2017.02.014
15. Panesar, A.S., Weaver, P.M.: Optimisation of blended bistable laminates for a morphing flap. Compos. Struct. **94**(10), 3092–3105 (2012)
16. Mattioni, F., Weaver, P.M., Friswell, M.: Multistable composite plates with piecewise variation of lay-up in the planform. Int. J. Solids Struct. **46**(1), 151–164 (2009)
17. Haldar, A., Reinoso, J., Jansen, E., Rolfes, R.: Multistable variable stiffness composites for adaptive structures: a parametric study. In: 2nd International Conference on Mechanics of Composites, Porto (2016)
18. Waldhart, C.: Analysis of tow-placed, variable-stiffness laminates. Ph.D. thesis, Virginia Polytechnic Institute and State University (1996)
19. Dano, M.L., Hyer, M.W.: Thermally-induced deformation behavior of unsymmetric laminates. Int. J. Solids Struct. **35**(17), 2101–2120 (1998)
20. Dano, M.L., Hyer, M.W.: Snap-through of unsymmetric fiber-reinforced composite laminates. Int. J. Solids Struct. **39**(1), 175–198 (2002)
21. Calladine, C.R.: Theory of Shell Structures, Cambridge University Press, Cambridge Books Online (1983)
22. Duran, A., Fasanella, N., Sundararaghavan, V., Waas, A.: Thermal buckling of composite plates with spatial varying fiber orientations. Compos. Struct. **124**, 228–235 (2015)
23. Diaconu, C.G., Weaver, P.M., Arrieta, A.F.: Dynamic analysis of bi-stable composite plates. J. Sound Vib. **322**(4–5), 987–1004 (2009)
24. Systèmes, D.: ABAQUS/Standard User's Manual, Version 6.9, Dassault Systemes Simulia Corporation Providence, RI, USA (2014)
25. Giddings, P.F., Bowen, C.R., Salo, A.I., Kim, H.A., Ive, A.: Bistable composite laminates: effects of laminate composition on cured shape and response to thermal load. Compos. Struct. **92**(9), 2220–2225 (2010). Fifteenth International Conference on Composite Structures
26. Hyer, M.W.: Some observations on the cured shape of thin unsymmetric laminates. J. Compos. Mater. **15**(2), 175–194 (1981)
27. Potter, K., Weaver, P., Seman, A.A., Shah, S.: Phenomena in the bifurcation of unsymmetric composite plates. Compos. A Appl. Sci. Manuf. **38**(1), 100–106 (2007)

Invariant-Based Finite Strain Anisotropic Material Model for Fiber-Reinforced Composites

Aamir Dean, José Reinoso, Shahab Sahraee, Benedikt Daum and Raimund Rolfes

Abstract Short fibre reinforced plastic (SFRP) materials are intensively used in several engineering sectors due to their excellent mechanical properties and production rates. In this investigation, an invariant-based transversely isotropic elasto-plastic model for finite strain applications and its corresponding numerical treatment are presented. The current model is based on the multiplicative decomposition of the deformation gradient. The main characteristic of the formulation is the mathematical realization of the incompressibility assumption with regard to the plastic behaviour in anisotropic finite strain setting. The proposed model is complying with thermodynamic restrictions and allows robust reliable numerical simulations. The accuracy of the model is verified by comparison against experimental data, showing a very satisfactory level of agreement.

1 Introduction

Short fiber-reinforced plastics (SFRPs) are materials which exhibit excellent specific strength and stiffness ratios. These materials are especially suitable for their

A. Dean (✉) · S. Sahraee · B. Daum · R. Rolfes
Institute of Structural Analysis, Leibniz Universität Hannover, Appelstr. 9A, 30167 Hannover, Germany
e-mail: a.dean@isd.uni-hannover.de

S. Sahraee
e-mail: s.sahraee@isd.uni-hannover.de

B. Daum
e-mail: b.daum@isd.uni-hannover.de

R. Rolfes
e-mail: r.rolfes@isd.uni-hannover.de

J.. Reinoso
Elasticity and Strength of Materials Group, School of Engineering, Universidad de Sevilla, Camino de Los Descubrimientos S/n, 41092 Seville, Spain
e-mail: jreinoso@us.es

J. Sorić et al. (eds.), *Multiscale Modeling of Heterogeneous Structures*, Lecture Notes in Applied and Computational Mechanics 86,
https://doi.org/10.1007/978-3-319-65463-8_5

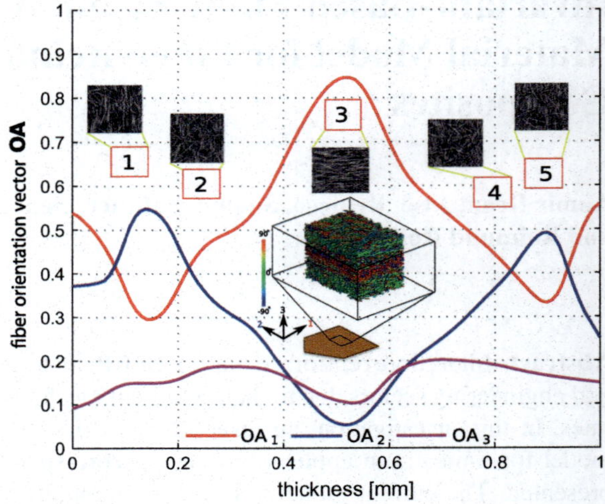

Fig. 1 Micro-computed tomography of SFRP PA6GF-30 processed by injection moulding procedure

incorporation into mass production, since they usually generate low manufacturing costs. In the last years, such materials have been extensively used in several industrial sectors, in particular in the automotive industry with special interest, among many others. In practical applications, one of the most relevant production techniques to manufacture engineering products made from SFRPs, is injection molding (IM), which leads to very complex internal arrangements of the reinforcing fibers, see Fig. 1. Of particular concern are SFRPs made of a polymer matrix with reinforcing short glass fibers, which are denominated as PAxGF-y, where x and y denote the polyamide-type and the fiber content, respectively.

Due to this intricate nature, the determination of the characteristic mechanical properties, which depend on the preferential fiber orientation, is of crucial importance. This characterization can be carried out using different experimental techniques such as optical observations, radiography procedures, CT scans, among others [3, 5, 30]. In this context, in the last three decades, several studies have been conducted in order to characterize the response of SFRP composites under different loading (static and fatigue) [4, 10, 11, 18, 21] and environmental scenarios [12].

From the mechanical point of view, as a consequence of the complex heterogeneous arrangement, the effective modeling of SFRP composites faces various notable difficulties. Original investigations in this area are due to Advani and Tucker [1, 2], who envisaged a tensorial formulation to approximate the probability function regarding the fiber orientation within the domain. Alternative methodologies regard multi-scale FE-based (FE^2) procedures using experimental data from 3D tomographies [15, 28, 34]. However, such FE^2-methods are tremendously expensive in large-scale simulations.

In order to avoid such computational demands, phenomenological anisotropic elasto-plastic models can be considered as a modeling alternative providing mechanical accuracy and numerical efficiency. In the related literature, a high number of investigations focused on the development of anisotropic elasto-plastic formulations under different modeling assumptions, especially within the finite deformation setting invoking the multiplicative decomposition of the deformation gradient between elastic and plastic counterparts [13, 17, 19, 20, 22, 24, 25, 32].

Inspired by these previous investigations, the current investigation presents the development of a novel phenomenological elasto-plastic invariant-based finite strain anisotropic material model for SFRP composites. Differing from previous studies [7, 8, 33], the current model incorporates the assumption of plastic incompressibility within the large deformation setting, accounting for the anisotropic character of SFRP composites through a structural tensorial representation. The proposed formulation is derived following a thermodynamic framework, which guarantees its consistency. On the computational side, specific aspects regarding the numerical integration of the evolution equations corresponding to the internal variables and the consistent elasto-plastic tangent moduli are outlined. Finally, the predictive capability of the model is examined through several applications.

The manuscript is organized as follows. Section 2 describes the basic arguments with regard to the continuous formulation. The constitutive model according to invariant-based formulation is given in Sect. 3. The numerical treatment of the proposed model within the context of a fully implicit nonlinear Finite Element Method (FEM) is addressed in Sect. 4. The applicability of the material model is confirmed through the examination of the experimental-numerical correlation regarding several applications (Sect. 5). Finally, the main conclusions of the current investigation are drawn in Sect. 6.

2 Continuous Formulation

This section presents the fundamental aspects of the novel finite deformation model for SFRP composites within the finite deformation setting. The current formulation uses an invariant-based formulation to account for the directional character of SFRPs composites using a tensorial representation in line with [7–9].

2.1 Basic Kinematics

Following the standard setting of finite inelasticity, consider a continuous three dimensional body which occupies the reference placement $\mathscr{B}_0 \subset \mathbb{R}^3$, where $\mathbf{X} \in \mathscr{B}_0$ denotes an arbitrary material point in this configuration. At time $t \in \mathbb{R}_+$, the corresponding spatially deformed configuration is denoted by $\mathscr{B}_t \subset \mathbb{R}^3$. An individual material point at time t is located at the position $\mathbf{x} \in \mathscr{B}_t$. Both configurations are related via the nonlinear deformation mapping $\boldsymbol{\varphi} : \mathscr{B}_0 \times [0, t] \rightarrow \mathbb{R}^3$, where $[0, t]$ denotes the time interval elapsed. This operator maps the reference material points

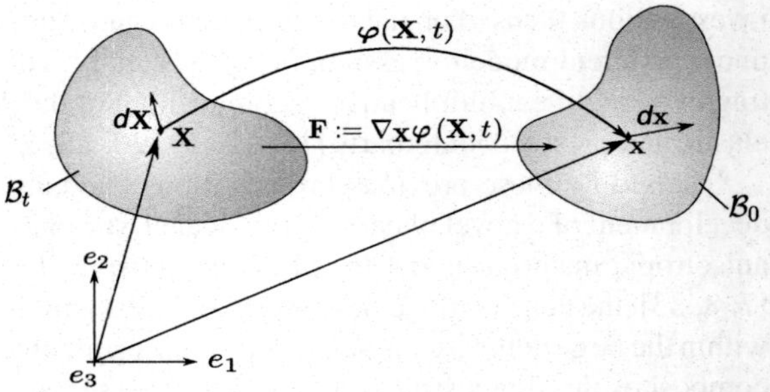

Fig. 2 Motion of a continuum body: reference and current configurations, the nonlinear mapping concept, deformation gradient

($\mathbf{X} \in \mathscr{B}_0$) onto the current material points ($\mathbf{x} \in \mathscr{B}_t$), i.e. $\mathbf{x} = \boldsymbol{\varphi}(\mathbf{X}, t)$, see Fig. 2. Therefore, the displacement vector at material point level is given by: $\mathbf{u} := \mathbf{x} - \mathbf{X}$.

As a measure of the deformation process experienced by the body, we consider the deformation gradient $\mathbf{F}$ that represents the linear mapping between tangent vectors in the reference and current configurations:

$$\mathbf{F} := \frac{\partial \boldsymbol{\varphi}(\mathbf{X}, t)}{\partial \mathbf{X}} = \mathbf{1} + \mathbf{H}(\mathbf{X}, t); \quad \mathbf{H}(\mathbf{X}, t) = \nabla_{\mathbf{X}} \mathbf{u}(\mathbf{X}, t), \tag{1}$$

where $\mathbf{1}$ is the second-order identity tensor and $\nabla_{\mathbf{X}}[\bullet]$ identifies the gradient of the quantity $\bullet$ with respect to the reference setting. The Jacobian of the transformation $J = \det[\mathbf{F}]$ has to satisfy $J \geq 0$. The polar decomposition of the deformation gradient is given by $\mathbf{F} = \mathbf{R}\mathbf{U}$, where $\mathbf{R}$ and $\mathbf{U}$ respectively denote the rotation tensor and the stretch tensor.

The definition of the symmetric right and left Cauchy-Green deformation tensors, $\mathbf{C}$ and $\mathbf{b}$, respectively, and the Green-Lagrange strain tensor, $\mathbf{E}$, is given by:

$$\mathbf{C} := \mathbf{F}^{\mathrm{T}}\mathbf{F}; \quad \mathbf{b} := \mathbf{F}\mathbf{F}^{\mathrm{T}}; \quad \mathbf{E} := \frac{1}{2}\left[\mathbf{C} - \mathbf{1}\right], \tag{2}$$

A central point of the proposed constitutive model for SFRPs is the adoption of the classical multiplicative decomposition of the deformation gradient into elastic and plastic counterparts considering a stress-free intermediate configuration $\bar{\mathscr{B}}$ [27], see Fig. 3:

$$\mathbf{F} = \mathbf{F}^e\mathbf{F}^p. \tag{3}$$

Based on Eq. (3), the elastic part $\mathbf{E}^e$ of the Green-Lagrange strain tensor in the intermediate configuration is defined as:

$$\mathbf{E}^e = \frac{1}{2}\left[\mathbf{C}^e - \mathbf{1}\right], \tag{4}$$

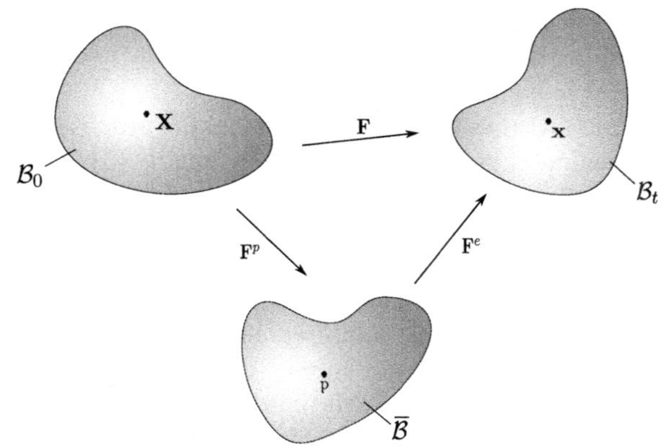

Fig. 3 Intermediate configuration definition: multiplicative decomposition of the deformation gradient $\mathbf{F} = \mathbf{F}^e\mathbf{F}^p$ for large deformation elasto-plasticity

where $\mathbf{C}^e := \mathbf{F}^{eT}\mathbf{F}^e$ is the elastic right Cauchy-Green strain tensor. Finally, the assumption of plastic incompressibility requires the satisfaction of the following constraint:

$$\det\mathbf{F}^p = 1. \tag{5}$$

2.2 Balance Laws

The local form of the balance of linear momentum, which governs the initial boundary value problem (IBVP) of the body's deformation process, takes the following form neglecting the inertia terms:

$$\mathrm{DIV}\mathbf{P}(\mathbf{X}, t) + \bar{\boldsymbol{\Upsilon}}(\mathbf{X}) = \mathbf{0}, \tag{6}$$

where the operator DIV[$\bullet$] stands for the divergence of the tensor field $\bullet$ with respect to the reference frame, $\bar{\boldsymbol{\Upsilon}}$ denotes the body forces (per unit reference volume) of the continuum. In Eq. (6), $\mathbf{P}$ is the first Piola-Kirchhoff stress tensor, which can be associated with the symmetric second Piola-Kirchhoff stress tensor $\mathbf{S}$ as follows: $\mathbf{P} = \mathbf{FS}$. The complete (IBVP) is defined with the suitable boundary conditions in terms of displacements $\bar{\mathbf{u}} : \Gamma_u \times [0, t] \rightarrow \mathbb{R}^3$ and nominal tractions $\bar{\mathbf{T}} : \Gamma_t \times [0, t] \rightarrow \mathbb{R}^3$ which are prescribed on the corresponding subsets of the body boundary.

The balance of angular momentum implies the symmetry condition of the second Piola-Kirchhoff stress tensor in the reference configuration $\mathbf{S} = \mathbf{S}^{\mathrm{T}}$. The balance of energy (first law of thermodynamics) postulates the energy preservation during the deformation process. The material version of the balance of energy reads:

$$\rho_0\dot{e} = \mathbf{S} : \dot{\mathbf{E}} + R - \mathrm{DIV}\,[\mathbf{Q}]\,, \tag{7}$$

where e denotes the specific internal energy, ρ_0 is the reference density, $\mathbf{Q}$ is the reference heat flux, and finally R refers to the internal heat source measured per unit reference volume.

The thermodynamic consistency of the proposed formulation is assessed via the evaluation of the Clausius-Planck inequality [31]:

$$\mathscr{D}_{\text{int}} = \mathbf{S} : \dot{\mathbf{E}} - \dot{e} + \vartheta \dot{\eta} \geq 0, \tag{8}$$

where $\mathscr{D}_{\text{int}}$ is the local dissipation per unit of volume, $\dot{\mathbf{E}}$ stands for the material time derivative of the Green-Lagrange strain tensor. The symbol $\dot{e}$ identifies the time derivative of the specific internal energy, whereas ϑ and $\dot{\eta}$ denote the temperature and the time derivative of the entropy of the system η. Under isothermal conditions and recalling the Legendre transformation [16], Eq. (8) is reduced to:

$$\mathscr{D}_{\text{int}} = \mathbf{S} : \dot{\mathbf{E}} - \dot{\Psi} \geq 0, \tag{9}$$

where Ψ is the Helmholtz free energy function that characterizes the material response.

3 Constitutive Model: Invariant-Based Formulation

3.1 Fundamental Aspects

The mechanical performance of SFRP composites exhibits relevant nonlinear effects along the deformation process prior to failure with a pronounced anisotropic character. This complex behavior arises from the molding flow production process which is employed for manufacturing purposes leading to nonuniform fiber distribution within the specimen. From the modeling standpoint, this directional dependency can be accounted for by means of a purely phenomenological anisotropic plasticity model at finite strains [7–9]. Assuming a tensorial representation of such anisotropic effects, we define a second-order structural tensor $\mathbf{A}$ in the reference configuration:

$$\mathbf{A} := \mathbf{a} \otimes \mathbf{a}, \tag{10}$$

where $\mathbf{a}$ is the direction with the highest aligned fiber content, coinciding with the molding direction. Consequently, the material response is invariant (symmetry transformations) with respect to: (i) arbitrary rotations around $\mathbf{a}$, (ii) reflections at planes parallel to $\mathbf{a}$ and, (iii) planes whose normal vector is aligned with $\mathbf{a}$ [6, 29].

Relying on the previous considerations, the Helmholtz free energy function Ψ that characterizes the mechanical response of SFRPs is assumed to allow the following decomposition:

$$\Psi(\mathbf{E}^e, \varsigma, \mathbf{A}) = \Psi^e(\mathbf{E}^e, \mathbf{A}) + \Psi^p(\varsigma, \mathbf{A}), \tag{11}$$

where $\Psi^e(\mathbf{E}^e, \mathbf{A})$ and $\Psi^p(\boldsymbol{\varsigma}, \mathbf{A})$ identify the elastic and plastic counterparts, respectively, and $\boldsymbol{\varsigma}$ stands for the vector of internal variables that trigger the evolution of the inelastic response. The anisotropic mechanical behavior is modeled through the consideration of the following irreducible integrity basis of invariants $\mathscr{P} := [J_1, ..., J_4]$. The invariants J_1 and J_2 are given by:

$$J_1 := \mathrm{tr}\left[\mathbf{E}^e\right], \quad J_2 := \mathrm{tr}\left[\left(\mathbf{E}^e\right)^2\right], \tag{12}$$

whereas the mixed invariants J_3 and J_4 render:

$$J_3 := \mathrm{tr}\left[\mathbf{A}\mathbf{E}^e\right], \quad J_4 := \mathrm{tr}\left[\mathbf{A}\left(\mathbf{E}^e\right)^2\right]. \tag{13}$$

Then, assuming a quadratic form, the elastic free energy function can be expressed as [8]:

$$\Psi^e(\mathbf{E}^e, \mathbf{A}) = \frac{\lambda}{2}J_1^2 + \mu_T J_2 + \alpha J_3 J_1 + 2\left(\mu_L - \mu_T\right) J_4 + \frac{\beta}{2}J_3^2 = \frac{1}{2}\mathbf{E}^e : \mathbb{C}^e : \mathbf{E}^e, \tag{14}$$

where λ, μ_L, μ_T, α, β identify the elastic constants [33].

The second Piola-Kirchhoff stress tensor $\mathbf{S}$ and its corresponding elasticity tensor $\mathbb{C}^e$ adopt the form:

$$\begin{aligned}\mathbf{S}(\mathbf{E}^e, \mathbf{A}) := \partial_{\mathbf{E}^e}\Psi = \lambda\mathrm{tr}[\mathbf{E}^e]\mathbf{1} + 2\mu_T\mathbf{E}^e + \alpha\left(\mathrm{tr}[\mathbf{A}\mathbf{E}^e] + \mathrm{tr}[\mathbf{E}^e\mathbf{A}]\right)\mathbf{1} + \\ 2(\mu_L - \mu_T)\left(\mathbf{E}^e\mathbf{A} + \mathbf{A}\mathbf{E}^e\right) + \beta\mathrm{tr}[\mathbf{A}\mathbf{E}^e]\mathbf{A}\end{aligned} \tag{15}$$

$$\mathbb{C}^e := \frac{\partial^2\Psi^e}{\partial\mathbf{E}^e \otimes \partial\mathbf{E}^e} = \lambda\mathbf{1}\otimes\mathbf{1} + 2\mu_T\mathbb{I} + \alpha\left(\mathbf{1}\otimes\mathbf{A} + \mathbf{A}\otimes\mathbf{1}\right) + 2\left(\mu_L - \mu_T\right)\mathbb{I}_{\mathbf{A}} + \beta\mathbf{A}\otimes\mathbf{A}, \tag{16}$$

where $\mathbb{I}$ stands for the fourth-order symmetric identity tensor and $\mathbb{I}_{\mathbf{A}}$ takes the form:

$$\mathbb{I}_{\mathbf{A}} \Rightarrow \mathbb{I}_{\mathbf{A},ijkl} = \mathbf{A}_{im}\mathbb{I}_{jmkl} + \mathbf{A}_{jm}\mathbb{I}_{mikl}. \tag{17}$$

Exploiting the multiplicative decomposition of the deformation gradient, Eq. (3), the internal dissipation under isothermal conditions yields:

$$\mathscr{D}_{\mathrm{int}} = \mathbf{S} : \left(\frac{1}{2}\mathbf{F}^{pT}\dot{\mathbf{C}}^e\mathbf{F}^p + \mathbf{F}^{pT}\left(\mathbf{C}^e\mathbf{L}^p\right)_{\mathrm{sym}}\mathbf{F}^p\right) - \dot{\Psi} \geq 0, \tag{18}$$

where the operator $(\bullet)_{\mathrm{sym}}$ stands for the symmetric part of the tensor field $\bullet$. The symbol $\mathbf{L}^p = \dot{\mathbf{F}}^p\mathbf{F}^{p-1}$ identifies the plastic velocity gradient, which can be split into its symmetric $\mathbf{D}^p$ (plastic deformation rate) and skew-symmetric $\mathbf{W}^p$ (plastic material spin) parts:

$$\mathbf{L}^p = \mathbf{D}^p + \mathbf{W}^p. \tag{19}$$

The insertion of the previous definitions into Eq. (18) yields:

$$\mathscr{D}_{\text{int}} = \frac{1}{2}\bar{\mathbf{S}} : \dot{\mathbf{C}}^e + \bar{\mathbf{S}} : \left(\mathbf{C}^e\mathbf{L}^p\right)_{\text{sym}} - \dot{\Psi}^e - \dot{\Psi}^p \geq 0, \tag{20}$$

where $\bar{\mathbf{S}} = \mathbf{F}^p\mathbf{S}\mathbf{F}^{pT}$ identifies the second Piola-Kirchhoff stress tensor counterpart in the intermediate configuration. The result is:

$$\mathscr{D}_{\text{int}} = \left(\frac{1}{2}\bar{\mathbf{S}} - \frac{\partial \Psi^e}{\partial \mathbf{C}^e}\right) : \dot{\mathbf{C}}^e + \left(\mathbf{C}^e\bar{\mathbf{S}}\right) : \mathbf{L}^p - \dot{\Psi}^p \geq 0. \tag{21}$$

Based on the previous procedure, the restriction with regard to the local internal dissipation in order to fulfill the second law of thermodynamics reads:

$$\mathscr{D}_{\text{int}} = \bar{\mathbf{\Sigma}} : \mathbf{L}^p - \dot{\Psi}^p \geq 0, \tag{22}$$

where $\bar{\mathbf{\Sigma}} = \mathbf{C}^e\bar{\mathbf{S}}$ identifies the so-called Mandel stress tensor.

3.2 Transversely Isotropic Yield Function

This section outlines the construction of the transversely isotropic yield function which characterizes the plastic locus of the current anisotropic finite strain elasto-plastic model [7, 8]. The elastic domain $\mathbb{E}$ is defined in terms of the symmetric part of Mandel stress tensor $\bar{\mathbf{\Sigma}}_{\text{s}}$ as follows:

$$\mathbb{E} = \left\{\left(\varsigma, \bar{\varepsilon}^p\right) | f\left(\bar{\mathbf{\Sigma}}_{\text{s}}, \mathbf{A}, \bar{\varepsilon}^p\right) \leq 0\right\}, \tag{23}$$

where $\bar{\varepsilon}^p$ identifies the equivalent plastic strain (hardening variable). The evolution equation of $\bar{\varepsilon}^p$ reads:

$$\dot{\bar{\varepsilon}}^p = \sqrt{\frac{2}{3}\mathbf{D}^p : \mathbf{D}^p}. \tag{24}$$

The proposed pressure-dependent, transversely isotropic and asymmetric yield surface $f\left(\bar{\mathbf{\Sigma}}_{\text{s}}, \mathbf{A}, \bar{\varepsilon}^p\right) \leq 0$ follows a quadratic construction, which can be expressed in terms of the invariant set as:

$$f\left(\bar{\mathbf{\Sigma}}_{\text{s}}, \mathbf{A}, \bar{\varepsilon}^p\right) = \zeta_1 I_1 + \zeta_2 I_2 + \zeta_3 I_3 + \zeta_4 I_3^2 + \zeta_5 I_4 + \zeta_6 I_4^2 - 1 \leq 0, \tag{25}$$

where I_i $(i = 1, ..., 4)$ denote the integrity basis (invariants) taking the form:

$$I_1 := \text{tr}\left[\left(\bar{\mathbf{\Sigma}}_{\text{s}}^{\text{pind}}\right)^2\right] - \text{tr}\left[\mathbf{A}\left(\bar{\mathbf{\Sigma}}_{\text{s}}^{\text{pind}}\right)^2\right]; \quad I_2 := \text{tr}\left[\mathbf{A}\left(\bar{\mathbf{\Sigma}}_{\text{s}}^{\text{pind}}\right)^2\right]; \tag{26}$$

$$I_3 := \text{tr}\left[\bar{\mathbf{\Sigma}}_s\right] - \text{tr}\left[\mathbf{A}\bar{\mathbf{\Sigma}}_s\right]; \quad I_4 := \frac{3}{2}\text{tr}\left[\mathbf{A}\bar{\mathbf{\Sigma}}_s^{\text{dev}}\right]. \tag{27}$$

In Eqs. (26)–(27), $\bar{\mathbf{\Sigma}}_s^{\text{dev}}$ denotes the deviatoric part of the symmetric Mandel stress tensor and $\bar{\mathbf{\Sigma}}_s^{\text{pind}}$ is the basic stress that induces plasticity [33]:

$$\bar{\mathbf{\Sigma}}_s^{\text{pind}} = \bar{\mathbf{\Sigma}}_s - \frac{1}{2}\left(\text{tr}\left[\bar{\mathbf{\Sigma}}_s\right] - \text{tr}\left[\mathbf{A}\bar{\mathbf{\Sigma}}_s\right]\right)\mathbf{1} + \frac{1}{2}\left(\text{tr}\left[\bar{\mathbf{\Sigma}}_s\right] - 3\text{tr}\left[\mathbf{A}\bar{\mathbf{\Sigma}}_s\right]\right)\mathbf{A}. \tag{28}$$

In condensed format, the yield function renders:

$$f\left(\bar{\mathbf{\Sigma}}_s, \mathbf{A}, \bar{\varepsilon}^p\right) = \frac{1}{2}\bar{\mathbf{\Sigma}}_s : \mathbb{K} : \bar{\mathbf{\Sigma}}_s + \mathbf{L} : \bar{\mathbf{\Sigma}}_s - 1 \leq 0, \tag{29}$$

with

$$\mathbb{K} := \zeta_1 \mathbb{P}^{\text{pind}} + (\zeta_2 - \zeta_1)\,\mathbb{P}_{\mathbf{A}}^{\text{pind}} + 2\zeta_4\,(\mathbf{1} - \mathbf{A}) \otimes (\mathbf{1} - \mathbf{A}) + \frac{9}{2}\zeta_6 \mathbf{A}^{\text{dev}} \otimes \mathbf{A}^{\text{dev}}, \tag{30}$$

$$\mathbf{L} := \zeta_3\,(\mathbf{1} - \mathbf{A}) + \frac{3}{2}\zeta_5 \mathbf{A}^{\text{dev}}, \tag{31}$$

$$\mathbb{P}^{\text{pind}} := \mathbb{I} - \frac{1}{2}\,(\mathbf{1} \otimes \mathbf{1}) + \frac{1}{2}\,(\mathbf{1} \otimes \mathbf{A} + \mathbf{A} \otimes \mathbf{1}) - \frac{3}{2}\mathbf{A} \otimes \mathbf{A}, \tag{32}$$

$$\mathbb{P}_{\mathbf{A}}^{\text{pind}} \Rightarrow \mathbb{P}_{\mathbf{A},ijkl}^{\text{pind}} = \mathbf{A}_{im}\mathbb{P}_{mjkl}^{\text{pind}} + \mathbf{A}_{mj}\mathbb{P}_{imkl}^{\text{pind}}, \tag{33}$$

$\mathbf{A}^{\text{dev}}$ being the deviatoric part of $\mathbf{A}$.

Figure 4 portraits a schematic 3D representation of the previous yield function in the principle stress and invariant space where an appropriate convex form can be observed.

Finally, the six parameters $\zeta_i\left(\bar{\varepsilon}^p\right), (i = 1, ..., 6)$ and their corresponding invariants are correlated with different loading states. In particular, the following physical interpretation of these parameters can be regarded [7, 8]: (1) ζ_1 concerns transverse shear loading states, (2) ζ_2 is associated with in-plane shear loadings, (3) ζ_3 and ζ_4 account for loading states transverse to the fiber direction, and finally (4) ζ_5 and ζ_6 involve the material response subjected to longitudinal loading aligned with the fiber direction.

3.3 Plastic Potential Function

Recalling the plastic incompressibility assumption, Eq. (5), the current model introduces the definition of a non-associative flow rule. The use of a non-associative flow rule results from the need for an accurate capturing of plastic deformations [8].

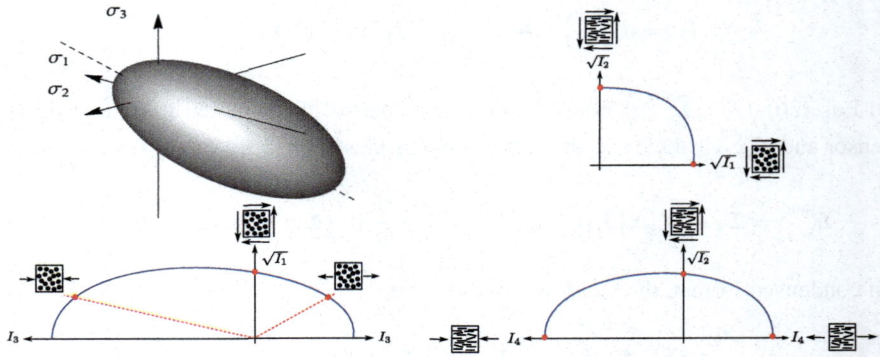

Fig. 4 Transversely isotropic yield function: 3D representation in the principal stress space and cross sections in the invariant space

Accordingly, the following anisotropic plastic potential function $g = g\left(\bar{\mathbf{\Sigma}}_s, \mathbf{A}\right)$ is formulated:

$$g\left(\bar{\mathbf{\Sigma}}_s, \mathbf{A}\right) = \iota_1 \bar{I}_1 + \iota_2 \bar{I}_2 - 1 \leq 0, \tag{34}$$

where $\bar{I}_1$ and $\bar{I}_2$ are the integrity basis (invariants):

$$\bar{I}_1 := \mathrm{tr}\left[\left(\bar{\mathbf{\Sigma}}_s^{dev}\right)^2\right]; \quad \bar{I}_2 := \mathrm{tr}\left[\mathbf{A}\left(\bar{\mathbf{\Sigma}}_s^{dev}\right)^2\right], \tag{35}$$

where ι_1 and ι_2 denote the plastic potential parameters [7]. In condensed format, g yields:

$$g\left(\bar{\mathbf{\Sigma}}_s, \mathbf{A}\right) = \frac{1}{2}\bar{\mathbf{\Sigma}}_s : \mathbb{M} : \bar{\mathbf{\Sigma}}_s - 1 \leq 0, \tag{36}$$

with

$$\mathbb{M} := 2\iota_1 \mathbb{I}^{dev} + \iota_2 \mathbb{I}_{\mathbf{A}}^{dev}; \quad \mathbb{I}^{dev} := \mathbb{I} - \frac{1}{3}\mathbf{1} \otimes \mathbf{1}; \quad \mathbb{I}_{\mathbf{A}}^{dev} \Rightarrow \mathbb{I}_{\mathbf{A},klmn}^{dev} = \mathbf{A}_{ij}\left(\mathbb{I}_{jsmn}^{dev}\mathbb{I}_{sikl}^{dev} + \mathbb{I}_{jskl}^{dev}\mathbb{I}_{simn}^{dev}\right). \tag{37}$$

Figure 5 depicts a cross section of the plastic potential in the invariant space and a 3D representation in the principal stress space.

3.4 Evolution Equations of the Internal Variables

Recalling the maximum energy dissipation principle [27] and using the non-associative flow rule introduced in Sect. 3.3, the evolution equations of the internal variables, namely the plastic velocity gradient $\mathbf{L}^p$ and the hardening variable $\bar{\varepsilon}^p$, are defined in the following.

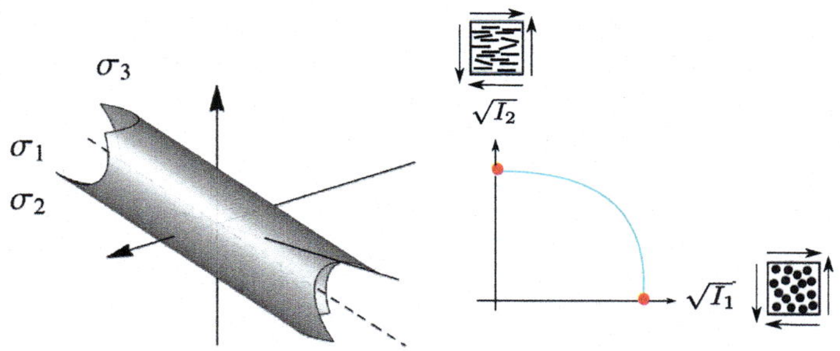

Fig. 5 Plastic potential function: 3D representation in the principal stress space and cross sections in the invariant space

Regarding the plastic velocity gradient, its corresponding evolution equation reads:

$$\mathbf{L}^p = \mathbf{D}^p = \dot{\gamma}\frac{\partial g\left(\bar{\boldsymbol{\Sigma}}_s, \mathbf{A}\right)}{\partial \bar{\boldsymbol{\Sigma}}_s} = \dot{\gamma}\mathbf{n}_g \quad \text{with} \quad \mathbf{n}_g = \mathbb{M} : \bar{\boldsymbol{\Sigma}}_s. \tag{38}$$

Referring to the equivalent plastic strain, the evolution equation takes the form:

$$\dot{\bar{\varepsilon}}^p = \sqrt{\frac{2}{3}\mathbf{D}^p : \mathbf{D}^p} = \dot{\gamma}\sqrt{\frac{2}{3}}\left\|\mathbf{n}_g\right\|, \tag{39}$$

$\dot{\gamma}$ identifying the so-called plastic multiplier.

It is worth mentioning that the symmetric part of the Mandel stress tensor is the unique operator that enters into the plastic potential function, and, therefore, evolution of the plastic material spin $\mathbf{W}^p$ vanishes. Consequently, the constitutive model is invariant with respect to any arbitrary rigid body rotation $\bar{\mathbf{Q}}$.

Finally, the standard Kuhn-Tucker loading/unloading conditions, which ensure the coherence of the model, take the form:

$$\dot{\gamma} \geq 0; \quad f\left(\bar{\boldsymbol{\Sigma}}_s, \mathbf{A}, \bar{\varepsilon}^p\right) \leq 0; \quad \dot{\gamma} f\left(\bar{\boldsymbol{\Sigma}}_s, \mathbf{A}, \bar{\varepsilon}^p\right) = 0. \tag{40}$$

Finally, the consistency condition is given by:

$$\dot{\gamma}\dot{f}\left(\bar{\boldsymbol{\Sigma}}_s, \mathbf{A}, \bar{\varepsilon}^p\right) = 0. \tag{41}$$

3.5 *Parameter Identification*

The yield function Eq. (25) and the plastic potential Eq. (34) are matched to actual materials via the coefficients of the invariants ζ_i and ι_i, respectively. A detailed description of the procedure to adjust these coefficients to experimental data is given in [7, 33], and the main points are outlined below.

The coefficients ζ_i control the size and shape of the elastic region as a function of the equivalent plastic strain variable $\bar{\epsilon}^p$. For each coefficient, the relation $\zeta_i(\bar{\epsilon}^p)$ should be determined from an independent experiment. Typically, this relation is obtained from experiments realizing a simple, controlled stress state with only one nonzero stress component $Y_j(\bar{\epsilon}^p)$ while yielding. In the present case, the following tests can be employed for this purpose: (i) in-plane shear test, (ii) transverse shear test, (iii) uniaxial longitudinal tension and (iv) compression tests, and (v) uniaxial transverse tension and (vi) compression tests. The relation $\zeta_i(Y_{\text{ft}}, Y_{\text{fc}}, Y_{\text{tt}}, Y_{\text{tc}}, Y_{\text{is}}, Y_{\text{ts}})$ can then be derived from inserting the stress tensor corresponding to the test in Eq. (25) and setting $f = 0$. The symbols Y_{ft}, Y_{fc}, Y_{tt} and Y_{tc} represent the uniaxial yield stresses in fiber direction, first index 'f', and transverse direction, first index 't'. The second index indicates tension ('t') or compression ('c'). The symbols Y_{is} and Y_{ts} stand for the transverse and in-plane shear yield stresses, respectively.

To comply with the maximum dissipation principle, the yield surface must be convex, which imposes a restriction to the allowable relations $\zeta_i(\bar{\epsilon}^p)$ which can be used in Eq. (25). Convexity is ensured, if the quadratic term in Eq. (29) is positive definite, and this requirement can be reduced to an inequality in terms of the yield stresses, c.f. Eq. (42), which must hold for any $\bar{\epsilon}^p$.

$$Y_{\text{ft}}Y_{\text{fc}}\left(4Y_{\text{ts}}^2 - Y_{\text{tt}}Y_{\text{tc}}\right) \geq Y_{\text{tc}}Y_{\text{ts}}^2Y_{\text{tt}} \tag{42}$$

The main motivation to adopt a non-associated plasticity scheme is the ability to optimize the plastic deformation behaviour independently of the yield strengths. The form of the plastic potential adopted in Eq. (34) has two adjustable coefficients ι_i. However, one of them is a scaling parameter associated with the size of the potential surface. The size of the plastic potential has no inherent meaning and can be set at will. This leaves only one remaining parameter to match with experimental data in the present case, but, if needed, extra parameters could be introduced by choosing a more complex form of g. Here, ι_1 is arbitrarily set to unity and ι_2 is used to enforce a certain plastic Poisson's ratio $\nu_{23}^p = \epsilon_{22}^p/\epsilon_{33}^p$ for uniaxial transverse tension.

$$\iota_1 = 1, \qquad \nu_{23}^p := \frac{\frac{\partial g}{\partial \sigma_{22}}}{\frac{\partial g}{\partial \sigma_{33}}}\|_{\sigma_{33}=Y_{\text{tt}}} = \frac{-3\iota_1 + \iota_2}{6\iota_1 + \iota_2} \Rightarrow \iota_2 = \frac{3 + 6\nu_{23}^p}{-1 + \nu_{23}^p}$$

Unlike the yield function coefficients, usually no evolution of ι_i with respect to the equivalent plastic strain is considered.

4 Numerical Treatment

This section presents the numerical treatment of the constitutive model given in Sect. 3. The construction of a numerical scheme for the solution of the initial boundary value problem (IBVP) associated involves two general steps [8]: (i) the local integration of the transversely isotropic elasto-plastic model via the corresponding return mapping algorithm, (ii) the introduction of the resulting stress and constitutive elasto-plastic operator into the weak form of the IBVP, which is discretized in space by means of standard brick elements and solved through a standard incremental-iterative Newton-Raphson scheme, see Appendix.

4.1 Numerical Time Integration: General Return Mapping Algorithm

The classical backward Euler scheme is the most extensively used implicit algorithm for the integration of the evolution equations into elasto-plastic constitutive models. This numerical procedure is carried out at integration point level within a standard nonlinear FE code.

The basic integration scheme comprises two fundamental stages: (1) an initial elastic predictor phase, and (2) a subsequent corrector step using a general return mapping [14, 27]. Let us consider a time interval $[t_n, t_{n+1}^{(i)}]$, with $t \in \mathbb{R}_+$, where t_n and $t_{n+1}^{(i)}$ identify the previously converged time step and the current prospective time step at the global FE Newton-Raphson iteration i, respectively. In the sequel, the superscript i is omitted in order to alleviate the notation. Additionally, we assume that all variables of the problem at t_n are known, denoting the incremental time step as $\Delta t = t_{n+1}^{(i)} - t_n$. According to this scheme, the temporal rates of the plastic deformation gradient and the equivalent plastic strain renders:

$$\dot{\mathbf{F}}^p = \frac{\mathbf{F}_{n+1}^p - \mathbf{F}_n^p}{\Delta t}; \quad \dot{\bar{\varepsilon}}^p = \frac{\bar{\varepsilon}_{n+1}^p - \bar{\varepsilon}_n^p}{\Delta t}. \tag{43}$$

For time integration of the evolution equations (Sect. 3.4), the discrete incremental forms according to the backward Euler algorithm take the form:

$$\mathbf{F}_{n+1}^p = \mathbf{F}_n^p + \gamma_{n+1}\mathbf{n}_{g,n+1}\mathbf{F}_{n+1}^p, \tag{44}$$

$$\bar{\varepsilon}_{n+1}^p = \bar{\varepsilon}_n^p + \gamma_{n+1}\sqrt{\frac{2}{3}}\left\|\mathbf{n}_{g,n+1},\right\| \tag{45}$$

$$f_{n+1} = f\left(\bar{\mathbf{\Sigma}}_{\mathrm{s},n+1}, \mathbf{A}, \bar{\varepsilon}_{n+1}^p\right) = 0, \tag{46}$$

where γ_{n+1} identifies the plastic multiplier.

To start the predictor-corrector procedure discussed above, within the predictor phase, initial purely elastic trial increment (denoted by the superscript 'tr' in the sequel) is assumed. Then, the trial elastic deformation gradient $\mathbf{F}_{n+1}^{e,\mathrm{tr}}$ reads:

$$\mathbf{F}_{n+1}^{e,\mathrm{tr}} = \mathbf{F}_{n+1}\mathbf{F}_n^{p-1}. \tag{47}$$

Accordingly, the following operators can be computed:

$$\mathbf{C}^{e,\mathrm{tr}}_{n+1} = \mathbf{F}^{e,\mathrm{tr}T}_{n+1}\mathbf{F}^{e,\mathrm{tr}}_{n+1}; \quad \mathbf{E}^{e,\mathrm{tr}}_{n+1} = \frac{1}{2}\left[\mathbf{C}^{e,\mathrm{tr}}_{n+1} - \mathbf{1}\right], \tag{48}$$

$$\bar{\mathbf{S}}^{\mathrm{tr}}_{n+1} = \mathbb{C}^{e} : \mathbf{E}^{e,\mathrm{tr}}_{n+1}; \quad \bar{\boldsymbol{\Sigma}}^{\mathrm{tr}}_{\mathrm{n+1}} = \mathbf{C}^{e,\mathrm{tr}}_{n+1}\bar{\mathbf{S}}^{\mathrm{tr}}_{n+1}. \tag{49}$$

Then, the trial elastic deformation gradient renders:

$$\mathbf{F}^{e}_{n+1} = \mathbf{F}^{e,\mathrm{tr}}_{n+1}\left(\mathbf{1} - \gamma_{n+1}\mathbf{n}_{g,n+1}\right). \tag{50}$$

The corresponding trial yield function reads:

$$f^{\mathrm{tr}}_{n+1}\left(\bar{\boldsymbol{\Sigma}}^{\mathrm{tr}}_{\mathrm{s},n+1}, \mathbf{A}, \bar{\varepsilon}^{p,\mathrm{tr}}_{n+1}\right) = \frac{1}{2}\bar{\boldsymbol{\Sigma}}^{\mathrm{tr}}_{\mathrm{s},n+1} : \mathbb{K}^{\mathrm{tr}}_{n+1} : \bar{\boldsymbol{\Sigma}}^{\mathrm{tr}}_{\mathrm{s},n+1} + \mathbf{L}^{\mathrm{tr}}_{n+1} : \bar{\boldsymbol{\Sigma}}^{\mathrm{tr}}_{\mathrm{s},n+1} - 1 \leq 0, \tag{51}$$

where the operators $\mathbb{K}^{\mathrm{tr}}$ and $\mathbf{L}^{\mathrm{tr}}$ depend on the parameters $\zeta^{\mathrm{tr}}_i = \zeta_i\left(\bar{\varepsilon}^{p,\mathrm{tr}}_{n+1}\right)$, $(i = 1, ..., 6)$.

If the predictor elastic trial state lies within the elastic domain $\mathbb{E}$, i.e. $f^{\mathrm{tr}}_{n+1} \leq 0$ (where we omit the explicit dependencies), this state is a solution of the constitutive problem stated above. Conversely, a plastic corrector step is required for $f^{\mathrm{tr}}_{n+1} > 0$, which is constructed as follows:

$$\mathbf{F}^{e}_{n+1} = \mathbf{F}^{e,\mathrm{tr}}_{n+1}\left(\mathbf{1} - \gamma_{n+1}\mathbf{n}_{g,n+1}\right), \tag{52}$$

$$\bar{\varepsilon}^{p}_{n+1} = \bar{\varepsilon}^{p}_{n} + \gamma_{n+1}\sqrt{\frac{2}{3}}\left\|\mathbf{n}_{g,n+1},\right\| \tag{53}$$

$$f_{n+1} = f\left(\bar{\boldsymbol{\Sigma}}_{\mathrm{s,n+1}}, \mathbf{A}, \bar{\varepsilon}^{p}_{n+1}\right) = 0, \tag{54}$$

where Eq. (54) stands for the yield criterion. Eqs. (52)–(54) identifies a discrete system of 11 nonlinear equations with 11 unknowns, which are solved simultaneously using a standard local Newton-Raphson procedure at integration point level. Thus, the corresponding residual equations $\mathbf{R}_{n+1} = \left\{\mathbf{R}_{\mathbf{F}^{e}_{n+1}}, \mathrm{R}_{\bar{\varepsilon}^{p}_{n+1}}, \mathrm{R}_{f_{n+1}}\right\}$ are arranged as follows:

$$\mathbf{R}_{\mathbf{F}^{e}_{n+1}} = \mathbf{F}^{e}_{n+1} - \mathbf{F}^{e,\mathrm{tr}}_{n+1}\left(\mathbf{1} - \gamma_{n+1}\mathbf{n}_{g,n+1}\right) = \mathbf{0}, \tag{55}$$

$$\mathrm{R}_{\bar{\varepsilon}^{p}_{n+1}} = \bar{\varepsilon}^{p}_{n+1} - \left(\bar{\varepsilon}^{p}_{n} + \gamma_{n+1}\sqrt{\frac{2}{3}}\left\|\mathbf{n}_{g,n+1}\right\|\right) = 0, \tag{56}$$

$$\mathrm{R}_{f_{n+1}} = f_{n+1} = f\left(\bar{\boldsymbol{\Sigma}}_{\mathrm{s,n+1}}, \mathbf{A}, \bar{\varepsilon}^{p}_{n+1}\right) = 0, \tag{57}$$

which are solved for the variables $\boldsymbol{\chi}_{n+1} = \left\{\mathbf{F}^{e}_{n+1}, \bar{\varepsilon}^{p}_{n+1}, \gamma_{n+1}\right\}$. The linearization of the residual equations $\mathbf{R}_{n+1}$ for the nonlinear solution procedure with respect to corresponding unknowns $\boldsymbol{\chi}_{n+1}$ can be computed as follows:

$$\mathbf{R}_{n+1}\left(\boldsymbol{\chi}_{n+1}^{k+1}\right) := \mathbf{R}_{n+1}\left(\boldsymbol{\chi}_{n+1}^{k}\right) + \mathbf{J}\left(\boldsymbol{\chi}_{n+1}^{k}\right)\left[\boldsymbol{\chi}_{n+1}^{k+1} - \boldsymbol{\chi}_{n+1}^{k}\right] = \mathbf{0}, \tag{58}$$

where the superscript k identifies the Newton-Raphson iteration index corresponding to the plastic corrector step of the present procedure. The Jacobian $\mathbf{J}$ matrix takes the form:

$$\mathbf{J} = \begin{bmatrix} \frac{\partial \mathbf{R}_{\mathbf{F}^e}}{\partial \mathbf{F}^e} & \frac{\partial \mathbf{R}_{\mathbf{F}^e}}{\partial \bar{\varepsilon}} & \frac{\partial \mathbf{R}_{\mathbf{F}^e}}{\partial \gamma} \\ \frac{\partial \mathrm{R}_{\bar{\varepsilon}}}{\partial \mathbf{F}^e} & \frac{\partial \mathrm{R}_{\bar{\varepsilon}}}{\partial \bar{\varepsilon}} & \frac{\partial \mathrm{R}_{\bar{\varepsilon}}}{\partial \gamma} \\ \frac{\partial \mathrm{R}_f}{\partial \mathbf{F}^e} & \frac{\partial \mathrm{R}_f}{\partial \bar{\varepsilon}} & \frac{\partial \mathrm{R}_f}{\partial \gamma} \end{bmatrix}. \tag{59}$$

The increment of the unknowns can be computed as:

$$\Delta\boldsymbol{\chi}_{n+1}^{k+1} := \boldsymbol{\chi}_{n+1}^{k+1} - \boldsymbol{\chi}_{n+1}^{k} = -\mathbf{J}^{-1}\left(\boldsymbol{\chi}_{n+1}^{k}\right)\mathbf{R}_{n+1}\left(\boldsymbol{\chi}_{n+1}^{k}\right), \tag{60}$$

where the initial values for the plastic corrector procedure correspond to the results of the elastic predictor phase:

$$\boldsymbol{\chi}_{n+1}^{k=0} = \begin{bmatrix} \mathbf{F}_{n+1}^{e,\mathrm{tr}} \\ \bar{\epsilon}_n \\ 0 \end{bmatrix}. \tag{61}$$

A representation of the current return mapping algorithm is shown in Fig. 6.

The closed form of the derivatives in Eq. (59) is outlined in the sequel. Thus, if $\mathbf{F}^e = \mathbf{F}^{e,\mathrm{tr}}\left(\mathbf{1} - \gamma\mathbf{n}_g\right)$, the increment of the elastic part of the deformation gradient $\Delta\mathbf{F}^e$ yields:

$$\Delta\mathbf{F}^e = \mathbf{F}^{e,\mathrm{tr}}\left(-\Delta\gamma\mathbf{n}_g - \gamma\frac{\partial\mathbf{n}_g}{\partial\mathbf{F}^e} : \Delta\mathbf{F}^e - \gamma\underbrace{\frac{\partial\mathbf{n}_g}{\partial\bar{\epsilon}}}_{0}\Delta\bar{\epsilon}\right). \tag{62}$$

Consequently, the entries of the first row of $\mathbf{J}$ can be expressed as:

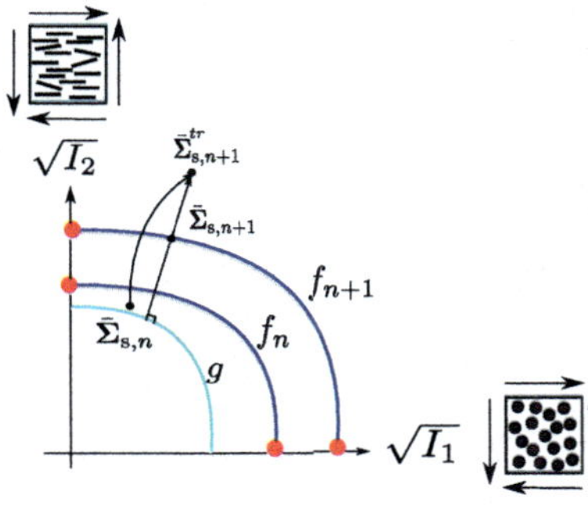

Fig. 6 Return mapping algorithm: graphical description in the invariant space

$$\frac{\partial \mathbf{R}_{\mathbf{F}^e}}{\partial \mathbf{F}^e} = \mathbb{V}^*, \tag{63}$$

with

$$\mathbb{V}^* \Rightarrow \mathbb{V}^*_{ijkl} = \left(\delta_{ik}\delta_{jl} + \gamma \mathbf{F}^{e,\text{tr}}_{im} \frac{\partial \mathbf{n}_{g,mj}}{\mathbf{F}^e_{kl}} \right). \tag{64}$$

Furthermore, it is noting that:

$$\frac{\partial \mathbf{R}_{\mathbf{F}^e}}{\partial \bar{\varepsilon}^p} = 0; \quad \frac{\partial \mathbf{R}_{\mathbf{F}^e}}{\partial \gamma} = \mathbf{F}^{e,\text{tr}} \mathbf{n}_g. \tag{65}$$

Therefore, the increment of the equivalent plastic strain $\Delta\bar{\varepsilon}^p$ reads:

$$\Delta\bar{\varepsilon}^p = \sqrt{\frac{2}{3}} \Delta\gamma \left\| \mathbf{n}_g \right\| + \sqrt{\frac{2}{3}} \gamma \frac{\partial \left\| \mathbf{n}_g \right\|}{\partial \mathbf{F}^e} : \Delta \mathbf{F}^e. \tag{66}$$

Then, the corresponding entries of the second row of $\mathbf{J}$ renders:

$$\frac{\partial \mathrm{R}_{\bar{\varepsilon}^p}}{\partial \mathbf{F}^e} = -\sqrt{\frac{2}{3}} \gamma \frac{\partial \left\| \mathbf{n}_g \right\|}{\partial \mathbf{F}^e}, \tag{67}$$

where

$$\frac{\partial \left\| \mathbf{n}_g \right\|}{\partial \mathbf{F}^e} = \frac{\mathbf{n}_g}{\left\| \mathbf{n}_g \right\|} : \frac{\partial \mathbf{n}_g}{\partial \mathbf{F}^e}. \tag{68}$$

Moreover, since $\mathbf{n}_g = \mathbb{M} : \bar{\mathbf{\Sigma}}_\text{s}$, this leads:

$$\frac{\partial \mathbf{n}_g}{\partial \mathbf{F}^e} = \mathbb{M} : \frac{\partial \bar{\mathbf{\Sigma}}_\text{s}}{\partial \mathbf{F}^e} \Rightarrow \left(\frac{\partial \mathbf{n}_g}{\partial \mathbf{F}^e} \right)_{ijkl} = \mathbb{M}_{ijab} \left(\frac{\partial \bar{\mathbf{\Sigma}}_\text{s}}{\partial \mathbf{F}^e} \right)_{abkl}, \tag{69}$$

where

$$\frac{\partial \bar{\mathbf{\Sigma}}_\text{s}}{\partial \mathbf{F}^e} = \frac{1}{2} \frac{\partial \left(\bar{\mathbf{\Sigma}} + \bar{\mathbf{\Sigma}}^T \right)}{\partial \mathbf{F}^e} \Rightarrow \left(\frac{\partial \bar{\mathbf{\Sigma}}_\text{s}}{\partial \mathbf{F}^e} \right)_{ijkl} = \frac{1}{2} \mathbb{C}^*_{imkl} \bar{\mathbf{S}}_{mj} + \frac{1}{2} \mathbb{C}^*_{mjkl} \bar{\mathbf{S}}_{im} + \frac{1}{4} \mathbf{C}^e_{im} \mathbb{C}^e_{mjab} \mathbb{C}^*_{abkl} + \frac{1}{4} \mathbf{C}^e_{mj} \mathbb{C}^e_{imab} \mathbb{C}^*_{abkl}. \tag{70}$$

In index notation yields:

$$\mathbb{C}^* \Rightarrow \mathbb{C}^*_{ijkl} = \left(\frac{\partial \mathbf{C}^e}{\partial \mathbf{F}^e} \right)_{ijkl} = \delta_{il} \mathbf{F}^e_{kj} + \delta_{jl} \mathbf{F}^e_{ki}. \tag{71}$$

Operating in a similar way:

$$\frac{\partial \mathrm{R}_{\bar{\varepsilon}^p}}{\partial \bar{\varepsilon}^p} = 1; \quad \frac{\partial \mathrm{R}_{\bar{\varepsilon}^p}}{\partial \gamma} = -\sqrt{\frac{2}{3}} \left\| \mathbf{n}_g \right\| . \tag{72}$$

Finally, the increment of the yield function Δf can be computed as:

$$\Delta f = \frac{\partial f}{\partial \mathbf{F}^e} : \Delta \mathbf{F}^e + \frac{\partial f}{\partial \bar{\varepsilon}^p} \Delta \bar{\epsilon} + \overset{0}{\cancel{\frac{\partial f}{\partial \gamma}}} \Delta \gamma = 0. \tag{73}$$

Thus, the corresponding derivatives read (which concern entries of the third row of $\mathbf{J}$) yield:

$$\frac{\partial f}{\partial \mathbf{F}^e} = \left(\mathbb{K} : \bar{\boldsymbol{\Sigma}}_{\mathrm{s}} + \mathbf{L} \right) : \frac{\partial \bar{\boldsymbol{\Sigma}}_{\mathrm{s}}}{\partial \mathbf{F}^e}, \tag{74}$$

$$\frac{\partial f}{\partial \bar{\varepsilon}^p} = \frac{1}{2} \bar{\boldsymbol{\Sigma}}_{\mathrm{s}} : \frac{\partial \mathbb{K}}{\partial \bar{\varepsilon}^p} : \bar{\boldsymbol{\Sigma}}_{\mathrm{s}} + \frac{\partial \mathbf{L}}{\partial \bar{\varepsilon}^p} : \bar{\boldsymbol{\Sigma}}_{\mathrm{s}}. \tag{75}$$

This novel material model is implemented into extended versions of the FE code FEAP and ABAQUS through user-defined material models, which operate at integration point level.

4.2 Algorithmic-Consistent Tangent Moduli

For fully implicit FE computations, the calculation of the tangential stiffness matrix at element level requires the derivation of the algorithmic tangent moduli which guarantees the quadratic convergence along the incremental-iterative solution process. We commence the derivation through the exploitation of Eq. (52). Accordingly, this results in:

$$\mathbf{F}^e = \mathbf{F}^{e,\mathrm{tr}} \left(\mathbf{1} - \gamma \mathbf{n}_g \right) = \mathbf{F}^{e,\mathrm{tr}} \mathbf{F}^p_*; \quad \text{with} \quad \mathbf{F}^{p-1}_* \mathbf{F}^p_n = \mathbf{F}^p. \tag{76}$$

Then, the incremental form of the plastic counterpart of the deformation gradient $\Delta \mathbf{F}^p_*$ renders:

$$\Delta \mathbf{F}^p_* = -\Delta \gamma \mathbf{n}_g - \gamma \frac{\partial \mathbf{n}_g}{\partial \vartheta} \Delta \vartheta - \gamma \frac{\partial \mathbf{n}_g}{\partial \mathbf{F}^e} : \Delta \mathbf{F}^e. \tag{77}$$

The increment of the elastic part of the deformation gradient can be expressed as:

$$\Delta \mathbf{F}^e = \Delta \mathbf{F} \mathbf{F}^{p-1} - \Delta \gamma \mathbf{F}^{e,\mathrm{tr}} \mathbf{n}_g - \gamma \mathbf{F}^{e,\mathrm{tr}} \frac{\partial \mathbf{n}_g}{\partial \mathbf{F}^e} : \Delta \mathbf{F}^e, \tag{78}$$

and the consistency condition renders:

$$\Delta f = \frac{\partial f}{\partial \mathbf{F}^e} : \Delta \mathbf{F}^e + \frac{\partial f}{\partial \bar{\varepsilon}^p} \Delta \bar{\varepsilon}^p = 0. \tag{79}$$

with

$$\Delta\bar{\varepsilon}^p = \sqrt{\frac{2}{3}}\Delta\gamma \left\|\mathbf{n}_g\right\| + \sqrt{\frac{2}{3}}\gamma \frac{\partial \left\|\mathbf{n}_g\right\|}{\partial \mathbf{F}^e} : \Delta\mathbf{F}^e. \tag{80}$$

The increment of the yield function takes the form:

$$\Delta f = \left(\frac{\partial f}{\partial \bar{\varepsilon}^p}\sqrt{\frac{2}{3}}\left\|\mathbf{n}_g\right\|\right)\Delta\gamma + \left(\frac{\partial f}{\partial \bar{\varepsilon}^p}\sqrt{\frac{2}{3}}\gamma\frac{\partial \left\|\mathbf{n}_g\right\|}{\partial \mathbf{F}^e} + \frac{\partial f}{\partial \mathbf{F}^e}\right) : \Delta\mathbf{F}^e = 0. \tag{81}$$

Based on the previous expression, the increment of the plastic multiplier is computed as:

$$\Delta\gamma = \underbrace{-\frac{\frac{\partial f}{\partial \bar{\varepsilon}^p}\sqrt{\frac{2}{3}}\gamma\frac{\partial \left\|\mathbf{n}_g\right\|}{\partial \mathbf{F}^e} + \frac{\partial f}{\partial \mathbf{F}^e}}{\frac{\partial f}{\partial \bar{\varepsilon}^p}\sqrt{\frac{2}{3}}\left\|\mathbf{n}_g\right\|}}_{\frac{\partial\gamma}{\partial\mathbf{F}^e}} : \Delta\mathbf{F}^e = \frac{\partial\gamma}{\partial\mathbf{F}^e} : \Delta\mathbf{F}^e. \tag{82}$$

Based on the previous derivations, $\Delta\mathbf{F}^p_*$ yields:

$$\Delta\mathbf{F}^p_* = \underbrace{-\left(\mathbf{n}_g \otimes \frac{\partial\gamma}{\partial\mathbf{F}^e} + \gamma\frac{\partial\mathbf{n}_g}{\partial\mathbf{F}^e}\right)}_{\mathbb{H}^p} : \Delta\mathbf{F}^e = \mathbb{H}^p : \Delta\mathbf{F}^e. \tag{83}$$

To accomplish the following steps, recalling $\bar{\mathbb{I}} \Rightarrow \bar{\mathbb{I}}_{ijkl} = \delta_{ik}\delta_{jl}$, we define:

$$\mathbb{N} = \bar{\mathbb{I}} + \gamma\mathbf{F}^{e,\mathrm{tr}}\frac{\partial\mathbf{n}_g}{\partial\mathbf{F}^e}; \quad \mathbb{N}^* = \mathbb{N} + \left(\mathbf{F}^{e,\mathrm{tr}}\mathbf{n}_g\right) \otimes \frac{\partial\gamma}{\partial\mathbf{F}^e}. \tag{84}$$

Inserting Eq. (82) into Eq. (78) yields:

$$\mathbb{N} : \Delta\mathbf{F}^e = \Delta\mathbf{F}\mathbf{F}^{p-1} - \mathbf{F}^{e,\mathrm{tr}}\mathbf{n}_g\left(\frac{\partial\gamma}{\partial\mathbf{F}^e} : \Delta\mathbf{F}^e\right). \tag{85}$$

Then, it is possible to obtain:

$$\Delta\mathbf{F}^e = \mathbb{N}^{*-1} : \left(\Delta\mathbf{F}\mathbf{F}^{p-1}\right). \tag{86}$$

An additional computation that should be performed is the increment of $\Delta\bar{\mathbf{S}}$:

$$\Delta\bar{\mathbf{S}} = \left(\frac{\partial\bar{\mathbf{S}}}{\partial\mathbf{E}^e} : \frac{\partial\mathbf{E}^e}{\partial\mathbf{F}^e}\right) : \Delta\mathbf{F}^e = \underbrace{\left(\mathbb{C}^e : \frac{\partial\mathbf{E}^e}{\partial\mathbf{F}^e}\right)}_{\mathbb{C}^e_{\mathbf{F}}} : \Delta\mathbf{F}^e, \tag{87}$$

Then, $\Delta\mathbf{F}^{p-1}$ takes the form:

$$\Delta \mathbf{F}^{p-1} = \Delta \left(\mathbf{F}_n^{p-1} \mathbf{F}_*^p \right) = \mathbf{F}_n^{p-1} \Delta \mathbf{F}_*^p, \tag{88}$$

and correspondingly:

$$\Delta \mathbf{F}^{p-\mathrm{T}} = \Delta \left(\mathbf{F}_*^p \mathbf{F}_n^{p-\mathrm{T}} \right) = \Delta \mathbf{F}_*^p \mathbf{F}_n^{p-\mathrm{T}}. \tag{89}$$

The increment of the first Piola-Kirchhoff stress tensor ($\mathbf{P} := \mathbf{FS}$) can be computed as:

$$\Delta \mathbf{P} = \Delta \left(\mathbf{FS} \right), \tag{90}$$

with $\mathbf{P}$ standing for the first Piola-Kirchhoff stress tensor. Expanding the previous expression yields:

$$\Delta \left(\mathbf{FS} \right) = \Delta \mathbf{FS} + \mathbf{F} \Delta \mathbf{S} = \Delta \mathbf{F} \left(\mathbf{F}^{p-1} \bar{\mathbf{S}} \mathbf{F}^{p-T} \right) + \mathbf{F} \Delta \left(\mathbf{F}^{p-1} \bar{\mathbf{S}} \mathbf{F}^{p-T} \right), \tag{91}$$

where

$$\Delta \left(\mathbf{FS} \right) = \Delta \mathbf{F} \left(\mathbf{F}^{p-1} \bar{\mathbf{S}} \mathbf{F}^{p-T} \right) + \mathbf{F} \Delta \mathbf{F}^{p-1} \bar{\mathbf{S}} \mathbf{F}^{p-T} + \mathbf{F} \mathbf{F}^{p-1} \Delta \bar{\mathbf{S}} \mathbf{F}^{p-T} + \mathbf{F} \mathbf{F}^{p-1} \bar{\mathbf{S}} \Delta \mathbf{F}^{p-T}, \tag{92}$$

and

$$\Delta \mathbf{F}^{p-T} = \Delta \left(\mathbf{F}_*^p \mathbf{F}_n^{p-T} \right) = \Delta \mathbf{F}_*^p \mathbf{F}_n^{p-T}. \tag{93}$$

Eq. (92) can be expanded as:

$$\begin{aligned} \Delta \left(\mathbf{FS} \right) = {} & \Delta \mathbf{F} \left(\mathbf{F}^{p-1} \bar{\mathbf{S}} \mathbf{F}^{p-T} \right) + \\ & \mathbf{F} \mathbf{F}_n^{p-1} \left(\mathbb{H}^p : \Delta \mathbf{F}^e \right) \bar{\mathbf{S}} \mathbf{F}^{p-T} + \\ & \mathbf{F} \mathbf{F}^{p-1} \left(\mathbb{C}_{\mathbf{F}}^e : \Delta \mathbf{F}^e \right) \mathbf{F}^{p-T} + \\ & \mathbf{F} \mathbf{F}^{p-1} \bar{\mathbf{S}} \left(\mathbb{H}^p : \Delta \mathbf{F}^e \right) \mathbf{F}_n^{p-T}. \end{aligned} \tag{94}$$

Inserting the results from Eq. (86) into Eq. (93) yields:

$$\begin{aligned} \Delta \left(\mathbf{FS} \right) = {} & \Delta \mathbf{F} \left(\mathbf{F}^{p-1} \bar{\mathbf{S}} \mathbf{F}^{p-T} \right) + \\ & \mathbf{F} \mathbf{F}_n^{p-1} \left(\mathbb{H}^p : \left(\mathbb{N}^{*-1} : \left(\Delta \mathbf{F} \mathbf{F}^{p-1} \right) \right) \right) \bar{\mathbf{S}} \mathbf{F}^{p-T} + \\ & \mathbf{F} \mathbf{F}^{p-1} \left(\mathbb{C}_{\mathbf{F}}^e : \left(\mathbb{N}^{*-1} : \left(\Delta \mathbf{F} \mathbf{F}^{p-1} \right) \right) \right) \mathbf{F}^{p-T} + \\ & \mathbf{F} \mathbf{F}^{p-1} \bar{\mathbf{S}} \left(\mathbb{H}^p : \left(\mathbb{N}^{*-1} : \left(\Delta \mathbf{F} \mathbf{F}^{p-1} \right) \right) \right) \mathbf{F}_n^{p-T}, \end{aligned} \tag{95}$$

which in index notation reads:

$$\Delta\left(\mathbf{FS}\right)_{ij} = \left(\delta_{ik}\mathbf{F}^{p-1}_{ln}\bar{\mathbf{S}}_{ns}\mathbf{F}^{p-T}_{sj}\right)\Delta\mathbf{F}_{kl} + \tag{96}$$

$$\left(\mathbf{F}_{im}\mathbf{F}^{p-1}_{n,mx}\mathbb{H}^{p}_{xnab}\mathbb{N}^{*-1}_{abky}\mathbf{F}^{p-1}_{ly}\bar{\mathbf{S}}_{ns}\mathbf{F}^{p-T}_{sj}\right)\Delta\mathbf{F}_{kl} + \tag{97}$$

$$\left(\mathbf{F}_{im}\mathbf{F}^{p-1}_{mn}\mathbb{C}^{e}_{\mathbf{F},nsab}\mathbb{N}^{*-1}_{abky}\mathbf{F}^{p-1}_{ly}\mathbf{F}^{p-T}_{sj}\right)\Delta\mathbf{F}_{kl} + \tag{98}$$

$$\left(\mathbf{F}_{im}\mathbf{F}^{p-1}_{mn}\bar{\mathbf{S}}_{ns}\mathbb{H}^{p}_{sxab}\mathbb{N}^{*-1}_{abky}\mathbf{F}^{p-1}_{ly}\mathbf{F}^{p-T}_{n,xj}\right)\Delta\mathbf{F}_{kl}. \tag{99}$$

Finally, in condensed format, it can be expressed:

$$\Delta\left(\mathbf{FS}\right) = \mathbb{C}^{ep} : \Delta\mathbf{F} \Rightarrow \Delta\left(\mathbf{FS}\right)_{ij} = \mathbb{C}^{ep}_{ijkl}\Delta\mathbf{F}_{kl}, \tag{100}$$

which closed form in index notation renders:

$$\begin{aligned}\mathbb{C}^{ep} \Rightarrow \mathbb{C}^{ep}_{ijkl} = {} & \delta_{ik}\mathbf{F}^{p-1}_{ln}\bar{\mathbf{S}}_{ns}\mathbf{F}^{p-T}_{sj} + \\ & \mathbf{F}_{im}\mathbf{F}^{p-1}_{n,mx}\mathbb{H}^{p}_{xnab}\mathbb{N}^{*-1}_{abky}\mathbf{F}^{p-1}_{ly}\bar{\mathbf{S}}_{ns}\mathbf{F}^{p-T}_{sj} + \\ & \mathbf{F}_{im}\mathbf{F}^{p-1}_{mn}\mathbb{C}^{e}_{\mathbf{F},nsab}\mathbb{N}^{*-1}_{abky}\mathbf{F}^{p-1}_{ly}\mathbf{F}^{p-T}_{sj} + \\ & \mathbf{F}_{im}\mathbf{F}^{p-1}_{mn}\bar{\mathbf{S}}_{ns}\mathbb{H}^{p}_{sxab}\mathbb{N}^{*-1}_{abky}\mathbf{F}^{p-1}_{ly}\mathbf{F}^{p-T}_{n,xj},\end{aligned} \tag{101}$$

To finish the current derivations, the following computations are also required:

$$\frac{\partial\left\|\mathbf{n}_g\right\|}{\partial\mathbf{F}^e} = \frac{\partial\left\|\mathbf{n}_g\right\|}{\partial\mathbf{n}_g} : \frac{\partial\mathbf{n}_g}{\partial\mathbf{F}^e} = \frac{\mathbf{n}_g}{\left\|\mathbf{n}_g\right\|} : \frac{\partial\mathbf{n}_g}{\partial\mathbf{F}^e}. \tag{102}$$

Given that $\mathbf{n}_g = \mathbb{M} : \bar{\boldsymbol{\Sigma}}_{\mathrm{s}}$, so:

$$\Delta\mathbf{n}_g = \mathbb{M} : \Delta\bar{\boldsymbol{\Sigma}}_{\mathrm{s}} = \mathbb{M} : \frac{\partial\bar{\boldsymbol{\Sigma}}_{\mathrm{s}}}{\partial\mathbf{F}^e} : \Delta\mathbf{F}^e, \tag{103}$$

and therefore:

$$\frac{\partial\mathbf{n}_g}{\partial\mathbf{F}^e} = \mathbb{M} : \frac{\partial\bar{\boldsymbol{\Sigma}}_{\mathrm{s}}}{\partial\mathbf{F}^e}. \tag{104}$$

5 Applications

In this section, several numerical results are presented in order to examine the performance of the constitutive model herein developed. The applications henceforth presented are: (i) a verification case concerning dog-bone specimen types under tensile loading with different preferential fibre orientations (Sect. 5.1), and (ii) a validation through a three-point bending test (Sect. 5.2).

5.1 Model Verification and Validation

In the first application, dog-bone specimen-types with different preferential fiber orientations are subjected to uniform tensile loading. These specimens are manufactured from the short fiber-reinforced thermoplastic PA6GF-30 and were experimentally investigated at the Institute of Forming Technology and Machines (IFUM, Hannover) [10]. The corresponding mechanical properties are given in Table 1. Note that averaged fiber distribution over the cross section of the specimen is considered complying with the so-called Equivalent Single Layer (ESL) approach [23].

The geometry of the specimen is shown in Fig. 7a, identifying the zero-degree reference material orientation. The specimen is discretized using 4860 first-order solid

Table 1 PA6GF-30: mechanical properties

E_{11} (MPa)	E_{22} (MPa)	G_{12} (MPa)	ν_{12}	ν_{23}
7893.550	3348.17	1800.94	0.23	0.62

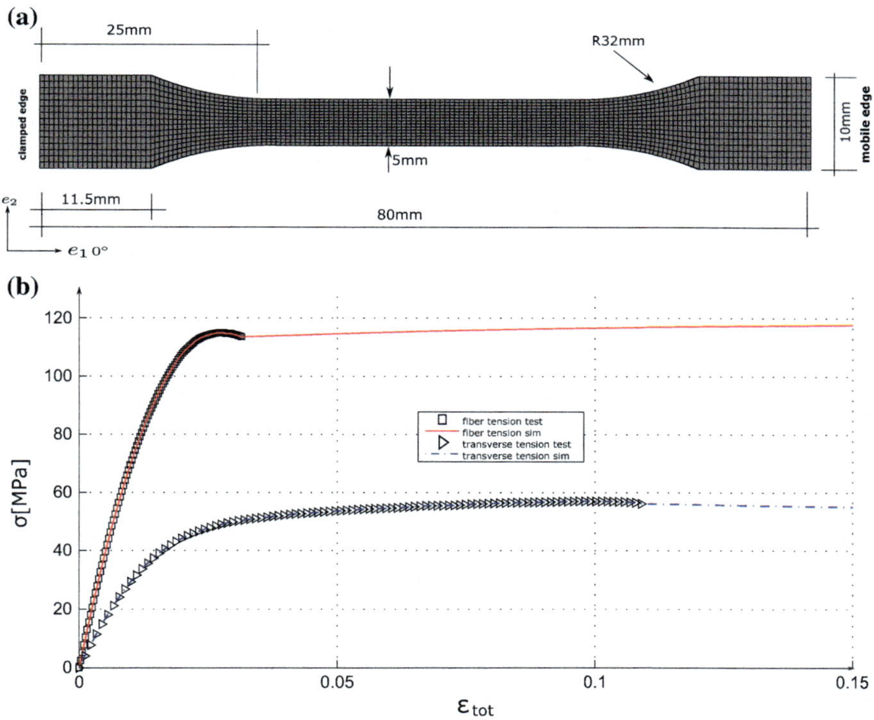

Fig. 7 Dog-bone specimens PA6GF-30 under uniaxial loading conditions. **a** Specimen definition **b** Experimental–numerical correlation for different preferential fiber orientations (0° and 90°)

elements. To reproduce the experimental conditions, the following boundary conditions are defined [7]: (i) fully restrained displacements at the clamped edge, and (ii) constrained displacements at the loaded edge, except the longitudinal displacement coinciding with the 0°-direction.

The initial yielding parameters, ζ_i, are reported in Table 2, whereas the plastic potential parameters ι_i are listed in Table 3. The geometry of the specimen is shown in Fig. 7a, identifying the zero-degree reference material orientation. Fig. 8 depicts the convex form of the yield surface.

Figures 7b shows the experimental–numerical correlation between the current simulations and the data reported in [10], whereby a satisfactory agreement can be observed.

Table 2 PA6GF-30: initial yielding parameters ζ_i

ζ_1	ζ_2	ζ_3	ζ_4	ζ_5	ζ_6
2.648×10^{-4}	2.648×10^{-4}	3.272×10^{-3}	2.523×10^{-5}	1.338×10^{-3}	2.588×10^{-4}

Table 3 PA66GF-35: plastic potential parameters ι_i

ι_1	ι_2
1.0	-17.684

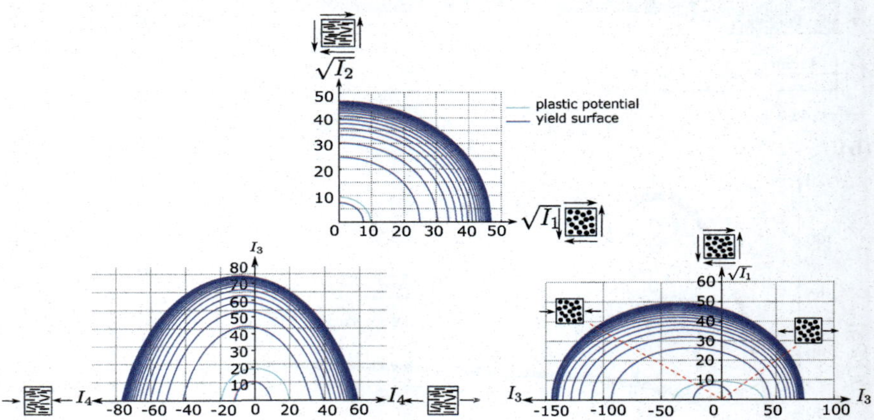

Fig. 8 PA6GF30: Characterization of the yield surface and plastic potential

Table 4 PA6GF-30: Mechanical properties

E_{11} (MPa)	E_{22} (MPa)	G_{12} (MPa)	ν_{12}	ν_{23}
5211.43	2262.86	1601.7	0.39	0.6

Table 5 PA6GF-30: initial yielding parameters ζ_i

ζ_1	ζ_2	ζ_3	ζ_4	ζ_5	ζ_6
4.521×10^{-4}	4.521×10^{-4}	6.624×10^{-3}	9.429×10^{-5}	3.899×10^{-3}	2.196×10^{-4}

Table 6 PA6GF-30: plastic potential parameters ι_i

ι_1	ι_2
1.0	-17.684

5.2 *Structural Application: Three-Point Bending Test*

The second application studied concerns the 3-point bending test previously reported in [26]. In particular, we restrict our analysis to the case of a loading velocity of 1.0 m/s to prevent incongruities with the quasi-static character of the current formulation. The material properties corresponding to the present case are listed in Table 4, whereas the plastic data are reported in Tables 5 and 6, respectively, complying with the ISO standard value corresponding to this material. Similarly to the previous application, we exploit the ESL approach to compute the corresponding mechanical properties over the plate thickness. This example is of special interest to characterize the mechanical performance and to trigger the fiber orientation along the loading procedure. Therefore, this application is herein used to assess the proposed formulation.

Figure 9a shows the geometric description of the current application, identifying the preferential fibre orientation with the longitudinal direction of the specimen and with the following geometric dimensions: (i) length $L = 50$ mm, (ii) width $B = 5$ mm, and (iii) thickness $t = 2$ mm. The plate is discretized using 7200 first-order solid elements. The pin for the loading application is meshed using 2100 elements with the same interpolation order and setting very high mechanical properties to prevent its deformation. The computations are performed prescribing the downward vertical displacement at the central pin equal to 9 mm using 1000 equal pseudo-time increments. It should be noted that ABAQUS simulations are performed using automatic time stepping. Figure 9b shows the longitudinal stress distribution over the plate thickness, featuring a nonuniform distribution over the thickness due to the imposed loading.

Simulations are conducted imposing a prescribed vertical displacement downwards at the central pin equal to 9 mm, see Fig. 9a ,b. In each increment, the global solution scheme is employed. Figure 9b shows the stress distribution due to the prescribed loading, where, as expected, a nonuniform strain distribution over the plate

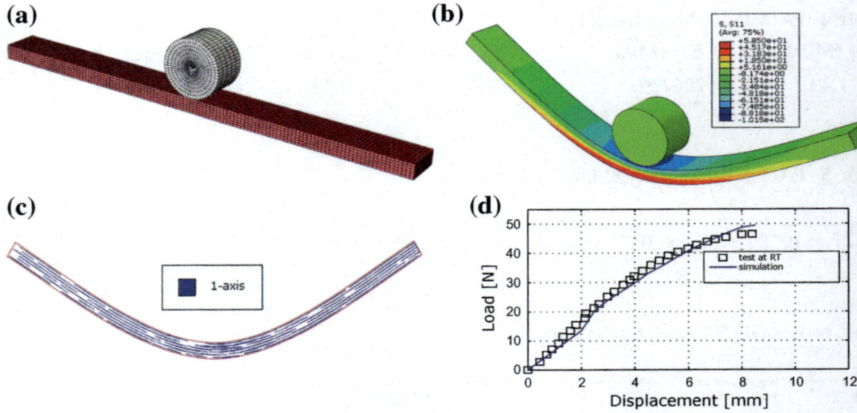

Fig. 9 Three-point bending test of PA6GF-30. **a** Specimen definition **b** Stress distribution **c** Mapping of the preferential material direction after the computation **d** Experimental–numerical correlation

thickness is estimated. Figure 9c displays the preferential material orientation along the thickness direction, whereby the mapping of the fibre alignments along the deformation process is mapped.

Finally, Fig. 9d shows the experimental-numerical correlation corresponding to the load–central displacement evolution curve of the application. In this graph, the experimental data are represented through discrete square symbols, whereas the simulation results are plotted using a solid line. Examining this evolution, it is interesting to see that a very good correlation is obtained along the whole loading procedure. In this respect, note that the mechanical performance of the systems is characterized by an initial linear evolution followed by a subsequent stage where notable nonlinear effects become appreciable.

6 Concluding Remarks

In this investigation, a new elasto-plastic invariant-based finite strain anisotropic material model for SFRP composites has been presented. The proposed formulation is suitable for arbitrarily large elastic and plastic deformations, assuming plastic incompressibility.

On the theoretical side, the model incorporates a non-associate flow rule to characterize the plastic evolution, which relies on the multiplicative decomposition of the deformation gradient. The constitutive equations are derived in a thermodynamically consistent format. On the computational side, the current investigation provides a comprehensive presentation of the numerical treatment within the context of nonlinear FEM. In particular, a closed form of the algorithmic tangent moduli is derived.

The reliability of the current formulation has been examined by verification and validation examples, showing a very satisfactory level of accuracy with respect to the experimental data.

Finally, further research activities will comprise the application of the proposed formulation to hybrid metal-composite clinching manufacturing processes.

Acknowledgements The authors gratefully acknowledge the financial support of the German Research Foundation (DFG) through the program SPP 1640 (joining by plastic deformation) under the contract No. RO 706/6-2. JR is also grateful to the Spanish Ministry of Economy and Competitiveness (Projects MAT2015-71036-P and MAT2015-71309-P) and the Andalusian Government (Project of Excellence No. TEP-7093). AD gratefully acknowledges the support of Mr. and Mrs. Dean.

Appendix

This appendix addresses the weak formulation of the IBVP presented in Eq. (6) (Sect. 2), which represents the most convenient setting to formulate the corresponding numerical approximation based on FEM (Finite Element Method) through the exploitation of the standard Galerkin procedure.

Assume that the reference body boundary $\partial\mathscr{B}_0$ is subdivided into the disjointed parts $\partial\mathscr{B}_{0,u} \subset \partial\mathscr{B}_0$ and $\partial\mathscr{B}_{0,t} \subset \partial\mathscr{B}_0$, with $\partial\mathscr{B}_0 = \partial\mathscr{B}_{0,u} \cup \partial\mathscr{B}_{0,t}$ and $\partial\mathscr{B}_{0,u} \cap \partial\mathscr{B}_{0,t} = \emptyset$. As customary, appropriate boundary conditions must be defined in order to guarantee the well-posedness of the IBVP. The weak form of the balance of linear momentum reads:

$$\begin{aligned} G^u(\mathbf{u}, \delta\mathbf{u}) &= \int_{\mathscr{B}_0} \left(\mathrm{DIV}\left[\mathbf{P}\right] + \bar{\mathbf{\Upsilon}}\right) \delta\mathbf{u}\mathrm{dV} = \int_{\mathscr{B}_0} \left(\mathrm{DIV}\left[\delta\mathbf{u}\mathbf{P}\right] - \mathbf{P} : \nabla_{\mathbf{X}}\delta\mathbf{u} + \bar{\mathbf{\Upsilon}}\delta\mathbf{u}\right) \mathrm{dV} \\ &= \int_{\mathscr{B}_0} \mathbf{P} : \delta\mathbf{F}\mathrm{dV} - \int_{\partial\mathscr{B}_0} \mathbf{T}\delta\mathbf{u}\mathrm{dA} - \int_{\mathscr{B}_0} \bar{\mathbf{\Upsilon}}\delta\mathbf{u}\mathrm{dV} = G^u_{int} + G^u_{ext} = 0, \end{aligned} \quad (105)$$

where $\delta\mathbf{u}$ renders the virtual displacement and $\delta\mathbf{F} = \nabla_{\mathbf{X}}\delta\mathbf{u}$ and $\mathbf{T} = \mathbf{PN}$ denotes the first Piola-Kirchhoff traction vector. Note that to achieve the present form of Eq. 105, the following rules are used:

$$\mathrm{DIV}\left[\mathbf{P}\right]\delta\mathbf{u} = \mathrm{DIV}\left[\delta\mathbf{u}\mathbf{P}\right] - \mathbf{P} : \nabla_{\mathbf{X}}\delta\mathbf{u}, \quad (106)$$

and the Gauss-Green theorem:

$$\int_{\mathscr{B}_0} \mathrm{DIV}\left[\delta\mathbf{u}\mathbf{P}\right] \mathrm{dV} = \int_{\partial\mathscr{B}_0} (\mathbf{PN})\, \delta\mathbf{u}\mathrm{dA}. \quad (107)$$

The virtual internal work G^u_{int} and the virtual work of external actions G^u_{ext} are given by:

$$G^u_{int}(\mathbf{u}, \delta\mathbf{u}) = \int_{\mathscr{B}_0} \mathbf{P} : \delta\mathbf{F}\mathrm{dV}, \quad (108)$$

$$G^u_{ext}(\mathbf{u}, \delta\mathbf{u}) = -\int_{\partial\mathscr{B}_0} \mathbf{T}\delta\mathbf{u}\mathrm{dA} - \int_{\mathscr{B}_0} \bar{\boldsymbol{\Upsilon}}\delta\mathbf{u}\mathrm{dV}. \tag{109}$$

The resulting set of nonlinear equations of the mechanical problem, Eq. 105, can be solved numerically through the use of the incremental and iterative Newton-Raphson solution scheme, which shows a quadratic convergence near the solution point. The consistent linearization of the given time integration algorithm, also called stress-update algorithm, leads to the derivation of the consistent tangent moduli, which describes in an incremental manner the stress sensitivity with respect to the deformation gradient increment. Following the directional derivative concept [16], the consistent linearization of Eq. 105 takes the following representation:

$$\mathrm{Lin}\left[G^u\left(\bar{\mathbf{u}}, \delta\mathbf{u}, \Delta\mathbf{u}\right)\right] = G^u\left(\bar{\mathbf{u}}, \delta\bar{\mathbf{u}}\right) + DG^u\left(\bar{\mathbf{u}}, \delta\mathbf{u}\right)\Delta\mathbf{u}. \tag{110}$$

In Eq. 105, the term $\mathbf{P} : \delta\mathbf{F}$ has to be linearized yielding:

$$\Delta\left(\mathbf{P} : \delta\mathbf{F}\right) = \Delta\mathbf{P} : \delta\mathbf{F}, \tag{111}$$

where $\Delta\mathbf{P}$ is derived in Sect. 4.2, with:

$$\Delta\mathbf{P} = \Delta\left(\mathbf{FS}\right) = \mathbb{C}^{ep} : \Delta\mathbf{F}, \tag{112}$$

where $\mathbb{C}^{ep}$ denotes the algorithmic elasto-plastic constitutive operator.

References

1. Advani, S.G., Tucker, C.L.: The use of tensors to describe and predict fiber orientation in short fiber composites. J. Rheol. **31**, 751–784 (1987)
2. Advani, S.G., Talreja, R.: A continuum approach to determination of elastic properties of short fibre composites. Mech. Compos. Mater. **29**(2), 171–183 (1993)
3. Arif, M.F., Saintier, N., Meraghni, F., Fitoussi, J., Chemisky, Y., Robert, G.: Multiscale fatigue damage characterization in short glass fiber reinforced polyamide-66. Comp. Part B **61**, 55–65 (2014)
4. Bellenger, V., Tcharkhtchi, A., Castaing, P.: Thermal and mechanical fatigue of a PA66/glass fibers composite material. Int. J. Fatigue **28**(10), 1348–1352 (2006)
5. Bernasconi, A., Cosmi, F., Hine, P.J.: Analysis of fibre orientation distribution in short fibre reinforced polymers: a comparison between optical and tomographic methods. Compos. Sci. Technol. **72**(16), 2002–2008 (2012)
6. Boehler, J.P.: Applications of tensor functions in solid mechanics. In: CISM Courses and Lectures, vol. 292. Springer (1987)
7. Dean, A., Reinoso, J., Sahraee, S., Rolfes, R.: An invariant-based anisotropic material model for short fiber-reinforced thermoplastics: coupled thermo-plastic formulation. Compos. Part A Appl. Sci. Manufact. **90**, 186–199 (2016)
8. Dean, A., Sahraee, S., Reinoso, J., Rolfes, R.: Finite deformation model for short fiber reinforced composites: application to hybrid metal-composite clinching joints. Compos. Struct. **151**, 162–171 (2016)

9. Dean, A., Sahraee, S., Reinoso, J., Rolfes, R.: A New invariant-based thermo-plastic model for finite deformation analysis of short fibre reinforced composites: development and numerical aspects. Compos. Part B Eng. **125**, 241–258 (2017)
10. De Monte, M., Moosbugger, E., Quaresimin, M.: Inuence of temperature and thickness on the off-axis behaviour of short glass bre reinforced polyamide 6.6Quasi-static loading. Compos. Part A **41**, 859–871 (2010)
11. De Monte, M., Moosbugger, E., Quaresimin, M.: Inuence of temperature and thickness on the off-axis behaviour of short glass bre reinforced polyamide 6.6Cyclic loading. Compos. Part A **41**, 1368–79 (2010)
12. Eftekhari, M., Fatemi, A.: Tensile, creep and fatigue behaviors of short fiber reinforced polymer composites at elevated temperatures: a literature survey. Fatigue Fract. Eng. Mater. Struct. **38**, 1395 (2015)
13. Eidel, B., Gruttmann, F.: Elastoplastic orthotropy at finite strains: multiplicative formulation and numerical implementation. Comput. Mater. Sci. **28**, 732–742 (2003)
14. Eidel, B.: Anisotropic Inelasticity: Modelling, Simulation, Validation. Ph.D. Dissertation, Technischen Universität Darmstadt, Darmstadt, Germany (2004)
15. Feyel, F., Chaboche, J-L.: FE2 multiscale approach for modelling the elastoviscoplastic behaviour of long fibre SiC/Ti composite materials, Computer Meth. in Appl. Mech. Eng. **183**, 309–330 (2000)
16. Holzapfel, G. A.: Nonlinear Solid Mechanics. Wiley (2000). ISBN: 978-0-471-82319-3
17. Lu, S.C.H., Pister, K.D.: Decomposition of deformation and representation of the free energy function for isotropic thermoelastic solids. Int. J. Solids Struct. **11**, 927–934 (1975)
18. Marco, Y., Le Saux, V., Jégou, L., Launay, A., Serrano, L., Raoult, I., Callot, S.: Dissipation analysis in SFRP structural samples: thermomechanical analysis and comparison to numerical simulations. Int. J. Fatigue **67**, 142–150 (2014)
19. Menzel, A., Steinmann, P.: On the spatial formulation of anisotropic multiplicative elasto-plasticity. Comput. Methods Appl. Mech. Eng. **192**, 3431–3470 (2003)
20. Miehe, C., Apel, N., Lambrecht, M.: Anisotropic additive plasticity in the logarithmic strain space: modular kinematic formulation and implementation based on incremental minimization principles for standard materials. Comput. Methods Appl. Mech. Eng. **191**, 5383–5425 (2002)
21. Mlekusch, B.: Thermoelastic properties of short-fibre-reinforced thermoplastics. Compos. Sci. Technol. **59**, 911–923 (1999)
22. Reese, S.: Meso-macro modelling of fibre-reinforced rubber-like composites exhibiting large elastoplastic deformation. Int. J. Solids Struct. **40**(4), 951–980 (2003)
23. Reinoso, J., Blázquez, A.: Application and finite element implementation of 7-parameter shell element for geometrically nonlinear analysis of layered CFRP composites. Compos. Struct. **139**, 263–276 (2016)
24. Sansour, C., Karsaj, I., Soric, J.: On a formulation for anisotropic elastoplasticity at finite strains invariant with respect to the intermediate configuration. J. Mech. Phys. Sol. **55**, 2406–2426 (2007)
25. Schröder, J., Gruttmann, F., Löblein, J.: A simple orthotropic finite elastoplasticity model based on generalized stress strain measures. Comput. Mech. **30**, 48–64 (2002)
26. Schröpfer, J.: Spritzgussbauteile aus kurzfaserverstärkten Kunststoffen: Methoden der Charakterisierung und Modellierung zur nichtlinearen Simulation von statischen und crashrelevanten Lastfällen. Dissertation Technischen Universit ät Kaiserslautern (2011)
27. Simo, J.C., Miehe, C.: Associative coupled thermoplasticity at finite strains: Formulation numerical analysis and implementation. Comput. Methods Appl. Mech. Eng. **98**, 41–104 (1992)
28. Spahn, J., Andrä, H., Kabel, M., Muller, R.: A multiscale approach for modeling progressive damage of composite materials using fast Fourier transforms Comput Method Appl. Mech. Eng. **268**, 871–883 (2014)
29. Spencer, A.J.M.: Theory of invariants, In: Eringen, A.C. (Ed.), Continuum Physics, vol. 1, Academic Press, New York, 239–353 (1971)

30. Thi, T.B.N., Morioka, M., Yokoyama, A., Hamanaka, S., Yamashita, K., Nonomura, C.: Measurement of fiber orientation distribution in injection-molded short-glass- fiber composites using X-ray computed tomography. J. Mater. Process. Technol. **219**, 1–9 (2015)
31. Truesdell, C., Noll, W.: The nonlinear field theories of mechanics, In: Flgge, S. (Ed.) Encyclopedia of Physics, vol. III/3. Springer, Berlin (1965)
32. Vladimirov, I. N., Pietryga, M. P., Reese, S.: Anisotropic finite elastoplasticity with nonlinear kinematic and isotropic hardening and application to sheet metal forming. Int. J. Plasticit. 0749–6419, **26**, 659–687 (2010)
33. Vogler, M., Rolfes, R., Camanho, P.P.: Modeling the inelastic deformation and fracture of polymer composites part I: plasticity model. Mech. Mater. **59**, 50–64 (2013)
34. Zohdi, T., Wriggers, P.: Aspects of the computational testing of properties of microheterogeneous material samples. Inter. J. Meth. Eng. **50**, 2573–2599 (2001)

Part II
Computational Solution Approaches

Unified Approach to Sensitivity Analysis Based Automation of Multi-scale Modelling

N. Zupan and J. Korelc

Abstract Use of different kinds of multi-scale methods is limited with specifications of the problem to be solved. Standard two-level finite element homogenization approach FE^2 is appropriate for problems with weakly coupled scales. If the difference between two scales is finite, or in the region of high gradients the FE^2 multi-scale approach fails, then some sort of domain decomposition method can be applied. Our motivation was to create computational environment, where the multi-scale code is automatically derived and various types of multi-scale approaches can be freely mixed. The described approach uses an advanced feature of software tools AceGen and AceFEM, that is automatic generation of the finite element codes for analytical first and second order sensitivity analysis with respect to prescribed essential boundary conditions as a unifying factor. The automatic-differentiation-based formulation (ADB) enables unification and automation of various multi-scale approaches for an arbitrary nonlinear, time dependent, coupled problem (e.g. general finite strain plasticity).

1 Multi-scale Methods

Multi-scale methods are nowadays widespread in computational mechanics. They are a good alternative, when for computing on one scale, very refined mesh is needed. Many of the multi-scale methods originate from the demand to model heterogeneous materials, like fiber reinforced composites, particle reinforced adhesives, concrete and even metal. A short overview of different multi-scale methods with their pros, cons and possible applications can be found in [2], while fundamentals of the computational micromechanics are described in [16]. Use of different kinds of methods is limited by the characteristics of the problem to be solved. Roughly, we can

N. Zupan (✉) · J. Korelc
Faculty of Civil and Geodetic Engineering, Jamova cesta 2, 1000 Ljubljana, Slovenia
e-mail: nina.zupan@fgg.uni-lj.si

J. Korelc
e-mail: joze.korelc@fgg.uni-lj.si

J. Sorić et al. (eds.), *Multiscale Modeling of Heterogeneous Structures*,
Lecture Notes in Applied and Computational Mechanics 86,
https://doi.org/10.1007/978-3-319-65463-8_6

separate multi-scale methods in two groups: on methods that are based on homogenization techniques and on domain decomposition methods. A basic hypothesis of homogenisation techniques is the complete separation of scales, where the size of heterogeneities is assumed to be infinitely smaller than the structural dimensions. Homogenized material behaviour of representative volume elements (RVEs), which contain microstructure, is considered to be representative for the entire or part of the structure. This approach is not correct in case of localisation, because homogenised material behaviour is influenced by cracks or instabilities within the RVE. In recent years some specialised multi-scale methods for fracture applications have been developed [10].

Standard two-level finite element homogenization approach FE^2 is appropriate for the problems where scales are separated far enough and are only weakly coupled, see [1, 13]. In this approach we have one FE model for the macro scale and the second one at each material integration point. The material response is obtained from the micro level FE analysis. If the difference between two scales is finite, or in the region of high gradients the FE^2 multi-scale approach fails, then some sort of domain decomposition method can be applied. One such method is mesh-in-element (MIEL) scheme described by Markovič and Ibrahimbegović in [11]. The aim of the paper is to present a unified approach to the development and implementation of FE^2 and MIEL multi-scale schemes based on boundary condition sensitivity analysis.

2 Automation of Multi-scale Methods

Our motivation was to create computational environment, where the multi-scale program code is automatically derived and various types of multi-scale and single-scale approaches can be freely mixed, while retaining quadratic convergence of the Newton-Raphson procedure. The described method uses an advanced feature of software tool [6], that is automatic generation of the finite element codes for analytical first and second order sensitivity analysis based on a specialized implementation of automatic differentiation technique. The automatic-differentiation-based (ADB) formulation enables unification and automation of various multi-scale approaches for an arbitrary nonlinear, time dependent coupled problem (e.g. general finite strain plasticity).

2.1 Automation of Boundary Condition Sensitivity Analysis

Automation of primal and sensitivity analysis is achieved through the hybrid symbolic-numeric approach to automation of finite element method that combines symbolic and algebraic capabilities of a general computer algebra system, e.g. Mathematica [12], an automatic differentiation technique (AD) and an automatic code generation with the general purpose finite element environment. The AD method is used for the evaluation of the exact derivatives of any arbitrary complex function via chain rule and represents an alternative solution to the numerical differentiation and symbolic differentiation. The result of the AD procedure is called "computational derivative" and is written as $\frac{\hat{\delta} f(\mathbf{a})}{\hat{\delta}\mathbf{a}}$. The AD operator $\frac{\hat{\delta} f(\mathbf{a})}{\hat{\delta}\mathbf{a}}$ represents partial differentiation of a function $f(\mathbf{a})$ with respect to variables $\mathbf{a}$. If, for example, alternative or additional dependencies for a set of intermediate variables $\mathbf{b}$ have to be considered for differentiation, then the AD exception is indicated by the following formalism

$$\left.\frac{\hat{\delta} f(\mathbf{a},\mathbf{b})}{\hat{\delta}\mathbf{a}}\right|_{\frac{D\mathbf{b}}{D\mathbf{a}}=\mathbf{M}}, \tag{1}$$

which indicates that during the AD procedure, the total derivatives of variables $\mathbf{b}$ with respect to variables $\mathbf{a}$ are set to be equal to matrix $\mathbf{M}$. The automatic differentiation exceptions are the basis for the automatic differentiation or ADB formulation of computational problem. The ADB notation can be directly translated to the AceFEM code and is part of numerically efficient code automation. Details of the method and of the corresponding software AceGen can be found in [4, 5, 7].

The automation of multi-scale analysis requires the automation of primal and sensitivity analysis. In primal analysis the response of the system is evaluated, while in sensitivity analysis the derivatives of the response, e.g. displacements, strains, stresses or work, with respect to arbitrary design parameter ϕ_i are sought. Design parameter can be any parameter that influences the response, e.g. material constants, load intensity and distribution, shape parameters or boundary conditions. For the automation of the multi-scale methods sensitivity analysis with respect to prescribed essential boundary conditions is important.

The procedures for solving the primal and sensitivity analysis are for an arbitrary coupled path dependent problem presented in detail in [5]. Here, a summary of the primal and sensitivity analysis of hyperelastic problems is given. Let us define a primal problem with the residual equation $\mathbf{R}(\mathbf{p}) = \mathbf{0}$, where $\mathbf{p}$ represents a set of nodal unknowns of the problem. The primal problem is solved by the standard Newton-Raphson iterative procedure (see e.g. [5]). For the boundary condition sensitivity analysis we define the residuals and the vectors of unknown as a function of a vector of design parameters $\boldsymbol{\phi}$ by

$$\mathbf{R}(\mathbf{p}(\boldsymbol{\phi}), \bar{\mathbf{p}}(\boldsymbol{\phi})) = \mathbf{0} \tag{2}$$

where $\bar{\mathbf{p}}$ represents a set of nodal unknowns with prescribed essential boundary conditions. The sensitivity problem can be obtained from the primal problem by differentiating (2) with respect to design parameters (3). Equation (3) represents a system of linear equations (5) for the unknown sensitivities of the primal unknowns of the problem $\frac{D\mathbf{p}}{D\phi_i}$. The right hand side is called "first order sensitivity pseudo load vector". The vector $\frac{D\bar{\mathbf{p}}}{D\phi_i}$ represents the rate of the change of the prescribed essential boundary conditions with respect to the change of design parameter and is called "prescribed boundary condition velocity field".

$$\frac{\partial \mathbf{R}}{\partial \mathbf{p}} \frac{D\mathbf{p}}{D\phi_i} + \frac{\partial \mathbf{R}}{\partial \bar{\mathbf{p}}} \frac{D\bar{\mathbf{p}}}{D\phi_i} = \mathbf{0} \tag{3}$$

$$^{I}\tilde{\mathbf{R}} = -\frac{\partial \mathbf{R}}{\partial \bar{\mathbf{p}}} \frac{D\bar{\mathbf{p}}}{D\phi_i} \tag{4}$$

$$\mathbf{K}\frac{D\mathbf{p}}{D\phi_i} = -{}^{I}\tilde{\mathbf{R}} \tag{5}$$

The sensitivity problem is solved after the convergence of the primal problem has been reached. For the automation we need only the ADB form of pseudo load vector $^{I}\tilde{\mathbf{R}}$ evaluated at the integration point of the individual finite element $^{I}\tilde{\mathbf{R}}_g$. The global pseudo load vector is then obtained by the standard integration over the element domain and the standard finite element assembly procedure of element contributions to global vector $^{I}\tilde{\mathbf{R}}_g$.

In the ADB form partial derivative is replaced with computational derivative and AD exceptions for the indirect dependencies $\bar{\mathbf{p}}(\boldsymbol{\phi})$ have to be added, leading to

$$^{I}\tilde{\mathbf{R}}_g = \left.\frac{\hat{\delta}\mathbf{R}_g}{\hat{\delta}\phi_i}\right|_{\frac{D\hat{\mathbf{p}}_e}{D\phi_i} = D_{\phi_i}\bar{\mathbf{p}}_e} \tag{6}$$

where $D_{\phi_i}\bar{\mathbf{p}}_e$ represents data structure of the first order prescribed boundary condition velocity field.

The second order sensitivity problem is obtained from the first order problem, by differentiating (3) with respect to design parameter ϕ_i. It results in

$$\frac{\partial^2 \mathbf{R}}{\partial \mathbf{p}^2}\frac{D\mathbf{p}}{D\phi_i}\frac{D\mathbf{p}}{D\phi_j} + \frac{\partial^2 \mathbf{R}}{\partial \mathbf{p}\partial \bar{\mathbf{p}}}\frac{D\bar{\mathbf{p}}}{D\phi_j}\frac{D\mathbf{p}}{D\phi_i} + \frac{\partial \mathbf{R}}{\partial \mathbf{p}}\frac{D^2\mathbf{p}}{D\phi_i D\phi_j} + \frac{\partial^2 \mathbf{R}}{\partial \bar{\mathbf{p}}\partial \mathbf{p}}\frac{D\mathbf{p}}{D\phi_j}\frac{D\bar{\mathbf{p}}}{D\phi_i} + \\ + \frac{\partial^2 \mathbf{R}}{\partial \bar{\mathbf{p}}^2}\frac{D\bar{\mathbf{p}}}{D\phi_i}\frac{D\bar{\mathbf{p}}}{D\phi_j} + \frac{D\mathbf{R}}{D\bar{\mathbf{p}}}\frac{\partial^2 \bar{\mathbf{p}}}{\partial\phi_i\partial\phi_j} = \mathbf{0} \tag{7}$$

$$\mathbf{K}\frac{D^2\mathbf{p}}{D\phi_i D\phi_j} = -{}^{II}\tilde{\mathbf{R}} \tag{8}$$

where $^{II}\tilde{\mathbf{R}}$ represents the second order sensitivity pseudo load vector. The ADB form of the integration point contribution to $^{II}\tilde{\mathbf{R}}$ can be derived from (7) as

$$^{II}\tilde{\mathbf{R}}_g = \left.\frac{\hat{\delta}}{\hat{\delta}\phi_j}\left(\left.\frac{\hat{\delta}\mathbf{R}_g}{\hat{\delta}\phi_i}\right|_{\frac{D\hat{\mathbf{p}}_e}{D\phi_i} = D_{\phi_i}\hat{\mathbf{p}}_e}\right)\right|_{\frac{D\hat{\mathbf{p}}_e}{D\phi_j} = D_{\phi_j}\hat{\mathbf{p}}_e,\ \frac{D(D_{\phi_i}\hat{\mathbf{p}}_e)}{D\phi_j} = D_{\phi_i\phi_j}\bar{\mathbf{p}}_e} \tag{9}$$

where $D_{\phi_i\phi_j}\bar{\mathbf{p}}_e$ represents data structure of the second order prescribed boundary condition velocity field. All first order sensitivities have to be calculated in order to be able to calculate the second order sensitivities. Equation (3) represents a system of linear equations (8) for the unknown sensitivities of the primal unknowns of the problem $\frac{D^2\mathbf{p}}{D\phi_i D\phi_j}$.

2.2 Automation of FE^2

The FE^2 method is a two-level scheme in which all information about micro-structure is obtained from computations on RVE-level by averaging the material response characterized by an appropriate stress measure and constitutive tangent matrix over RVE. RVE is attached to each integration point of the macroscopic FE problem, as shown in Fig. 1. For a typical finite strain problem that leads to $\boldsymbol{P}_M = \{\boldsymbol{P}_m\}$ and $\mathbf{A}_M = \{\frac{\partial \boldsymbol{P}_m}{\partial \boldsymbol{F}_M}\}$, where $\boldsymbol{P}_M$ and $\boldsymbol{P}_m$ are the first Piola-Kirchoff stress tensors at macro and micro level, $\boldsymbol{F}_M$ is the macroscopic deformation gradient and $\mathbf{A}_M$ a macroscopic constitutive matrix. This information is then used at the macro level for the evaluation of integration point contribution to element residual $\mathbf{R}_{Mg}$ and tangent matrix $\mathbf{K}_{Mg}$ at the macro level, as follows

$$\mathbf{R}_{Mg} = \boldsymbol{P}_M : \frac{\partial \boldsymbol{F}_M}{\partial \mathbf{p}_{Me}} \tag{10}$$

$$\mathbf{K}_{Mg} = \frac{\partial \mathbf{R}_M}{\partial \mathbf{p}_{Me}} + \frac{\partial \mathbf{R}_M}{\partial \boldsymbol{P}_M} \frac{D\boldsymbol{P}_M}{D\boldsymbol{F}_M} \frac{\partial \boldsymbol{F}_M}{\partial \mathbf{p}_{Me}} \tag{11}$$

where $\mathbf{p}_{Me}$ represents a set of nodal unknowns of macro element. Prescribed displacements $\bar{\mathbf{u}}_m$ at the corner nodes of the RVE are computed from macroscopic deformation gradient $\boldsymbol{F}_M$ by

$$\bar{\mathbf{u}}_m = (\boldsymbol{F}_M - \mathbf{I})\,\boldsymbol{X}_m. \tag{12}$$

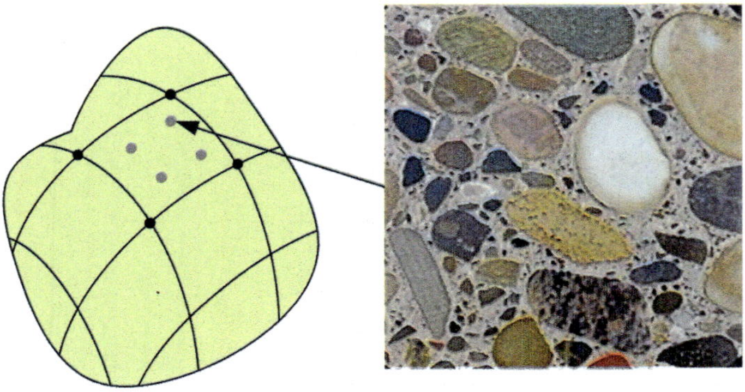

Fig. 1 Macroscopic finite element problem and RVE attached to each integration point

For the unconstrained boundary nodes periodicity of boundary conditions is adopted with the use of Lagrange multipliers.

The FE^2 method can be implemented in different ways (see e.g. [8]). For efficiency of the method efficient calculation of the macroscopic constitutive matrix $\mathbf{A}_M$ is important. In a conventional way of computing macroscopic constitutive tangent, computation of a Schur complement of RVEs tangent matrix is needed. Some optimisations of efficiency were already made with a tangent computation technique that is based on perturbation approach [15].

An alternative for the calculation of macroscopic tangent is calculation of consistent macroscopic stiffness matrix via direct differentiation sensitivity analysis of the micro-structure with respect to macro strain measure used to impose boundary conditions on RVE. It was described in [9] and [14] for time-independent problems. Sensitivity parameters of the problem are components of the macro deformation gradient $\boldsymbol{F}_M$, thus $\boldsymbol{\phi} = \{F_{M,11}, F_{M,12}, F_{M,13}, F_{M,21}, \ldots\}$. For the complete formulation of the prescribed boundary condition sensitivity problem, as presented in Sect. 2.1, we also need prescribed boundary condition velocity field $D_{\phi_i}\bar{\mathbf{p}}_e$, see Table 1. The components of $D_{\phi_i}\bar{\mathbf{p}}_e$ are obtained by the differentiation of (12) with respect to $\boldsymbol{\phi}$ (13)

Table 1 Comparison between FE^2 and MIEL

Method	Micro-scale	Characteristic velocity field
FE^2 periodic BC	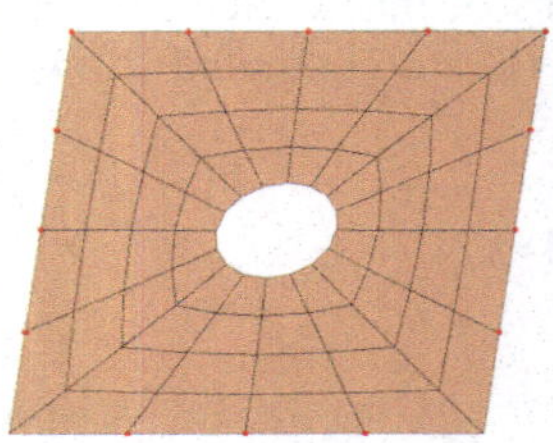	$\frac{\partial \bar{u}_{mi}}{\partial F_{Mjk}} = \delta_{ij} X_{mk}$ 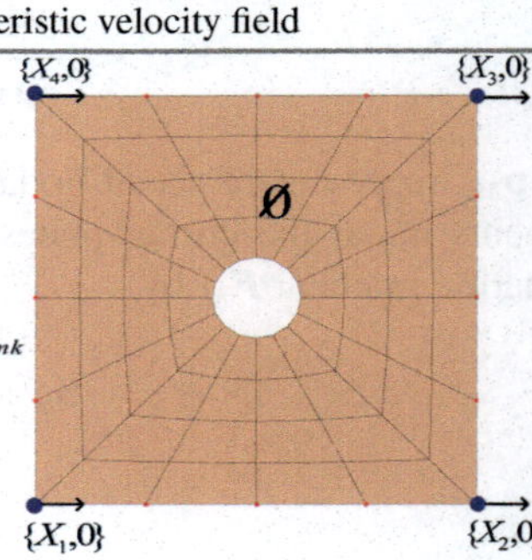
MIEL	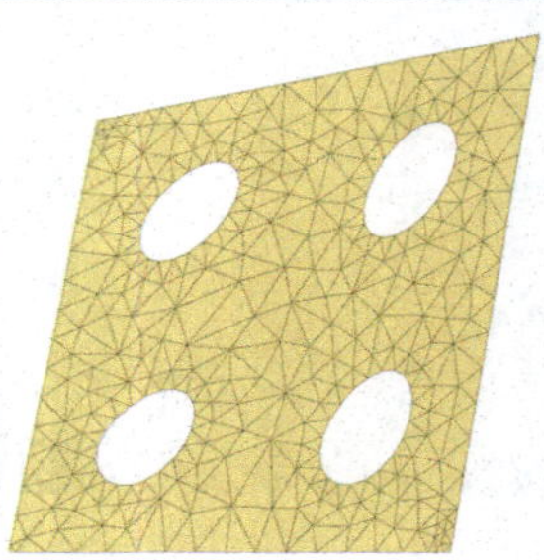	$\frac{\partial \sum N_i(\Xi) u_{mi}}{\partial u_{Mj}} = N_j(\Xi)$ 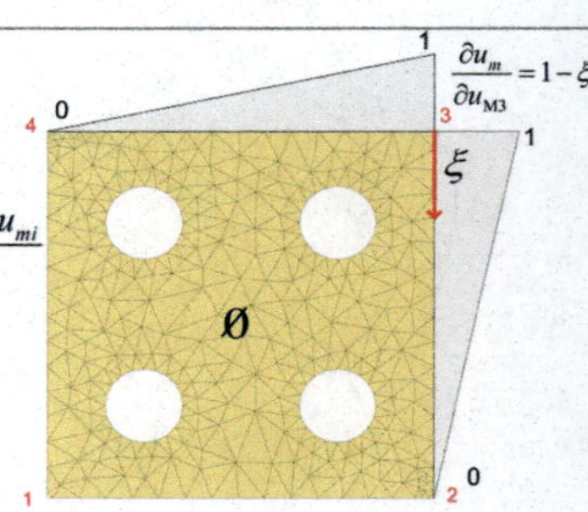

$$\frac{\partial \bar{u}_{m\,i}}{\partial F_{M\,j\,k}} = \delta_{ij}\, X_{m\,k} \tag{13}$$

and evaluated at the RVEs corner nodes. The result of the sensitivity analysis is derivatives $\frac{D\mathbf{p}}{D\boldsymbol{\phi}}$ that can be used to evaluate an integration point contribution to macroscopic constitutive matrix

$$\mathbf{A}_g = \frac{\partial \boldsymbol{P}_m}{\partial \boldsymbol{F}_M} = \left.\frac{\hat{\delta} \boldsymbol{P}_m}{\hat{\delta} \boldsymbol{F}_M}\right|_{\frac{D\hat{\mathbf{p}}_{m\,e}}{D\boldsymbol{\phi}} = D_{\boldsymbol{\phi}}\hat{\mathbf{p}}_{m\,e}} \tag{14}$$

At the end the automation of the integration point contribution to the residual and tangent matrix of macro element leads from (10) and (11) to

$$\mathbf{R}_{Mg} = \left.\frac{\hat{\delta}\mathbf{W}_M}{\hat{\delta}\mathbf{p}_{Me}}\right|_{\boldsymbol{P}_M = const.} \tag{15}$$

$$\mathbf{K}_{Mg} = \left.\frac{\hat{\delta}\mathbf{R}_{Mg}}{\hat{\delta}\mathbf{p}_{Me}}\right|_{\frac{D\boldsymbol{P}_M}{D\boldsymbol{F}_M} = \mathbf{A}_M} \tag{16}$$

where a pseudo-potential scalar function $W_M = \boldsymbol{P}_M : \boldsymbol{F}_M$ was introduced in order to make automatic differentiation more efficient (see e.g. [5]). Output of the Newton-Raphson iteration at the macro level is a new macro deformation gradient $\boldsymbol{F}_M$ that is used within the next iteration at the macro level. The analytical sensitivity analysis ensures consistent tangent matrix and a quadratic convergence rates on both scales, RVE-level and macro-level.

In [14] a step forward was made with the introduction of symmetric stretch tensor $\boldsymbol{U}_M$ as strain measure at macro level instead of asymmetric deformation gradient $\boldsymbol{F}_M$, to determine boundary conditions on embedded micro-structure. Stretch tensor $\boldsymbol{U}_M$ can be calculated as matrix square root of Cauchy-Green tensor $\boldsymbol{C}_M$, for which efficient, automated way of evaluation together with its derivatives can be found in [3]. Use of symmetric stretch tensor $\boldsymbol{U}_M$ that has only 6 components instead of $\boldsymbol{F}_M$ with 9, significantly reduces computational cost of boundary condition related sensitivity analysis of micro-structure and with it the evaluation of local macroscopic stress tensors and tangent matrices. The following equations summarize the alternative formulation.

$$\boldsymbol{C}_M = \boldsymbol{F}_M^T \cdot \boldsymbol{F}_M, \quad \boldsymbol{U}_M = \sqrt{\boldsymbol{C}_M} \tag{17}$$

$$\boldsymbol{\phi} = \{U_{M,11}, U_{M,12}, U_{M,13}, U_{M,22}, \ldots\} \tag{18}$$

$$\bar{\mathbf{u}}_m = (\boldsymbol{U}_M - \mathbf{I})\, \boldsymbol{X}_m, \quad \frac{\partial \bar{u}_{m\,i}}{\partial U_{M\,j\,k}} = \delta_{ij}\, X_{m\,k} \tag{19}$$

$$\mathbf{A}_g = \frac{\partial \boldsymbol{P}_m}{\partial \boldsymbol{U}_M} = \left.\frac{\hat{\delta} \boldsymbol{P}_m}{\hat{\delta} \boldsymbol{U}_M}\right|_{\frac{D\hat{\mathbf{p}}_{m\,e}}{D\boldsymbol{\phi}} = D_{\boldsymbol{\phi}}\hat{\mathbf{p}}_{m\,e}} \tag{20}$$

$$\mathbf{R}_{Mg} = \boldsymbol{P}_M : \frac{\partial \boldsymbol{U}_M}{\partial \mathbf{p}_{Me}} = \left.\frac{\hat{\delta} \mathbf{W}_M}{\hat{\delta} \mathbf{p}_{Me}}\right|_{\boldsymbol{P}_M = const.} \tag{21}$$

$$\mathbf{K}_{Mg} = \frac{\partial \mathbf{R}_M}{\partial \mathbf{p}_{Me}} + \frac{\partial \mathbf{R}_M}{\partial \boldsymbol{P}_M} \frac{D\boldsymbol{P}_M}{D\boldsymbol{U}_M} \frac{\partial \boldsymbol{U}_M}{\partial \mathbf{p}_{Me}} = \left.\frac{\hat{\delta} \mathbf{R}_{Mg}}{\hat{\delta} \mathbf{p}_{Me}}\right|_{\frac{D\boldsymbol{P}_M}{D\boldsymbol{F}_M} = \mathbf{A}_M} \tag{22}$$

Again, only the first order sensitivity analysis of the micro problem unknowns $\mathbf{p}_m$ with respect to the components of the macro stretch tensor $\frac{D\mathbf{p}_m}{D\boldsymbol{\phi}}$ is needed and it can be automatically obtained using procedures from Sect. 2.1.

2.3 Automation of MIEL

When the difference between two scales is finite, or in the region of high gradients the FE^2 multi scale approach fails, then we need to use some sort of domain decomposition method. One possibility is the mesh-in-element or MIEL scheme described e.g. by Markovič and Ibrahimbegović in [11]. We used it as a starting point for the development of an automatized multi-scale method. The FE models of different scales communicate between each other through degrees of freedom of the finite element at the macro-scale. The residual and tangent matrix are for each macro element obtained directly from micro-scale problem. Each macro element thus represents one micro problem, see Fig. 2. Macro element performs only proper transfer of components of the macro element residual vector and tangent matrix from micro scale to macro scale finite element assembly procedure. At the macro level residual and tangent are assembled from individual macro elements and macro response is calculated. Originally the Schur complement of the micro problem global tangent matrix is used to calculate components of the macro element tangent matrix. Here the boundary condition sensitivity analysis will be used again to obtain the same. For densely meshed micro-structure calculations of the Schur complement inflicts high memory allocation and is time consuming, which is not the case for the sensitivity analysis based implementation of the MIEL method. This is due to the fact that the number of sensitivity parameters remains the same, regardless of the density of the micro mesh, whereas the size of the Schur complement grows with the number of the nodes on the boundary of the micro problem.

Let $\mathbf{p}_{M\,e}$ be a vector of unknowns in the nodes of the macro element, $\mathbf{p}_{m\,e}$ a vector of unknowns in the nodes of the characteristic micro problem element and W strain energy function. The outer shape of the micro problem is the same as the shape of the corresponding macro element. The prescribed essential boundary

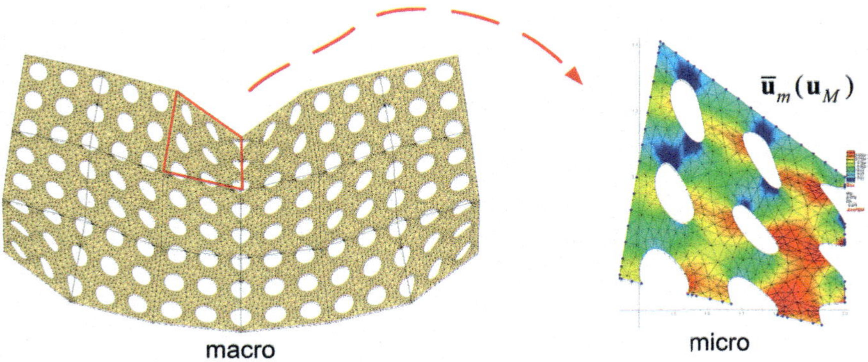

Fig. 2 Macro and micro scale for MIEL

conditions (displacements) are identical to the displacements at the boundary of the corresponding macro element. The integration point contribution (g-th integration point in the e-th element of the micro mesh) to the macro residual and macro tangent matrix is then

$$\mathbf{R}_{Mg} = \frac{\partial W(\mathbf{p}_{m\,e}(\mathbf{p}_{M\,e}))}{\partial \mathbf{p}_{M\,e}} = \frac{\partial W}{\partial \mathbf{p}_{m\,e}} \frac{D\mathbf{p}_{m\,e}}{D\mathbf{p}_{M\,e}} \tag{23}$$

$$\mathbf{K}_{Mg} = \frac{\partial \mathbf{R}_{Mg}}{\partial \mathbf{p}_{M\,e}} = \frac{\partial^2 W}{\partial \mathbf{p}_{m\,e}{}^2} \frac{D\mathbf{p}_{m\,e}}{D\mathbf{p}_{M\,e}} + \frac{\partial W}{\partial \mathbf{p}_{m\,e}} \frac{D^2\mathbf{p}_{m\,e}}{D\mathbf{p}_{M\,e}^2} \tag{24}$$

The implicit dependencies $\frac{D\mathbf{p}_{m\,e}}{D\mathbf{p}_{M\,e}}$ and $\frac{D^2\mathbf{p}_{m\,e}}{D\mathbf{p}_{M\,e}^2}$ are obtained by the first and second order sensitivity analysis. Thus, the sensitivity analysis based automation of the MIEL scheme requires the second order sensitivity analysis for a set of sensitivity parameters $\mathbf{p}_{M\,e}$, as presented in Sect. 2.1. The ADB form of (23) and (24) then leads to

$$\mathbf{R}_{Mg} = \left.\frac{\hat{\delta}\mathbf{W}}{\hat{\delta}\mathbf{p}_{M\,e}}\right|_{\frac{D\mathbf{p}_{m\,e}}{D\mathbf{p}_{M\,e}} = D_{\mathbf{p}_{M\,e}}\mathbf{p}_{m\,e}} \tag{25}$$

$$\mathbf{K}_{Mg} = \left.\frac{\hat{\delta}\mathbf{R}_{Mg}}{\hat{\delta}\mathbf{p}_{M\,e}}\right|_{\frac{D\mathbf{p}_{m\,e}}{D\mathbf{p}_{M\,e}} = D_{\mathbf{p}_{M\,e}}\mathbf{p}_{m\,e},\ \frac{D(D_{\mathbf{p}_{M\,e}}\mathbf{p}_m)}{D\mathbf{p}_{M\,e}} = D_{\mathbf{p}_{M\,e}\mathbf{p}_{M\,e}}\mathbf{p}_{m\,e}} \tag{26}$$

where data structures $D_{\mathbf{p}_{M\,e}}\mathbf{p}_{m\,e} = \frac{D\mathbf{p}_{m\,e}}{D\mathbf{p}_{M\,e}}$ and $D_{\mathbf{p}_{M\,e}\mathbf{p}_{M\,e}}\mathbf{p}_{m\,e} = \frac{D^2\mathbf{p}_{m\,e}}{D\mathbf{p}_{M\,e}^2}$ are the results of the first and second order sensitivity analysis.

For the complete formulation of the prescribed boundary condition sensitivity problem, as presented in Sect. 2.1, we need the first and second order prescribed boundary condition velocity fields $D_{\phi_i}\bar{\mathbf{p}}_e$ and $D_{\phi_i\phi_j}\bar{\mathbf{p}}_e$. Let $\bar{\mathbf{p}}_m$ be a vector of unknowns at the boundary of micro problems with prescribed essential boundary conditions, thus $\bar{\mathbf{p}}_m = \bar{\mathbf{p}}_m(\mathbf{p}_{M\,e})$. The set of sensitivity parameters of the MIEL problem is $\boldsymbol{\phi} = \mathbf{p}_{M\,e}$. The components of $D_{\phi_i}\bar{\mathbf{p}}_e$ are obtained by the differentiation of $\bar{\mathbf{p}}_m(\mathbf{p}_{M\,e})$ with respect to $\mathbf{p}_{M\,e}$. Let us assume the standard interpolation of the unknown field u on the boundary of the macro element $u = \sum N_i(\Xi)\,u_i$, where $N_i(\Xi)$ is the shape functions and u_i the nodal unknowns and $\frac{\partial u}{\partial u_i} = N_i(\Xi)$. Thus, the components of the first order boundary condition velocity field $D_{\phi_i}\bar{\mathbf{p}}_e$ are the values of the macro element shape functions at the position of the boundary nodes of the micro mesh, see Table 1. The second order velocity field is in this case zero $D_{\phi_i\phi_j}\bar{\mathbf{p}}_e = \mathbf{0}$.

3 Unified Approach

The automatic-differentiation-based (ADB) formulation enables unification and automation of various multi-scale approaches for an arbitrary nonlinear, time dependent coupled problem (e.g. general finite strain plasticity). For all methods we need individual finite element codes that support the first and second order sensitivity analysis that is used for the evaluation of implicit derivatives, that is derivatives of unknowns of the problem. Sensitivity related codes are general, thus problem undependent. Additional problem dependent user subroutines are required to evaluate homogenized constitutive matrix and macro stress for FE^2 and residual and macro tangent matrix for MIEL.

Differences between the methods are in essential boundary conditions at micro mesh and in essential boundary conditions velocity fields needed for the sensitivity analysis. Macro element used in FE^2 evaluates residual and tangent matrix (see Eqs. (15) and (16)), whereas macro element used for MIEL is used just for the transformation of data. The implemented FE^2 and MIEL schemes together represent unified approach to the automation of multi-scale modelling. The use of this approach is presented on an example.

Implementation of the presented multi-scale computational approach in AceFEM is fully parallelized for multi-core processors. Micro problems are distributed on kernels by evaluating each individual micro problem always at the same kernel. For FE^2 each RVE is associated with individual Gauss point and can be calculated on individual kernel. The same goes for MIEL, where each micro problem can be distributed to individual kernel. With parallelized computation, computational time for complex problems can be significantly reduced. The setup is also appropriate for the implementation on clusters.

4 Two-Level Path-Following Procedure

In AceFEM solving of nonlinear problems is done implicitly with a Newton-Raphson type iterative solution procedure. Different path-following procedures can be used

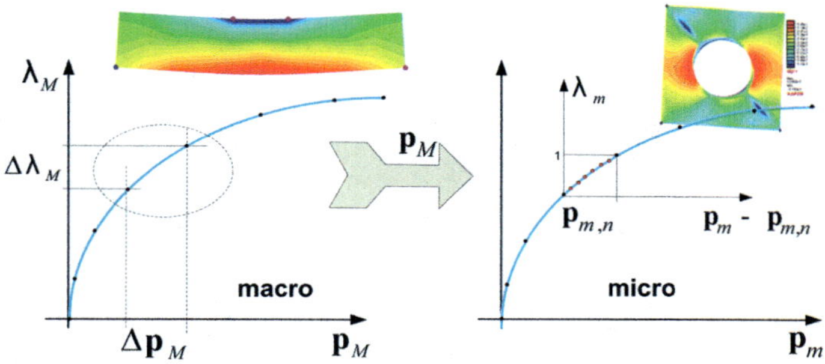

Fig. 3 Two-level path-following procedure

Table 2 Comparison of simulations

	FE^2	MIEL	Mixed
	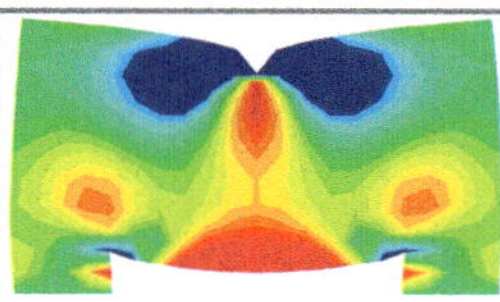	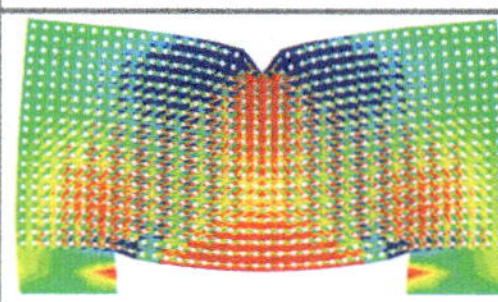	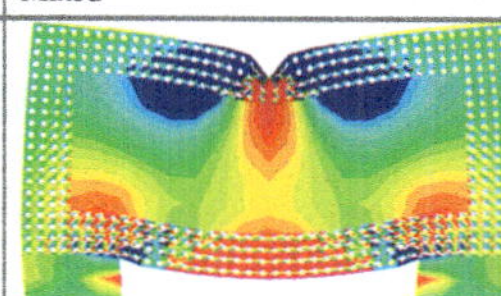
No. micro problems	800	200	488
No. macro elements	216	216	216
Total DOF	1.030.881	132.881	563.921
Total time (s)	192.4	50.2	127.3

for implicit solution procedure of the parametrized system of nonlinear equations. In the paper the path is parametrized by load level (λ). For every load step at macro level ($\Delta\lambda_M$), several substeps can be done at micro level ($\Delta\lambda_m$). Since we have two scales, we have in general a path following procedure at both levels, resulting in two-level path following procedure presented in Fig. 3. Traditionally, each step at macro level is followed by only one step at micro level. Sensitivity analysis based multi-scale analysis allows extension to more general case, where each macro step can be followed by an arbitrary number of micro substeps (Fig. 3).

5 Numerical Example

As a numerical example, bending of a beam with enforced vertical displacement was investigated for various multi-scale methods and material models. In Table 2 different combinations of multi-scale schemes are presented. The first presented in Fig. 4, is mixed, where both FE^2 and MIEL are used. In the next two examples only

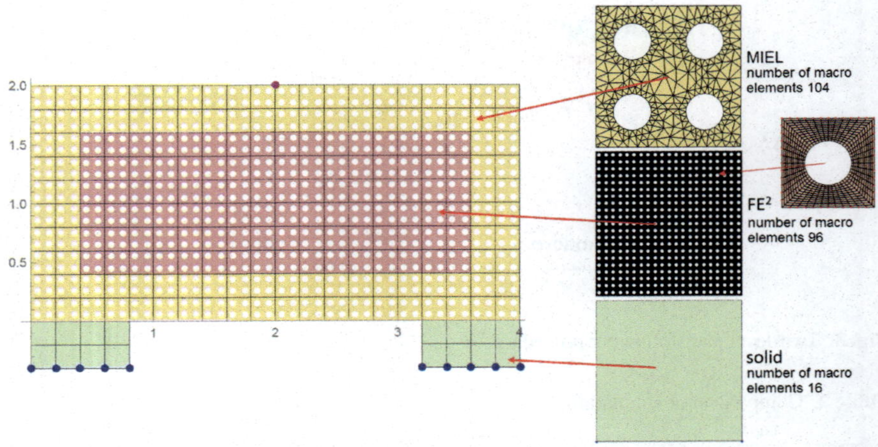

Fig. 4 Mixed multi-scale model

one method, either FE^2 or MIEL, is used. Supports are in all cases modelled with 16 macro solid elements. For these three combinations in Table 2 the numbers of micro and macro elements and total DOF are compared. In case of MIEL, the number of micro problems is equal to that of macro elements, whereas for FE^2 the number of micro problems for one macro element is equal to that of integration points, in this case 4. Total DOF represents the number of equations that need to be solved and is the biggest for sole use of FE^2 method. The example was run for two material models. First, a Neo-Hookean type hyperelastic material was used. The computational times for different schemes are compared in Table 2 and the distribution of strain Exx for mixed case is shown in Fig. 5.

Next, a finite strain JC plasticity material was used. Since the elasto-plastic problems are path dependent, the influence of the implementation of the two-level path following procedure on the convergence rate of the Newton-Raphson iterative procedure was additionally investigated.

In the two-level path following algorithm it is important that the sensitivity analysis is carried out correctly. For path dependent problems only correct sensitivity analysis at micro level leads to macro tangent matrix, which is algorithmically consistent and leads to quadratically convergent scheme. Sensitivity analysis is done correctly, if it is integrated along the whole path of micro sub-increment. For every substep at micro level the sensitivity has to be updated.

In Table 3 convergence rate of the NR method is compared for the FE^2 scheme: HY 1/1—hyperelastic material, each macro load step is followed by one micro load step; HY 1/5 uncons. —hyperelastic material, each micro increment is divided into 5 substeps, sensitivity is not updated; PL 1/1—elasto-plastic material, each macro load step is followed by one micro load step; PL 1/5—elasto-plastic material, each micro increment is divided

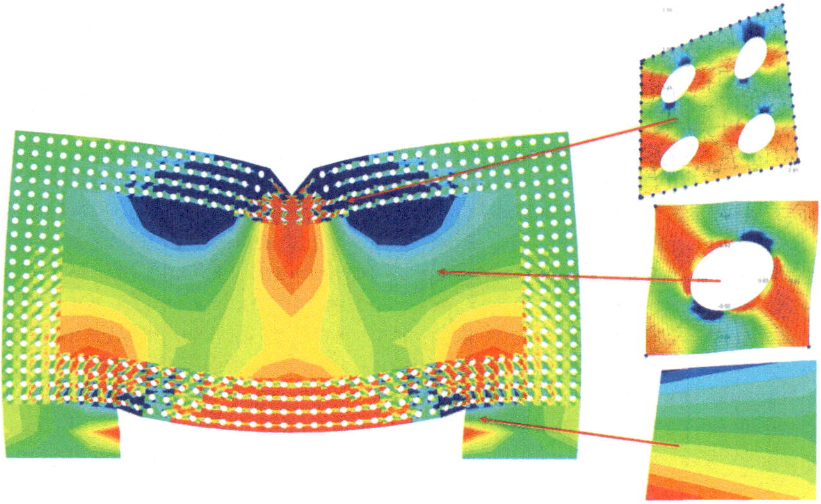

Fig. 5 Results for strains Exx for mixed scheme

Table 3 Comparison of convergences for FE^2 scheme

it.	HY 1/1	HY 1/5 uncons.	PL 1/1	PL 1/5	PL 1/5 uncons.
1	9.79×10^{-4}	9.79×10^{-4}	8.92×10^{-4}	8.92×10^{-4}	8.92×10^{-4}
2	2.18×10^{-7}	2.18×10^{-7}	2.30×10^{-4}	2.30×10^{-4}	2.30×10^{-4}
3	8.45×10^{-14}	8.45×10^{-14}	2.01×10^{-5}	1.76×10^{-5}	1.81×10^{-5}
4	/	/	2.21×10^{-8}	2.10×10^{-8}	2.57×10^{-6}
5	/	/	1.45×10^{-13}	4.68×10^{-14}	6.11×10^{-7}
6	/	/	/	/	1.53×10^{-7}
–	–	–	–	–	–
15	/	/	/	/	7.37×10^{-13}

into 5 substeps, sensitivity is updated; PL 1/5 uncons.—elasto-plastic material, each micro increment is divided into 5 substeps, sensitivity is not updated.

For hyperelastic material convergence rate is quadratic also for inconsistent sensitivity analysis, regardless of the number of micro substeps, because the problem is not path dependent. For elasto-plastic material the quadratic convergence is lost, unless the sensitivity is properly updated (PL 1/5 uncons.). The same is valid also for MIEL and mixed schemes. Results for MEIL and mixed models are presented in Tables 4 and 5.

6 Conclusions

Unified approach to the automation of various multi-scale approaches through the automatic-differentiation-based (ADB) formulation enables mixed use of different multi-scale and one-scale methods in one model for an arbitrary nonlinear, time dependent coupled problem (e.g. general finite strain plasticity). Codes of the finite element for analytical first and second order sensitivity analysis are generated automatically. Multi-scale FE^2 and MIEL schemes are implemented based on boundary condition sensitivity analysis. The differences between the FE^2 and MIEL are in essential boundary condition at micro mesh and in essential boundary condition velocity fields needed for the sensitivity analysis. For FE^2 first order sensitivity is enough, whereas for MIEL second order sensitivity is needed.

In a conventional way of computing macroscopic tangent matrix for MIEL a Schur complement is needed. We used the boundary condition sensitivity analysis to obtain macroscopic constitutive tangent matrix, which is numerically more efficient

Table 4 Comparison of convergences for MIEL scheme

it.	HY 1/1	HY 1/5 uncons.	PL 1/1	PL 1/5	PL 1/5 uncons.
1	9.80×10^{-4}	9.80×10^{-4}	1.76×10^{-3}	1.76×10^{-3}	1.76×10^{-3}
2	2.21×10^{-7}	2.21×10^{-7}	1.48×10^{-4}	1.52×10^{-4}	2.03×10^{-4}
3	8.23×10^{-14}	8.20×10^{-14}	1.12×10^{-4}	1.12×10^{-4}	1.67×10^{-4}
4	/	/	5.78×10^{-7}	8.86×10^{-7}	6.03×10^{-5}
5	/	/	5.47×10^{-11}	9.52×10^{-11}	1.83×10^{-5}
6	/	/	/	/	1.53×10^{-7}
–	–	–	–	–	–
16	/	/	/	/	4.33×10^{-11}

Table 5 Comparison of convergences for mixed FE^2 and MIEL model

it.	HY 1/1	HY 1/5 uncons.	PL 1/1	PL 1/5	PL 1/5 uncons.
1	1.01×10^{-3}	1.01×10^{-3}	1.54×10^{-3}	1.54×10^{-3}	1.54×10^{-3}
2	2.38×10^{-7}	2.38×10^{-7}	24.10×10^{-4}	4.12×10^{-4}	5.00×10^{-4}
3	1.47×10^{-13}	1.48×10^{-13}	1.95×10^{-4}	1.92×10^{-4}	1.38×10^{-4}
4	/	/	4.26×10^{-6}	3.98×10^{-6}	1.38×10^{-5}
5	/	/	1.33×10^{-9}	81.51×10^{-9}	5.51×10^{-7}
6	/	/	1.25×10^{-16}	9.74×10^{-17}	2.99×10^{-8}
7	/	/	/	/	1.51×10^{-9}
–	–	–	–	–	–
10	/	/	/	/	2.02×10^{-13}

for densely meshed micro-structures. This is due to the fact that the number of sensitivity parameters remains the same regardless of the density of the micro mesh, while the size of the Schur complement grows with the number of the nodes on the boundary of the micro problem.

Traditionally in multi-scale methods one macro time step is followed by one micro time step. Sensitivity analysis based multi-scale analysis allows that each macro step can be followed by an arbitrary number of micro substeps. For path-dependent problems only correctly done sensitivity analysis at micro level leads to algorithmically consistent macro tangent matrix and to quadratic convergence. For every substep at micro level, sensitivity has to be updated, otherwise for path dependent problems the quadratic convergence is lost.

Acknowledgements The financial support for this work was obtained from the Slovenian Research Agency within the PhD Grant Agreement (annex No: 630-34/2015-7).

References

1. Fèyel, F.: Multiscale FE^2 elastoviscoplastic analysis of composite structures. Comput. Mat. Sci. **16**, 344–354 (1999)
2. Geers, M.G.D., Kouznetsova, V.G., Brekelmans, W.A.M.: Multi-scale computational homogenization: trends and challenges. J. Comput. Appl. Math. **234**, 2175–2182 (2010)
3. Hudobivnik, B., Korelc, J.: Closed-form representation of matrix functions in the formulation of nonlinear material models. Finite Elem. Anal. Des. **111**, 19–32 (2016)
4. Korelc, J.: Automatic generation of finite-element code by simultaneous optimization of expressions. Theor. Comput. Sci. **187**, 231–248 (1997)
5. Korelc, J.: Automation of primal and sensitivity analysis of transient coupled problems. Comput. Mech. **44**, 631–649 (2009)
6. Korelc, J., Wriggers, P.: Automation of finite element methods. Springer, Switzerland (2016)
7. Korelc, J.: AceFEM and AceGen user manuals (2017). http://symech.fgg.uni-lj.si/
8. Kouznetsova, V., Brekelmans, W.A.M., Baaijens, F.P.T.: An approach to micro-macro modelling of heterogeneous materials. Comput. Mech. **27**, 37–48 (2001)
9. Lamut, M., Korelc, J., Rodič, T.: Multiscale modelling of heterogeneous materials. Mat. Tech. **45**, 421–426 (2011)
10. Loehnert S., Mueller-Hoeppe D.S.: Multiscale methods for fracturing solids. In: Reddy B.D. (ed) IUTAM Symposium on Theoretical, Computational and Modelling Aspects of Inelastic Media. IUTAM BookSeries, vol. 11. Springer, Dordrecht (2008)
11. Markovič, D., Ibrahimbegović, A.: On micro-macro interface conditions for micro scale based FEM for inelastic behaviour of heterogeneous materials. Comput. Methods Appl. Mech. Eng. **193**, 5503–5523 (2004)
12. Mathematica 11, Wolfram Research Inc., (2016). http://www.wolfram.com
13. Miehe, C., Schotte, J., Schröder, J.: Computational micro-macro transitions and overall moduli in the analysis of polycrystals at large strains. Comput. Mat. Sci. **16**, 372–382 (1999)
14. Šolinc, U., Korelc, J.: A simple way to improved formulation of FE^2 analysis. Comput. Mech. **56**, 905–915 (2015)
15. Temizer, I., Wriggers, P.: On the computation of the macroscopic tangent for multiscale volumetric homogenization problems. Comput. Methods Appl. Mech. Eng. **198**, 495–510 (2008)
16. Zohdi, T.I., Wriggers, P.: An introduction to computational micromechanics. Springer, Berlin, Heidelberg (2008)

Efficient Multiscale FE-FFT-Based Modeling and Simulation of Macroscopic Deformation Processes with Non-linear Heterogeneous Microstructures

Julian Kochmann, Lisa Ehle, Stephan Wulfinghoff, Joachim Mayer, Bob Svendsen and Stefanie Reese

Abstract The purpose of this work is the prediction of micromechanical fields and the overall material behavior of heterogeneous materials using an efficient and robust two-scale FE-FFT-based computational approach. The macroscopic boundary value problem is solved using the finite element (FE) method. The constitutively dependent quantities such as the stress tensor are determined by the solution of the local boundary value problem. The latter is represented by a periodic unit cell attached to each macroscopic integration point. The local algorithmic formulation is based on fast Fourier transforms (FFT), fixed-point and Newton-Krylov subspace methods (e.g. conjugate gradients). The handshake between both scales is defined

J. Kochmann (✉) · S. Wulfinghoff · S. Reese
Institute of Applied Mechanics, RWTH Aachen University,
52074 Aachen, Germany
e-mail: julian.kochmann@rwth-aachen.de

S. Wulfinghoff
e-mail: stephan.wulfinghoff@rwth-aachen.de

S. Reese
e-mail: stefanie.reese@rwth-aachen.de

L. Ehle · J. Mayer
Central Facility for Electron Microscopy, RWTH Aachen University,
52074 Aachen, Germany
e-mail: ehle@gfe.rwth-aachen.de

J. Mayer
e-mail: mayer@gfe.rwth-aachen.de

B. Svendsen
Material Mechanics, RWTH Aachen University,
52062 Aachen, Germany
e-mail: b.svendsen@mpie.de

B. Svendsen
Microstructure Physics and Alloy Design, Max-Planck-Institut
für Eisenforschung GmbH, 40237 Düsseldorf, Germany

J. Sorić et al. (eds.), *Multiscale Modeling of Heterogeneous Structures*,
Lecture Notes in Applied and Computational Mechanics 86,
https://doi.org/10.1007/978-3-319-65463-8_7

through the Hill-Mandel condition. In order to ensure accurate results for the local fields as well as feasible overall computation times, an efficient solution strategy for two-scale full-field simulations is employed. As an example, the local and effective mechanical behavior of ferrit-perlit annealed elasto-viscoplastic 42CrMo4 steel is studied for three-point-bending tests. For simplicity, attention is restricted to the geometrically linear case and quasi-static processes.

1 Introduction

The macroscopic mechanical behavior of most materials of technological importance (e.g. polycrystalline materials, fiber-reinforced composites, high-strength ceramics) is to a large extent dictated by their microstructure which varies in the distribution, orientation, size and morphology of individual grains and different phase constituents. The derivation of appropriate phenomenological constitutive laws for such complex materials is often difficult or even not possible.

Instead, assuming scale separation, the phenomenological model might be replaced by an additional boundary value problem (BVP) which is attached as a representative volume element (RVE) (e.g. [10, 11]) to each macroscopic integration point (e.g. Gauss point). The handshake between both scales is then given by the mean RVE response in the sense of Hill-Mandel [11]. Most common in the context of computational scale bridging methods is the multilevel finite element or FE^2 method (e.g. [6, 18, 33, 35]) which is based on computational homogenization (e.g. [19, 32, 41]) and finite element-based algorithmic formulations at both scales. Compared to finite element (FE) methods, the computational efficiency of the local solution procedure can be significantly increased (e.g. [4, 27]) using fast Fourier transforms (FFT), fixed-point and Green's function methods (e.g. [23, 24]). First, employing FFT, or more generally speaking spectral methods, periodic boundary conditions are implicitly given by global higher-order periodic shape functions. Compared to other types of boundary conditions (e.g. displacement, traction, mixed), periodic boundary conditions have been proven to be optimal for RVE-based computational homogenization methods (see e.g. [25]). Second, such meshfree methods operate on regular voxel grids which makes the discretization of complex and fine scale microstructures 'easily' possible (see e.g. [4, 12]). Third, the fixed-point-based iterative solution scheme avoids the assembly procedure and the computation and inversion of the tangent stiffness matrix which are classical features of implicit FE methods and often represent time-consuming numerical operations. Fourth, efficient and powerful FFT software libraries (e.g. FFTW) and simulation kits (e.g. DAMASK, OpenPhase) are available that make the implementation and application of new and complex material models 'straightforward'. However, there are of course drawbacks to be mentioned, as well. First, spectral methods are accompanied by non-physical spurious oscillations that negatively affect the spatial resolution and convergence behavior (e.g. [39]). Second, the basic fixed-point scheme by Moulinec and Suquet [23, 24] is only conditionally convergent depending on the choice of a homogeneous reference material and suffers from 'many' fixed-point iterations for high contrasts in mechanical fields (e.g. stiffness). Different computational schemes have been developed to tackle these problems (e.g. [1, 13, 22, 31, 38]). Among other methods (e.g. [13, 31]), using finite difference (FD) approximations of the divergence and gradient operators (e.g. [39]) leads to a tremendous improvement of the spatial resolution by reducing the strength of higher-order frequencies. Replacing the fixed-point-based scheme by Newton-Krylov subspace methods (e.g. conjugate gradients [7, 12, 38, 42], GMRES [34]) leads to quadratic convergence independently of the choice of the homogeneous reference material [42].

Despite such recent improvements, the two-scale FE-FFT method has only been addressed in a few works. First, Spahn et al. [36] proposed such a two-scale computational approach for the modeling of progressive damage in fiber-reinforced composite materials. Recently, Kochmann et al. [14] presented an FE-FFT- and phase-field-based method to model martensitic phase transformations in elastic polycrystalline materials. In both cases, the macroscopic BVP was rather simple and fine local discretizations lead to unsatisfactory computation times. The purpose of this work is the extension to viscoplastic polycrystals, realistic EBSD-based virtual grain structures and the establishment of process-microstructure-property relations (e.g. yield strength, hardness) for virtual three-point-bending tests. An efficient solution strategy for two-scale full-field simulations [15] is adapted which ensures both, feasible overall CPU-times and accurate micromechanical fields.

2 Material Model Formulation

In this section the two-scale material model is presented in a nutshell, more details can be found e.g. in [14]. The continuum and scale briding relations are summarized in Sect. 2.1 and local field relations in Sect. 2.2. In what follows, macroscopic quantities are denoted by the index 'M' and local fields by no index. Quantities which are referred to the current time step are denoted by the superscript $t+1$. Further notations will be introduced when needed.

2.1 Macroscopic Relations and Scale-Bridging

Let us consider the macroscopic structure Ω_{M} and boundary $\partial\Omega_{\mathrm{M}}$ on which displacement $\partial\Omega_{\mathrm{M}}^{\mathbf{u}}$ and/or traction boundary conditions (BC) $\partial\Omega_{\mathrm{M}}^{\mathbf{t}}$ are applied such that $\partial\Omega_{\mathrm{M}} = \partial\Omega_{\mathrm{M}}^{\mathbf{u}} \cup \partial\Omega_{\mathrm{M}}^{\mathbf{t}}$ and $\partial\Omega_{\mathrm{M}}^{\mathbf{u}} \cap \partial\Omega_{\mathrm{M}}^{\mathbf{t}} = \emptyset$. The macroscopic linearized strain tensor

$$\varepsilon_{\mathrm{M}}(\mathbf{x}_{\mathrm{M}}) = \nabla^{\mathrm{sym}}\mathbf{u}_{\mathrm{M}}(\mathbf{x}_{\mathrm{M}}) = \frac{1}{2}\left(\nabla\mathbf{u}_{\mathrm{M}}(\mathbf{x}_{\mathrm{M}}) + (\nabla\mathbf{u}_{\mathrm{M}}(\mathbf{x}_{\mathrm{M}}))^{\top}\right) \tag{1}$$

is introduced at each material point $\mathbf{x}_M \in \Omega_M$ in terms of the displacement vector $\mathbf{u}_M(\mathbf{x}_M)$ to define the macroscopic stress tensor $\boldsymbol{\sigma}_M(\mathbf{x}_M)$ which is assumed to fullfill the quasi-static linear momentum balance

$$\operatorname{div} \boldsymbol{\sigma}_M(\mathbf{x}_M) = \mathbf{0}. \tag{2}$$

The solution of (2) is directly correlated to the local problem which is embedded as a periodic unit cell (UC) of the domain Ω at each $\mathbf{x}_M \in \Omega_M$. Let $\mathbf{g}(\mathbf{x}; \mathbf{x}_M)$ represent an arbitrary local field (e.g. strain, stress) corresponding to the macroscopic counterpart $\mathbf{g}_M(\mathbf{x}_M)$. The fluctuation field

$$\tilde{\mathbf{g}}(\mathbf{x}) := \mathbf{g}(\mathbf{x}; \mathbf{x}_M) - \mathbf{g}_M(\mathbf{x}_M) \tag{3}$$

is defined as the deviation from $\mathbf{g}_M(\mathbf{x}_M)$ at each $\mathbf{x}_M \in \Omega_M$. Per definition, $\mathbf{g}_M(\mathbf{x}_M) = \frac{1}{\Omega} \int_\Omega \mathbf{g}(\mathbf{x}; \mathbf{x}_M)\, d\Omega(\mathbf{x})$ holds which implies satisfaction of

$$\frac{1}{\Omega} \int_\Omega \tilde{\mathbf{g}}(\mathbf{x}; \mathbf{x}_M)\, d\Omega(\mathbf{x}) = 0 \tag{4}$$

and represents the energy consistency Hill-Mandel [11] condition. In particular, this means that the handshake between both scales is solely accomplished through

$$\boldsymbol{\varepsilon}_M(\mathbf{x}_M) := \frac{1}{\Omega} \int_\Omega \boldsymbol{\varepsilon}(\mathbf{x}; \mathbf{x}_M)\, d\Omega(\mathbf{x}) \tag{5}$$

$$\boldsymbol{\sigma}_M(\mathbf{x}_M) := \frac{1}{\Omega} \int_\Omega \boldsymbol{\sigma}(\mathbf{x}; \mathbf{x}_M)\, d\Omega(\mathbf{x}). \tag{6}$$

2.2 Local Problem

As alluded above, the material behavior at the macro scale is dictated by the solution of the local problem

$$\operatorname{div} \boldsymbol{\sigma}(\mathbf{x}, \boldsymbol{\varepsilon}_e) = \mathbf{0} \tag{7}$$

which is formulated in terms of the elastic strain field $\boldsymbol{\varepsilon}_e(\mathbf{x}; \mathbf{x}_M) = \boldsymbol{\varepsilon}(\mathbf{x}; \mathbf{x}_M) - \boldsymbol{\varepsilon}_p(\mathbf{x})$. Irreversible deformations due to dislocation slip are characterized by the plastic strain tensor $\boldsymbol{\varepsilon}_p(\mathbf{x})$. Note that internal variables such as $\boldsymbol{\varepsilon}_p(\mathbf{x})$ or the accumulated plastic slip $\gamma_{acc}(\mathbf{x})$ are purely local, whereas strains $\boldsymbol{\varepsilon}(\mathbf{x}; \mathbf{x}_M)$ and stresses $\boldsymbol{\sigma}(\mathbf{x}; \mathbf{x}_M)$ vary on both length scales.

Since $\boldsymbol{\varepsilon}_M(\mathbf{x}_M)$ is prescribed by the kinematic relations (1) at the continuum level, the fluctuation field $\tilde{\boldsymbol{\varepsilon}}(\mathbf{x})$ (see Fig. 1) represents the primary unknown of the non-linear equilibrium equations (7). For better readability, the dependence on $\mathbf{x}$ and $\mathbf{x}_M$ is omitted henceforth.

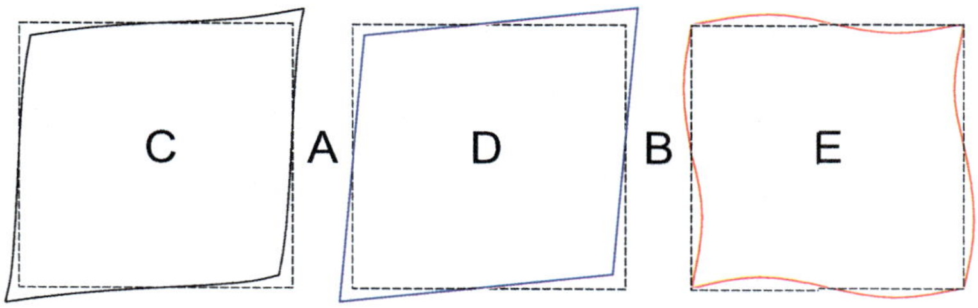

Fig. 1 Additive decomposition of local strain field (black) into volume average part (blue) and fluctuation field (red)

3 Crystal Plasticity Constitutive Law

Restricting ourselves to elasto-viscoplastic polycrystalline aggregates (e.g. [3, 20, 28, 30]), the stress-strain relation is defined as:

$$\boldsymbol{\sigma} = \mathbb{C}\boldsymbol{\varepsilon}_e = \mathbb{C}(\boldsymbol{\varepsilon} - \boldsymbol{\varepsilon}_p). \tag{8}$$

The evolution of the plastic strain

$$\dot{\boldsymbol{\varepsilon}}_p = \sum_{\alpha}^{n_{\text{slip}}} \dot{\gamma}_\alpha \, \mathbf{m}_\alpha^{\text{sym}} \tag{9}$$

is assumed to be governed by the following flow rule:

$$\dot{\gamma}_\alpha = \text{sgn}(\tau_\alpha)\, \dot{\gamma}_0 \, f_\alpha(\boldsymbol{\sigma}, \tau_\alpha^c). \tag{10}$$

The plastic shear rate $\dot{\gamma}_\alpha$, the critical resolved shear stress (CRSS) and the symmetric Schmid tensor $\mathbf{m}_\alpha^{\text{sym}} = \frac{1}{2}\left(\mathbf{d}_\alpha \otimes \mathbf{n}_\alpha + \mathbf{n}_\alpha \otimes \mathbf{d}_\alpha\right)$ are defined on the slip systems $\alpha = 1, \ldots, n_{\text{slip}}$ which are characterized by the slip directions $\mathbf{d}_\alpha$ and plane normals $\mathbf{n}_\alpha$. The specific form of (10) is adapted from [40]

$$f_\alpha(\boldsymbol{\sigma}, \tau_\alpha^c) = \begin{cases} \left\langle \dfrac{|\tau_\alpha| - \tau_\alpha^c}{\tau^D} \right\rangle^p & |\tau_\alpha| \leq \tau^R \\ \beta\left(|\tau_\alpha| - \tau^R\right) + \left\langle \dfrac{\tau^R - \tau_\alpha^c}{\tau^D} \right\rangle^p & |\tau_\alpha| > \tau^R, \end{cases} \tag{11}$$

where $\langle \bullet \rangle = (|\bullet| + \bullet)/2$, τ^D and τ_0^c define the McAuley brackets drag stress and initial CRSS, respectively. The specific form of (11) makes use of a linear regularized problem with slope β which serves as an improved initial guess for the non-regularized problem yielding robust numerical computations. The influence of the loading speed on the local material response is described by the rate sensitivity parameter p. Using a backward Euler time discretization, the accumulated plastic slip $\gamma_{\rm acc} = \sum_\alpha \int |\dot{\gamma}_\alpha| dt$, the material constant $\tau^R = \tau^c_\alpha(\gamma_{\rm acc}) + \Delta\tau$ and yield stress $\tau^c_\alpha(\gamma_{\rm acc}) = \tau^c_0 + g(\gamma_{\rm acc})$ are defined. The hardening function is assumed to be of Voce-type behavior

$$g(\gamma_{\rm acc}) = \tau^c_0 + (\tau_\infty - \tau^c_0)\left(1 - \exp\left[-\frac{H\gamma_{\rm acc}}{\tau_\infty - \tau^c_0}\right]\right) + \theta_\infty \gamma_{\rm acc} \tag{12}$$

where τ_∞, H and θ_∞ are the saturation stress (for $\theta_\infty = 0$), hardening modulus and parameter, respectively. The constitutive relation (8) is given by the solution of the following set of non-linear equilibrium equations in terms of the unknows $(\boldsymbol{\sigma}, \tau^c_\alpha)$

$$\mathbf{r}_\xi = \begin{pmatrix} \mathbf{r}_\sigma \\ r_{\tau^c} \end{pmatrix} = \begin{pmatrix} \mathbb{C}^{-1}\boldsymbol{\sigma} - \boldsymbol{\varepsilon} + \boldsymbol{\varepsilon}^t_p + \Delta t\, \dot{\boldsymbol{\varepsilon}}_p(\boldsymbol{\sigma}, \tau^c) \\ (g(\gamma_{\rm acc}) - \tau^c_\alpha)/\tau_0 \end{pmatrix} = \mathbf{0}. \tag{13}$$

The parameter τ_0 is introduced to obtain a dimensionless residual $\mathbf{r}_\xi$.

4 Algorithmic Formulation

The algorithmic formulation at the macro scale is based on the FE method which is rather standard and thus not presented in this work. Details can be found e.g. in [14]. The local solution algorithm is based on fast Fourier transforms, fixed-point and Green's function methods (e.g. [23, 24]) and derived in Sect. 4.1.

4.1 Fast Fourier Transforms

The transform of an arbitrary local field $\mathbf{g}(\mathbf{x})$ in real space $\mathbf{x} \in \Omega$ to Fourier space $\mathbf{k} \in \hat{\Omega}$ is denoted by

$$\hat{\mathbf{g}}(\mathbf{k}) = \mathscr{F}\{\mathbf{g}(\mathbf{x})\} = \sum_{\mathbf{x}\in\Omega} \mathbf{g}(\mathbf{x}) \exp(-\imath \mathbf{k} \cdot \mathbf{x}) \tag{14}$$

and its inverse transform by

$$\mathbf{g}(\mathbf{x}) = \mathscr{F}^{-1}\{\hat{\mathbf{g}}(\mathbf{k})\} = \frac{1}{|\Omega|} \sum_{\mathbf{k}\in\hat{\Omega}} \hat{\mathbf{g}}(\mathbf{k}) \exp(\imath \mathbf{k} \cdot \mathbf{x}) \tag{15}$$

in terms of the complex number $\imath = \sqrt{-1}$ and wave vector $k_i = (2\pi\imath/N)/\Delta x$ $(i = -N/2, \ldots, N/2 - 1$ for N even and $i = -(N-1)/2, \ldots, (N-1)/2-$

1 for odd N). The number of grid points is denoted by N and grid spacing by Δx, respectively. The Fourier representation of the gradient and divergence operators is defined as:

$$\mathscr{F}\{\nabla\,\mathbf{g}(\mathbf{x})\} = \hat{\mathbf{g}}(\mathbf{k}) \otimes \imath\,\mathbf{k}, \qquad \mathscr{F}\{\operatorname{div}\mathbf{g}(\mathbf{x})\} = \hat{\mathbf{g}}(\mathbf{k}) \cdot \imath\,\mathbf{k}. \tag{16}$$

Let us introduce the so-called 'polarization stress' (see e.g. [8, 9, 37])

$$\boldsymbol{\tau} = \boldsymbol{\sigma}(\boldsymbol{\varepsilon}^e) - \mathbb{C}^{(0)}\nabla^{\mathrm{sym}}\mathbf{u} \tag{17}$$

relative to an homogeneous reference material with elastic stiffness $\mathbb{C}^{(0)}$ to rewrite (7) in the following form:

$$\operatorname{div}\mathbb{C}^{(0)}\nabla\mathbf{u} + \operatorname{div}\boldsymbol{\tau} = \mathbf{0}. \tag{18}$$

Introducing the Green function tensor

$$\hat{G}^{(0)}_{jm} = (\mathbb{C}^{(0)}_{jkmn}k_k k_n)^{-1} \qquad \text{for} \quad \mathbf{k} \neq \mathbf{0}, \tag{19}$$

Fourier transform of (18) leads to

$$\hat{\mathbf{u}} = \hat{\mathbf{G}}^{(0)}\hat{\boldsymbol{\tau}}\,\imath\,\mathbf{k} \qquad \text{for} \quad \mathbf{k} \neq \mathbf{0}. \tag{20}$$

Assuming an isotropic reference material with Lamé parameters $\mu^{(0)}$ and $\lambda^{(0)}$, the symmetric fourth order Lippmann-Schwinger operator

$$\begin{aligned} \hat{\Gamma}^{(0)}_{ijkl} = {} & \frac{1}{4\mu^{(0)}|\mathbf{k}|^2}(\delta_{ki}k_l k_j + \delta_{li}k_k k_j + \delta_{kj}k_l k_i + \delta_{lj}k_k k_i) \\ & - \frac{\lambda^{(0)} + \mu^{(0)}}{\mu^{(0)}(\lambda^{(0)} + 2\mu^{(0)})}\frac{k_i k_j k_k k_l}{|\mathbf{k}|^4} \end{aligned} \tag{21}$$

is introduced to transform (20) into

$$\hat{\varepsilon} = -\hat{\boldsymbol{\Gamma}}^{(0)}\hat{\boldsymbol{\tau}} \qquad \text{for} \quad \mathbf{k} \neq \mathbf{0} \tag{22}$$

which represents the fundamental equation of the basic-fixed point scheme by Moulinec and Suquet [23, 24]. Since the right-hand side of (22) depends on ε via (17), an iterative solution scheme is required, in general (see e.g. [12, 24]). Note that $\mathbf{k} = \mathbf{0}$ represents the mean value in Fourier space and is exploited to apply ε_{M} to the UC which means in turn that $\mathbf{k} = \mathbf{0}$ defines the interface between the local and continuum scale.

Algorithm 1 Basic algorithm for fixed-point-based iterative solution of the microscopic mechanical BVP

1. **initialization** ($i = 0$): $\boldsymbol{\varepsilon}^{(i)} = \boldsymbol{\varepsilon}_{\mathrm{M}}$;
2. **repeat** until convergence
   a. $\boldsymbol{\tau}^{(i)} = \boldsymbol{\sigma}(\boldsymbol{\varepsilon}_e^{(i)}) - \mathbb{C}^{(0)}\boldsymbol{\varepsilon}^{(i)}$ from (17);
   b. $\hat{\boldsymbol{\tau}}^{(i)} = \mathscr{F}\left\{\boldsymbol{\tau}^{(i)}\right\}$ from (14);
   c. $\hat{\boldsymbol{\varepsilon}}^{(i+1)} = \begin{cases} -\hat{\boldsymbol{\Gamma}}^{(0)}\hat{\boldsymbol{\tau}}^{(i)} & \text{for } \mathbf{k} \neq \mathbf{0} \\ \boldsymbol{\varepsilon}_{\mathrm{M}} & \text{for } \mathbf{k} = \mathbf{0} \end{cases}$ from (22);
   d. $\boldsymbol{\varepsilon}^{(i+1)} = \mathscr{F}^{-1}\left\{\hat{\boldsymbol{\varepsilon}}^{(i+1)}\right\}$ from (15);
   e. *constitutive relation*:
   solve (13) at each $\mathbf{x} \in \Omega$ with accuracy tol_{ξ};
   f. *convergence check*:
   if $||\boldsymbol{\varepsilon}^{(i+1)} - \boldsymbol{\varepsilon}^{(i)}||/||\boldsymbol{\varepsilon}_{\mathrm{M}}|| \leqslant \mathrm{tol}_{\varepsilon}$,
   g. else i++, repeat starting at a.;

The fixed-point algorithm for the basic scheme is given by Algorithm 1. As indicated above, the basic fixed-point scheme is only conditionally convergent and suffers from 'many' fixed-point iterations for high stiffness contrasts. Using Newton-Krylov solvers (e.g. conjugate gradients [7, 12, 42], GMRES [34]) instead, has the possibility of quadratic convergence independent of the choice of $\mathbb{C}^{(0)}$ [42]. To this end, $\boldsymbol{\sigma}^{(i+1)}$ is linearized around $\boldsymbol{\varepsilon}^{(i)}$ yielding

$$\boldsymbol{\sigma}^{(i+1)} = \boldsymbol{\sigma}^{(i)}(\boldsymbol{\varepsilon}^{(i)}) + \mathbb{C}_{\mathrm{alg}}^{(i)}\Delta\boldsymbol{\varepsilon}^{(i+1)}, \tag{23}$$

where $\mathbb{C}_{\mathrm{alg}}^{(i)} = \partial_{\boldsymbol{\varepsilon}^{(i)}}\boldsymbol{\sigma}^{(i)}$ represents the algorithmic tangent modulus. Inserting (23) and $\boldsymbol{\varepsilon}^{(i+1)} = \boldsymbol{\varepsilon}^{(i)} + \Delta\boldsymbol{\varepsilon}^{(i+1)}$ into (22) leads to the following system of equations

$$\Delta\hat{\boldsymbol{\varepsilon}}^{(i+1)} + \hat{\boldsymbol{\Gamma}}^{(0)}\left(\widehat{[\mathbb{C}_{\mathrm{alg}}^{(i)} - \mathbb{C}^{(0)}]\Delta\boldsymbol{\varepsilon}^{(i+1)}}\right) = \hat{\boldsymbol{\Gamma}}^{(0)}\hat{\boldsymbol{\sigma}}^{(i)} \tag{24}$$

which has to be solved numerically for $\Delta\boldsymbol{\varepsilon}^{(i+1)}$.[1] In this work, a conjugate gradient (Polak-Ribière) solver with accuracy tol_{ω} is used for the solution of (24). The small strain algorithm for the solution of (24) was formulated based on the finite strain documentation by [12] and implemented analogously.

5 Solution Procedure

Within the two-scale model (see Fig. 2) three different 'computational levels' can be identified:

[1]Note that the identity $\boldsymbol{\varepsilon}^{(i)} = \boldsymbol{\varepsilon}_{\mathrm{M}} + \hat{\boldsymbol{\Gamma}}^{(0)} * [\mathbb{C}^{(0)}\boldsymbol{\varepsilon}^{(i)}]$ was used to arrive at (24), where $\mathbf{f} * \mathbf{g} = \int_{-\infty}^{\infty} \mathbf{f}(\boldsymbol{\xi})\,\mathbf{g}(\mathbf{x} - \boldsymbol{\xi})\,d\boldsymbol{\xi}$ denotes the convolution integral of two arbitrary fields $\mathbf{f}$ and $\mathbf{g}$.

- The **continuum (FE)** level,
- the **local (FFT)** level,
- and the **local material point** level.

At the FE-level (1), an implicit solution scheme and reduced FE formulation with hourglass stabilization (see e.g. [29]) is employed yielding one integration point $\mathbf{x}_\mathrm{M}$ per element e and locking free element behavior. The overall consistent algorithmic tangent modulus $\mathbb{C}_\mathrm{M} := \partial_{\varepsilon_\mathrm{M}} \boldsymbol{\sigma}_\mathrm{M}$ is required for the linearization procedure and computed by numerical differentiation for simplicity. This means that at the FFT-level (2) the local problem has to be solved (i) once for the evaluation of the stress-strain relation (8) and (ii) three times for the computation of $\mathbb{C}_\mathrm{M}$ by perturbing each coef-

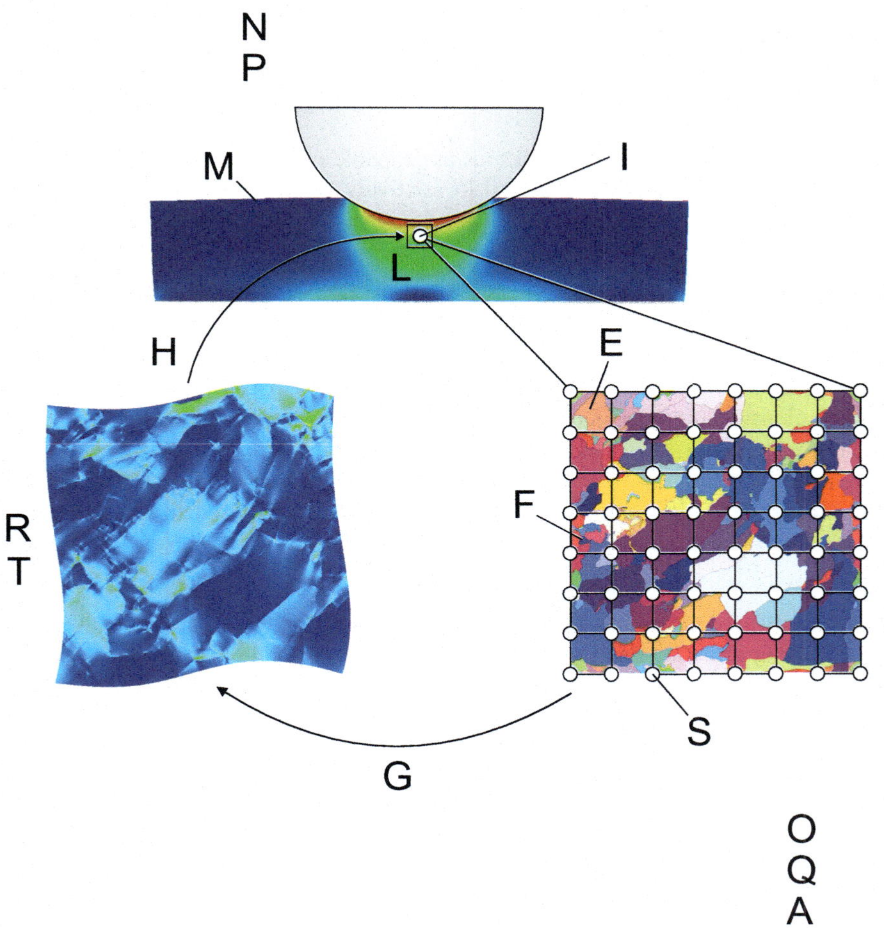

Fig. 2 Schematic of FE-FFT-based micro-to-macro transition with polycrystalline microstructure

ficient of ε_M (2D, plane strain). In order to improve the local convergence rate and reduce the overall computation time, the local fields $\boldsymbol{\sigma}$ and ε which are available from (i), are taken as local starting solution for (ii). By doing this, the FFT solver only requires 1–2 Newton iterations to achieve convergence. In order to reduce the number of Newton iterations at the FFT-level, n_{pre} presteps are performed before each Newton iteration which improves the initial guess for the local field ε yielding robust computations and quadratic convergence. At the lowest level, the local material point level (3), the stabilization technique which was introduced in Sect. 3, ensures convergence for relatively large time increments Δt and rate sensitivity parameter p. The original power law is replaced by a regularized linear approximation which serves as an improved starting solution for the numerical solution of (13). The slope of the regularized problem is adapted depending on the convergence behavior. If the finally converged solution does not coincide with the original power law, the slope is increased and an additional local iteration is performed. If no converged solution is achieved, the slope is decreased. This procedure is repeated until convergence is obtained which ensures that the finally converged solution coincides with the original power law solution.

Since coarse unit cell discretizations are capable of resolving the basic features of complex microstructures and lead to relatively small errors in terms of effective properties (e.g. stress) (see e.g. [14, 15]) the following solution procedure is employed:

- **Pre-processing**: mesh convergence analysis in terms of effective properties (e.g. macroscopic stress);
- **Processing**: run two-scale full-field simulation with coarse unit cell discretization and save macroscopic strain tensor $\varepsilon_M^{(i+1)}(\mathbf{x}_M)$ for each converged load step i at each Gauss point of particular interest $\mathbf{x}_M$; use time step increment Δt;
- **Post-processing**: solve local problem (stand-alone computation) for each i by reading and applying $\varepsilon_M^{(i+1)}(\mathbf{x}_M)$ to unit cell which is characterized by a fine local mesh discretization with $\Delta t^* = \Delta t$; if $\Delta t^* < \Delta t$ is required to get converged solution, decrease $\Delta t^* = \Delta t/2$ and linearly interpolate between $\varepsilon_M^{(i+1)}(\mathbf{x}_M)$ and $\varepsilon_M^{(i)}(\mathbf{x}_M)$.

In a nutshell the solution procedure is based on non-converged unit cell discretizations for two-scale simulations. A mesh convergence analysis beforehand ensures that the coarse discretization still leads to an error in the effective quantities which is in an acceptable range. The processing and post-processing lead to feasible overall CPU-times and accurate micromechanical fields. A detailed and systematic analysis as well as relevant associated examples can be found in [15].

6 Numerical Examples

The present two-scale model is implemented into the finite element program FEAP (http://www.ce.berkeley.edu/projects/feap) as a user element routine. Local and macroscopic fields are visualized using Paraview (http://www.paraview.org). At the continuum level, the residual force-based convergence tolerance is set to $\text{tol}_{\mathbf{f}} = 10^{-6}$ N. At the local level, the tolerance for the Newton algorithm is chosen as $\text{tol}_{\varepsilon} = 10^{-8}$ and for the conjugate gradient solver as $\text{tol}_{\omega} = 10^{-14}$. The Newton-type solution of (13) is considered to be converged based on $\text{tol}_{\xi} = 10^{-10}$. If not stated otherwise, square grids ($N_1 = N_2 = N$) with equidistant grid spacing $\Delta x_1 = \Delta x_2 = \Delta x$ are considered. The material is assumed to consist of body-centered cubic (BCC) unit cell structures, where two families of slip systems which are described by almost identical activation energies, are taken into account: the $\mathbf{d}_\alpha = \{110\}$, $\mathbf{s}_\alpha = [111]$ and the $\mathbf{d}_\alpha = \{210\}$, $\mathbf{s}_\alpha = [111]$ systems. The material parameters used in this work are fitted to experimental data of ferrit-perlit annealed 42CrMo4 steel and are summarized in Table 1.

In order to reduce the strength of spurious oscillations, a central finite difference approximation of the gradient and divergence operator is employed [39]. The effective Lamé constants $\mu^{(0)}$ and $\lambda^{(0)}$ are determined based on the arithmetic averages

$$\mu^{(0)} = \frac{1}{2}\left(\min\{\mu(\mathbf{x})\} + \max\{\mu(\mathbf{x})\}\right), \quad \lambda^{(0)} = \frac{1}{2}\left(\min\{\lambda(\mathbf{x})\} + \max\{\lambda(\mathbf{x})\}\right). \quad (25)$$

EBSD maps were digitized and edited (e.g. reflection at edges and insertion of additional virtual grains) to generate periodic boundary conditions and square computational domains. The physical side length of the unit cell was set to $L_1 = L_2 = L = 250\,\mu\text{m}$. For simplicity the Eulerian angles $\{\theta_1, \Theta, \theta_2\}$ which are shown for

Table 1 Used elasto-viscoplastic material parameters; also, $\tau^R = \tau^D = \tau_0 = 1$ MPa are set

$\mathbb{C}_{11}$ (GPa)	$\mathbb{C}_{12}$ (GPa)	$\mathbb{C}_{44}$ (GPa)	τ_0^c (MPa)	H (MPa)	$\dot{\gamma}_0$ (1/s)	p (–)	τ_∞ (MPa)	θ_∞ (MPa)
237	141	58	150	1200	0.001	20	200	120

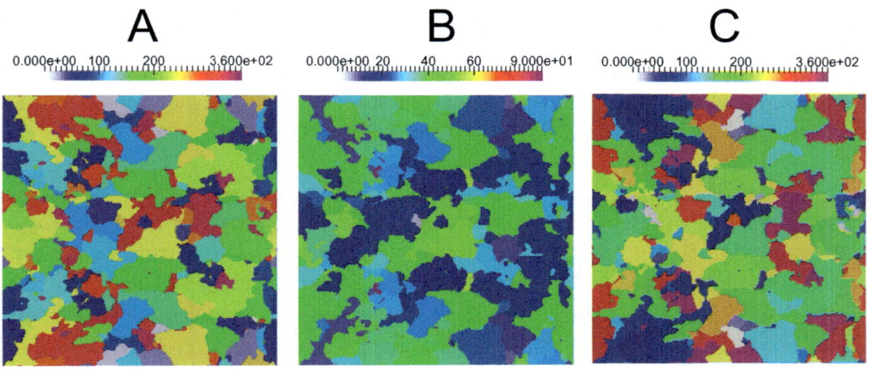

Fig. 3 Visualization of Eulerian angles θ_1, Θ and θ_2 obtained from digitized and edited EBSD maps of ferrit-perlit annealed 42CrMo4 steel

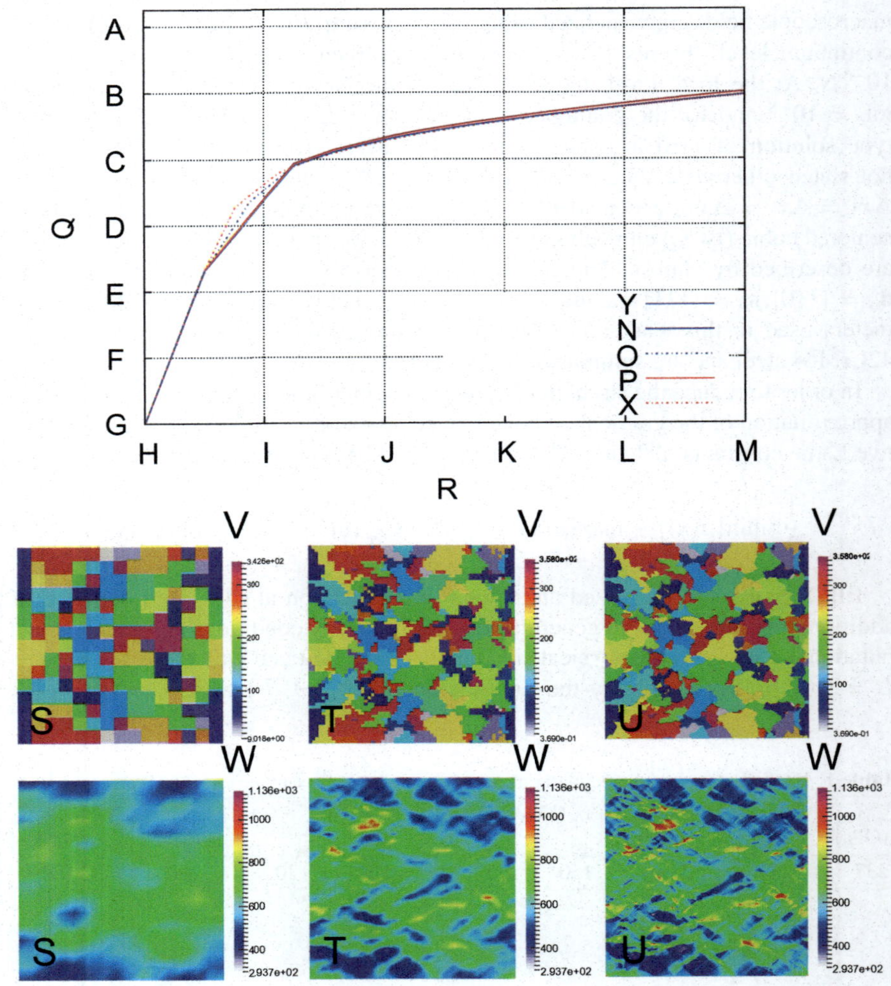

Fig. 4 Top: effective stress-strain response for uniaxial tension virtual experiments with different UC discretizations. Bottom: visualization of grain structure and local equivalent plastic strain distribution for $N = 15, 63$ and $N = 255$ grid points at final load step $\varepsilon_{M\,11} = 1\%$

a $N = 255$ unit cell (UC) discretization in Fig. 3 are prescribed at each macroscopic integration point $\mathbf{x}_M$. First, uniaxial tension virtual experiments are performed with one element ($e = 1$) at the macro scale and different UC discretizations ($N = 15, 63, 255$) at the Gauss point level are considered. Displacement controlled boundary conditions are applied at constant displacement rate $\dot{u}_{M\,11} = 1$ mm/s until $\varepsilon_{M\,11} = 1\%$ is reached. The effective stress-strain response of the two-scale full-field simulations which are performed with $N = 15$ and $N = 255$ UC discretizations, are visualized in Fig. 4 by red curves. The stand-alone computations with $N = 15, 63$

and $N = 255$ at the Gauss-point level based on the strain history of the two-scale simulation with $N = 15$ are described by blue curves in Fig. 4. The Eulerian angle θ_1 and local equivalent plastic strain distribution $\varepsilon^p_{\text{eq}}$ are visualized for all discretizations in the bottom part of Fig. 4. The coarsest discretization $N = 15$ is not even capable of resolving the correct distribution and number of grains. However, this coarse representation leads to a qualitatively quite good behavior in terms of the effective stress $\sigma_{\text{M}\,11}$. Considering a discretization with $N = 511$ grid points as converged solution, the error is approximately 7.9% for $N = 15$. In a post-processing step, local fields are generated based on $N = 255$ yielding an effective error of 0.87%. This solution strategy leads to micromechanical fields which are as accurate as the ones computed in the two-scale simulation with $N = 255$. At the same time an overall speed-up of about 97 is obtained. Different UC discretizations and loading cases (e.g. tension, shear, mixed loading) have been investigated in [15]. For all cases, the magnitude of the error made by this solution procedure is the approximately the same. Thus, the following two-scale simulations are performed with $N = 15$ grid points and the post-processing generation of local fields is based on $N = 255$.

6.1 *Three Point Bending*

Next, a virtual three-point-bending test is considered. The beam with physical dimensions $L_1 L_2 = 25 \times 3\,\text{mm}^2$ is discretized using 300 bilinear quadrilateral elements with one integration point per element in each of which the EBSD-based grain structure (see Fig. 3) is embedded. The contact between the rollers and the beam is described based on an augmented Lagrangean formulation. The hardmetal rollers are modeled as linear elastic and isotropic with elastic constants $E = 640\,\text{GPa}$ and $\nu = 0.23$. The left roller is fixed in all spatial directions, symmetry conditions in x_1-direction are applied on the right edge of the beam and the right roller is supposed to move $u_{\text{M}\,11} = 3\,\text{mm}$ upwards at fixed displacement rate $\dot{u}_{\text{M}\,11} = 1\,\text{mm/s}$. Local fields and effective properties are visualized and investigated in the element which is highlighted in black in Fig. 5 and denoted as e^*, in what follows. Using automatic time stepping, 200 loadsteps are required to move the right roller $u_{\text{M}\,11} = 3\,\text{mm}$ upwards which corresponds to an equivalent macroscopic strain of $\varepsilon_{\text{M}\,\text{eq}}{=}7\%$. Note that the number of loadsteps correlates to the size of Δt which in turn is limited mostly by the contact formulation.

On both scales quadratic convergence is observed and the overall CPU-time for the coarse two-scale full-field simulation + fine post-processing (see Sect. 5) is approximately 12 h. The effective stress distribution $\sigma_{\text{M}\,\text{eq}}$ is visualized in Fig. 5 at loadstep 10, 100 and 200 as well as the local equivalent plastic strain $\varepsilon^p_{\text{eq}}$ and stress distribution σ_{eq}. The most critical points in the structure can be found in regions where the curvature is highest. Looking at the micromechanical fields, shear bands are observed in $\varepsilon^p_{\text{eq}}$ and stress concentrations at grain boundaries in σ_{eq}.

Assuming that the hardness is proportional to the yield strength of the material, e.g.

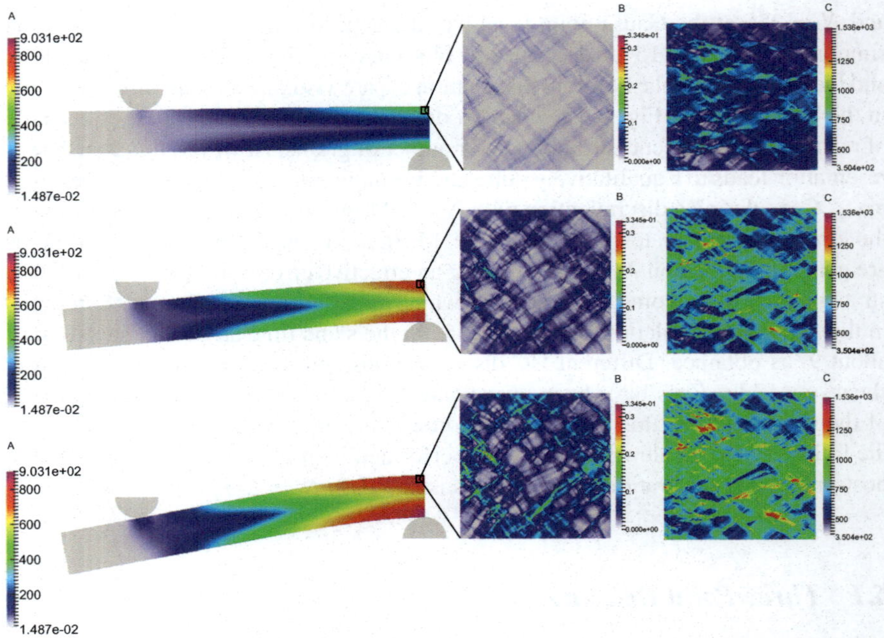

Fig. 5 Visualization of effective equivalent stress, local equivalent stress, plastic strain fields at first loadstep 10, 100 and 200

$$HV \sim 3\sigma_y \sim 3 \cdot 2\tau^c \tag{26}$$

let us analyze the influence of the curvature of the beam on the hardness evolution in e^*. To this end, two different beams are considered which are defined by the two different thicknesses $L_2^1 = 1.5$ mm and $L_2^2 = 3$ mm (see Fig. 6). In order to get a geometry- and process-independent analysis, the curvature κ which is approximated based on displacement differences at nodal positions, is multiplied with the thickness L_2 and plotted versus the hardness based on (26). For both thicknesses, similar hardness-curvature curves are obtained which are visualized in Fig. 6. The difference between both curves could have been expected since the simple macroscopic FE discretization does not lead to a fully converged solution. Since this work is about the qualitative analysis of local and effective properties, this fact is acceptable at this point. The results represent a first step towards the solution of an inverse integrity problem, namely which process parameters have to be set to get a desired surface layer property. In this case, Fig. 6 can be considered to identify the required curvature to achieve a desired surface layer property which is the hardness distribution in this case. Furthermore, an extension of the model to three dimensions allows the direct comparison of experimentally measured structural (e.g. force-displacement curves) and local information (e.g. EBSD, REM, TEM data) which leads to a better

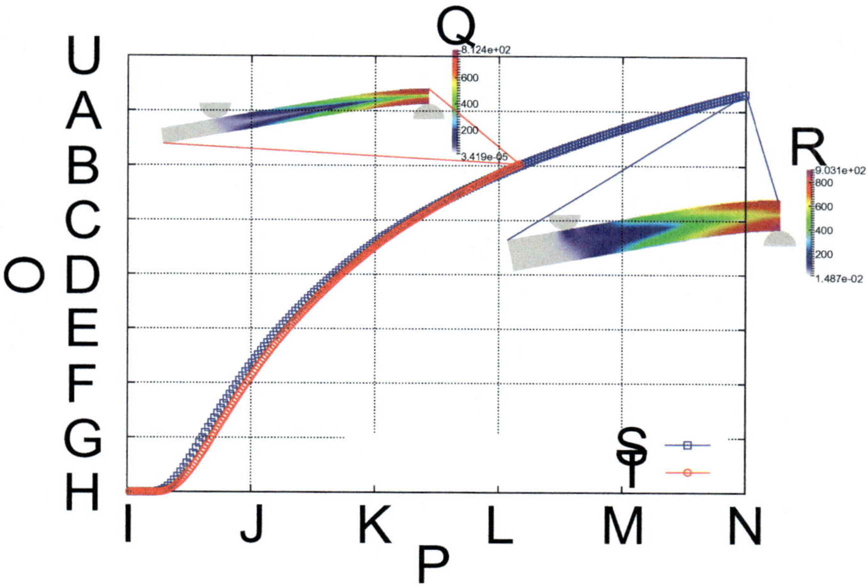

Fig. 6 Hardness versus κL_2 for $L_2^1 = 3$ mm (blue) and $L_2^2 = 1.5$ mm (red) beam

understanding of the underlying physics and mechanisms. This will be discussed in future developments of this work as well as the application of more complex macroscopic processes (e.g. deep rolling process).

7 Discussion

In this work an efficient two-scale model was presented which couples FE modeling of the structural material behavior to FFT and Newton-Krylov-based modeling of the local mechanical behavior of elasto-viscoplastic polycrystalline aggregates. An efficient but simple solution strategy is adapted from [15] which ensures both, accurate micromechanical fields as well as feasible overall CPU-times. This solution procedure is characterized by a pre-processing, processing and post-processing step and leads to a tremendous increase of computational efficiency. As an example, EBSD maps of 42CrMo4 steel are digitized, edited and embedded in each macroscopic integration point of a three-point-bending specimen. The numerical results allow the investigation of structural properties (e.g. force-displacement curves) and microstructural information (e.g. local stress and strain distribution). Assuming that the hardness of the material is proportional to the yield strength of the phase constituents at the micro scale, the hardness distribution at the surface is studied depending on the curvature of the bending beam. To this end, two different beams with different thicknesses are considered. Plotting the hardness versus the curvature at the work-piece surface helps to find the required curvature to achieve a desired surface layer property, in this case the hardness of the material. The model allows the prediction of the yield strength and hardness of the material as well as the state of residual stresses, the prediction of which is still barely possible in manufacture processing as well as simulation sciences. Future developments will focus on the extension to finite strain crystal plasticity, texture evolution and three-dimensional problems. Furthermore, the numerical examples in this work have been performed on a single CPU. Therefore, it is reasonable to employ parallelization techniques on both scales to decrease the overall computation time and make the simulation of more complex macroscopic boundary value problems (e.g. deep rolling) possible in future work.

Acknowledgements Financial support of Subprojects M03 and C02 of the Transregional Collaborative Research Center SFB/TRR 136 by the German Science Foundation (DFG) is gratefully acknowledged.

References

1. Brisard, S., Dormieux, L.: FFT-based methods for the mechanics of composites: a general variational framework. Comput. Mater. Sci **49**(3), 663–671 (2010)
2. Brisard, S., Dormieux, L.: Combining Galerkin approximation techniques with the principle of Hashin and Shtrikman to derive a new FFT-based numerical method for the homogenization of composites. Comput. Methods Appl. Mech. Eng. **217**(220), 197–212 (2012)
3. Diard, O., Leclerq, S., Rousselier, G., Cailletaud, G.: Evaluation of finite element based analysis of 3D multicrystalline aggregates plasticity—application to crystal plasticity model identification and the study of stress and strain fields near grain boundaries. Int. J. Plast. **21**, 691–722 (2005)
4. Eisenlohr, P., Diehl, M., Lebensohn, R.A., Roters, F.: A spectral method solution to crystal elasto viscoplasticity at finite strains. Int. J. Plast. **46**, 37–53 (2013)
5. Eyre, D.J., Milton, G.W.: A fast numerical scheme for computing the response of composites using grid refinement. Eur. Phys. J. Appl. Phys. **6**, 41–47 (1999)
6. Geers, M.G.D., Kouznetsova, V., Brekelmans, W.A.M.: Multi-scale computational homogenization: trends and challenges. J. Comput. Appl. Math. **234**, 2175–2182 (2010)
7. Gélébart, L., Mondon-Cancel, R.: Non-linear extension of FFT-based methods accelerated by conjugated gradients to evaluate the mechanical behavior of composite materials. Comput. Mater. Sci **77**, 430–439 (2013)
8. Hashin, Z., Shtrikman, H.: On some variational principles in anisotropic and nonhomogeneous elasticity. J. Mech. Phys. Solids **10**, 335–342 (1962)
9. Hashin, Z., Shtrikman, S.: A variational approach to the theory of the elastic behavior of multiphase materials. J. Mech. Phys. Solids **11**, 127–140 (1963)
10. Hashin, Z.: Analysis of composite materials—a survey. J. Appl. Mech. **36**, 481–505 (1983)
11. Hill, R.: Elastic properties of reinforced solids: some theoretical principles. J. Mech. Phys. Solids **11**, 357–372 (1963)
12. Kabel, M., Böhlke, T., Schneider, M.: Efficient fixed point and Newton-Krylov solvers for FFT-based homogenization of elasticity at large deformations. Comput. Mech. **54**, 1497–1514 (2014)
13. Kabel, M., Merkert, D., Schneider, M.: Use of composite voxels in FFT-based homogenization. Comput. Methods Appl. Mech. Eng. **294**, 168–188 (2015)

14. Kochmann, J., Rezaei Mianroodi, J., Wulfinghoff, S., Reese, S., Svendsen, B.: Two-scale, FE-FFT- and phase-field based computational modeling of bulk microstructure evolution and macroscopic material behavior. Comput. Methods Appl. Mech. Eng. **305**, 89–110 (2016)
15. Kochmann, J., Wulfinghoff, S., Ehle, L., Mayer, J., Svendsen, B., Reese, S.: Efficient and accurate two-scale FE-FFT-based prediction of the effective material behaviour of elasto-viscoplastic polycrystals. Comput. Mech., in press (2017). https://doi.org/10.1007/s00466-017-1476-2
16. Michel, J.C., Moulinec, H., Suquet, P.: A computational scheme for linear and non-linear composites with arbitrary phase contrast. Int. J. Numer. Methods Eng. **52**, 139–160 (2001)
17. Michel, J.C., Moulinec, H., Suquet, P.: A computational method based on augmented Lagrangians and Fast Fourier Transforms for composites with high contrast. Comput. Model. Eng. Sci. **1**, 79–88 (2000)
18. Miehe, C., Schotte, J., Schröder, J.: Computational micro-macro transitions and overall moduli in the analysis of polycrystals at large strains. Comput. Mater. Sci **16**, 372–382 (1999)
19. Miehe, C., Schröder, J., Schotte, J.: Computational homogenization analysis in finite plasticity. Simulation of texture development in polycrystalline materials. Comput. Methods Appl. Mech. Eng. **171**, 387–418 (1999)
20. Mika, D.P., Dawson, P.R.: Effects of grain interaction on deformation in polycrystals. Mater. Sci. Eng. A **257**, 62–76 (1998)
21. Monchiet, V., Bonnet, G.: A polarization-based FFT iterative scheme for computing the effective properties of elastic composites with arbitrary contrast. Int. J. Numer. Methods Eng. **89**, 1419–1436 (2012)
22. Moulinec, H., Silva, F.: Comparison of three accelerated FFT-based schemes for computing the mechanical response of composite materials. Int. J. Numer. Methods Eng. **97**, 960–985 (2014)
23. Moulinec, H., Suquet, P.: A fast numerical method for computing the linear and nonlinear mechanical properties of composites. C.R. Acad. Sci. Ser. IIb: Mech. Phys. Chim. Astron. **318**, 1417–1423 (1994)
24. Moulinec, H., Suquet, P.: A numerical method for computing the overall response of nonlinear composites with complex microstructures. Comput. Methods Appl. Mech. Eng. **157**(1), 69–94 (1998)
25. Nemat-Nasser, S., Hori, M.: Micromechanics: Overall Properties of Heterogeneous Materials. Amsterdam (1993)
26. Castaneda, P.P., Suquet, P.: Advances in Applied Mechanics, vol. 34, pp. 171–302. Elsevier (1997)
27. Prakash, A., Lebensohn, R.A.: Simulations of micromechanical behavior of polycrystals: finite element versus Fast Fourier Transforms. Model. Simul. Mater. Sci. Eng. **17**, 16pp (2009)
28. Raabe, D., Sachtleber, M., Zhao, Z., Roters, F.: Micromechanical and macromechanical effects in grain scale polycrystal plasticity experimentation and simulation. Acta Mater. **49**, 3433–3441 (2001)
29. Reese, S., Wriggers, P.: A stabilization technique to avoid hourglassing in finite elasticity. Int. J. Numer. Methods Eng. **48**, 79–109 (2000)
30. Roters, F., Eisenlohr, P., Hantcherli, L., Tjahjanto, D.D., Bieler, T.R., Raabe, D.: Overview on constitutive laws, kinematics, homogenization and multiscale methods in crystal plasticity finite element modeling: theory, experiments, applications. Acta Mater. **58**, 1152–1211 (2010)
31. Schneider, M., Ospald, F., Kabel, M.: FFT-based homogenization for microstructures discretized by linear hexahedral elements. Int. J. Numer. Methods Eng. (2016). https://doi.org/10.1002/nme.5336
32. Schröder, J.: Homogenisierungsmethoden der nichtlinearen Kontinuumsmechanik unter Beachtung von Instabilitäten. Universität Stuttgart, Habilitationsschrift (2000)
33. Schröder, J.: A numerical two-scale homogenization scheme: the FE^2-method. CISM Int. Centre Mech. Sci. **550**, 1–64 (2014)
34. Shanthraj, P., Eisenlohr, P., Diehl, M., Roters, F.: Numerically robust spectral methods for crystal plasticity simulations of heterogeneous materials. Int. J. Plast. **66**, 31–45 (2015)

35. Smit, R.J.M., Brekelmans, W.A.M., Meijer, H.E.H.: Prediction of the mechanical behavior of nonlinear heterogeneous systems by multi-level finite element modeling. Comput. Methods Appl. Mech. Eng. **155**, 181–192 (1998)
36. Spahn, J., Andrae, H., Kabel, M., Müller, R.: A multiscale approach for modeling progressive damage of composite materials using fast Fourier transforms. Comput. Methods Appl. Mech. Eng. **268**, 871–883 (2004)
37. Talbot, D.R.S., Willis, J.R.: Variational principles for inhomogeneous nonlinear media. Int. J. Appl. Math. **35**, 39–54 (1985)
38. Vinogradov, V., Milton, G.W.: An accelerated FFT algorithm for thermoelastic and non linear composites. Int. J. Numer. Methods Eng. **76**, 1678–1695 (2008)
39. Willot, F.: Fourier-based schemes with modified Green operator for computing the electrical response of heterogeneous media with accurate local fields. C.R. Acad. Sci. Ser. IIb: Mech. Phys. Chim. Astron. **323**(3), 232–245 (2014)
40. Wulfinghoff, S., Böhlke, T.: Equivalent plastic strain gradient crystal plasticity—enhanced power law subroutine. GAMM-Mitteilunge **36**(2), 134–148 (2013)
41. Wulfinghoff, S., Reese, S.: Efficient computational homogenization of simple elasto-plastic microstructures using a shear band approach. Comput. Methods Appl. Mech. Eng. **298**, 350–372 (2016)
42. Zeman, J., Vodrejc, J., Novak, J., Marek, I.: Accelerating a FFT-based solver for numerical homogenization of a periodic media by conjugate gradients. J. Comput. Phys. **229**(21), 8065–8071 (2010)
43. Zeman, J., de Geuss, T.W.J., Vondejc, J., Peerlings, R.H.J., Geers, M.G.D.: A finite element perspective on non-linear FFT-based micromechanical simulations. Int. J. Numer. Methods Eng. (2016). https://doi.org/10.1002/nme.5481

Experimental-Numerical Validation Framework for Micromechanical Simulations

Ante Buljac, Modesar Shakoor, Jan Neggers, Marc Bernacki, Pierre-Olivier Bouchard, Lukas Helfen, Thilo F. Morgeneyer and François Hild

Abstract A combined experimental-numerical framework is presented in order to validate computations at the microscale. It is illustrated for a flat specimen with two holes, which is made of cast iron and imaged via in situ synchrotron laminography at micrometer resolution during a tensile test. The region in the reconstructed volume between the two holes is analyzed via Digital Volume Correlation (DVC) to measure displacement fields. Finite Element (FE) simulations, whose mesh is made consistent with the studied material microstructure, are driven by measured Dirichlet boundary conditions. Damage levels and gray level residuals for DVC measurements *and* FE simulations are assessed for validation purposes.

1 Introduction

The prediction of forming processes and in-service life of metals and alloys raises important issues for ductile fracture, which have led researchers to investigate advanced damage models. A first type of damage models, which is known as macro-

A. Buljac · J. Neggers · F. Hild (✉)
LMT, ENS Paris-Saclay/CNRS/Université Paris-Saclay,
61 avenue du Président Wilson, 94235 Cachan cedex, France
e-mail: hild@lmt.ens-cachan.fr

A. Buljac · T.F. Morgeneyer
MINES ParisTech, PSL Research University, Centre des Matériaux,
CNRS UMR 7633, BP 87, 91003 Evry, France

M. Shakoor · M. Bernacki · P.-O. Bouchard
MINES ParisTech, PSL - Research University, CEMEF - Centre de mise en forme des matériaux, CNRS UMR 7635, CS 10207 rue Claude Daunesse, 06904 Sophia Antipolis Cedex, France

L. Helfen
ANKA/Institute for Photon Science and Synchrotron Radiation Karlsruhe Institute of Technology (KIT), 76131 Karlsruhe, Germany

L. Helfen
European Synchrotron Radiation Facility (ESRF), 38043 Grenoble, France

J. Sorić et al. (eds.), *Multiscale Modeling of Heterogeneous Structures*,
Lecture Notes in Applied and Computational Mechanics 86,
https://doi.org/10.1007/978-3-319-65463-8_8

scopic postulates [1–3], is used to predict not only damage inception but also the softening and transition to fracture. Due to their macroscopic nature, they are known to have limited predictive capabilities and are usually calibrated and applied for specific loading conditions. For applications such as material forming, where loading may be complex and non proportional, these limitations become problematic [4, 5].

Microscopic models [6, 7] are an alternative where the macroscopic response is derived from averaged microscale calculations. This scale transition may be purely analytical [6] or performed via computations on ideal microstructures [7]. The predictive capacities of such models are also limited for arbitrary loading conditions [4, 5] because of restrictive assumptions used in their derivations [6, 7]. Further, the calibration of these models is challenging since they usually require advanced identification techniques [8–10]. It is worth noting that some damage variables such as porosity can now be observed experimentally thanks to X-ray imaging techniques [11–14]. Inclusions and voids can be studied individually based on manual [13, 15] or automatic [14] procedures.

Simulations allow experimentally observed quantities such as porosity and number of fractured/debonded inclusions to be related to internal variables such as plastic strain and stress-based criteria. These microscale computations are usually driven with idealistic microstructures, constitutive behavior, and simplified kinematic or static boundary conditions that do not capture local strain and stress states that inclusions and voids are subjected to [11, 14, 16, 17]. The principal aim of the present work is to develop reliable simulations at the microscale using validated models to describe the three steps of ductile damage (i.e., nucleation, growth and coalescence). The first step then consists of developing an experimental-numerical framework, which enables numerical models to be probed with respect to experimental data.

The material of interest is nodular graphite cast iron made of a ferritic matrix, graphite nodules, and no significant initial porosity. Upon loading, ductile fracture is caused by nodule/matrix debonding, void growth and coalescence [18–20]. Literature data [18, 19, 21, 22] show that the nodules can be modeled as voids since their stress-carrying capacity is very small in tension. Such hypothesis will be made herein. One of the present challenges is to test this type of assumption with local error estimators (i.e., at the microscale). It will also allow microscopic models to be developed in order to better capture the final stages of failure via calibrated criteria associated with different mechanisms [23, 24].

The framework followed herein, which was first applied to another test case [25], quantitatively compares experimental bulk data with 3D computations. It consists of the following steps (Fig. 1):

- X-ray laminography, which is a non-destructive 3D imaging technique for laterally extended 3D objects [26–30], to acquire radiographs and subsequently reconstruct 3D volumes of different steps of a mechanical test. By post-processing such bulk data, the morphology of the two-phase microstructure can be revealed and its changes can be analyzed.
- Digital volume correlation (DVC) to measure 3D displacement fields [31–34]. Small interrogation volumes are *independently* registered in the considered Region

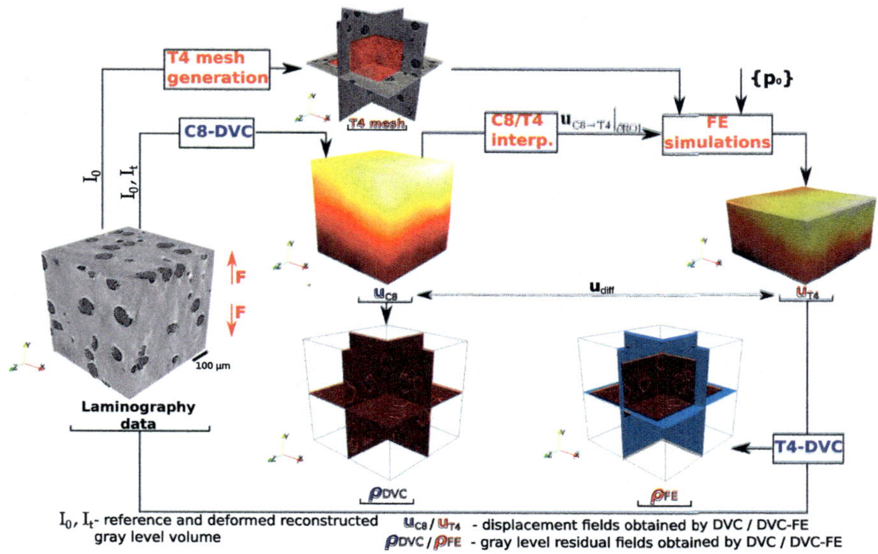

Fig. 1 Schematic representation of the methods used in the present chapter for validating numerical simulations at the microscale (after Ref. [25])

of Interest (ROI). The only information that is kept is the mean displacement assigned to each analyzed Zone of Interest (ZOI) center. In the following, FE-based approaches [35] will be considered. Registrations are performed over the *whole* ROI using FE discretizations. Such DVC approaches can be directly linked with numerical simulations of mechanical tests [36–38]. In particular, DVC measurements serve as Dirichlet boundary conditions to the Finite Element (FE) computations at the microscale.

- FE simulations to explicitly model the actual morphology of cast iron thanks to laminography data (see e.g., Refs. [39, 40]). The Level-Set (LS) procedure [41, 42], which is used herein, enables interfaces to be described in FE simulations under large deformations and complex topological events [43–45]. It is worth noting that regularity [46] and conservation [47] issues have to be handled with care.
- FE computations are run with an elastoplastic law to describe the nonlinear behavior of the ferritic matrix. The nodules are modeled as elastic media with very low Young's modulus.
- Comparisons between experiments (i.e., DVC measurements) and 3D FE computations driven by measured displacements (i.e., DVC-FE) are performed for displacement fields and, more importantly, gray level residuals, which were shown to be very powerful error estimators [25].
- The change of the mean volume fraction of pores is also compared by analyzing the reconstructed volumes and the predictions with DVC-FE.

The chapter is structured as follows. The experimental setup and laminography are first discussed. Digital Volume Correlation is summarized next. Uncertainty quantifications are performed. FE computations including the microstructure meshing procedure are then described. Last, the results from both methods are compared relatively via kinematic field subtractions and absolutely by computing gray level residuals. The predictions of the damage state are also confronted with experimental evidence.

2 Experimental and Numerical Framework

2.1 *Experiments*

The studied material is commercial nodular graphite cast iron (serial code: EN-GJS-400). Figure 2a shows the sample geometry, which is inspired by Ref. [48]. The holes have been machined via Electrical Discharge Machining (EDM). The load is manually applied to the sample by controlling the global relative displacement via screw rotation.

After applying each loading step, a set of radiographs is acquired while the sample is rotated about the laminographic axis (i.e., parallel to the specimen thickness direction). This axis is inclined with respect to the X-ray beam direction

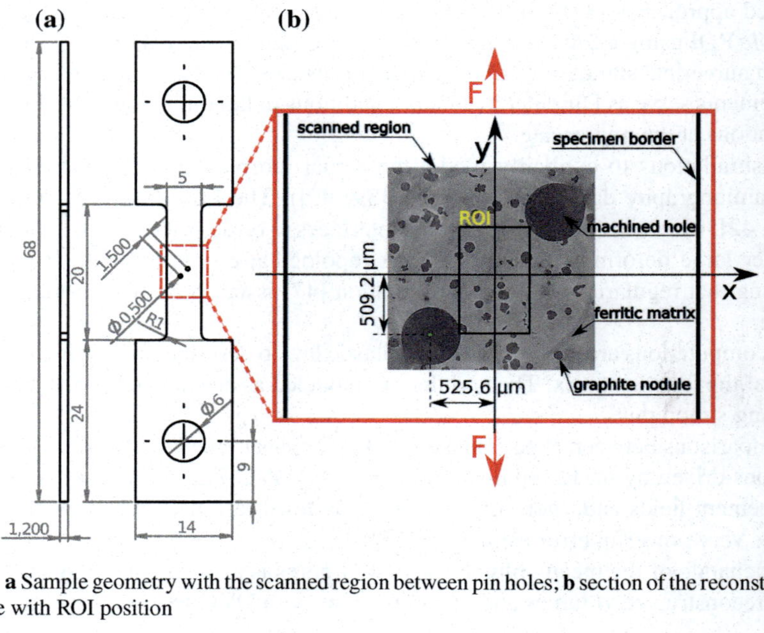

Fig. 2 **a** Sample geometry with the scanned region between pin holes; **b** section of the reconstructed volume with ROI position

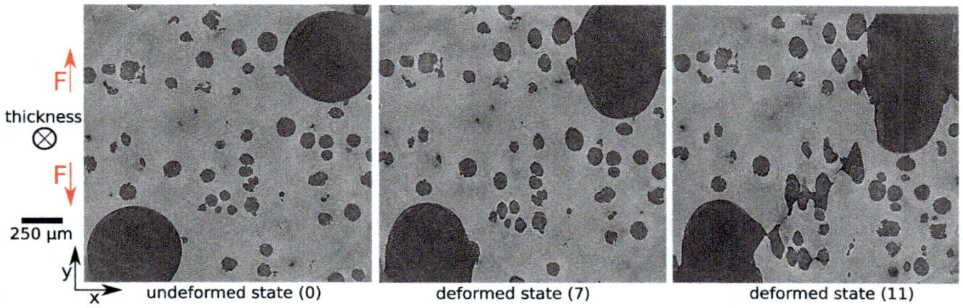

Fig. 3 Mid-thickness section of the reconstructed volume for three different loading steps

by an angle $\theta \approx 60°$. The series of radiographs is then used to reconstruct 3D volumes via filtered-back projection [49]. A GPU-accelerated implementation of this algorithm [50] has been utilized herein. The reconstructed volume size is $1600 \times 1600 \times 1600$ voxels (each voxel has a physical length of 1.1 μm). After scanning the undeformed state (0) three times, 12 additional scans are performed upon stepwise loading. The last scan corresponds to the final crack.

The scanned zone encompasses the two holes. The selected ROI for DVC *and* FE calculations mainly focuses on the ligament between the two holes (Fig. 2b). The two machined holes are 500 μm in diameter and the nodule population, which is assumed to behave as voids in the FE computations, has a characteristic diameter of 60 μm. It is considered as secondary void population, which can be observed at micrometer resolutions. Figure 3 shows mid-thickness sections of the reconstructed volume for three different load stages. Classical void coalescence mechanisms are accompanied by sheet coalescence between the two machined holes in the last loading step (deformed state (11)).

2.2 Digital Volume Correlation

Global DVC, which is used herein, is an extension of global 2D DIC [51, 52]. Reconstructed volumes are described by discrete gray level fields of spatial (voxel) coordinate $\mathbf{x}$. DVC consists in registering the gray levels I_0 in the reference configuration with those of the deformed volume I_t such that their conservation is obtained

$$I_0(\mathbf{x}) = I_t[\mathbf{x} + \mathbf{u}(\mathbf{x})] \tag{1}$$

where $\mathbf{u}$ is the Lagrangian displacement field. In experiments gray level conservation (1) is never satisfied in laminography due to acquisition noise and reconstruction artifacts [53]. Therefore the gray level residual $\rho(\mathbf{x}) = I_0(\mathbf{x}) - I_t[\mathbf{x} + \mathbf{u}(\mathbf{x})]$ is globally minimized by considering its L2-norm with respect to kinematic unknowns, which parameterize the measured displacement field. For global DVC, the whole ROI is considered and the global residual Φ_c^2

$$\Phi_c^2(\{\mathbf{u}\}) = \sum_{ROI} \rho^2(\mathbf{x}, \{\mathbf{u}\}) \tag{2}$$

is minimized with respect to the unknown degrees of freedom u_p gathered in the column vector $\{\mathbf{u}\}$ when the displacement field is written as

$$\mathbf{u}(\mathbf{x}, \{\mathbf{u}\}) = \sum_{p} u_p \boldsymbol{\Psi}_p(\mathbf{x}) \tag{3}$$

where $\boldsymbol{\Psi}_p(\mathbf{x})$ are selected displacement fields associated with the parameterization of $\mathbf{u}(\mathbf{x}, \{\mathbf{u}\})$. Finite element shape functions are of particular interest since they provide direct links between measured displacement fields and numerical simulations. DVC based on hexahedral finite elements with trilinear shape functions [35] is utilized herein. Only a part of the reconstructed volume, which is referred to as DVC ROI, is considered (Fig. 2b). To keep large ROI sizes, the reconstructed volumes are coarsened (i.e., each 8 neighboring voxels are averaged to form one supervoxel).

The measurement uncertainties are quantified by registering two volumes of the unloaded sample (0) with (coined "rbm") and without (i.e., "bis") rigid body motion (RBM) applied between acquisitions. Noise and reconstruction artifacts make these two volumes non identical. The corresponding displacement fields account for laminography and DVC effects on the measurement uncertainties [54]. The measurement uncertainties are assessed by the standard deviation of displacement fields. Figure 4 shows the standard displacement uncertainties for different element sizes ℓ. Decreasing the element size induces an increase of the displacement uncertainty [55, 56]. The element size used hereafter is set to $\ell = 16$ supervoxels and corresponds to a standard displacement uncertainty of 0.25 supervoxel. This level is the limit below which the estimated displacement levels are no longer trustworthy.

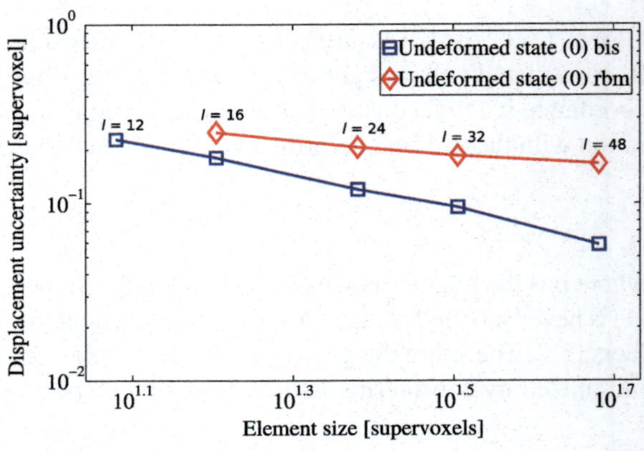

Fig. 4 Standard displacement uncertainties as functions of the element size ℓ expressed in supervoxels for two different acquisitions of the reference configuration

Table 1 Elastoplastic properties of the ferritic matrix

E (GPa)	ν	σ_y (MPa)	K (MPa)	n
210	0.30	290	382	0.35

Successful DVC registrations were achieved for the first 9 incremental calculations (i.e., registrations between step $n-1$ and step n). The measured displacement fields will serve as DVC-FE boundary conditions. The measured displacement fields are interpolated for each loading step onto the FE mesh of the ROI using the shape functions of the DVC mesh).

2.3 Simulations

To perform microscale FE simulations the numerical framework discussed in Refs. [44, 46, 47, 57, 58] is followed. The ROI selected for the FE simulations has to belong to all DVC ROIs for each analyzed loading step and to be made as large as possible [25]. To model the experimental microstructure, standard image processing operations are carried out [59, 60], namely, smoothing the data, applying a gray value threshold to separate matrix and voids, and then converting these binary data into signed distance function. The latter is interpolated onto a first mesh of uniform size of 10 μm of the FE ROI via trilinear interpolation. The corresponding signed distance function is then regularized with a parallel reinitialization algorithm [46], and used to locate the interfaces [25, 47]. An adaption step is added to control the local maximum curvature of the interface [25, 58]. These different steps are exemplified in Fig. 5 for a 2D laminography section. The final mesh has a size of 10 μm close to matrix/void interfaces and 50 μm at a distance of 100 μm from any interface with a linear transition. As shown in Fig. 5 the FE discretization of the microstructure is very close to the experimental observation.

The graphite nodules are modeled as zones with very low Young's modulus [18, 19, 21, 22, 25], while the ferritic matrix is considered as an elastoplastic medium with power law hardening

$$\sigma_0(p) = \sigma_y + Kp^n \tag{4}$$

where p is the equivalent plastic strain, σ_y the initial yield stress, K the plastic modulus and n the hardening exponent. The properties of the matrix (Table 1) are deduced from tensile experiments on pure ferrite [21].

The satisfaction of equilibrium equations is obtained with a mixed velocity-pressure formulation solved with $P1^+/P1$ elements to avoid locking [61]. The non-linear behavior of the matrix requires Newton-Raphson schemes to be implemented

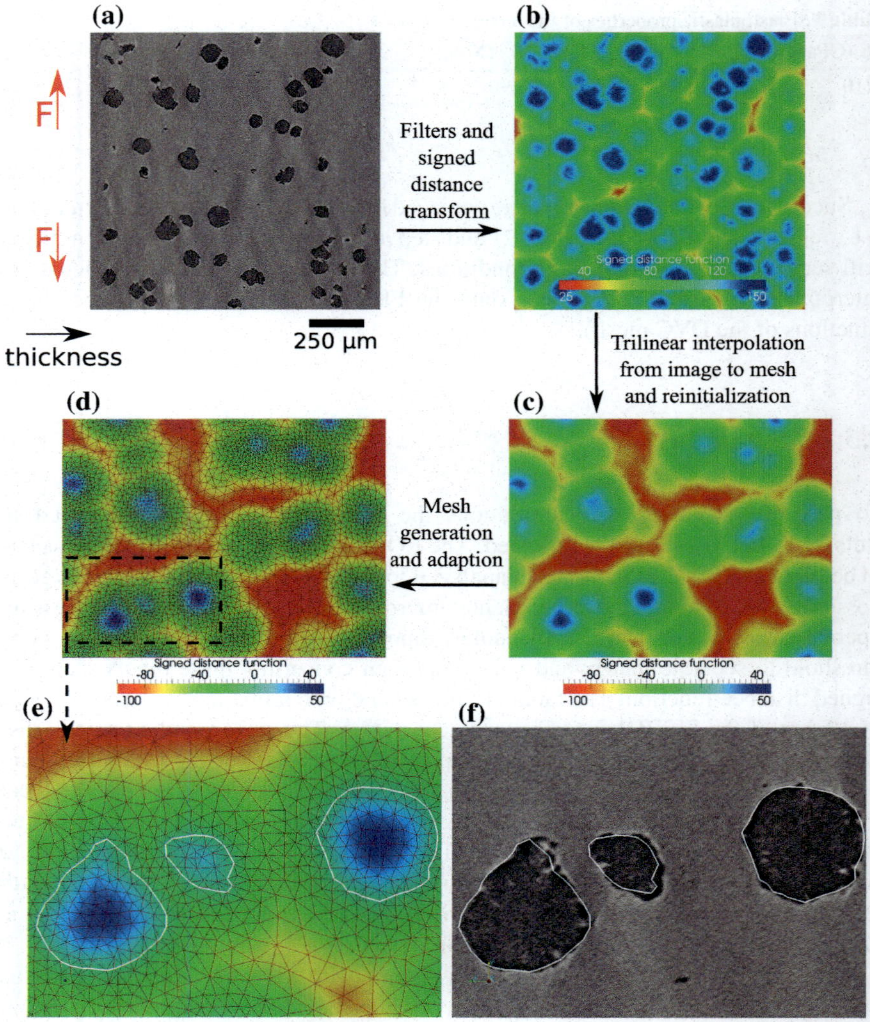

Fig. 5 Image immersion and meshing. **a** Initial laminography 2D section. **b** Signed distance function computed thanks to image processing. **c** Signed distance function interpolated and reinitialized on the FE mesh [46]. **d** Conforming FE mesh generated and adapted to interfaces and local maximum curvature, **e** Zoom on the FE mesh. **f** Comparison between initial laminography 2D section and interfaces in the final FE mesh (in white)

locally and globally [62]. An updated Lagrangian scheme is used to handle large deformations. Further, large distortions and possible flip of elements are avoided with automatic mesh motion and adaption [47].

3 Results

The numerical results using DVC-FE are illustrated in Fig. 6. This computation considers 100 voids meshed with ≈1 million elements. Void growth and equivalent plastic strains develop as more load is applied.

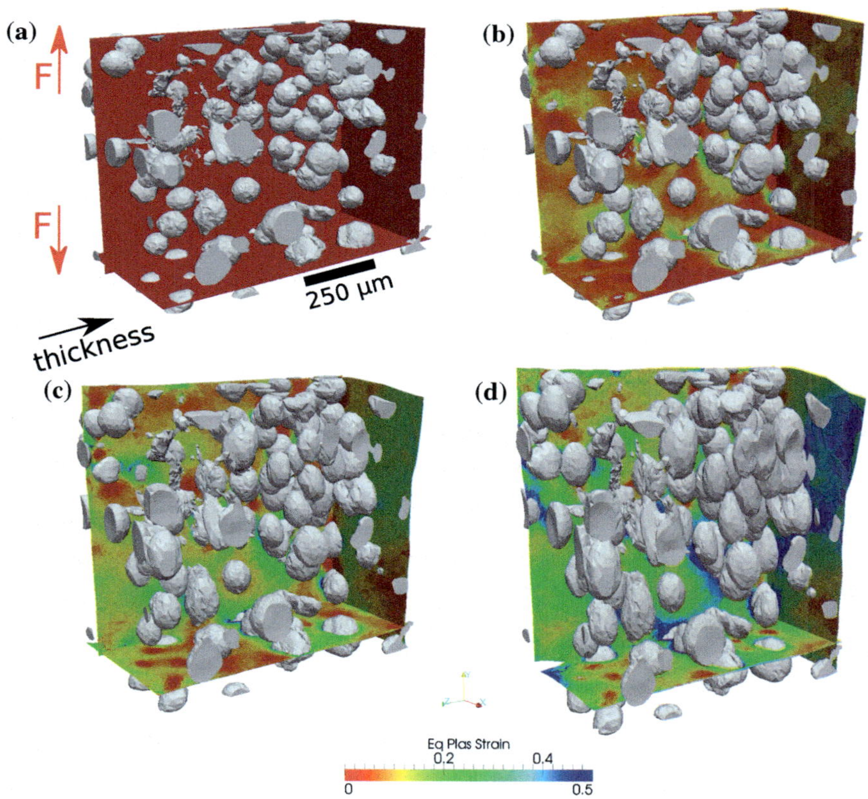

Fig. 6 ROI calculation results using DVC-FE showing the 3D meshed voids and the equivalent plastic strain on sections when: **a** $u = 0$ (undeformed state), **b** $u = 83.4$ μm, **c** $u = 192.2$ μm, **d** $u = 320.8$ μm

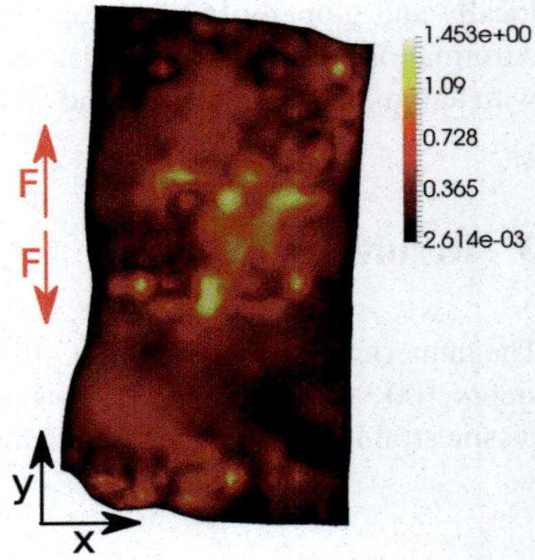

Fig. 7 Mid-section normal to z-direction showing absolute difference between DVC and DVC-FE displacement fields. The displacement difference is expressed in supervoxels (1 supervoxel $\longleftrightarrow$ 2.2 μm)

3.1 *Error Estimators*

Relative displacement comparisons are first reported. Measured displacement fields (via DVC) are applied to the boundaries of the FE ROI. They are also available within the whole ROI. Thus, DVC and DVC-FE displacement fields can be interpolated on the same mesh and directly compared as reported in Fig. 7. The main differences are concentrated around debond zones between the matrix and the nodules, while those close to the boundaries are mostly zero. The fact that the differences become significantly larger than the displacement uncertainty is a first indication of model error.

The errors in terms of gray level residuals are now discussed. For each pair of consecutive loading steps, the volume reconstructed for the second step can be deformed back with the measured or computed displacement field. This corrected volume can be compared *voxelwise* with the volume of the first step. With a newly developed tetrahedral-DVC code [38, 63] FE computations with tetrahedral meshes can be imported in the reconstructed volumes frame where the displacement fields are interpolated voxelwise. The deformed volume $I_t(\mathbf{x})$ is corrected by the computed displacement field $\mathbf{u}_{FE}(\mathbf{x})$, i.e., $I_t(\mathbf{x} + \mathbf{u}_{FE}(\mathbf{x}))$ is obtained. The gray level residuals, namely, differences between the volume of the reference configuration $I_0(\mathbf{x})$ and the corrected deformed volume $I_t(\mathbf{x} + \mathbf{u}(\mathbf{x}))$ are assessed for DVC *and* FE computations. Quantitative and local error measurements are evaluated for DVC and DVC-FE procedures. Figure 8 shows the standard deviation of residual fields that are normalized by the dynamic range of the volume (i.e., 256 gray levels). The DVC residuals remain close to those observed in the uncertainty analysis for which no strains occurred. Therefore the DVC results are deemed trustworthy.

The errors produced by the micromechanical models inside the DVC-FE domain also remain low and slightly increase at later loading steps (from $\approx$15% initially to $\approx$20% in last loading step). However they are always higher than the DVC residuals. This observation confirms model errors that become more significant as coalescence

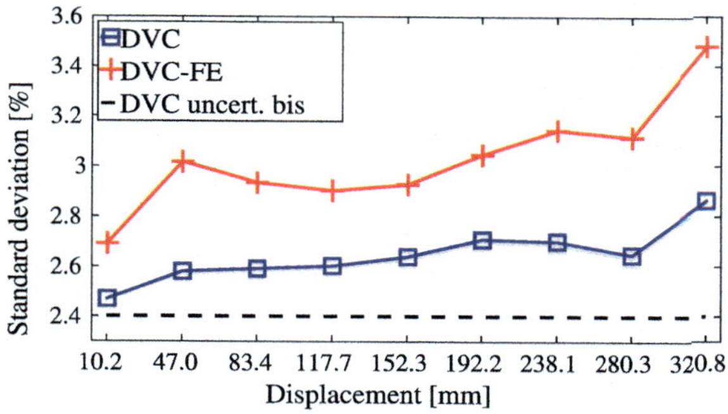

Fig. 8 Standard deviation for the dimensionless gray level residual fields for all loading steps. For comparison purposes, the dashed line corresponds to the uncertainty analysis for the so-called "bis" case (see Sect. 2.2)

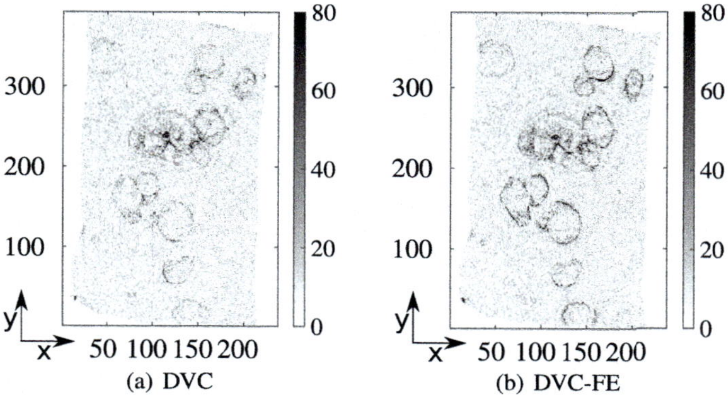

Fig. 9 Absolute gray level differences at the z midsection after correction with DVC (**a**) and DVC-FE (**b**) displacements for the ninth loading step

sets in. Figure 9 confirms that these differences between DVC-FE simulations and experiments are mostly concentrated around interfaces.

3.2 *Damage Analysis*

Damage predictions of DVC-FE are qualitatively compared studying the x-midsection of the ROI with experimental images in Fig. 10. Since measured boundary conditions are expected to follow experimental images at the spatial resolution of DVC, the matrix/void interfaces in the simulation (in white in the figure) are superimposed. The interfaces are very accurately meshed on average and tracked during the simulation up to the last loading step. Quantitatively void growth is defined by

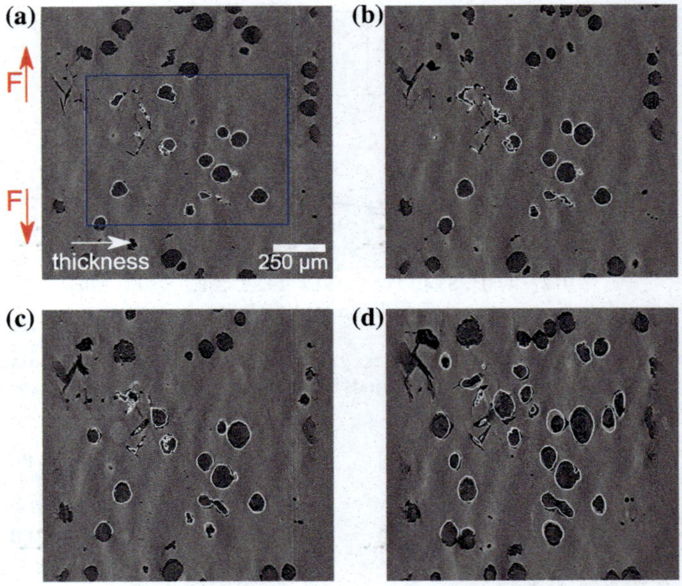

Fig. 10 ROI (blue line) calculation results using DVC-FE comparing the numerical matrix/void interface (white line) with experimental images for the x-midsection. **a** $u = 0$ (undeformed state), **b** $u = 83.4$ μm, **c** $u = 192.2$ μm, **d** $u = 320.8$ μm

Fig. 11 Void volume change observed experimentally and predicted within the present framework

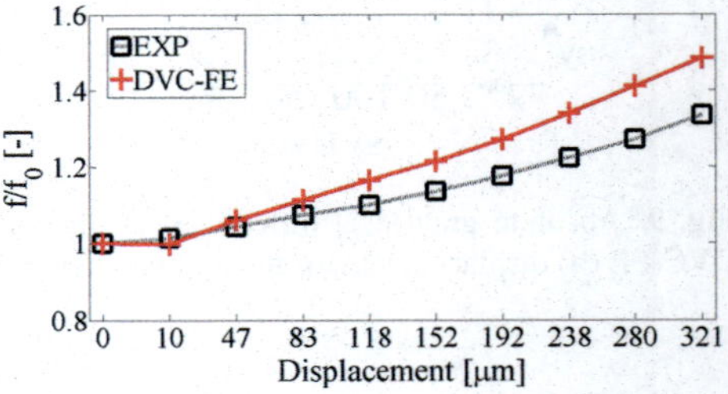

$$f = \frac{\text{void volume}}{\text{ROI volume}}, \quad \text{void growth} = \frac{f}{f_0} \tag{5}$$

where f_0 is the initial void volume fraction. Void growth plots are shown in Fig. 11 in which the 'EXP' curve is obtained in processed laminography volumes (i.e., images with smooth signed distance functions as shown in Fig. 5b).

The numerical results show a small decrease of porosity p at the first loading step. This is not observed experimentally. This first loading step is bigger than the subsequent ones, which asks for extensive remeshing in the computations. Consequently interfaces are slightly smoothened and void volume can be diffused. For the other loading steps, void growth is overestimated numerically. This may be due to the fact that nodules are considered as very soft media in the computations, while in reality only the voids grow after nodule/matrix interface debonding (Fig. 10).

4 Discussion

Although the results using DVC-FE look very promising, several issues need to be addressed. There still are gaps between FE-DVC and DVC results (see Figs. 9 and 10). This gap increases when reaching the final loading steps. Similarly, the displacement difference (Fig. 7) is significantly higher than the displacement uncertainty reported in Fig. 4. The differences are mainly concentrated around matrix/nodule interfaces. This observation calls for better models of the nodules and interface debonding. Further, the increase of the error at later loading steps proves the inability of the constitutive law used for the ferritic matrix to fully capture the acceleration of void growth and subsequent coalescence. Better calibrated and more advanced plasticity models may be considered at the microscale to better capture the multiscale plastic flow. These additional developments will extensively rely on DVC-FE and its ability to provide experimentally measured boundary conditions for micromechanical simulations. The extension of Integrated-DVC to 4D analyses [38] will be utilized to conduct inverse analyses based on these error measurements and calibrate material parameters at the microscale.

Acknowledgements This work was performed within the COMINSIDE project funded by the French Agence Nationale de la Recherche (ANR-14-CE07-0034-02 grant). We also acknowledge the European Synchrotron Radiation Facility for provision of beamtime at beamline ID15, experiment ME 1366. It is also a pleasure to acknowledge the support of BPI France ("DICCIT" project), and of the Carnot M.I.N.E.S institute ("CORTEX" project). M. Kuna and L. Zybell from IMFD, TU Freiberg, are thanked for materials supply and machining as well as for scientific discussions.

References

1. Kachanov, L.: Bull. SSR Acad. Sci. Division of Technical Sciences (in Russian) **8**, 26 (1958)
2. Rabotnov, Y.: On the Equations of State for Creep, pp. 307–315. McMillan, New York, USA (1963)
3. Lemaitre, J.: A Course on Damage Mechanics. Springer-Verlag, Berlin (1992)
4. Cao, T.S., Maire, E., Verdu, C., Bobadilla, C., Lasne, P., Montmitonnet, P., Bouchard, P.O.: Comput. Mat. Sci. **84**, 175 (2014)
5. Cao, T.S., Bobadilla, C., Montmitonnet, P., Bouchard, P.O.: J. Mat. Proc. Technol. **216**, 385 (2015)
6. Gurson, A.: ASME J. Eng. Mat. Technol. **99**, 2 (1977)
7. Needleman, A., Tvergaard, V.: J. Mech. Phys. Solids **32**(6), 461 (1984)
8. Geers, M., De Borst, R., Peijs, T.: Compos. Sci. Tech. **59**, 1569 (1999)
9. Claire, D., Hild, F., Roux, S.: C. R. Mécanique **330**, 729 (2002)
10. Roux, S., Hild, F.: Exp. Mech. **48**(4), 495 (2008)

11. Babout, L., Bréchet, Y., Maire, E., Fougères, R.: Acta Mat. **52**(15), 4517 (2004)
12. Bouchard, P.O., Bourgeon, L., Lachapèle, H., Maire, E., Verdu, C., Forestier, R., Logé, R.: Mat. Sci. Eng. A **496**(1–2), 223 (2008)
13. Ueda, T., Helfen, L., Morgeneyer, T.F.: Acta Mat. **78**, 254 (2014)
14. Hannard, F., Pardoen, T., Maire, E., Le Bourlot, C., Mokso, R., Simar, A.: Acta Mat. **103**, 558 (2016)
15. Morgeneyer, T., Taillandier-Thomas, T., Buljac, A., Helfen, L., Hild, F.: J. Mech. Phys. Solids **96**, 550 (2016)
16. Morgeneyer, T., Besson, J., Proudhon, H., Starink, M., Sinclair, I.: Acta Mat. **57**(13), 3902 (2009)
17. Tang, S., Kopacz, A.M., Chan O'Keeffe, S., Olson, G.B., Liu, W.K.: J. Mech. Phys. Solids **61**(11), 2108 (2013)
18. Dong, M.J., Prioul, C., François, D.: Metall. Mat. Trans. A **28**(11), 2245 (1997)
19. Hütter, G., Zybell, L., Kuna, M.: Eng. Fract. Mech. **144**, 118 (2015)
20. Tomičević, Z., Kodvanj, J., Hild, F.: Eur. J. Mech. A/Solids **59**, 140 (2016)
21. Zhang, K., Bai, J., François, D.: Int. J. Solids Struct. **36**(23), 3407 (1999)
22. Bonora, N., Ruggiero, A.: Int. J. Solids Struct. **42**(5–6), 1401 (2005)
23. Hütter, G., Zybell, L., Mühlich, U., Kuna, M.: Comput. Mat. Sci. **80**, 61 (2013)
24. Hütter, G., Zybell, L., Kuna, M.: Int. J. Solids Struct. **51**(3–4), 839 (2014)
25. Buljac, A., Shakoor, M., Neggers, J., Helfen, L., Bernacki, M., Bouchard, P.O., Morgeneyer, T.F., Hild, F.: Comput. Mech. **59**(3), 419 (2017)
26. Helfen, L., Myagotin, A., Pernot, P., DiMichiel, M., Mikulík, P., Berthold, A., Baumbach, T.: Nucl. Inst. Meth. Phys. Res. B **563**, 163 (2006)
27. Helfen, L., Morgeneyer, T., Xu, F., Mavrogordato, M., Sinclair, I., Schillinger, B., Baumbach, T.: Int. J. Mat. Res. **2012**(2), 170 (2012)
28. Maurel, V., Helfen, L., N'Guyen, F., Köster, A., Di Michiel, M., Baumbach, T., Morgeneyer, T.: Scripta Mat. **66**, 471 (2012)
29. Bull, D., Spearing, S., Sinclair, I., Helfen, L.: Compos. Part A **52**, 62 (2013)
30. Reischig, P., Helfen, L., Wallert, A., Baumbach, T., Dik, J.: Appl. Phys. A **111**, 983 (2013)
31. Bay, B., Smith, T., Fyhrie, D., Saad, M.: Exp. Mech. **39**, 217 (1999)
32. Smith, T., Bay, B., Rashid, M.: Exp. Mech. **42**(3), 272 (2002)
33. Bornert, M., Chaix, J., Doumalin, P., Dupré, J., Fournel, T., Jeulin, D., Maire, E., Moreaud, M., Moulinec, H.: Inst. Mes. Métrol. **4**, 43 (2004)
34. Verhulp, E., van Rietbergen, B., Huiskes, R.: J. Biomech. **37**(9), 1313 (2004)
35. Roux, S., Hild, F., Viot, P., Bernard, D.: Comp. Part A **39**(8), 1253 (2008)
36. Rannou, J., Limodin, N., Réthoré, J., Gravouil, A., Ludwig, W., Baïetto, M., Buffière, J., Combescure, A., Hild, F., Roux, S.: Comp. Meth. Appl. Mech. Eng. **199**, 1307 (2010)
37. Bouterf, A., Roux, S., Hild, F., Adrien, J., Maire, E.: Strain **50**(5), 444 (2014)
38. Hild, F., Bouterf, A., Chamoin, L., Mathieu, F., Neggers, J., Pled, F., Tomičević, Z., Roux, S.: Adv. Mech. Simul. Eng. Sci. **3**(1), 1 (2016)
39. Zhang, Y., Bajaj, C., Sohn, B.S.: Comput. Meth. Appl. Mech. Eng. **194**(48–49), 5083 (2005)
40. Young, P.G., Beresford-West, T.B.H., Coward, S.R.L., Notarberardino, B., Walker, B., Abdul-Aziz, A.: Phil. Trans. A **366**(1878), 3155 (2008)
41. Osher, S., Sethian, J.A.: J. Comput. Phys. **79**(1), 12 (1988)
42. Kimmel, R., Shaked, D., Kiryati, N., Bruckstein, A.M.: Comput. Vis. Image Underst. **62**(3), 382 (1995)
43. Sukumar, N., Chopp, D., Moës, N., Belytschko, T.: Comput. Meth. Appl. Mech. Eng. **190**(46–47), 6183 (2001)
44. Roux, E., Bernacki, M., Bouchard, P.O.: Comput. Mat. Sci. **68**, 32 (2013)
45. Quan, D.L., Toulorge, T., Marchandise, E., Remacle, J.F., Bricteux, G.: Comp. Meth. Appl. Mech. Eng. **268**, 65 (2014)
46. Shakoor, M., Scholtes, B., Bouchard, P.O., Bernacki, M.: Appl. Math. Model. **39**(23–24), 7291 (2015)
47. Shakoor, M., Bouchard, P.O., Bernacki, M.: Int. J. Num. Meth. Eng. **109**(4), 555 (2017)

48. Weck, A., Wilkinson, D.: Acta Materialia **56**(8), 1774 (2008)
49. Myagotin, A., Voropaev, A., Helfen, L., Hänschke, D., Baumbach, T., Trans, I.E.E.E.: Image Process. **22**(12), 5348 (2013)
50. Vogelgesang, M., Farago, T., Morgeneyer, T.F., Helfen, L., dos Santos Rolo, T., Myagotin, A., Baumbach, T.: J. Synchrotron Rad. **23**, 1254 (2016)
51. Besnard, G., Hild, F., Roux, S.: Exp. Mech. **46**, 789 (2006)
52. Hild, F., Roux, S.: Exp. Mech. **52**(9), 1503 (2012)
53. Xu, F., Helfen, L., Baumbach, T., Suhonen, H.: Optics Exp. **20**, 794 (2012)
54. Morgeneyer, T., Helfen, L., Mubarak, H., Hild, F.: Exp. Mech. **53**(4), 543 (2013)
55. Limodin, N., Rthor, J., Adrien, J., Buffire, J., Hild, F., Roux, S.: Exp. Mech. **51**(6), 959 (2011)
56. Leclerc, H., Périé, J., Hild, F., Roux, S.: Mech. Indust. **13**, 361 (2012)
57. Roux, E., Shakoor, M., Bernacki, M., Bouchard, P.O.: Model. Simul. Mat. Sci. Eng. **22**(7), 075001 (2014)
58. Shakoor, M., Bernacki, M., Bouchard, P.O.: Eng. Fract. Mech. **147**, 398 (2015)
59. Schindelin, J., Arganda-Carreras, I., Frise, E., Kaynig, V., Longair, M., Pietzsch, T., Preibisch, S., Rueden, C., Saalfeld, S., Schmid, B., Tinevez, J.Y., White, D.J., Hartenstein, V., Eliceiri, K., Tomancak, P., Cardona, A.: Nature Meth. **9**(7), 676 (2012)
60. Schneider, C.A., Rasband, W.S., Eliceiri, K.W.: Nature Meth. **9**(7), 671 (2012)
61. Boffi, D., Brezzi, F., Demkowicz, L.F., Durán, R.G., Falk, R.S., Fortin, M.: Mixed Finite Elements, Compatibility Conditions, and Applications. Lecture Notes in Mathematics, vol. 1939. Springer, Berlin (2008)
62. Wagoner, R.H., Chenot, J.L.: Metal Forming Analysis. Cambridge University Press (2001)
63. Leclerc, H., Neggers, J., Mathieu, F., Roux, S., Hild, F.: Correli 3.0. IDDN.FR.001.520008.000.S.P.2015.000.31500, Agence pour la Protection des Programmes, Paris, France (2015)

Stochastic Upscaling via Linear Bayesian Updating

Sadiq M. Sarfaraz, Bojana V. Rosić, Hermann G. Matthies and Adnan Ibrahimbegović

Abstract In this work we present an upscaling technique for multi-scale computations based on a stochastic model calibration technique. We consider a coarse scale continuum material model described in the framework of generalised standard materials. The model parameters are considered uncertain in this approach, and are approximated using random variables. The update or calibration of these random variables is performed in a Bayesian framework where the information from a deterministic fine scale model computation is used as observation. The proposed approach is independent w.r.t. the choice of models on coarse and fine scales. Simple numerical examples are shown to demonstrate the ability of the proposed approach to calibrate coarse-scale elastic and inelastic material parameters.

1 Introduction

Heterogeneous microstructures occur in many naturally existing or man-made materials: rocks/soils and concrete being two well-known examples. The spatial scales on which these heterogeneities occur are orders of magnitudes smaller than the ones for which one would like to do response predictions. This means that it is computationally not possible to resolve the small scales (micro-scale) on the predictive scale (macro-scale).

S.M. Sarfaraz (✉) · B.V. Rosić · H.G. Matthies
Institute of Scientic Computing, TU Braunschweig, Braunschweig, Germany
e-mail: m.sarfaraz@tu-bs.de

B.V. Rosić
e-mail: bojana.rosic@tu-bs.de

H.G. Matthies
e-mail: wire@tu-bs.de

A. Ibrahimbegović
Laboratoire Roberval, UTC Compiègne, Compiègne, France
e-mail: adnan.ibrahimbegovic@utc.fr

J. Sorić et al. (eds.), *Multiscale Modeling of Heterogeneous Structures*,
Lecture Notes in Applied and Computational Mechanics 86,
https://doi.org/10.1007/978-3-319-65463-8_9

Additionally, models at the micro-scale may be of an entirely different nature than what is desired on the macro-scale—typically a continuum model—as, for example, the micro-scale model may be a discrete model. Hence it becomes clear that to obtain continuum material models on the macro-scale, the pure "homogenisation" is insufficient. This situation is more broadly described in [18, 26], see also the references therein.

Thus we are looking at a stochastic upscaling approach. Much of the literature to this subject has been reviewed in the monograph [31], as well as in [2, 4, 5, 9–11, 23, 37, 40], and [6], to which we refer for the sake of brevity in this exposition. One way to achieve the coupling between scales with possibly completely different descriptions is to use concepts of machine learning as in [22], the concept of which is often, at least conceptually, grounded in Bayesian ideas. Here we directly take a Bayesian approach.

More information on general Bayesian calibration can be found in [15, 20, 21]. The evaluation of the posterior probability density function (pdf) is often based on statistically based sampling methods such as Markov Chain Monte Carlo (MCMC) [8]. To avoid this computationally expensive procedure, we want to take a faster filtering approach, which allows various approximations. One of the most well-known and simplest methods in this approach is the *Kalman filter* (KF) and its extensions. Therefore, here we use a numerical strategy based on the (generalised) polynomial chaos expansion (gPCE) applied to a new extension of the KF and approximate Bayesian estimation, which can estimate non-Gaussian distributions without sampling [32, 34, 36].

As is well-known in Bayesian analysis, one generally speaking only obtains answers in relation to things one asks. Therefore a frame of reference has to be established a priori within which the identification will take place. The general set-up we propose is therefore as follows: on the macro-scale a continuum material model is derived which not only covers the mean (i.e. homogenised) behaviour, but also the possible deviations from it. As the micro-scale mechanical behaviour we have in mind involves both reversible (i.e. elastic) as well as irreversible (i.e. inelastic) behaviour, this has to be reflected also in the constitutive models considered on the macro-scale. Here we only want to show a proof-of-concept, so we will limit ourselves to a simple but characteristic case. For the sake of simplicity we limit ourselves to isothermal conditions and we shall exclude strain-rate dependent behaviour, and for the inelastic or irreversible part we only consider ductile non-softening behaviour, i.e. strain-rate independent plasticity with hardening.

As this is to be a model for possibly more complex behaviour, we shall assume that the macro-scale continuum model can be described as a generalised standard material model [12, 13, 30]. This has the advantage that these materials are completely characterised by the specification of two scalar functions, the stored energy resp. Helmholtz free energy, and the dissipation pseudo-potential. In this way the simple case chosen here can be generalised to more complex material behaviour. In our view this description is also a key for the connection with the micro-scale behaviour. No matter how the physical and mathematical/computational description on the micro-scale has been chosen, in all cases where the description is based on physical principles it will be possible to define the stored (Helmholtz free) energy and the dissipation (entropy production). These two thermodynamic functions will thus be used to connect the micro—and the macro-structure models.

In a little more detail, the identification of the macro-structure generalised standard material constitutive model proceeds as follows: the micro-structure is exposed to some external action resp. stimulus, in this purely mechanical case this is large scale homogeneous deformation. The response is measured in the change of the two thermodynamic functions alluded to: the stored resp. Helmholtz free energy and the dissipation resp. entropy production. The main purpose of this note is to show that this idea is computationally feasible.

The content of this note is as follows: In Sect. 2 the stochastic upscaling will be described. The identification resp. calibration will use Bayesian ideas [32, 34, 36]. As this topic is only a computational tool, it will only be sketched here briefly in Sect. 2.1, and more detailed accounts can be found in the literature. The coarse—and fine-scale model used in this note will be described in Sect. 3 and Sect. 4 respectively. These theoretical concepts will be numerically applied to a simple but illustrative example in Sect. 5.

2 Stochastic Upscaling

Bayesian ideas are connected with a probabilistic description of our knowledge, in this case the parameters which describe a particular instance of a generalised standard material, and additional information is reflected in conditional probabilities, conditioned on this additional information. Here this additional information is the information from the fine-scale model. Our presentation follows [27].

2.1 Bayesian Inverse Problems

Conditional probabilities are rigorously derived from conditional expectations. Here that foundation will also be taken as the computational starting point, by numerical approximation of conditional expectation. For nonlinear models, further simplifications are needed, which give a computationally efficient algorithm, leading via a generalisation of the well-known Gauss-Markov theorem to something which may be seen as a substantial extension of the Kalman filter. The resulting filter is therefore termed the *Gauss-Markov-Kalman* filter (GMKF). Here we give a short description of the connection of the GMK-filter with Bayesian updating via conditional expectation.

2.1.1 Mathematical Set-Up

Assume that one has a computational model—here that will be our coarse-scale model—symbolically written as

$$A_c(u_c, \boldsymbol{q}) = f_c, \tag{1}$$

where the variable $u_c \in \mathcal{U}_c$ represents a the state of the system in a vector space $\mathcal{U}_c$, the variables $\boldsymbol{q} = [q_1, \ldots, q_n]^T$ are parameters to calibrate the model, f_c stands for the external influences—the loading, action, initial conditions, experimental set-up—and operator A_c describes the system under consideration. Although written as a stationary system, for the sake of simplicity we will tacitly assume that Eq. (1) covers also time-evolution problems. The parameters $\boldsymbol{q}$ may actually include the state u_c or parts of it, or the initial conditions in case of a time-evolution problem. To keep things simple, only one update step is described.

In addition there is a second system—in this case the fine-scale model—something we can evaluate at possibly high cost, but which does not need any parameters for calibration

$$A_f(u_f) = f_f, \tag{2}$$

with state $u_f \in \mathcal{U}_f$, where it is assumed that f_c and f_f describe the same situation. Again it is written in a simple stationary form, although it may also cover evolutionary problems. The model Eq. (2) is to be used to calibrate the parameters $\boldsymbol{q}$ in such a way that the predictions of Eq. (1) match those of Eq. (2) as well as possible.

As $\mathcal{U}_c \neq \mathcal{U}_f$, the states u_c and u_f can not be directly compared, and the two models are to be compared by some observables or measurements $y \in \mathcal{Y}$, where $\mathcal{Y}$ is typically some vector space like $\mathbb{R}^m$. This measurement function for Eq. (1)

$$y_c = Y_c(\boldsymbol{q}, u_c(\boldsymbol{q}, f_c)) \tag{3}$$

will also shortly be denoted as $y_c = Y_c(\boldsymbol{q})$. We also assume a similar function

$$y_f = Y_f(u_f(f_f)), \tag{4}$$

which models the same observation in relation to Eq. (2). Further assume that y_c is not observed directly, but rather $y_c + \epsilon$, where ϵ is a random variable, which in the case of Eq. (2) being reality models the errors of the measurement device, and in case of Eq. (2) being a computational model can represent the *model error* of Eq. (2), i.e. the difference between it and *reality*. The observation model is hence

$$z = y_c + \epsilon = Y_c(\boldsymbol{q}) + \epsilon. \tag{5}$$

The goal of calibration is now to estimate $\boldsymbol{q}$ such that y_c and y_f resp. z and y_f deviate as little as possible.

2.1.2 Synopsis of Bayesian Estimation

The idea is that the observation z, depending on the unknown parameters $\boldsymbol{q}$, should give an indication on what $\boldsymbol{q}$ should be. But in general the mapping $\boldsymbol{q} \mapsto Y_c(\boldsymbol{q})$ is not invertible, i.e. z does not contain information to uniquely determine $\boldsymbol{q}$, or there are many $\boldsymbol{q}$ which give a good fit. Therefore the inverse problem of determining $\boldsymbol{q}$ from observing z is termed an ill-posed problem.

In the Bayesian view, see e.g. [39], the unknown resp. uncertain parameter $\boldsymbol{q}$ is modelled as a random variable (RV)—also called the prior model—and additional information on the system through measurement or observation changes the probabilistic description to the so-called posterior model. It is well-known that such a Bayesian update is in fact closely related to conditional expectation [3], and this will be the basis of the method presented. For these and other probabilistic notions see for example [33] and the references therein. We also show computational procedures for this update through methods based on *functional approximation* or *spectral representation* of stochastic problems [29]. These approximations are in the simplest case known as Wiener's so-called homogeneous or polynomial chaos expansion, which are polynomials in independent Gaussian RVs—the "chaos"—and which can also be used numerically in a Galerkin procedure [29]. Since the parameters of the model to be estimated are uncertain, all relevant information may be obtained via their stochastic description.

Formally, assume that the uncertain parameters are given by a random variable (RV)

$$\boldsymbol{q} : \Omega \to \mathbb{R}^n \text{ as a RV on a probability space } (\Omega, \mathfrak{A}, \mathbb{P}), \tag{6}$$

where the set of elementary events is Ω, $\mathfrak{A}$ a σ-algebra of measurable events, and $\mathbb{P}$ a probability measure. The *expectation* corresponding to $\mathbb{P}$ will be denoted by $\mathbb{E}\,()$, e.g. $\bar{\boldsymbol{x}} := \mathbb{E}\,(\boldsymbol{q}) := \int_\Omega \boldsymbol{q}(\omega)\,\mathbb{P}(\mathrm{d}\omega)$.

As formally $\boldsymbol{q}$ is a RV, so is the state u_c, and also the prediction of the "true" measurement y_c Eq. (1). Also assume that the error $\varepsilon(\omega)$ is a $\mathcal{Y}$-valued RV, and in total the prediction of the observation or measurement Eq. (5) $z(\omega) = y_c(\omega) + \varepsilon(\omega)$ therefore becomes a RV as well; i.e. we have a *probabilistic* model of the observation.

2.1.3 The Theorem of Bayes and Conditional Expectation

We recall Bayes's theorem as formulated by Laplace, commonly accepted as a consistent way to incorporate new knowledge into a probabilistic description [39]. The elementary textbook statement of the theorem is

$$\mathbb{P}(\mathcal{I}_{\boldsymbol{q}}|\mathcal{M}_z) = \frac{\mathbb{P}(\mathcal{M}_z|\mathcal{I}_{\boldsymbol{q}})}{\mathbb{P}(\mathcal{M}_z)}\,\mathbb{P}(\mathcal{I}_{\boldsymbol{q}}), \quad \text{if } \mathbb{P}(\mathcal{M}_z) > 0, \tag{7}$$

where $\mathcal{I}_{\boldsymbol{q}}$ is some subset of possible $\boldsymbol{q}$ on which we would like to gain some information, and $\mathcal{M}_z$ is the information provided by the measurement. The term $\mathbb{P}(\mathcal{I}_{\boldsymbol{q}})$ is the so-called *prior*, it is what we know before the observation $\mathcal{M}_z$. The quantity $\mathbb{P}(\mathcal{M}_z|\mathcal{I}_{\boldsymbol{q}})$ is the so-called *likelihood*, the conditional probability of $\mathcal{M}_z$ assuming that $\mathcal{I}_{\boldsymbol{q}}$ is given. The term $\mathbb{P}(\mathcal{M}_z)$ is the so called *evidence*, the probability of observing $\mathcal{M}_z$ in the first place. It is necessary to make the right hand side of Eq. (7) into a real probability—summing to unity—and hence the conditional probability $\mathbb{P}(\mathcal{I}_{\boldsymbol{q}}|\mathcal{M}_z)$, the *posterior*, reflects our knowledge on $\mathcal{I}_{\boldsymbol{q}}$ *after* observing $\mathcal{M}_z$. This statement Eq. (7) faces problems if the set of observations $\mathcal{M}_z$ has vanishing measure, $\mathbb{P}(\mathcal{M}_z) = 0$, as is often the case when we observe *continuous* random variables, and the theorem would have to be formulated in *densities*, or more precisely in probability density functions (pdfs), which may be possible under some additional assumptions.

To avoid the difficulties with conditional probabilities, *Kolmogorov* defined as a first and fundamental notion *conditional expectation*, from which conditional probabilities may easily be recovered. It has to be defined w.r.t sub-σ-algebras $\mathfrak{B} \subset \mathfrak{A}$ of the underlying σ-algebra $\mathfrak{A}$. The σ-algebra may be loosely seen as the collection of subsets of Ω on which we can make statements about their probability. The sub-σ-algebra $\mathfrak{B}$ may be seen as the sets on which we learn something through the observation.

The simplest, although slightly restricted, way to define the conditional expectation [3] is to just consider RVs with *finite variance*, i.e. the Hilbert-space

$$\mathcal{S} := \mathrm{L}_2(\Omega, \mathfrak{A}, \mathbb{P}) := \{r : \Omega \to \mathbb{R} \;:\; r \text{ measurable w.r.t. } \mathfrak{A},\ \mathbb{E}\left(|r|^2\right) < \infty\}.$$

If $\mathfrak{B} \subset \mathfrak{A}$ is a sub-σ-algebra, the space

$$\mathcal{S}_{\mathfrak{B}} := \mathrm{L}_2(\Omega, \mathfrak{B}, \mathbb{P}) := \{r :\in \mathcal{S} \;:\; r \text{ measurable w.r.t. } \mathfrak{B}\}$$

is a *closed* subspace, and hence has a well-defined continuous orthogonal projection $P_{\mathfrak{B}} : \mathcal{S} \to \mathcal{S}_{\mathfrak{B}}$. The *conditional expectation* (CE) of a RV $r \in \mathcal{S}$ w.r.t. a sub-σ-algebra $\mathfrak{B}$ is then defined as that orthogonal projection

$$\mathbb{E}\left(r|\mathfrak{B}\right) := P_{\mathfrak{B}}(r) \in \mathcal{S}_{\mathfrak{B}}. \tag{8}$$

As the CE is an orthogonal projection, it minimises the squared error

$$\mathbb{E}\left(|r - \mathbb{E}\left(r|\mathfrak{B}\right)|^2\right) = \min\{\mathbb{E}\left(|r - \tilde{r}|^2\right) \;:\; \tilde{r} \in \mathcal{S}_{\mathfrak{B}}\}, \tag{9}$$

from which one obtains the *variational equation* or orthogonality relation

$$\forall \tilde{r} \in \mathcal{S}_{\mathfrak{B}} : \quad \mathbb{E}\left(\tilde{r}(r - \mathbb{E}\left(r|\mathfrak{B}\right))\right) = 0. \tag{10}$$

In our case of an observation of a RV z, the sub-σ-algebra $\mathfrak{B}$ will be the one generated by the *observation* z, i.e. $\mathfrak{B} = \sigma(z)$, and the corresponding CE will simply be denoted as $\mathbb{E}\left(r|z\right) := \mathbb{E}\left(r|\sigma(z)\right)$. According to the *Doob-Dynkin* lemma [3], $\mathcal{S}_{\sigma(z)}$

is given by

$$\mathcal{S}_{\sigma(z)} := \{r \in \mathcal{S} \ : \ r(\omega) = \phi(z(\omega)), \phi \text{ measurable}\}, \tag{11}$$

i.e. functions of the observation. This means intuitively that anything we learn from an observation is a function of the observation, and the subspace $\mathcal{S}_{\sigma(z)} \subset \mathcal{S}$ is where the information from the measurement lies. A RV r may be decomposed into its orthogonal components w.r.t. $\mathcal{S}_{\sigma(z)}$ by

$$\begin{aligned} r &= P_{\sigma(z)}(r) + (I_{\mathcal{S}} - P_{\sigma(z)})(r) \\ &= P_{\sigma(z)}(r) + \big(r - P_{\sigma(z)}(r)\big) = \mathbb{E}\,(r|z) + (r - \mathbb{E}\,(r|z))\,, \end{aligned} \tag{12}$$

where $(I_{\mathcal{S}} - P_{\sigma(z))}(r) \in \mathcal{S}^{\perp}_{\sigma(z)}$, the orthogonal complement of $\mathcal{S}_{\sigma(z)}$. Obviously $P_{\sigma(z)}(r)$ is the best estimator for r measured in the error norm squared $\|r - P_{\sigma(z)}(r)\|^2_{\mathcal{S}}$ from the subspace $\mathcal{S}_{\sigma(z)}$. From a measurement z we learn something about the component $P_{\sigma(z)}(r)$ in $\mathcal{S}_{\sigma(z)}$. Hence one simple approach is the least-squares approximation, which also underlies the Gauss-Markov theorem and its extensions [24].

If $\boldsymbol{q}_f$ is our prior knowledge before the measurement, the forecast, we define the filtered, analysed, or assimilated RV $\boldsymbol{q}_a$ after the observation $\check{y}$ from Eq. (12) as

$$\begin{aligned} \boldsymbol{q}_a &= \mathbb{E}(\boldsymbol{q}_f|\check{y}) + \big(\boldsymbol{q}_f - \mathbb{E}(\boldsymbol{q}_f|z)\big) \\ &= \boldsymbol{q}_f + \big(\mathbb{E}(\boldsymbol{q}_f|\check{y}) - \mathbb{E}(\boldsymbol{q}_f|z)\big) = \boldsymbol{q}_f + \boldsymbol{q}_i, \end{aligned} \tag{13}$$

where $\boldsymbol{q}_i = \mathbb{E}(\boldsymbol{q}_f|\check{y}) - \mathbb{E}(\boldsymbol{q}_f|z)$ is called the *innovation*, and as $\mathbb{E}(\boldsymbol{q}_a|\check{y}) = \mathbb{E}(\boldsymbol{q}_f|\check{y})$, it follows that $\mathbb{E}(\boldsymbol{q}_i|\check{y}) = 0$.

Equation (13) is the *nonlinear* CE-filter [27], but as $\mathbb{E}(\boldsymbol{q}_f|z)$ can be a complicated function of z, it may be difficult to compute. A simpler version results if in Eq. (11) one takes only the affine functions, i.e. a smaller subspace

$$\mathcal{S}_{\sigma(z),1} := \{r \in \mathcal{S} \ : \ r(\omega) = H(z(\omega)) + b, \ H \text{ linear}\} \subset \mathcal{S}, \tag{14}$$

and the minimisation Eq. (9) is performed over this smaller subspace, resulting in an optimal linear map $\boldsymbol{K}_q$ (the so-called Kalman-gain) [24, 27]—the constant b in Eq. (14) cancels in Eq. (13) and hence does not have to be computed. With this simplification in Eq. (13) one arrives at the *Gauss-Markov-Kalman*-filter (GMKF):

$$\begin{aligned} \boldsymbol{q}_a &= \boldsymbol{q}_f + (\boldsymbol{K}_{\boldsymbol{q}}(\check{y}) - \boldsymbol{K}_{\boldsymbol{q}}(z)) \\ &= \boldsymbol{q}_f + \boldsymbol{K}_{\boldsymbol{q}}(\check{y} - z) = \boldsymbol{q}_f + \boldsymbol{K}_{\boldsymbol{q}}(\check{y} - (Y_c(\boldsymbol{q}_f) + \epsilon)). \end{aligned} \tag{15}$$

This is an equation between RVs—$\boldsymbol{q}_f = \boldsymbol{q}_f(\omega)$ and $\boldsymbol{K}_{\boldsymbol{q}}(z) = \boldsymbol{K}_{\boldsymbol{q}}(z(\omega))$ are RVs, and hence so is $\boldsymbol{q}_a = \boldsymbol{q}_a(\omega)$—and for a computation it has to be discretised.

2.1.4 Spectral or Functional Approximation

Our starting point is Eq. (15). As it is a relation between RVs, it certainly also holds for samples of the RVs, and this is the basis of the *ensemble* Kalman filter, the EnKF [7]. The sampling points are sometimes also denoted as particles, and the EnKF is a simple version of a particle filter.

Here we want to pursue the more promising functional or spectral approximation [27, 29] for all the RVs in Eq. (15). This means that all RVs, say $\boldsymbol{q}(\omega)$, are described as functions of known RVs $\{\theta_1(\omega), \ldots, \theta_\ell(\omega), \ldots\}$. Often, when for example stochastic processes or random fields are involved, one has to deal here with *infinitely* many RVs, which for an actual computation have to be truncated to a finte vector $\boldsymbol{\theta}(\omega) = [\theta_1(\omega), \ldots, \theta_n(\omega)]$ of significant RVs. We shall assume that these have been chosen such as to be independent. As we want to later approximate $\boldsymbol{q} \in \mathbb{R}^n$, we do not need more than n RVs $\boldsymbol{\theta}$.

One further chooses a finite set of linearly independent functions $\{\Psi_\alpha\}_{\alpha\in\mathcal{J}_M}$ of the variables $\boldsymbol{\theta}(\omega)$, where the index α often is a multi-index, and the set $\mathcal{J}_M$ is a finite set of multi-indices with cardinality (size) M. Many different systems of functions can be used, classical choices are [29] multivariate polynomials—leading to the *polynomial chaos expansion* (PCE). We shall assume that the set $\{\Gamma_\alpha\}_{\alpha\in\mathcal{J}_M}$ includes all the *linear* functions of $\boldsymbol{\theta}$. Thus a RV $\boldsymbol{q}(\omega)$ would be replaced by a functional approximation

$$\boldsymbol{q}(\omega) = \sum_{\alpha\in\mathcal{J}_M} \boldsymbol{q}_\alpha \Psi_\alpha(\boldsymbol{\theta}(\omega)) = \sum_{\alpha\in\mathcal{J}_M} \boldsymbol{q}_\alpha \Psi_\alpha(\boldsymbol{\theta}) = \boldsymbol{q}(\boldsymbol{\theta}). \tag{16}$$

The argument ω will be omitted from here on, as we transport the probability measure $\mathbb{P}$ on Ω to $\boldsymbol{\Theta} = \Theta_1 \times \cdots \times \Theta_n$, the range of $\boldsymbol{\theta}$, giving $\mathbb{P}_\theta = \mathbb{P}_1 \times \cdots \times \mathbb{P}_n$ as a product measure, where $\mathbb{P}_\ell = (\theta_\ell)_*\mathbb{P}$ is the distribution measure of the RV θ_ℓ, as the RVs θ_ℓ are independent. All computations following this are performed on $\boldsymbol{\Theta}$, typically some subset of $\mathbb{R}^n$. Hence n is the dimension of our problem, and if n is large, one faces a high-dimensional problem.

The filter Eq. (15) then reads (see [27] for more details)

$$\boldsymbol{q}_a(\boldsymbol{\theta}) = \boldsymbol{q}_f(\boldsymbol{\theta}) + \boldsymbol{C}_{x_f z}\boldsymbol{C}_z^{-1}(\check{y} - z(\boldsymbol{\theta})) = \boldsymbol{q}_f(\boldsymbol{\theta}) + \boldsymbol{K}_{\boldsymbol{q}}(\check{y} - z(\boldsymbol{\theta})). \tag{17}$$

this has been termed—especially if the approximating functions are polynomials—as a *spectral Kalman filter* (SPKF). Inserting the functional approximations into Eq. (17), one obtains an explicit and easy to evaluate expression for the assimilated or updated variable in terms of the input.

3 The Coarse-Scale Model

The coarse scale model is a continuum model, and here we take the simplest case and assume the displacements and strains to be infinitesimal, and the material properties to be spatially constant. This is the concrete example of Eq. (1) in Sect. 2.1.1. The material is assumed to be of the standard generalised type [12, 13, 30]. The behaviour of such materials is completely charcterised by two functions: the stored resp. Helmholtz free energy density $\psi_c(\varepsilon, \boldsymbol{w}, \boldsymbol{q})$ for the reversible behaviour, and the dissipation pseudo-potential density $\varphi_c(\varepsilon, \dot{\varepsilon}, \boldsymbol{w}, \dot{\boldsymbol{w}}, \boldsymbol{q})$ for the irreversible behaviour, and the assumption of maximal dissipation. Here ε is strain, $\boldsymbol{w}$ is a collection of internal phenomenological variables (the memory of the material), and $\boldsymbol{q}$ is a collection of parameters specifing the detailled character of the functions ψ_c and φ_c.

For the identification the coarse scale model will occupy exactly the domain of a standard quadrilateral element, and the loading will consist of imposed displacements on the boundary.

We consider the pressure sensitive materials such as concrete and rocks described in a simplified manner using the Drucker-Prager yield criterion. Here, an associated rate-independent model with linear hardening is considered for which the Helmholz free energy and the yield function are defined as

$$\psi_c = \frac{1}{2}(\varepsilon - \varepsilon_p) : C : (\varepsilon - \varepsilon_p) + \frac{1}{2}K_p \nu^2, \tag{18}$$

$$f(\sigma, \chi) = \sqrt{\mathrm{dev}(\sigma) : \mathrm{dev}(\sigma)} - \frac{1}{3}\mathrm{tr}(\sigma)\tan(\alpha) - \sqrt{\frac{2}{3}}(\sigma_y - \chi), \tag{19}$$

respectively. In Eq. (18) K_p denotes the isotropic hardening and C is the elastic constitutive tensor assumed to be isotropic and homogeneous, and hence described by bulk and shear moduli κ and G, respectively. Similarly, in Eq. (19) σ_y denotes the yield stress, α is the friction angle which relates cohesion c and yield stress as $c = \tan(\alpha)\sigma_y$, $\sigma := C : (\varepsilon - \varepsilon_p)$ is the stress, and $\chi := K_p\nu$ is the backstress. The variable ν is the accumulated hardening, and ε_p is the total plastic strain; both are internal phenomenological variables and are components of $\boldsymbol{w}$.

The model as just described is deterministic, and is to be extended into its probabilistic counterpart as the material properties are unknown and are to be determined from the energy measurements. By collecting all unknown material parameters, i.e. the variables in the elastic constitutive tensor namely $\{\kappa, G\}$ and inelastic ones in the yield function:$\{\sigma_y, K_p, c\}$ into the vector

$$\boldsymbol{q} = [\log\kappa, \log G, \log\sigma_y, \log K_p, \log c]^T,$$

one may describe $\boldsymbol{q}$ a priori in a Bayesian manner via a corresponding probability distribution, e.g. chosen via the maximum entropy principle. As all the parameters are positive, we actually estimate their logarithms (which are unconstrained, at the same time producing the proper metric for these positive parameters), and hence take as prior distributions for these indepedent normally distributed random variables. This gives the original parameters a log-normal distribution as prior.

In the Bayesian identification the uncertain material parameters are random variables, and hence the coarse scale material is a *stochastic* one. This means that the above mentioned stored energy density ψ_c and the dissipation density φ_c become random variables, and additionally to their spatial dependence on the location x as densities, they are formally functions of the variable $\omega \in \Omega$, i.e. elementary probability events as described in Sect. 2.1.2. The computation of the coarse scale is hence a so-called stochastic problem which has similar form to the deterministic one [18, 26]. Computationally, the striking difference upon full discretisation lies in the problem dimension. Namely, the state variable lives in a space obtained as the tensor product of the corresponding deterministic space and the space of random variables $\mathcal{S}$. Thus, the stochastic problem requires a temporal, spatial, and stochastic discretisation, largely increasing the problem dimension and thus the computational effort. For a full discussion of such computations, see [35].

The probabilistic discretisation is done in a functional approximation setting as already described in Sect. 2.1.4, i.e. as ansatz are taken the polynomial chaos expansions for all describing variables, like ε, $\boldsymbol{w}$, and $\boldsymbol{q}$, as in Eq. (16). Additionally, the spatial dependence is modelled in a standard manner by finite element shape functions. Further detailed information on the mathematical description and on the solution of the whole problem may be found in [35]. Here the coefficients in the expressions like Eq. (16) were found by pseudo-spectral projection resp. regression.

The measurement for the coarse scale model as described in Eq. (3) in Sect.. 2.1 takes now the concrete form

$$Y_c(\omega) = \left[\ldots, \int \psi_c(x, \omega, \varepsilon, \boldsymbol{w}, \boldsymbol{q}) N_m(x)\, \mathrm{d}x, \int \varphi_c(x, \omega, \varepsilon, \dot{\varepsilon}, \boldsymbol{w}, \dot{\boldsymbol{w}}, \boldsymbol{q}) N_m(x)\, \mathrm{d}x\right], \tag{20}$$

where $\{N_m\}_m$ are the coarse scale element shape functions. Hence the measurements Eq. (20) are random variables describing certain spatial averages of the stored and dissipated energies in the domain—one quad element—of the coarse scale model.

4 The Fine-Scale Model

As the main purpose here is to show that and how the upscaling procedure works, the fine scale or "micro-scale model" is here taken to be the fine-discretised version of the coarse-scale continuum model of Sect. 3. But it will become obvious that any model which allows a "measurement" resp. computation of stored and dissipated energies could be used.

The notion of the standard generalised material is kept, only the same computational domain as for the coarse-scale model, which was there discretised by *only one* quad element, is here discretised by a high number (50 × 50 block) of finite elements. The description of the material parameters is not assumed to be constant as in the coarse-scale model in Sect. 3, but are taken to correspond to a realisation of random fields.

This means that the material description is similar as in Sect. 3, but the parameters which occur there in the vector $\boldsymbol{q}$ are spatially varying realisations of random fields. Also, here they are not random variables, but fixed though spatially varying; albeit unknown to the coarse-scale model in the identification procedure. They do not need to be modelled as variables. Hence the fine-scale stored resp. Helmholtz free energy density is given by $\psi_f(x, \varepsilon_f(x), \boldsymbol{w}_f(x))$, where ε_f and $\boldsymbol{w}_f$ are the corresponding fine-scale fields. Similar considerations hold for the fine scale dissipation pseudo-potential density $\varphi_f(x, \varepsilon_f(x), \dot{\varepsilon}_f(x), \boldsymbol{w}_f(x), \dot{\boldsymbol{w}}_f(x))$. This is the concrete realisation of the abstract model Eq. (2) in Sect. 2.1.1.

The fine-scale measurements—the concrete realisation of Eq. (4) in Sect.. 2.1—are hence given by

$$Y_f = \left[\ldots, \int \psi_f(x, \varepsilon_f, \boldsymbol{w}_f) N_m(x)\, \mathrm{d}x, \int \varphi_f(x, \varepsilon_f, \dot{\varepsilon}_f, \boldsymbol{w}_f, \dot{\boldsymbol{w}}_f) N_m(x)\, \mathrm{d}x \right], \tag{21}$$

where again the $\{N_m(x)\}_m$ are the *coarse-scale* shape functions.

5 Numerical Results

To demonstrate the proposed strategy, we consider the identification of the homogeneous material parameters of the coarse-scale Drucker-Prager model given energy measurements on the fine scale. The spatial domain is a two-dimensional square block of unit length consisting of one and 50×50 finite elements on the coarse and fine levels, respectively. Both the scales are modelled by the identical continuum model.

Furthermore we consider strain controlled experiments yielding different deformed configurations namely:

- case I: bi-axial compression

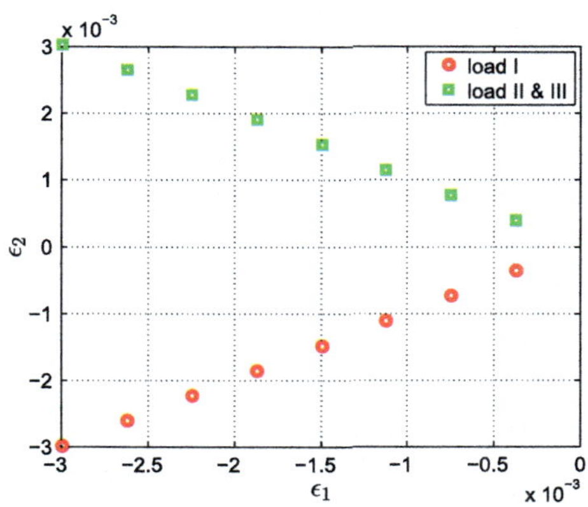

Fig. 1 Loading paths in principal strain space

Table 1 Deformation matrix for different loading cases

$\boldsymbol{H}_I$	$\boldsymbol{H}_{II}$	$\boldsymbol{H}_{III}$
$\begin{bmatrix} -0.003 & 0 \\ 0 & -0.003 \end{bmatrix}$	$\begin{bmatrix} 0.0 & 0.003 \\ 0.003 & 0.0 \end{bmatrix}$	$\begin{bmatrix} 0.003 & 0 \\ 0 & -0.003 \end{bmatrix}$

- case II: bi-axial compression-tension
- case III: pure shear

The corresponding boundary displacements are enforced by specifying the respective displacement gradient

$$\boldsymbol{u}_b = \boldsymbol{H}\boldsymbol{x}_b \tag{22}$$

where $\mathbf{u}_b$ and $\mathbf{x}_b$ are the boundary displacements and nodal coordinates respectively. The deformation gradient $\boldsymbol{H}$ defines the given loading case, with values specified in Table 1 and the loading paths depicted in Fig. 1. As we are dealing with different experiments, the update of the coarse-scale model parameters is performed in a sequential way such that the energy measurements from the first experiment are used to obtain the intermediate posterior which further serves as a prior for the second experiment.

5.1 The Homogeneous Case

This trivial case characterised by homogeneous Drucker-Prager material model on both scales serves as an example for the verification of the proposed strategy. Both scales are modelled using one element. The measurements used for upscaling are incremental stored energy and dissipation. The material properties on fine-scale are given in Table. 2.

Furthermore, we assume that only the value of the measurement is known and that the true parameter values are unavailable. Hence, as prior on the coarse-scale we adopt the lognormal distributions characterised by the second order statistics, see Table. 3.

Note that the prior mean and standard deviation correspond to 10% offset from fine truth and 5% coefficient of variation.

We first consider the update of the elastic parameters $\{\kappa, G\}$ using the first elastic step from loading configuration I (bi-axial compression) and II (pure shear) in a sequential manner. In the first loading step, the coarse-scale κ is updated to the fine

Table 2 Fine-scale material parameters

κ	G	σ_y	K_p	c
204 GPa	92 GPa	170 MPa	100 MPa	283.33 MPa

truth shown in Fig. 2a, followed by the update of G in the second step Fig. 2b. This is expected behaviour, as the measured elastic energy is a consequence of the volumetric change (characterising κ) in the first step, and shape distortion (characterising G) in the second one, respectively.

For the update of the inelastic parameters $\{\sigma_y, K_p, c\}$ we use incremental plastic dissipation and stored hardening energy as measurements. The update is performed sequentially in four plastic increments, the first two being from loading case II (pure shear) and the remaining ones from case III (compression-tension) hence the type of measurements is kept same during the increments. As can be seen in Fig. 3a–c, with each new information the posterior distributions of all of three parameters are getting more narrow, i.e. more information is brought to the update process and we

Table 3 Coarse-scale prior statistics

Property	κ	G	σ_y	K_p	c
μ	224.89 GPa	101.2 GPa	187 MPa	110 MPa	311.67 MPa
σ	11.2 GPa	5.06 GPa	9.35 MPa	5.5 MPa	15.58 MPa

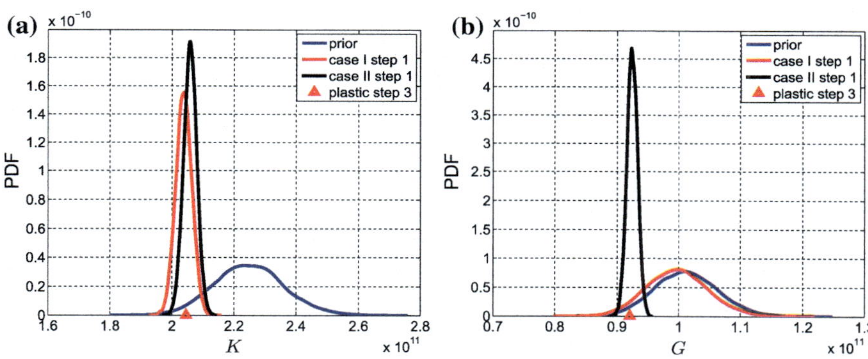

Fig. 2 Update of elastic material parameters $\{\kappa, G\}$ using elastic step of load case I (bi-axial compression) followed by elastic step of load case II (pure shear), shown in (**a**) and (**b**) respectively

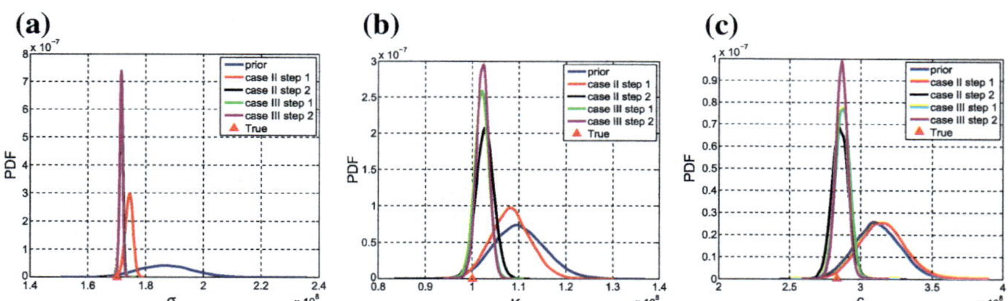

Fig. 3 Updates for plastic material parameters $\{\sigma_y, K_p, c\}$ for the first 2 plastic steps for loading II and the latter 2 with loading III shown in (**a**), (**b**) and (**c**) respectively

are less uncertain about the model parameter values. In addition, the cohesion c in Fig. 3c is characterised by a posterior mean which equals the true parameter values, whereas σy and K_p are slightly offset, see Fig. 3a–b.

5.2 The Heterogeneous Case

In order to simulate more realistic multiscale problem, here is considered an example in which the fine-scale is described by deterministic heterogeneous properties, i.e. a particular realisations of log-normal random fields described by Gaussian covariance functions.

We restrict our study on upscaling of plastic parameters, keeping elastic values as known and deterministic for the coarse-scale. The loading sequence is kept the same as in the homogeneous case. The fine-scale realisations of $\{\sigma_y, Kp, c\}$ are generated using values from Table 2 as mean and 5% coefficient of variation for varying correlation lengths $\ell_c = \{5\,\ell_e, 10\,\ell_e, 50\,\ell_e\}$, where $\ell_e = 1/50$ is the fine-scale element length (see Fig. 4a–c). The prior mean values for the coarse-scale inelastic parameters are taken 5% off from their fine-scale counterpart, with 5% as coefficient of variation.

The main goal here is to find the unknown equivalent coarse-scale homogeneous parameters for the given spatially heterogeneous material parameters on fine-scale. For each step of loading increment we obtain a new update for the coarse-scale values. For σ_y, shown in Fig. 5a–c, from the second increment onwards, the posterior gets significantly narrower and the shift in distribution is quite small, this becomes more prominent as one goes to maximum value of $\ell_c = 50\,\ell_e$ in Fig. 5c. This shows that after significant plastic deformation, σ_y becomes insensitive to the applied loading. For the cohesion c, referring to Fig. 5d–f, the posterior becomes narrower in the second increment, however there is no significant reduction with further updates, moreover the effect of increasing l_c is much less pronounced as compared to σ_y. In line with the cohesion c, the hardening modulus K_p (Fig. 5g–i) also exhibits a similar response, however with one striking difference: in the first update the shift in the distribution is quite significant for $\ell_c = \{5\,\ell_e, 10\,\ell_e\}$ cases in Fig. 5g–h. The

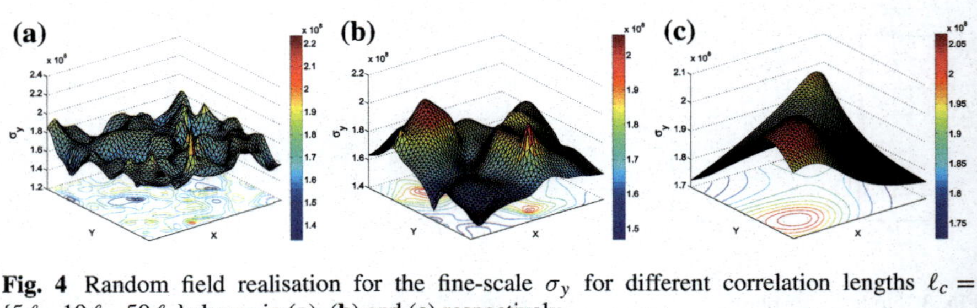

Fig. 4 Random field realisation for the fine-scale σ_y for different correlation lengths $\ell_c = \{5\,\ell_e, 10\,\ell_e, 50\,\ell_e\}$ shown in (**a**), (**b**) and (**c**) respectively

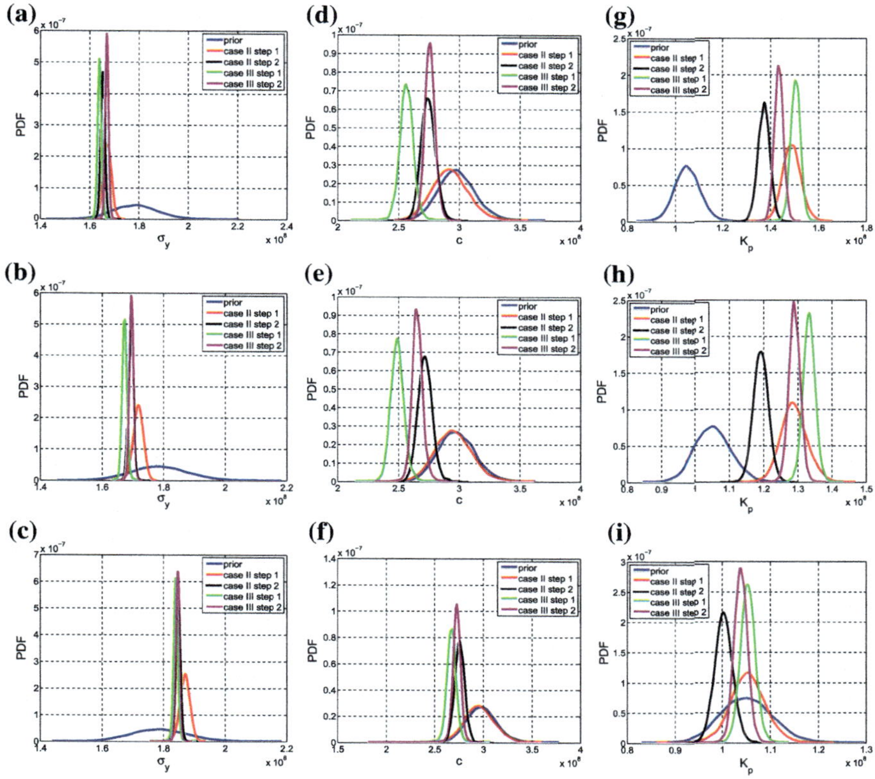

Fig. 5 Updates using first 2 plastic steps for load case II (pure shear) and the latter 2 with loading III (compression-tension) shown for yield stress σ_y in (**a**)–(**c**), cohesion c in (**d**)–(**f**) and hardening modulus K_p in (**g**)–(**i**) for $\ell_c = \{5\,\ell_e, 10\,\ell_e, 50\,\ell_e\}$ respectively

variability in the updates for coarse-scale K_p can be substantiated with the following argument: owing to the higher degree of spatial heterogeneities (small ℓ_c cases) at fine-scale, the imposed homogeneous strain field produces stress field which varies significantly from one element to the other.

Therefore on the coarse-scale, a linear isotropic hardening relation is apparently not sufficient to capture fine-scale heterogeneities, one may need a more elaborate hardening law (e.g. saturation hardening [17]) on coarse-scale. To conclude the discussion for the heterogeneous case, we show that the distributions of forward energies computed from the updated posterior on the coarse-scale narrow down on the fine-scale value. For the sake of brevity, we show two instances in Figs. 6a–b and 7a–b for the updated incremental hardening energy and dissipation as evidence.

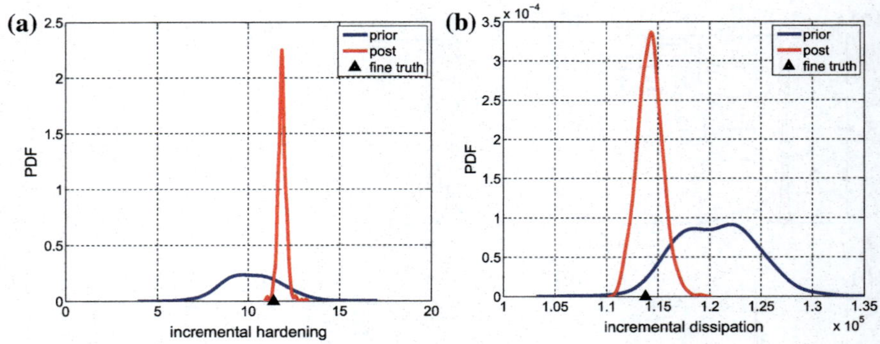

Fig. 6 Incremental hardening energy (**a**) and dissipation (**b**) for the load case II (pure shear) second increment and $\ell_c = 10\,\ell_e$ realisation

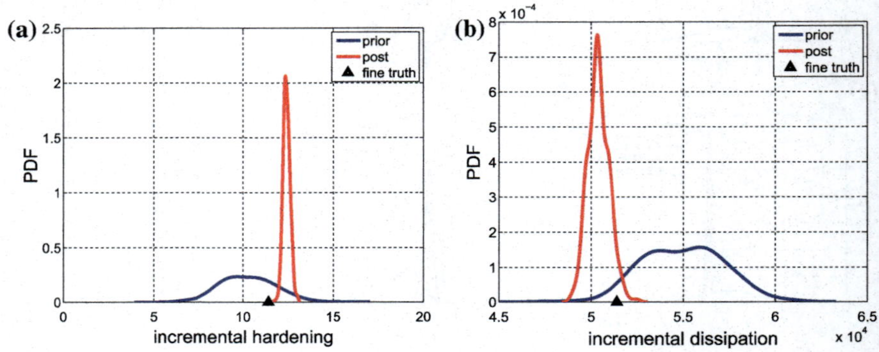

Fig. 7 Incremental hardening (**a**) and dissipation (**b**) for the load case III (compression tension) second increment and $\ell_c = 10\,\ell_e$ realisation

6 Conclusion

In this paper we have shown a novel stochastic strategy to upscale fine-scale information onto a coarse scale. The lack of knowledge about the coarse-scale renders it stochastic, which is then updated in a Bayesian framework. The output quantities of interest (QoI) are updated random variables which give us not just a mean (deterministic value), but a distribution over it, allowing us to estimate the inherent uncertainty. The proposed approach is shown to work for non-linear material models as demonstrated in illustrative numerical examples. The idea of using energy as measurement to update coarse-scale models provides a promising outlook for using more complicated fine-scale models. Moreover the method takes advantage of existing deterministic codes for the mechanical FE-computations and fast sampling-free Bayesian filters [32, 34, 36] to achieve its objective in a computationally efficient way.

Acknowledgements This paper is based upon work supported by DFG (Deutsche Forschungsgemeinschaft), Germany, and ANR (Agence Nationale de la Recherche), France, under the grant of the SELF-TUM project.

References

1. Arsigny, V., Fillard, P., Pennec, X., Ayache, N.: Geometric means in a novel vector space structure on symmetric positive-definite matrices. SIAM J. Matrix Anal. Appl. **29**(1), 328–347 (2006)
2. Asokan, B.V., Zabaras, N.: A stochastic variational multiscale method for diffusion in heterogeneous random media. J. Comput. Phys. **218**, 654–676 (2006). https://doi.org/10.1016/j.jcp.2006.02.026
3. Bobrowski, A.: Functional Analysis for Probability and Stochastic Processes. Cambridge University Press, Cambridge (2005)
4. Clément, A., Soize, C., Yvonnet, J.: Uncertainty quantification in computational stochastic multiscale analysis of nonlinear elastic materials. Comput. Methods Appl. Mech. Eng. **254**, 61–82 (2013). https://doi.org/10.1016/j.cma.2012.10.016
5. Das, S., Ghanem, R.: Stochastic upscaling for inelastic material behavior from limited experimental data. In: Ghosh, S., Dimiduk, D. (eds.) Computational Methods for Microstructure-Property Relationships, pp. 443–468. Springer, Berlin (2011). https://doi.org/10.1007/978-1-4419-0643-4_12
6. Demmie, P.N., Ostoja-Starzewski, M.: Local and nonlocal material models spatial randomness and impact loading. Arch. Appl. Mech. (2015). https://doi.org/10.1007/s00419-015-1095-3
7. Evensen, G.: Data Assimilation–the Ensemble Kalman Filter. Springer, Berlin (2009)
8. Gelman, A., Carlin, J.B., Stern, H.S., Rubin, D.B.: Bayesian Data Analysis. Taylor, Francis (2014)
9. Gorguluarslan, R., Choi, S.-K.: A simulation-based upscaling technique for multiscale modeling of engineering systems under uncertainty. Int. J. Multiscale Comput. Eng. **12**(2014), 549–566 (2014). https://doi.org/10.1615/IntJMultCompEng.011519
10. Graham-Brady, L., Arwade, S., Corr, D., Gutiérrez, M., Breysse, D., Grigoriu, M., Zabaras, N.: Probability and materials: from nano- to macro-scale: a summary. Probab. Eng. Mech. **21**, 193–199 (2006). https://doi.org/10.1016/j.probengmech.2005.10.005
11. Greene, M.S., Liu, Y., Chen, W., Liu, W.K.: Computational uncertainty analysis in multiresolution materials via stochastic constitutive theory. Comput. Methods Appl. Mech. Eng. **200**(1–4), 309–325 (2011). https://doi.org/10.1016/j.cma.2010.08.013
12. Halphen, B., Nguyen, Q.S.: Sur les matériaux standard généralisés. J. de Méc.**14**:39–63 (1975) Available from: http://hal.archives-ouvertes.fr/hal-00105514
13. Halphen, B., Nguyen, Q.S.: Plastic and visco-plastic materials with generalized potential. Mech. Res. Commun. **1**(1), 43–47 (1974). https://doi.org/10.1016/0093-6413(74)90034-2
14. Han, W., Reddy, B.D.: Plasticity mathematical theory and numerical analysis, 2nd edn. In: Interdisciplanary Applied Mathematics, Springer Verlag, New York (2013)
15. Hawkins-Daarud, A., Prudhomme, S., van der Zee, K.G., Oden, J.T.: Bayesian calibration validation and uncertainty quantification of diffuse interface models of tumor growth. J. math. biol. **67**(6–7), 1457–1485 (2013)
16. Ibrahimbegović, A., Fadi, G., Lotfi, C.: Classical plasticity and viscoplasticity models reformulated: theoretical basis and numerical implementation. Int. J. Numer Methods Eng. **42**(8), 1499–1535 (1998)
17. Ibrahimbegović, A., Markovič, D., Gatuingt, F.: Constitutive model of coupled damage-plasticity and its finite element implementation. Revue Eur. des Elém. **12**(4), 381–405 (2003)
18. Ibrahimbegovic, A., Matthies, H.G.: Probabilistic multiscale analysis of inelastic localized failure in solid mechanics. Comput. Assist Methods Eng Sci. **19**:277–304 (2012). Available from: http://cames.ippt.gov.pl/pdf/CAMES_19_3_5.pdf

19. Ibrahimbegović, A.: Nonlinear Solid Mechanics of Solid Mechanics and its Applications. vol. 160, Springer (2009)
20. Kaipio, J., Somersalo, E.: Statistical and Computational Inverse Problems (applied mathematical sciences) Vol. 160, 1st (ed.) Springer, Berlin (2004)
21. Kennedy, M.C., O'Hagan, A.: Bayesian calibration of computer models. J. Royal Statist. Series B **63**(3), 425–464 (2001)
22. Koutsourelakis, P.: Stochastic upscaling in solid mechanics: an excercise in machine learning. J. Comput. Phys. **226**, 301–325 (2007). https://doi.org/10.1016/j.jcp.2007.04.01
23. Liu, Y., Greene, S.M., Chen, W., Dikin, D.A., Liu, W. K.: Computational microstructure characterization and reconstruction for stochastic multiscale material design. Comput. Aided Des. **45**(1), 65–76 (2013). https://doi.org/10.1016/j.cad.2012.03.007
24. Luenberger, D.G.: Optimization by Vector Space Methods. Wiley, Chichester (1969)
25. Maso, D.G., de Simone, A., Mora, M.G.: Quasistatic evolution problems for linearly elastic? perfectly plastic materials. Arch. Rational Mech. Anal. **180**, 237–291 (2006)
26. Matthies, H.G., Ibrahimbegović, A.: Stochastic multiscale coupling of inelastic processes in solid mechanic. In: Papadrakakis, M., Stefanou, G. (eds.) Multiscale Modelling and Uncertainty Quantification of Materials and Structures, vol. 3, pp. 135–157. Springer, Berlin (2014). https://doi.org/10.1007/978-3-319-06331-7_9
27. Matthies, H.G., Zander, E., Rosić, B.V., Litvinenko, A., Pajonk, O.: Inverse problems in a bayesian setting. In: Ibrahimbegović, A. (ed.) Computational Methods for Solids and Fluids—Multiscale Analysis Probability Aspects and Model Reduction Computational Methods in Applied Sciences, vol. 3, pp. 245–286. Springer, Berlin (2016). https://doi.org/10.1007/978-3-319-27996-1
28. Matthies, H.G.: Computation of constitutive response in nonlinear computational mechanics. In: Wriggers, P., Wagner, W. (eds.) State of the Art. Springer, Berlin, Heidelberg (1991)
29. Matthies, H.G.: Uncertainty quantification with stochastic finite elements. In: Stein, E., de Borst, R., Hughes, T.J.R. (eds.) Encyclopaedia of Computational Mechanics. Wiley, Chichester (2007). https://doi.org/10.1002/0470091355.ecm071
30. Nguyen, Q.S.: On the elastic plastic initial-boundary value problem and its numerical integration. Int. J. Numer. Methods Eng. **11**(5), 817–832 (1977). https://doi.org/10.1002/nme.1620110505
31. Ostoja-Starzewski, M.: Microstructural Randomness and Scaling in Mechanics of Materials. Chapman and Hall, Boca Raton, FL (2008)
32. Pajonk, O., Rosić, B.V., Litvinenko, A., Matthies, H.G.: A deterministic filter for non-Gaussian Bayesian estimation-applications to dynamical system estimation with noisy measurements. Phys. D **241**, 775–788 (2012). https://doi.org/10.1016/j.physd.2012.01.001
33. Papoulis, A.: Probability Random Variables and Stochastic Processes, 3rd edn. McGraw-Hill Series in Electrical Engineering, McGraw-Hill, New York (1991)
34. Rosić, B., Sýkora, J., Pajonk, O., Kučerová, A., Matthies, H.G.: Comparison of numerical approaches to Bayesian updating computational methods for solids and fluids—multiscale analysis, probability aspects, and model reduction. In: Ibrahimbegović, A. (ed.) Computational Methods in Applied Sciences, vol. 41, pp. 427–461. Springer, Berlin (2016). https://doi.org/10.1007/978-3-319-27996-1
35. Rosić, B., Matthies, H.G.: Variational theory and computations in stochastic plasticity. Archi. Comput. Methods Eng. **22**(3), 457–509 (2014)
36. Rosić, B.V., Litvinenko, A., Pajonk, O., Matthies, H.G.: Sampling-free linear Bayesian update of polynomial chaos representations. J. Comput. Phys. **231**, 5761–5787 (2012). https://doi.org/10.1016/j.jcp.2012.04.044
37. Stefanou, G., Savvas, D., Papadrakakis, M.: Stochastic finite element analysis of composite structures based on material microstructure. Compos. Struct. **132**, 384–392 (2015). https://doi.org/10.1016/j.compstruct.2015.05.044
38. Suquet, P., Lahellec, N.: Elasto-plasticity of heterogeneous materials at different scale. Proced. IUTAM **10**, 247–262 (2014). https://doi.org/10.1016/j.piutam.2014.01.021

39. Tarantola, A.: Inverse Problem Theory and Methods for Model Parameter Estimation. SIAM, Philadelphia, PA (2005)
40. Yvonnet, J., Bonnet, G.: A consistent nonlocal scheme based on filters for the homogenization of heterogeneous linear materials with non-separated scales. Int. J. Solids Struct. **51**(1), 196–209 (2014). https://doi.org/10.1016/j.ijsolstr.2013.09.023

A Model Reduction Technique in Space and Time for Fatigue Simulation

Mainak Bhattacharyya, Amélie Fau, Udo Nackenhorst, David Néron and Pierre Ladevèze

Abstract The simulation of mechanical responses of structures subjected to cyclic loadings for a large number of cycles remains a challenge. The goal herein is to develop an innovative computational scheme for fatigue computations involving non-linear mechanical behaviour of materials, described by internal variables. The focus is on the Large Time Increment (LATIN) method coupled with a model reduction technique, the Proper Generalized Decomposition (PGD). Moreover, a multi-time scale approach is proposed for the simulation of fatigue involving large number of cycles. The quantities of interest are calculated only at particular pre-defined cycles called the "nodal cycles" and a suitable interpolation is used to estimate their evolution at the intermediate cycles. The proposed framework is exemplified for a structure subjected to cyclic loading, where damage is considered to be isotropic and micro-defect closure effects are taken into account. The combination of these techniques reduce the numerical cost drastically and allows to create virtual S-N curves for large number of cycles.

M. Bhattacharyya (✉) · A. Fau · U. Nackenhorst
IBNM, Leibniz Universität Hannover, Appelstraße 9a, 30167 Hannover, Germany
e-mail: mainak.bhattacharyya@ibnm.uni-hannover.de

A. Fau
e-mail: amelie.fau@ibnm.uni-hannover.de

U. Nackenhorst
e-mail: nackenhorst@ibnm.uni-hannover.de

D. Néron · P. Ladevèze
LMT, ENS Paris-Saclay, CNRS, Université Paris-Saclay,
61 avenue du Président Wilson, 94235 Cachan, France
e-mail: neron@lmt.ens-cachan.fr

P. Ladevèze
e-mail: ladeveze@lmt.ens-cachan.fr

J. Sorić et al. (eds.), *Multiscale Modeling of Heterogeneous Structures*,
Lecture Notes in Applied and Computational Mechanics 86,
https://doi.org/10.1007/978-3-319-65463-8_10

1 Introduction

The phenomenon of fatigue has been of great importance in the design of mechanical components and civil structures. Fatigue in the most generic case can be defined as the change in properties of a structure subjected to repetitive loading. The focus here is on mechanical fatigue involving fluctuations in externally applied load with respect to time. The nature of the loading however can be completely periodic, non-periodic, or random. Whatever the nature of fatigue maybe, the failure of the structure is composed of three stages [1]:

1. nucleation and growth of micro-voids into a macrocrack,
2. stable propagation of the macro-crack,
3. unstable crack propagation that leads to complete failure.

There are three main engineering approaches to ensure the safety of a structure [1]. The "safe-life" approach demands that the structure remains safe under a prescribed load for a certain number of cycles. The "fail-safe" approach requires that a structure is able to withstand damage without a catastrophic failure of the whole structure. The "damage tolerance" approach investigates the ability of a structure to survive with damage before reparation. Whatever the engineering goal, researchers have to represent the complex phenomena of cyclic fatigue or random fatigue with pertinent and flexible models.

1.1 Different Modelling Approaches for Fatigue Analysis

Historical approaches for fatigue analysis are based on the exploitation of some empirical curves introduced by Wöhler [2]. The stress range is plotted against the experimentally observed number of surviving cycles giving rise to S-N curves. This may be a straightforward tool for "safe-life" analysis if a perfect periodic loading is considered with the same mean stress as the one of the S-N curve. From the asymptotic behaviour of the curve, an endurance limit, below which the material will never fail, may be defined. For loading inducing appreciable plastic deformation e.g. low cycle fatigue, it is better to introduce a strain-life approach as proposed by Coffin [3] and Manson [4] for metallic materials. Similar to S-N curves that are used for stress-life approaches, ε_a-N or $\Delta\varepsilon$-N curves are generally used for strain-life approaches [5], where ε_a and $\Delta\varepsilon$ are the total strain amplitude and total strain range respectively. The empirical method may be extended to different sophisticated cases e.g.:

- taking into account the mean stress effect on fatigue as proposed by Gerber [6], Goodman [7] and Soderberg [8];
- characterising the fatigue life of a structure when subjected to blocks of cyclic stresses of different amplitudes by using a cumulative damage rule coupled with S-N curve. The accumulation damage rule may be linear as the Palmgren-Miner damage rule [9, 10], or non-linear accumulation rule as proposed by Marco and Starkey [11];

- for notched members using the Neuber rule [12], which takes into account the stress concentration factor and the fatigue notch factor;
- for multiaxial stresses, an effective stress based for example on von Mises criterion, Crossland criterion [13], Sines criterion [14], or Dang Van criterion [15] is used together with the Basquin relation. An effective strain or effective plastic strain may either be estimated according to the maximum shear strain theory or according to the distortional energy theory;
- for random loading, by cycle counting approach the random load history is representing as a discrete number of equivalent cycles, for example using rainflow counting.

The aforementioned methods, are phenomenological in nature and deal with the usage of empirical relations that are based on experimental findings. For example, they do not offer the flexibility to investigate the order in which different levels of load are applied to the structure. To overcome this limitation, the changes in a given structure can be described by internal variables using continuum mechanics approaches. Damage, defined as an internal variable, is used to quantify the initiation phase of a macro-crack that reduces the load carrying capacity of the material [5]. Continuum damage mechanics to predict fatigue life was introduced by Lemaitre and Lesne [16] where a non-linear continuous fatigue damage model was used to describe the different phases of the deterioration process. Various modifications and developments have been done over the years for the modelling of fatigue damage to incorporate the physical phenomena as far as possible. For instance, when the load is large, the structure undergoes appreciable plastic deformation, leading to a fatigue life less than 10^5 cycles, which is referred as low-cycle or oligocyclic fatigue. On the contrary, under high-cycle fatigue (HCF), the loading is much less than the yield stress. Therefore, no macro-plastic deformation is involved. The structure may handle a very large number of cycles. The two-scale damage model [17] is an important development for modelling HCF, and represents the macro-elastic behavior while the damage is evaluated only at the micro-scale. Different micromechanical models may be proposed to represent the micro-void growth [18].

To investigate macro-crack propagation, the premise of fracture mechanics is used knowing the pre-existence of a crack within the material. The most traditional law to describe the crack growth is the Paris-Erdogan law [19]. More information on classical techniques and recent developments on fatigue modelling and simulations can be found in the review articles [20, 21].

Here, a continuum damage approach is used for modelling the fatigue behaviour. This allows to consider the chronology of the different cycles or inertial effects due to high-frequency cycles. However, this approach may lead to very expensive computational cost. Developments have been done in the light of new and robust numerical techniques that reduce CPU cost. For example, this problem may be overcome by using model order reduction techniques.

1.2 Model Reduction Technique for Fatigue Computation

Model order reduction is a family of numerical strategies which has shown efficiency for many large size mechanical problems such as parametric studies or real-time computations [22]. The solution is approximated by solving the Galerkin problem within a reduced-order basis, whose dimension is much smaller than the size of the original high-dimensional model. Proper Orthogonal Decomposition (POD) as proposed e.g. in [23] is based on a training stage. From the solution of the full-order problem at some particular time instants and/or parameter values arbitrarily chosen, a reduced-order basis is built as the truncation of a singular value decomposition [24]. In Reduced-Basis approach a greedy algorithm is employed to define the most relevant calculations within the parametric space to optimise the enrichment of the reduced-order basis [25]. Using Proper Generalised Decomposition (PGD), the problem is also solved in a relevant reduced-order basis, but the basis is defined on-the-fly by a greedy algorithm [26–28].

As computations for continuum damage problems may face strain localisation, numerical response is highly sensible to any modification of the model. Therefore damage computational strategies using model order reduction are challenging and may be hazardous. Controlling the accuracy of POD computations with circumspection is recommended and adaptive schemes such as A Priori Hyper Reduction Method [29] or POD coupled with Newton-Krylov algorithms [30] are preferred. Delamination has been modelled by cohesive zone and PGD by Metoui et al. [31], PGD-based multi-scale computations for rate-dependent damage model have been recently proposed by El Halabi et al. [32].

The LATIN method [26] in its classical sense is employed such that an approximation of the solution on the whole time-space domain is defined at every iteration, and the gobal equilibrium is tackled as a linearised problem and the non-linear material behaviours are considered separately. Therefore, this approach offers a convenient framework to include model-order reduction techniques even for non-linear computations. It has been developed for solving plasticity and visco-plasticity problems and shown a drastic decrease of the computational cost compared to a classical approach [33–35]. Here an extended version is used for (visco-)plastic problems with damage that incorporates micro-defects closure effects [36].

For fatigue computation applied to a large number of cycles, advanced numerical strategies may be required to work around the computational cost due to the large time domain.

1.3 Efficient Time Schemes for Fatigue Computations

For high or very high cycle fatigue or for combined cycle fatigue in which the load is a combination of large amplitude low frequency and small amplitude high frequency loads [1], the computational cost of the time integration scheme may be extremely high. The computational cost may be decreased by benefiting from the cyclic loading to not compute explicitly the full number of cycles.

The jump cycle procedure is a very robust technique [5] which works around full blocks of cycles. After computing in details a set of cycles, they are used to establish a trend line and extrapolate the quantities of interest for further evolution. Then, the extrapolated state is used as initial condition for a further computation such that the whole time domain is spanned. A control function has been proposed such that the lengths of cycle jumps are monitored to ensure an accurate approximation [37]. This method is most suitable for quasi-linear systems. However the control function allows to also consider non-linear behaviour and cycles jumps are automatically shortened or even cancelled [38, 39].

For combined cycle fatigue, time homogenisation techniques are based on a hypothesis of time scale separation between a large time scale associated with the low-frequency load and a small time scale associated with the high-frequency load. The ratio between both of them is assumed to be small enough such that the two scales can be considered as independent, and the behaviour due to the high frequency is homogenised for a time discretisation scheme consistently defined for the low-frequency load. Developed initially for plasticity and quasi-static computations [40], this method has been extended to damage and dynamic effects [41].

In the framework of LATIN method, as the whole time domain is available at each iteration, a two-time scale in a "finite element like" discretisation of the time intervals has been proposed for visco-plasticity problems [26, 42]. Some cycles of interest, called "time nodes" or "nodal cycles", are computed in the classical time scheme. The evolution of the quantities of interest in the time slot between two time nodes is interpolated using shape functions as for space finite-element method.

As empirical methods are based on expensive and cumbersome experiment, the goal of this project is to provide virtual experiments based on continuum mechanics approach. To overcome the computational cost due to these models, a sophisticated model order reduction scheme is proposed which reduces drastically the cost and allows to compute virtual test even for a very large number of cycles. The proposed numerical framework is flexible concerning the damage model, the global framework and equations considered are summarised in Sect. 2. Then, the model order reduction based on the separation of variables is introduced in Sect. 3. Finally, the numerical treatment of the time problem, which allows to reduce the numerical cost, is detailed in Sect. 4.

2 Continuum Damage Mechanics Approach

The focus herein is on the methods that deal in quantifying the change in material properties due to fatigue loading using internal damage variable in a continuum mechanics framework.

A continuous reference structure for any quasi-static analysis can be considered in a spatial domain Ω. For simplistic case, the evolution of the state of the structure can be considered to be isothermal within time domain $[0, T]$. Such a reference structure in general is subjected to prescribed body forces $\underline{f}_d$, to traction forces $\underline{F}_d$ over a part $\partial_2\Omega$ of the boundary $\partial\Omega$, and to prescribed displacements $\underline{u}_d$ over the complementary part $\partial_1\Omega$. Thereby the compatibility of the applied forces with the internal stress is given by the equilibrium equation which in a weak form becomes the static admissibility condition, which is defined such that $\forall \underline{u}^* \in \mathcal{U}^*$

$$\int_{[0,T]\times\Omega} \boldsymbol{\sigma} : \boldsymbol{\varepsilon}\left(\underline{u}^*\right)\, \mathrm{d}\Omega\, \mathrm{d}t = \int_{[0,T]\times\Omega} \underline{f}_d \cdot \underline{u}^*\, \mathrm{d}\Omega\, \mathrm{d}t + \int_{[0,T]\times\partial_2\Omega} \underline{F}_d \cdot \underline{u}^*\, \mathrm{d}S\, \mathrm{d}t \tag{1}$$

with the stress tensor $\boldsymbol{\sigma}$ being statically admissible, and $\mathcal{U}^*$ being the homogeneous vector space associated to the space $\mathcal{U}$ of kinematically admissible field $\underline{u}$. Also, the compatibility of the prescribed displacement with the generated strain within the structure is given by the strain-displacement relationship which in weak form becomes the kinematic admissibility condition, which is defined such that $\forall \boldsymbol{\sigma}^* \in \mathcal{F}^*$

$$\int_{[0,T]\times\Omega} \boldsymbol{\sigma}^* : \boldsymbol{\varepsilon}\, \mathrm{d}\Omega\, \mathrm{d}t = \int_{[0,T]\times\partial_1\Omega} \boldsymbol{\sigma}^*\underline{n} \cdot \underline{u}_d\, \mathrm{d}S\, \mathrm{d}t \tag{2}$$

with $\boldsymbol{\varepsilon}$ being the total strain tensor which is kinematically admissible and can be split into an elastic part $\boldsymbol{\varepsilon}^e$ and a plastic part $\boldsymbol{\varepsilon}^p$ additively. $\mathcal{F}^*$ is the homogeneous vector space associated to the space $\mathcal{F}$ of statically admissible field $\boldsymbol{\sigma}$.

The mechanical properties of the materials are described by a set of constitutive relations. The equations of state for elasto-(visco)plastic materials subjected to unilateral damage are obtained from a free energy function, and are given as [5]

$$\varepsilon^e_{ij} = \frac{1+\nu}{E}\left[\frac{\langle\sigma\rangle^+_{ij}}{1-D} + \frac{\langle\sigma\rangle^-_{ij}}{1-hD}\right] - \frac{\nu}{E}\left[\frac{\langle\sigma_{kk}\rangle}{1-D} + \frac{\langle-\sigma_{kk}\rangle}{1-hD}\right]\delta_{ij}, \tag{3a}$$

$$\beta_{ij} = C\alpha_{ij}, \tag{3b}$$

$$R = g'(r), \tag{3c}$$

$$Y = \frac{1+\nu}{2E}\left[\frac{\langle\sigma\rangle^+_{ij}\langle\sigma\rangle^+_{ij}}{(1-D)^2} + h\frac{\langle\sigma\rangle^-_{ij}\langle\sigma\rangle^-_{ij}}{(1-hD)^2}\right] - \frac{\nu}{2E}\left[\frac{\langle\sigma_{kk}\rangle^2}{(1-D)^2} + h\frac{\langle-\sigma_{kk}\rangle^2}{(1-hD)^2}\right]. \tag{3d}$$

Here, Eq. (3a) represents the elastic state law, which because of damage is nonlinear in nature, as it is not described by a linear operator, with E and ν being the modulus of elasticity and Poisson ratio respectively. D is the isotropic damage variable and h is the closure parameter representing micro-defects closure effect, which indicates that the effect of damage is more predominant in tension than in compression. During compression, some of the micro-defects are closed, thereby the effective area is increased, the material regains some of the stiffness. This can be represented by the effective modulus of elasticity during tension $\tilde{E}^+$, which is given by $\tilde{E}^+ = E(1 - D)$, and by the effective elastic modulus during compression $\tilde{E}^-$ given by $\tilde{E}^- = E(1 - hD)$. The closure parameter h has values between 0 (complete stiffness recovery) and 1 (no stiffness recovery). Equations (3b) and (3c) give the relationships between the internal variables α_{ij} and r describing kinematic and isotropic hardening of the material and their corresponding thermodynamic forces β_{ij} and R, by means of material parameter C for kinematic hardening and through a function g' for isotropic hardening. Equation (3d) defines the strain energy release rate Y, which is the thermodynamic force corresponding to damage and is non-linear with respect to the stress tensor and the damage variable.

The evolution equations obtained from potential functions are written as

$$\dot{\varepsilon}_{ij}^{p} = \dot{\lambda}_p \frac{\partial F}{\partial \sigma_{ij}}, \tag{4a}$$

$$\dot{\alpha}_{ij} = \dot{\lambda}_p \frac{\partial F}{\partial \beta_{ij}}, \tag{4b}$$

$$\dot{r} = \dot{\lambda}_p \frac{\partial F}{\partial R}, \tag{4c}$$

$$\dot{D} = \dot{\lambda}_D \frac{\partial F_D}{\partial Y}. \tag{4d}$$

This is the normality rule for standard materials. Equation (4a) gives the evolution of the plastic strain with respect to time, with $\dot{\lambda}_p$ being the plastic multiplier which is measured from the consistency condition, and F is a potential which can be obtained from experimental findings. This function has to be convex, non-negative and should pass through the origin. $\frac{\partial F}{\partial \sigma_{ij}}$ is the flow vector indicating plastic flow is normal to the potential function F. This potential F, for associative plasticity is considered to be the yield function. Equations (4b) and (4c) similarly, give the evolution of the internal variables for kinematic and isotropic hardening respectively. Equation (4d) describes the evolution of damage with respect to time, where $\dot{\lambda}_D$ is the damage multiplier and F_D is a potential which is also identified from experimental findings. All the conditions required to formulate F must also be taken into account for F_D. These evolution equations, essentially, take into account the history dependency of the material.

The initiation of the macro-crack is indicated by the critical damage D_c and the material is assumed to fail when D_c is reached.

3 LATIN-Based Model Order Reduction Approach for Damage Computation

The LATIN approach tackles the set of equations on the whole time-space domain at every iteration. The equations are considered iteratively, namely the global equilibrium of the structure on one hand, the non-linear elastic law and the non-linear evolution equations on another hand. Model reduction is used in the sense of separation of time and space variables. LATIN-PGD approach allows for visco-plasticity case to define both quantities of interest which are stress and plastic strain using a unique time basis. Including damage, a new quantity of interest is added, which is the elastic strain, the total set of the solution field can be represented as $\boldsymbol{s} = \left\{\dot{\boldsymbol{\varepsilon}}^p, \boldsymbol{\varepsilon}^e, \dot{\boldsymbol{X}}, \dot{D}, \boldsymbol{\sigma}, \boldsymbol{Z}, Y\right\}$, where $\boldsymbol{X}$ represents the set of hardening variables (kinematic and isotropic) and $\boldsymbol{Z}$ is the conjugate variable of $\boldsymbol{X}$. The non-linear state law could be included in the global stage and then the separation of variables tackled by using different time functions for stress and strain. Otherwise, as presented here, the non-linearity due to the state law can be tackled in the local stage, and the stress can be decoupled in two parts, one defined from the local stage and the other one which can be written as a time-space decomposition form using the same time functions as the plastic strain one. The algorithm is only briefly overviewed in this contribution. Detailed explanation of every step and details about the numerical implementation may be found in [36].

The algorithm is initialised by solving the problem considering the boundary conditions of the exact problem but assuming that the material behaviour is perfectly elastic for the whole loading conditions. Then, plastic and damage corrections are added to the elastic solution at each subsequent iteration. The set of equations is divided in two sub-groups, one comprising the global and linear equations whereas the other one comprises the local and non-linear ones. One LATIN iteration consists of two parts:

- the global and linear problem is solved in the space $\boldsymbol{A_d}$ which belongs to the manifold of the admissibility conditions Eqs. (1) and (2), the linear state laws Eqs. (3b) and (3c), and the non-linear state law for damage Eq. (3d);
- the local and non-linear problem is solved in the space $\boldsymbol{\Gamma}$ which belongs to the manifold of the evolution equations Eq. (4) and the elastic state law Eq. (3a) which was not linearisable due to damage.

It can be noted that Eq. (3d) although being non-linear is tackled with the group of linear equations as a post-processing step from the knowledge of the stress tensor and the damage variable at the end of each iteration. The exact solution $\boldsymbol{s}_{ex}$ of the problem is defined as the intersection of the two manifolds by

$$\boldsymbol{s}_{ex} \in \boldsymbol{A_d} \cap \boldsymbol{\Gamma}. \tag{5}$$

The approximation of the solution is looked alternatively in the two manifolds until reaching convergence. From the knowledge of one step, the approximation in the following manifold is looked for by using certain linear operators called search direction operators.

3.1 Local Stage

In the local stage, evolution equations for internal variables, which are local in space and non-linear, are solved. The elastic state law, being non-linear, is also tackled in this stage. From the solution set $\boldsymbol{s}_i \in \boldsymbol{A}_d$ at LATIN iteration i, the approximation $\hat{\boldsymbol{s}}_{i+1/2} \in \boldsymbol{\Gamma}$ is estimated such that the local search directions are satisfied

$$\begin{bmatrix} \hat{\dot{\boldsymbol{\varepsilon}}}^p_{i+1/2} - \dot{\boldsymbol{\varepsilon}}^p_i \\ -\left(\hat{\dot{\boldsymbol{X}}}_{i+1/2} - \dot{\boldsymbol{X}}_i\right) \\ \hat{\boldsymbol{\varepsilon}}^e_{i+1/2} - \boldsymbol{\varepsilon}^e_i \\ \hat{\dot{D}}_{i+1/2} - \dot{D}_i \end{bmatrix} + \mathbf{B}^+ \begin{bmatrix} \hat{\boldsymbol{\sigma}}_{i+1/2} - \boldsymbol{\sigma}_i \\ \hat{\boldsymbol{Z}}_{i+1/2} - \boldsymbol{Z}_i \\ \hat{\boldsymbol{\sigma}}_{i+1/2} - \boldsymbol{\sigma}_i \\ \hat{Y}_{i+1/2} - Y_i \end{bmatrix} = 0. \tag{6}$$

Here, $\mathbf{B}^+$ is the direction of ascent. Following [26], the search direction is considered to be vertical such that

$$\left(\mathbf{B}^+\right)^{-1} = 0. \tag{7}$$

The solution of the search direction equation (Eq. 6) along with the evolution equations (Eq. 4) and the non-linear elastic law (Eq. 3a) constitute $\hat{\boldsymbol{s}}_{i+1/2}$. From the approximation at the local stage $\hat{\boldsymbol{s}}_{i+1/2}$, the solution set $\boldsymbol{s}_{i+1}$ is estimated in the global stage.

3.2 Global Stage Including Model Order Reduction

In the global stage, the solution set $\boldsymbol{s}_{i+1} \in \boldsymbol{A}_d$ satisfies the state laws, the admissibility conditions and the descent search directions

$$\begin{bmatrix} \dot{\boldsymbol{\varepsilon}}^p_{i+1} - \hat{\dot{\boldsymbol{\varepsilon}}}^p_{i+1/2} \\ -\left(\dot{\boldsymbol{X}}_{i+1} - \hat{\dot{\boldsymbol{X}}}_{i+1/2}\right) \\ \boldsymbol{\varepsilon}^e_{i+1} - \hat{\boldsymbol{\varepsilon}}^e_{i+1/2} \end{bmatrix} - \mathbf{B}^- \begin{bmatrix} \boldsymbol{\sigma}_{i+1} - \hat{\boldsymbol{\sigma}}_{i+1/2} \\ \boldsymbol{Z}_{i+1} - \hat{\boldsymbol{Z}}_{i+1/2} \\ \boldsymbol{\sigma}_{i+1} - \hat{\boldsymbol{\sigma}}_{i+1/2} \end{bmatrix} = 0, \tag{8a}$$

$$\left[\, \dot{D}_{i+1} - \hat{\dot{D}}_{i+1/2} \,\right] - \mathbf{b}^- \left[\, Y_i - \hat{Y}_{i+1/2} \,\right] = 0, \tag{8b}$$

where $\mathbf{B}^- = \begin{bmatrix} \mathbf{H}^- & 0 \\ 0 & \mathbf{C}^{-1} \end{bmatrix}$. The operator $\mathbf{H}^-$ belongs to the tangent space associated with the solution set $\hat{\boldsymbol{s}}_{i+1/2}$ in the manifold $\boldsymbol{\Gamma}$, and $\mathbf{C}$ is the undamaged Hooke tensor. Considering the damage variable is not updated in the linear stage, the search direction operator $\mathbf{b}^-$ is defined as zero.

The first step being the calculation of the hardening variables, the state equations are combined in the form

$$\boldsymbol{Z}_{i+1} = \boldsymbol{\Lambda} \boldsymbol{X}_{i+1}, \tag{9}$$

where $\mathbf{\Lambda}$ is a linear operator containing the sate law parameters. The search direction equation for hardening variables Eq. (8a) combined with the state equation (Eq. 9) can be written as

$$-\left(\dot{\boldsymbol{X}}_{i+1}-\hat{\dot{\boldsymbol{X}}}_{i+1/2}\right)=\mathbf{H}_Z(\mathbf{\Lambda}\boldsymbol{X}_{i+1}-\hat{\boldsymbol{Z}}_{i+1/2}), \tag{10}$$

with $\mathbf{H}_Z$ being the decoupled part of $\mathbf{H}^-$ that relates the internal variables with the corresponding associated variables. The hardening variables are then obtained by solving Eq. (8a) in time at each Gauss point.

The difficulty to calculate the stresses and strains, compared to former works with the LATIN method, is that the elastic state law Eq. (3a) is non-linear due to the presence of damage, leading to solve a non-linear problem at the global stage. This point is particularly tricky as it prevents the introduction of a model reduction strategy at this stage. The idea proposed herein is to transform this non-linear problem into separate linear equations by decomposing stress and total strain into two parts depending on plastic deformation and damage respectively.

The quantities of interest at this point $\boldsymbol{\sigma}_{i+1}$, $\boldsymbol{\varepsilon}^e_{i+1}$ and $\dot{\boldsymbol{\varepsilon}}^p_{i+1}$ are represented in a corrective form at iteration $i+1$ as

$$\Delta\boldsymbol{\sigma}_{i+1}=\boldsymbol{\sigma}_{i+1}-\boldsymbol{\sigma}_i,\quad \Delta\boldsymbol{\varepsilon}^e_{i+1}=\boldsymbol{\varepsilon}^e_{i+1}-\boldsymbol{\varepsilon}^e_i \quad\text{and}\quad \Delta\dot{\boldsymbol{\varepsilon}}^p_{i+1}=\dot{\boldsymbol{\varepsilon}}^p_{i+1}-\dot{\boldsymbol{\varepsilon}}^p_i. \tag{11}$$

The stress and total strain corrections in the global stage at iteration $i+1$ are separated into parts depending on plastic deformation $(\Delta\boldsymbol{\sigma}'_{i+1}, \Delta\boldsymbol{\varepsilon}'_{i+1})$ and on damage $(\Delta\tilde{\boldsymbol{\sigma}}_{i+1}, \Delta\tilde{\boldsymbol{\varepsilon}}_{i+1})$,

$$\Delta\boldsymbol{\sigma}_{i+1}=\Delta\boldsymbol{\sigma}'_{i+1}+\Delta\tilde{\boldsymbol{\sigma}}_{i+1}, \tag{12a}$$

$$\Delta\boldsymbol{\varepsilon}_{i+1}=\Delta\boldsymbol{\varepsilon}'_{i+1}+\Delta\tilde{\boldsymbol{\varepsilon}}_{i+1}. \tag{12b}$$

From these separations and the search direction equation along with the additive strain decomposition relation, it can be established that

$$\Delta\boldsymbol{\sigma}'_{i+1}+\Delta\tilde{\boldsymbol{\sigma}}_{i+1}=\mathbf{C}\left(\Delta\boldsymbol{\varepsilon}'_{i+1}-\Delta\boldsymbol{\varepsilon}^p_{i+1}\right)+\mathbf{C}\left(\Delta\tilde{\boldsymbol{\varepsilon}}_{i+1}-\Delta\boldsymbol{\varepsilon}^R_{i+1}\right), \tag{13}$$

where $\Delta\boldsymbol{\varepsilon}^R_{i+1}$ can be interpreted as a residual strain obtained from non-linear state law at iteration $i+1$. $\Delta\tilde{\boldsymbol{\sigma}}_{i+1}$ and $\Delta\tilde{\boldsymbol{\varepsilon}}_{i+1}$ are thereby obtained from the equilibrium equation, directly.

On the other hand, if only the plastic part is considered, the search direction can be re-written as

$$\Delta\dot{\boldsymbol{\varepsilon}}^p_{i+1}-\mathbf{H}_\sigma\Delta\boldsymbol{\sigma}'_{i+1}+\bar{\mathbf{\Delta}}_{i+1}=0. \tag{14}$$

with $\bar{\mathbf{\Delta}}_{i+1}$ is a plastic corrective term from the local stage and $\mathbf{H}_\sigma$ represents the decoupled part of $\mathbf{H}^-$ that relates stress to plastic strain rate.

The correction terms linked with the plastic behaviour $\Delta\sigma'_{i+1}$ and $\Delta\varepsilon'_{i+1}$ are then written in a separable form using Proper Generalised Decomposition.

3.2.1 Separation of Variables

The Proper Generalised Decomposition (PGD) is a flexible model order reduction technique, which is not based on a training stage. As at every LATIN iteration, the quantities of interest are approximated on the whole space-time domain by a linear form of the mechanical equilibrium, the usage of PGD coupled with LATIN is convenient. As any function dependent on several independent variables can be approximated as an infinite sum of products of one-variable functions [26, 43], PGD looks for an approximation of any quantity of interest as a finite sum of products of low-dimensional functions defined by a greedy algorithm. Therefore, this approximation includes an error due to the truncation of the series of separable forms. Here the plastic strain and stress part due to plastic deformation dependent on space and time variables are approximated as

$$\begin{aligned} \dot{\boldsymbol{\varepsilon}}^{p}\left(\underline{x}, t\right) &= \sum_{j=1}^{\mu} \dot{\lambda}_{j}\left(t\right) \bar{\boldsymbol{\varepsilon}}_{j}^{p}\left(\underline{x}\right), \\ \boldsymbol{\sigma}'\left(\underline{x}, t\right) &= \sum_{j=1}^{\mu} \lambda_{j}\left(t\right) \mathbb{C}\bar{\boldsymbol{\varepsilon}}_{j}^{p}\left(\underline{x}\right), \end{aligned} \tag{15}$$

where μ is the number of pairs involved in the decomposition, and $\mathbb{C}$ is a linear operator which relates the space functions of stress and plastic strain.

3.2.2 Updating Stage

The greedy algorithm is such that after defining on-the-fly a first pair of space and time functions, at every iteration a first decomposition is looked using the previously defined space functions and updating the time functions. This step is equivalent to a Proper Orthogonal Decomposition on the current space basis. Considering μ space-time modes have been generated to approximate the stress and plastic strain rate at iteration i, the corrections of stress and the plastic strain rate at iteration $i+1$ are given as

$$\begin{aligned} \Delta\dot{\boldsymbol{\varepsilon}}_{i+1}^{p}\left(\underline{x}, t\right) &= \sum_{j=1}^{\mu} \Delta\dot{\lambda}_{j}\left(t\right) \bar{\boldsymbol{\varepsilon}}_{j}^{p}\left(\underline{x}\right), \\ \Delta\boldsymbol{\sigma}'_{i+1}\left(\underline{x}, t\right) &= \sum_{j=1}^{\mu} \Delta\lambda_{j}\left(t\right) \mathbb{C}\bar{\boldsymbol{\varepsilon}}_{j}^{p}\left(\underline{x}\right). \end{aligned} \tag{16}$$

The updates of the time functions are calculated by minimising a mechanical residual which is defined as the norm of the search direction operator, i.e.

$$\{\Delta\lambda_j\}_{j=1}^{\mu} = \underset{\{\Delta\lambda_i\}_{j=1}^{\mu}}{\arg\min} \left\| \sum_{j=1}^{\mu} \Delta\dot{\lambda}_j \bar{\boldsymbol{\varepsilon}}_j^p - \mathbf{H}_\sigma \sum_{j=1}^{\mu} \Delta\lambda_j \mathbb{C} \bar{\boldsymbol{\varepsilon}}_j^p + \bar{\boldsymbol{\Delta}} \right\|_{\mathbf{H}_\sigma^{-1}} . \tag{17}$$

Then, if the improvement of the approximation is not efficient enough, a new pair is added to the decomposition.

3.2.3 Enrichment of Space-Time Bases

The objective of the enrichment phase is to add a new space-time pair. The corrections of stress and plastic strain for this case are written as

$$\begin{aligned} \Delta\dot{\boldsymbol{\varepsilon}}_{i+1}^p \left(\underline{x}, t\right) &= \dot{\lambda}_{\mu+1}(t)\, \bar{\boldsymbol{\varepsilon}}_{\mu+1}^p \left(\underline{x}\right), \\ \Delta\boldsymbol{\sigma}'_{i+1} \left(\underline{x}, t\right) &= \lambda_{\mu+1}(t)\, \mathbb{C} \bar{\boldsymbol{\varepsilon}}_{\mu+1}^p \left(\underline{x}\right), \end{aligned} \tag{18}$$

with the intention of calculating the separable quantities $\dot{\lambda}_{\mu+1}$ and $\bar{\boldsymbol{\varepsilon}}_{\mu+1}^p$.

A hybrid strategy is used, the space function $\bar{\boldsymbol{\varepsilon}}_{\mu+1}^p$ is calculated from a Galerkin formulation, by using the kinematic admissibility condition (Eq. 2) such that $\forall \boldsymbol{\sigma}^*$ which is statically admissible to zero

$$\int_{[0,T]\times\Omega} \Delta\dot{\boldsymbol{\varepsilon}}' : \boldsymbol{\sigma}^* \, \mathrm{d}\Omega \, \mathrm{d}t = 0 \tag{19}$$

and the static admissibility condition (Eq. 1) such that $\forall \underline{u}^*$ which is kinematically admissible to zero

$$\int_{[0,T]\times\Omega} \Delta\boldsymbol{\sigma}'_{i+1} : \boldsymbol{\varepsilon}\left(\underline{u}^*\right) \, \mathrm{d}\Omega \, \mathrm{d}t = 0 \tag{20}$$

with

$$\Delta\boldsymbol{\sigma}'_{i+1} = \mathbf{C}\left(\Delta\boldsymbol{\varepsilon}'_{i+1} - \Delta\boldsymbol{\varepsilon}_{i+1}^p\right). \tag{21}$$

Subsequently the time function $\lambda_{\mu+1}$ is solved similarly as in the update stage by minimising a mechanical residual

$$\lambda_{\mu+1} = \underset{\lambda_{\mu+1}}{\arg\min} \left\| \dot{\lambda}_{\mu+1} \bar{\boldsymbol{\varepsilon}}_{\mu+1}^p - \mathbf{H}_\sigma \lambda_{\mu+1} \mathbb{C} \bar{\boldsymbol{\varepsilon}}_{\mu+1}^p + \bar{\boldsymbol{\Delta}} \right\|_{\mathbf{H}_\sigma^{-1}} . \tag{22}$$

This fixed point iteration between space and time problems converges quickly.

Once the stress tensor is known at iteration $i+1$, the strain energy release rate for damage is calculated from Eq. (3d).

3.3 Numerical Example of a Plate Under Cyclic Loading

The proposed usage of LATIN-PGD for damage problem is exemplified for a two-dimensional problem as depicted in Fig. 1. Material considered is a Cr-Mo steel at 580 °C [5]. The variation in material property is represented by the yield stress, with $\sigma_y = 80$ MPa at $x = 0$, $\forall y \in [0, W]$ and $\sigma_y = 85$ MPa at $x = L$, $\forall y \in [0, W]$ and with a linear variation along the length of the structure.

The distribution of the damage variable D at $t = T$ is depicted in Fig. 2. It shows localisation, with maximum near $(x, y) = (L, W)$ and minimum near $(x, y) = (L, 0)$.

For fatigue computation, calculation of the time dependent quantities is expensive. Therefore a peculiar effort is done to reduce the cost of the time-computation.

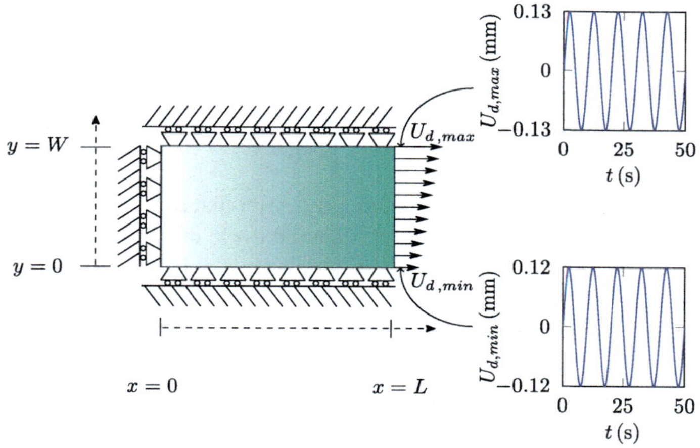

Fig. 1 A plate under traction with linearly distributed cyclic loading and variable material properties

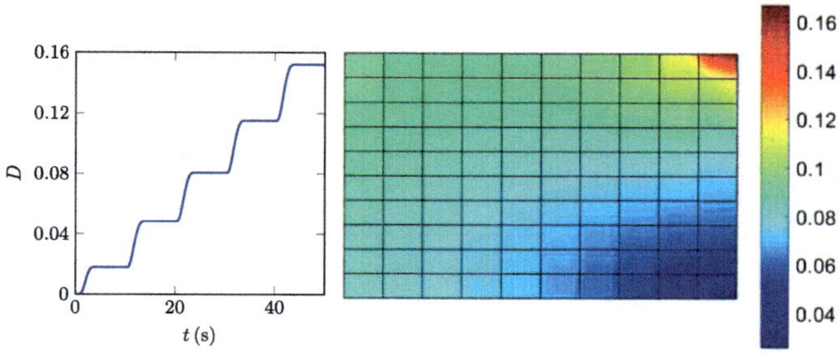

Fig. 2 Damage evolution for the weakest part of the plate and damage distribution at $t = T$ in the plate under linearly distributed cyclic loading

4 A Two-Time Scale Approach

In the case of a large number of cycles, the previous strategy can be enhanced by adding a multi-time scale feature. To describe the time dependent quantities defined on the whole time domain $[0, T]$, two time scales are introduced,

- a long time discretisation θ defined on the interval $[0, T]$ that represents the slow evolution along the cycles,
- a short time discretisation τ describing the rapid evolution within a cycle.

The idea is to introduce a finite element like description of the temporal quantities which are calculated only at certain chosen cycles called the "nodal cycles" (Fig. 3). For any time element $\left[\theta_m, \theta_{m+1}\right]$ once the nodal cycles m and $m+1$ are known a linear one-dimensional interpolation formula can be used [42]. If χ represents any temporal quantity over time element $\left[\theta_m, \theta_{m+1}\right]$, then

$$\chi\left(t\right) = \frac{\theta_{m+1} - \theta}{\theta_{m+1} - \theta_m}\chi_m\left(\tau\right) + \frac{\theta - \theta_m}{\theta_{m+1} - \theta_m}\chi_{m+1}\left(\tau\right), t \in \left[\theta_m, \theta_{m+1} + \Delta T\right], \quad (23)$$

where $\chi_m\left(\tau\right)$ and $\chi_{m+1}\left(\tau\right)$ are defined $\forall\tau \in \left[\theta_m, \theta_m + \Delta T\right]$ and $\forall\tau \in \left[\theta_{m+1}, \theta_{m+1} + \Delta T\right]$ respectively.

The first few cycles are computed classically. After that the nodal cycles are calculated progressively. The last classically computed cycle defined over $[\theta_0, \theta_0 + \Delta T]$ becomes the nodal cycle 0 and thereby the idea is to calculate nodal cycle 1, defined over $[\theta_1, \theta_1 + \Delta T]$. Thereafter knowing nodal cycle 1, nodal cycle 2 is calculated. This computation is continued till the last nodal cycle is calculated.

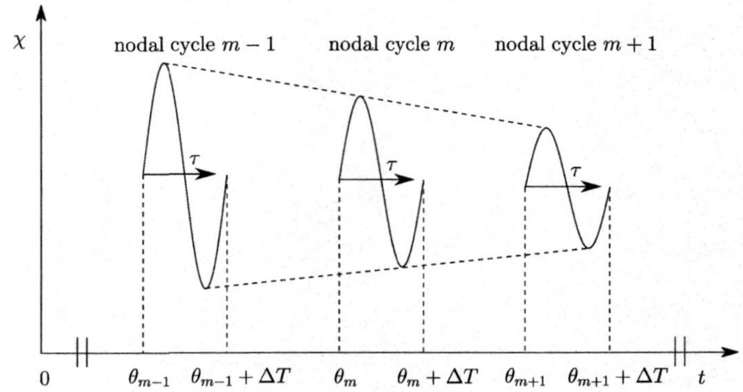

Fig. 3 Schematic of the two-time scale indicating the nodal cycles

4.1 Initialisation

The initialisation of the quantities of interest at nodal cycle m, knowing them at nodal cycle $m-1$ depends on the quantities that are being initialised. The quantities that are cyclic, namely the stress, elastic strain as well as the kinematic variables, are duplicated from the nodal cycle $m-1$ and a transformation is considered such that they are periodic, preserving the continuity from cycle $m-1$. The time functions representing the plastic strain $\{\lambda_j\}_{j=1}^{\mu}$ are also duplicated in a similar manner. The quantities that are non-cyclic and non-decreasing, namely damage and isotropic variables, are initialised as constant over nodal cycle m, with the magnitude being that obtained at $\theta_{m-1}+\Delta T$. The strain energy release rate for damage is calculated from the damage and the stress tensor. Thereafter the solution field over the nodal cycle m is calculated iteratively using the two-step algorithm till a convergence is obtained.

4.2 Local Stage

In the local stage, all the quantities of interest except damage do not need any time integration and can be calculated directly. The only problem in this stage is while integrating $\dot{D}$ to obtain the damage variable. To integrate Eq. (4d) over the nodal cycle m the initial condition at θ_m needs to be known. Considering a general first order ODE [26]

$$\frac{\mathrm{d}\chi}{\mathrm{d}t}+\kappa\chi=\upsilon, \tag{24}$$

defined over the complete time domain, with κ and υ being time-dependent known quantities. The idea is to calculate $\chi(\theta_m)$ from $\chi(\theta_{m-1})$. The time element $[\theta_{m-1},\theta_m]$ is discretised into certain instances Θ^k such that $\Theta^k=\theta_{m-1}+k\Delta T$, with $k=0,1,2,\cdots,p-1$, where p is the number of cycles in the time element $[\theta_{m-1},\theta_m]$. This provides $\Theta^0=\theta_{m-1}$ and $\Theta^{p-1}=\theta_m$. Knowing $\chi(\theta^k)$, Eq. (24) can be solved to obtain $\chi(\theta^{k+1})$ as

$$\chi\left(\Theta^{k+1}\right)=\check{\chi}\left(\Theta^{k+1},\Theta^k\right)+\Re\left(\Theta^{k+1},\Theta^k\right)\chi\left(\Theta^k\right), \tag{25}$$

where $\check{\chi}$ represents the solution of the ODE with zero initial condition, and $\Re$ represents the "resolvent" operator [26].

The challenge henceforth is to calculate $\check{\chi}$ and $\Re$ with minimum numerical cost. The easiest way is to calculate the quantities only at the nodal cycles $m-1$ and m and then use linear interpolations to obtain $\check{\chi}\left(\Theta^{k+1},\Theta^k\right)$ and $\Re\left(\Theta^{k+1},\Theta^k\right)$, i.e.

$$\check{\chi}\left(\Theta^{k+1},\Theta^k\right)=\nu_{m-1}\check{\chi}\left(\theta_{m-1}+\Delta T,\theta_{m-1}\right)+\nu_m\check{\chi}\left(\theta_m+\Delta T,\theta_m\right), \tag{26a}$$

$$\Re\left(\Theta^{k+1},\Theta^k\right)=\nu_{m-1}\Re\left(\theta_{m-1}+\Delta T,\theta_m\right)+\nu_m\Re\left(\theta_m+\Delta T,\theta_m\right), \tag{26b}$$

with ν being the linear shape functions defined as

$$\nu_{m-1} = \frac{\theta_m - \Theta^k}{\theta_m - \theta_{m-1}}, \nu_m = \frac{\Theta^k - \Theta_{m-1}}{\theta_m - \theta_{m-1}}. \tag{27}$$

These interpolated values can be used to rewrite Eq. (25) as

$$\begin{aligned} \chi\left(\Theta^{k+1}\right) = \nu_{m-1}\left[\check{\chi}\left(\theta_{m-1} + \Delta T, \theta_{m-1}\right) + \Re\left(\theta_{m-1} + \Delta T, \theta_{m-1}\right)\chi\left(\Theta^k\right)\right] \\ + \nu_m\left[\check{\chi}\left(\theta_m + \Delta T, \theta_m\right) + \Re\left(\theta_m + \Delta T, \theta_m\right)\chi\left(\Theta^k\right)\right]. \end{aligned} \tag{28}$$

Starting from $\chi\left(\Theta^0 = \theta_{m-1}\right)$, $\chi\left(\Theta^p = \theta_m\right)$ is calculated progressively, thereby Eq. (24) over the nodal cycle m.

The evolution equation of damage Eq. (4d) is solved over cycle m using the aforementioned technique.

4.3 Global Stage

The spatial modes, calculated for the initial cycles, are reused to compute the time functions of the PGD modes over the nodal cycle m. The initialisation of the time functions $\{\lambda_j\}_{j=1}^{\mu}$ is such that continuity is maintained with respect to the nodal cycle $m - 1$. Thereafter corrections of the time functions $\{\Delta\lambda_j\}_{j=1}^{\mu}$ are computed by solving Eq. (17) using zero initial conditions.

The next concern are the kinematic variables, which also being cyclic can be treated in a similar way. Equation (10) has to be solved for the kinematic variables over the nodal cycle m. However, an exact measurement of the initial conditions using the "resolvent" technique, being numerically expensive, is not necessary. The initialisation of the kinematic variables has been done to maintain continuity with respect to the nodal cycle $m - 1$. Thereby the quantities are calculated in terms of corrections by solving the ODE with zero initial condition.

The isotropic variable, however, being non-cyclic needs an accurate measurement of the initial condition for time integration of Eq. (10) over the nodal cycle m. This first order ODE is solved using the "resolvent" technique previously described.

This approach provides a drastic reduction in cost compared to classical LATIN technique.

4.4 A Numerical Example of a Bar Under Fatigue Loading to Build Virtual ε_a-N Curves

The one-dimensional numerical example considered here is a bar under traction to build virtual ε_a-N curves. The material considered is Cr-Mo steel at 25 °C with

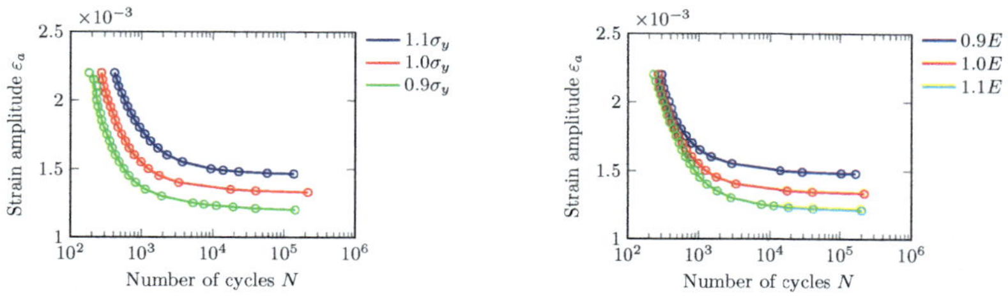

Fig. 4 ε_a-N curves for different yield stresses and moduli of elasticity

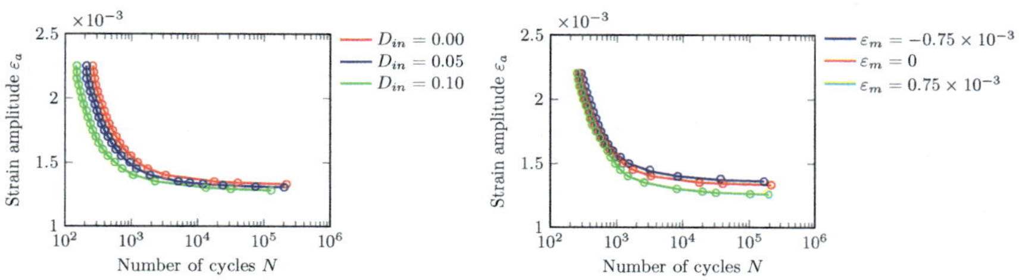

Fig. 5 ε_a-N curves for different initial damage conditions and total mean strains

kinematic hardening. The structure is clamped at one end and subjected to a sinusoidal displacement of the form $U_d(t) = U_0 \sin\left(\frac{2\pi t}{T}\right)$ at the other end. The most important material properties in this model are the yield stress $\sigma_y = 189$ MPa and the modulus of elasticity $E = 199.74$ GPa.

Experimental ε_a-N are generally obtained when a particular specimen is loaded under a given ε_a and the number of cycles needed for the specimen to rupture is measured. This experiment is repeated for several values of ε_a to obtain different values of N. For the numerical tests described here, a critical damage value of 0.2 is considered as a failure point. Similar to the physical experiments, several numerical tests are conducted by varying ε_a to obtain different values of N needed by the structure to reach the critical damage level. Some virtual ε_a-N curves are depicted in Figs. 4 and 5. The time domain for each numerical test is discretised uniformly, with a lesser number of cycles per time element for larger ε_a. The range of the number of cycles per element is between 10 and 200. The computation of each curve requires approximately 1 hr.

The influence of yield stress σ_y is depicted in Fig. 4. It is observed that, with the increase in σ_y, the structure becomes less susceptible to damage for a given strain amplitude. As the damage threshold considered in the model is directly proportional to the yield stress, more number of cycles are needed to reach the critical damage. The influence of elastic modulus E is also shown in Fig. 4. It is witnessed that, with the increase in E, the stress in the structure increases for a given strain amplitude, resulting in increased susceptibility to damage. Also, the damage threshold, considered in the model is inversely proportional to the modulus of elasticity, resulting in a higher damage for a higher value of E at a given ε_a. The aforementioned reasons culminate in the decrease in the number of cycles to failure with increase in E for a given strain amplitude. It is also noticed that this decrease is profound at lower strain amplitudes.

The next sets of tests investigate the influence of initial damage and mean strain on the ε_a-N curves, as shown in Fig. 5. The first set of tests investigates the influence of the presence of initial damage. It can be observed that, with the increase in magnitude of the initial damage, less number of cycles are needed to reach the critical damage for a given strain rate. The subsequent set of tests consists of different total mean strains ε_m. For a positive mean strain, the stress is more in the tensile part than in the compressive part, thereby a higher damage is obtained resulting into a lower N, for a given ε_a, than compared to the zero mean strain case. For a negative mean strain, the stress is more in the compressive region, thereby damage evolution is less compared to the zero mean stress case, resulting into a higher N for a given ε_a.

To evaluate the accuracy and the efficiency of the two-time scale approach a case with $U_0 = 1.40$ mm for 2000 cycles is investigated. A mono-scale LATIN-PGD computation obtained in a CPU time of 19 hrs is considered as a reference. The performance of the two-time scale approach is analysed using various uniform time discretisations.

The evolution of damage for different sizes of time elements is plotted in Fig. 6. For decreasing size of time elements the evaluation converges to the reference solution. The relative error of the two-time scale approach with respect to the reference solution is defined as

$$\text{Error} = \left[\frac{\int_{[0,T]\times\Omega} (D_{ms} - D_{ts}) \cdot (D_{ms} - D_{ts}) \, \mathrm{d}\Omega \, \mathrm{d}t}{\int_{[0,T]\times\Omega} (D_{ms} + D_{ts}) \cdot (D_{ms} + D_{ts}) \, \mathrm{d}\Omega \, \mathrm{d}t} \right]^{1/2}, \tag{29}$$

where D_{ms} and D_{ts} are the damage variables computed using the mono-scale computation and the two-scale methods respectively. The evolution of this error and the computational time with respect to the size of time elements is depicted in Fig. 6 too. The multi-scale computational times are represented as percentages of the mono-scale computational time in Fig. 6. Using only one time element containing 2000

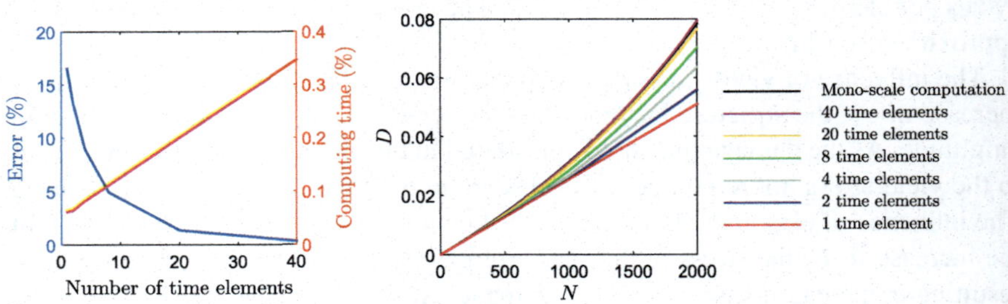

Fig. 6 Accuracy and numerical cost for different time discretisations

cycles, the computational cost has been decreased to 0.06% of the reference cost, but the error is 16% of the reference. By increasing the number of time elements, the computational cost increases but the error decreases. For 40 elements of 50 cycles, the error is only 0.36% for a computational cost 0.3% of the reference solution. The calculation time is drastically less compared to a mono-scale LATIN-PGD approach for acceptable accuracy.

5 Conclusion

An innovative numerical approach has been here presented for the computation of fatigue damage. A non-incremental technique has been used as numerical framework and the computational cost has been reduced by the usage of PGD that converts the actual high dimensional problem into much lower dimension. A multi-scale approach in time domain has been suggested to extend the LATIN-PGD method for the simulation of fatigue damage behaviour involving large number of cycles. By that, the increased numerical efficiency has been demonstrated which paves a way for physically based high cycle fatigue simulations.

References

1. Suresh, S.: Fatigue of Materials, 2nd edn. Cambridge University Press, Cambridge (2001)
2. Wöhler, A.: Über die Festigkeitsversuche mit Eisen und Stahl. Zeitschrift für Bauwesen **20**, 73–106 (1870)
3. Coffin, L.: The stability of metals under cyclic plastic strain. J. Basic Eng. **82**, 671 (1960)
4. Manson, S.S.: Fatigue: a complex subject-some simple approximations. Exp. Mech. **5**, 193–226 (1965)
5. Lemaitre, J., Desmorat, R.: Engineering Damage Mechanics: Ductile, Creep, Fatigue and Brittle Failures. Springer (2005)
6. Gerber, H.: Bestimmung der zulässigen Spannungen in Eisen-Constructionen. Zeitschrift des Bayerischen Architekten- und Ingenieur-Vereins **6**, 101–110 (1874)
7. Goodman, J.: Mechanics Applied to Engineering. Longman, Green and Company, London (1899)
8. Soderberg, C.R.: Factor of safety and working stress. Trans. Am. Soc. Mech. Eng. **52**, 13–28 (1939)
9. Palmgren, A.: Die Lebensdauer von Kugellagern. Zeitschrift des Vereins Deutscher Ingenieure **68**, 339–341 (1924)
10. Miner, M.A.: Cumulative damage in fatigue. J. Appl. Mech. **12**, 159–164 (1945)
11. Marco, S.M., Starkey, W.L.: A concept of fatigue damage. Trans. ASME **32**(76), 627 (1954)
12. Neuber, H.: Theory of stress concentration for shear-strained prismatical bodies with arbitrary nonlinear stress-strain law. J. Appl. Mech. **28**(4), 544–550 (1961)
13. Crossland, B.: Effect of large hydrostatic pressure on the torsional fatigue strength of an alloy steel. In: Proceedings of the International Conference on Fatigue of Metals, pp. 138–149. IME London (1956)
14. Sines, G.: Failure of materials under combined repeated stresses with superimposed static stresses. Technical report, National Advisory Committee for Aeronautics, Washington DC (1955)

15. Dang, V.K.: Sur la résistance à la fatigue des métaux. In: Sciences Technique Armement, vol. 47 (1973)
16. Chaboche, J.-L., Lesne, P.-M.: A non-linear continuous fatigue damage model. Fatigue Fract. Eng. Mater. Struct. **11**(1), 1–17 (1988)
17. Lemaitre, J., Sermage, J.P., Desmorat, R.: A two scale damage concept applied to fatigue. Int. J. Fract. **97**(1–4), 67–81 (1999)
18. Pirondi, A., Bonora, N., Steglich, D., Brocks, W., Hellmann, D.: Simulation of failure under cyclic plastic loading by damage models. Int. J. Plast. **22**, 2146–2170 (2006)
19. Paris, P.C., Erdogan, F.: A critical analysis of crack propagation laws. J. Basic Eng. **85**, 528–534 (1963)
20. Cui, W.: A state-of-the-art review on fatigue life prediction methods for metal structures. J. Mar. Sci. Technol. **7**(1), 43–56 (2002)
21. Santecchia, E., Hamouda, A.M.S., Musharavati, F., Zalnezhad, E., Cabibbo, M., El Mehtedi, M., Spigarelli, S.: A review on fatigue life prediction methods for metals. Adv. Mater. Sci. Eng. **2016**(26) (2016)
22. Prud'homme, C., Rovas, D., Veroyand, K., Machiels, L., Maday, Y., Patera, A., Turinici, G.: Reliable real-time solution of parametrized partial differential equations: Reduced-basis output bound methods. J. Fluids Eng. **124**(1), 70–80 (2002)
23. Chatterjee, A.: An introduction to the proper orthogonal decomposition. Curr. Sci. **78**(7), 808–817 (2000)
24. Kerschen, G., Golinval, J.-C., Vakakis, A.F., Bergman, L.A.: The method of proper orthogonal decomposition for dynamical characterization and order reduction of mechanical systems: an overview. Nonlinear Dyn. **41**(1), 147–169 (2005)
25. Rozza, G.: Reduced-basis methods for elliptic equations in sub-domains with a posteriori error bounds and adaptivity. Appl. Numer. Math. **55**, 403–424 (2004)
26. Ladevèze, P.: Nonlinear computational structural mechanics. In: Mechanical Engineering Series. Springer, New York (1999)
27. Ammar, A., Mokdad, B., Chinesta, F., Keunings, R.: A new family of solvers for some classes of multidimensional partial differential equations encountered in kinetic theory modeling of complex fluids. J. Non-Newto. Fluid Mech. **3**(139), 153–176 (2006)
28. Chinesta, F., Ladevèze, P., Cueto, E.: A short review on model order reduction based on proper generalized decomposition. Arch. Comput. Meth. Eng. **18**, 395–404 (2011)
29. Ryckelynck, D., Missoum Benziane, D., Cartel, S., Besson, J.: A robust adaptive model reduction method for damage simulations. Comput. Mater. Sci. **50**, 1597–1605 (2011)
30. Kerfriden, P., Gosselet, P., Adhikari, S., Bordas, S.P.A.: Bridging proper orthogonal decomposition methods and augmented Newton-Krylov algorithms: an adaptive model reduction for highly nonlinear mechanical problems. Comput. Meth. Appl. Mech. Eng. **200**, 850–866 (2011)
31. Metoui, S., Prulière, E., Ammar, A., Dau, F., Iordanoff, I.: The proper generalized decomposition for the simulation of delamination using cohesive zone model. Int. J. Numer. Meth. Eng. **99**(13), 1000–1022 (2014)
32. El Halabi, F., González, D., Sanz-Herrera, J.A., Doblaré, M.: A PGD-based multiscale formulation for non-linear solid mechanics under small deformations. Comput. Meth. Appl. Mech. Eng. **305**, 806–826 (2016)
33. Ladevèze, P., Nouy, A.: On a multiscale computational strategy with time and space homogenization for structural mechanics. Comput. Meth. Appl. Mech. Eng. **192**, 3061–3087 (2003)
34. Ladevèze, P., Néron,D., Passieux, J.-C.: On Multiscale Computational Mechanics with Time-space Homogenization. In: Fish, J. (ed.) pp. 247–282. Oxford University Press (2009)
35. Relun, N., Néron, D., Boucard, P.-A.: A model reduction technique based on the PGD for elastic-viscoplastic computational analysis. Comput. Mech. **51**, 83–92 (2013)
36. Bhattacharyya, M., Fau, A., Nackenhorst, U., Néron, D., Ladevèze, P.: A LATIN-based model reduction approach for the simulation of cycling damage. Comput. Mech. (2017) (submitted)
37. Cojocaru, D., Karlsson, A.M.: A simple numerical method of cycle jumps for cyclically loaded structures. Int. J. Fatigue **28**(12), 1677–1689 (2006)

38. Nesnas, K., Saanouni, K.: A cycle jumping scheme for numerical integration of coupled damage and viscoplastic models for cyclic loading path. Revue Européenne des Éléments Finis **9**, 865–891 (2000)
39. Lesne, P.-M., Savalle, S.: An efficient cycle jump technique for viscoplastic structures calculations involving large number of cycles. In: Proceedings of 2nd International Conference on Computational Plasticity, pp. 591–602 (1989)
40. Guennouni, T., Aubry, D.: Réponse homogénéisée en temps de structures sous chargements cycliques. Comptes rendus de l'Académie des sciences. Série 2, Mécanique, Physique, Chimie, Sciences de l'univers. Sciences de la Terre **303**(20), 1765–1768 (1986)
41. Oskay, C., Fish, J.: Fatigue life prediction using two-scale temporal asymptotic homogenization. Int. J. Numer. Meth. Eng. **61**, 329–359 (2004)
42. Cognard, J.-Y., Ladevèze, P.: A large time increment approach for cyclic viscoplasticity. Int. J. Plast. **9**(2), 141–157 (1993)
43. Ladevèze, P.: Separated representations and PGD-based model reduction. In: Fundamentals and Applications, Chapter PGD in Linear and Nonlinear Computational Solid Mechanics, pp. 91–152. Springer Vienna (2014)

Finite and Virtual Element Formulations for Large Strain Anisotropic Material with Inextensive Fibers

P. Wriggers, B. Hudobivnik and J. Schröder

Abstract Anisotropic material with inextensive or nearly inextensible fibers introduce constraints in the mathematical formulations of the underlying differential equations from mechanics. This is always the case when fibers with high stiffness in a certain direction are present and a relatively weak matrix material is supporting these fibers. In numerical solution schemes like the finite element method or the virtual element method the presence of constraints—in this case associated to a possible fiber inextensibility compared to a matrix—lead to so called locking-phenomena. This can be overcome by special interpolation schemes as has been discussed extensively for volume constraints like incompressibility as well as contact constraints. For anisotropic material behaviour the most severe case is related to inextensible fibers. In this paper a mixed method is developed for finite elements and virtual elements that can handle anisotropic materials with inextensive and nearly inextensive fibers. For this purpose a classical ansatz, known from the modeling of volume constraint is adopted leading stable elements that can be used in the finite strain regime.

Keywords Anisotropic material · Finite element analysis · Virtual element schemes · Mixed methods · Constraints

1 Introduction

Many different approaches were developed over the last decade to formulate finite elements for anisotropic material with nearly inextensive fibers. This is not the case for the virtual element method which is a relatively new discretization scheme and thus only limited applications of this method to anisotropic materials are known. The

P. Wriggers (✉) · B. Hudobivnik
Leibniz Universität Hannover, Hannover, Germany
e-mail: wriggers@ikm.uni-hannover.de

J. Schröder
Universität Duisburg-Essen, Duisburg, Germany
e-mail: j.schroeder@uni-due.de

J. Sorić et al. (eds.), *Multiscale Modeling of Heterogeneous Structures*,
Lecture Notes in Applied and Computational Mechanics 86,
https://doi.org/10.1007/978-3-319-65463-8_11

general problem related to the modeling of anisotropic materials is the high stiffness ratio between fiber and matrix material with the limit case of inextensible fibers where this ratio tends to infinity. This is physically related to the exact fulfilment of the kinematic constraint associated with the inextensibility of fibers in certain directions.

Generally the method of Lagrange multipliers provides a possibility to fulfil such constraints for small and and finite deformations. In this paper the Lagrange multiplier approach is employed to model anisotropic material behaviour at finite strains. Furthermore a relaxed version, i.e., the perturbed Lagrangian formulation, is used to model extensible fibers as well.

Boundary value problems that incorporate constraints usually cannot be solved using standard displacement interpolations. This leads to locking. Its main source is related to the fact that the constraints eventually restrict the deformation mode present in the displacement interpolation. This is also true when the constraints are fulfilled approximately, like in penalty methods. Locking phenomena are well-analyzed for geometrically linear problems in the case of volume constraints, see e.g. Hughes [19], Babuska and Suri [4], Zienkiewicz and Taylor [49] and Wriggers [44]. They were investigated in the mathematical community quite early, see Babuska [3] and Brezzi [11], and are now well understood. In general the Babuska-Brezzi (BB) condition can be employed to investigate the stability behaviour of mixed finite elements in the linear range. Within nonlinear problems the BB condition can only be used at certain stages of the analysis, see e.g. Chapelle and Bathe [13].

Different strategies were pursued in computational mechanics over the last years in order to circumvent locking effects. It became evident that element ansatz functions that interpolate the deformation or displacement field within an element with first order shape functions (bi- or tri-linear interpolation) do not converge properly when applied to problems with constraints like incompressibility or anisotropic material behaviour. Thus different variational formulations were explored in order to construct finite elements that can be used for problems with constraints. Approaches include reduced integration and stabilization, see e.g. Zienkiewicz et al. [50] for the linear case. Many variants can be found in the literature. It was shown that the reduced integration has to be used together with stabilization and can be extended to nonlinear problems, see e.g. Belytschko et al. [9] and Reese and Wriggers [33] leading to elements that are in general locking free for incompressible deformations. Additionally these elements are very efficient due to reduced integration. However stabilized elements rely on artificial stabilization parameters and thus the numerical solution can depend on theses parameters in certain cases.

Formulations, based on the mixed variational principle of HU- WASHIZU, were developed, e.g. see Simo and co-workers who introduced the enhanced strain elements first for the geometrically linear and nonlinear problems, e.g. see Simo and Rifai [40] and then for large deformations, Simo and Armero [38] and Simo et al. [39]. However, these elements depict non-physical instabilities at certain deformation states.

Other mixed finite element formulations, that are stable, perform well in the framework of small deformations and isotropy, e.g. see Bathe [5] and Brezzi and

Fortin [12]. Extensions to problems undergoing finite deformations are discussed in Auricchio et al. [1, 2] for the case of incompressibility. For finite strain anisotropic material behavior it is even more complex to find good finite element formulations. Many classical approaches that were designed for fiber-reinforced materials depict non-physical behavior, see e.g. Weiss et al. [43] and Holzapfel et al. [18]. Discussions related to the correct formulations of the mathematical model for anisotropic behaviour can be found in e.g. Sansour [34] and Helfenstein et al. [17]. These authors state that all fiber-related terms have to be provided in the energy by the complete deformation tensor and not by its isochoric part.

Reduced integration schemes using a special stabilization have been successfully applied to the simulation of composite reinforced material, see Hamila and Boisse [16]. Also special interpolations eliminated locking behaviour for composite materials, see ten Thjie and Akkerman [42]. Still many researchers use Hu-Washizu-based displacement, dilatation and pressure formulations, early introduced for incompressible materials by Simo et al. [41], for nearly incompressible materials with highly stiff fibers (like in arterial walls), see Zdunek et al. [48] and the references therein. However for strongly anisotropic material with inextensive fibers these approaches have limited performance, especially at finite strains.

A new formulation was presented in Schröder et al. [37] who introduced a novel finite element formulation that is developed especially for anisotropic materials, based on strain energy functions as discussed in Schröder and Neff [36] and Schröder [35]. There the constraints, associated with the anisotropy, are controlled by an additional deformation measure. A second-order tensorial Lagrange-multiplier was introduced via a discontinuous ansatz. This approach offers the opportunity to reduce the interpolation order of the anisotropic part and thus is able to relax the constraints due to anisotropy. This formulation leads to a stable methods for the solution of problems with anisotropic materials undergoing large strains. Another approach that introduces finite element formulations for anisotropy was discussed in Wriggers et al. [47] who used a mixed approach for the inclusion of the constraint related to the inextensive fibers. In this paper the latter approach modeled by two different different discretizations schemes. One is related to classical Taylor-Hood finite elements, see also Wriggers et al. [47] while the other approach is using the virtual element techniques where a formulation developed in Wriggers et al. [45] is employed. In both approaches a constraint equation is introduced on the basis of a Lagrange multiplier method. This allows to select ansatz functions as well for the displacement field in fiber direction as for the fiber forces. Additionally a perturbed Lagrangian formulation is introduced to relax the constraint condition and to be able to introduce real fiber stiffnesses.

The performance of the developed element formulations is compared to existing formulations using benchmark problems. All numerical results were obtained with the AceGen/AceFEM system developed in Korelc [20–22], see also Korelc and Wriggers [24].

2 Anisotropic Material with Inextensive Fibers at Large Strain

In this section a summary of the continuum mechanics background is provided for the formulation of problems exhibiting anisotropic response in finite elasticity. The formulation is reduced to the necessary equations that are needed to formulate the problem in AceGen. This omits many derivations since automatic differentiation is used. The formulation accounts for transversely isotropic material behaviour by using a mixed approach. It is assumed that the material is not extendable or only nearly extansable in a given fiber direction **a**.

2.1 Continuum Mechanics

All formulations are presented with respect to the initial configuration. For this we introduce the deformation $\boldsymbol{\varphi}(\mathbf{X}, t)$ which maps points in the initial configuration to the current or deformed configuration. The deformation can be computed using the coordinates of the initial configuration and the displacement field: $\boldsymbol{\varphi}(\mathbf{X}\,,t) = \mathbf{X} + \mathbf{u}(\mathbf{X}\,,t)$.
Using this deformation map, the deformation gradient can be computed as

$$\mathbf{F} = \text{Grad}\,\boldsymbol{\varphi}(\mathbf{X}\,,t) = \text{Grad}\,(\mathbf{X} + \mathbf{u}(\mathbf{X}\,,t)) = \mathbf{1} + \mathbf{H} \tag{1}$$

where $\mathbf{H} = \text{Grad}\,\mathbf{u}(\mathbf{X}\,,t)$. Note that the the volume change J is defined as the determinant of the deformation gradient: $J = \det \mathbf{F}$.
Based on the deformation gradient the Cauchy-Green tensor can be formulated as

$$\mathbf{C} = \mathbf{F}^T\,\mathbf{F} \tag{2}$$

Based on these kinematical quantities one can formulate a strain energy function for hyperelastic materials. The following isotropic strain energy function W^{iso} can be used to describe the behaviour of the isotropic part of the material:

$$W^{iso}(\mathbf{u}) = \frac{\mu}{2}(\,\text{tr}\,\mathbf{C} - 3 - 2\,\log J\,) + \frac{\lambda}{4}(\,J^2 - 1 - 2\log J\,) \tag{3}$$

where μ and λ are the Lame constants, see e.g. Wriggers [44]. Any other strain energy function that describes hyperelastic material behaviour can be selected as well.

2.2 Kinematical Anisotropic Constraint

The enforcement of the constraint that ensures that the material does not extend in the direction $\mathbf{a}$ leads to the following condition

$$\mathbf{a} \cdot \mathbf{E}\,\mathbf{a} = 0 \tag{4}$$

where $\mathbf{E}$ is the Green-Lagrangian strain tensor

$$\mathbf{E} = \frac{1}{2}\,(\mathbf{F}^T\mathbf{F} - \mathbf{1}) = \frac{1}{2}\,(\mathbf{C} - \mathbf{1})\,. \tag{5}$$

Since it is simpler to work with the right Cauchy Green tensor, see (2), the constraint can be written as

$$2\,\mathbf{a} \cdot \mathbf{E}\,\mathbf{a} = \mathbf{a} \cdot (\mathbf{C} - \mathbf{1})\,\mathbf{a} = \mathbf{a} \cdot \mathbf{C}\,\mathbf{a} - 1 \quad \text{for } \|\mathbf{a}\| = 1 \tag{6}$$

Furthermore we can write

$$\mathbf{a} \cdot \mathbf{C}\,\mathbf{a} = \mathbf{C} \cdot \mathbf{M} = \text{tr}[\,\mathbf{CM}\,] \quad \text{with } \mathbf{M} = \mathbf{a} \otimes \mathbf{a}\,. \tag{7}$$

Since Eq. (7) defines the stretch in direction of $\mathbf{a}$ we have

$$\lambda_c^2 = \text{tr}[\,\mathbf{CM}\,] \tag{8}$$

which in case of a fiber constraint in the direction of $\mathbf{a}$ leads to $\lambda_c^2 = 1$.

2.3 Lagrange Multiplier Formulation

Based on these kinematical relations different constraints and associated forms of a Lagrange multiplier approach can be formulated:

- **One or two constraints**. The Lagrange multiplier term related to the constraints of a material that is not extendable in the directions $\mathbf{a}_i$ $(i = 1, 2)$ yields with (7) and the associated structure tensors $\mathbf{M}_i$

$$W^{tiL}(\mathbf{C}\,,\sigma_{c\,i}) = \sum_{i=1}^{2} \sigma_{c\,i}\,(\text{tr}[\,\mathbf{CM}_i\,] - 1) \tag{9}$$

  where $\sigma_{c\,i}$ are the Lagrangian multipliers that physically represent the fiber stress related to the constraint.
- **Constraints for tension only**. In case that the response of the fiber system only occurs in tension states (9) can be re-written by using the Macauley bracket:

$\langle x \rangle = \frac{1}{2}(x + \|x\|)$. This choice yields

$$W^{tiL}(\mathbf{C}\,,\sigma_c) = \sum_{i=1}^{2} \sigma_{c\,i}\,\langle \mathrm{tr}[\,\mathbf{CM}_i\,] - 1\rangle \tag{10}$$

Now one of the two variants of $W^{tiL}(\mathbf{C}\,,\sigma_{c\,i})$, discussed above, can be added to (3) which leads to the final form of the strain energy

$$W(\mathbf{C}(\mathbf{u})\,,\sigma) = W^{iso}(\mathbf{C}(\mathbf{u})) + W^{tiL}(\mathbf{C}(\mathbf{u})\,,\sigma_{c\,i})\,. \tag{11}$$

All different forms of $W^{tiL}(\mathbf{C}\,,\sigma_{c\,i})$ lead to a pure mixed form since unknowns are the displacement field $\mathbf{u}$ and the fiber stresses σ_c.

2.4 *Perturbed Lagrangian and Penalty Formulation*

Additionally there is the possibility to use a so called perturbed Lagrangian formulation which can be stated as follows

$$W(\mathbf{C}\,,\sigma) = W^{iso}(\mathbf{C}) + W^{tiPL}(\mathbf{C}\,,\sigma_{c\,i})\,. \tag{12}$$

with

$$W^{tiPL}(\mathbf{C}\,,\sigma_c) = \sum_{i=1}^{2}\left[\sigma_{c\,i}\,(\mathrm{tr}[\,\mathbf{CM}_i\,] - 1) - \frac{1}{2\,C_{c\,i}}\,\sigma_{c\,i}^2\right] \tag{13}$$

here again $C_{c\,i}$ is a penalty parameter. For $C_{c\,i} \to \infty$ (13) reduces to (11). The perturbed Lagrangian formulation leads in the continuous version to a penalty method, but for different ansatz spaces for $\sigma_{c\,i}$ and the displacement field $\mathbf{u}$ it can lead to a different finite element scheme. Note, that the perturbed Lagrangian formulation can also be used to introduce of a fiber stiffness that is related to the physical behaviour of the anisotropic material. In that case C_c has a physical meaning.

Another possibility is to enforce the constraint via the penalty method. In this way the constraints are approximated by introducing a penalty term related to the constraint. The associated formulations includes the constraint (8) in the strain energy as follows

$$W^p(\mathbf{C}) = W^{iso}(\mathbf{C}) + W^{pen}(\mathbf{C})\,. \tag{14}$$

with

$$W^{pen}(\mathbf{C}) = \sum_{i=1}^{2}\left[\frac{C_{c\,i}}{2}\,(\mathrm{tr}[\,\mathbf{CM}_i\,] - 1)^2\right] \tag{15}$$

here $C_{c\,i}$ are a penalty parameters in fiber direction. For $C_{c\,i} \to \infty$ (13) it can be shown that the constraint is fulfilled exactly.[1] The penalty formulation can also be used to introduce a certain fiber stiffness that is related to the physical behaviour of the fiber. In that case $C_{c\,i}$ has a physical meaning.

3 Mixed Element Formulation

For the mixed interpolation tetrahedral and hexahedral elements are selected and compared. For both element formulations a quadratic interpolation for the displacement field $\mathbf{u}$ and a linear interpolation for the mixed variables $\sigma_{c\,i}$ is selected. This choice is motivated by the classical mixed formulation, known as Taylor-Hood element, for the incompressibility constraint. For anisotropic material with inextensive fibers the variables $\sigma_{c\,i}$ are the stress component related to the constraint, e.g. $\sigma_{c\,i}$ are the stresses in direction of $\mathbf{a}_i$.
Note that in the mixed form for the incompressibility with the constraint $(J-1)$, that is related to the determinant of $\mathbf{F}$, a cubic function of the components of the deformation gradient describes this constraint. In the case of the constraint (9) for anisotropic materials this function is only a quadratic form of the components of the deformation gradient. Thus it is not obvious that the same choice for the interpolation of $\sigma_{c\,i}$ will be sufficient.[2]

Now ansatz functions for the displacement field and the Lagrangian multiplier (fiber stresses) $\sigma_{c\,i}$ have to be formulated. The quadratic shape functions that approximate the displacement field

$$\mathbf{u}_e = \sum_{I=1}^{n_u} N_I(\xi\,,\eta\,,\zeta)\,\mathbf{u}_I \tag{16}$$

are given below for the a tetrahedron with 10 nodes ($n_u = 10$)

$$\begin{aligned} N_1 &= (2\xi-1)\xi\,,\ N_2 = (2\eta-1)\eta\,,\ N_3 = (2\zeta-1)\zeta\,,\\ N_4 &= (2\kappa-1)\kappa\,, N_5 = 4\xi\eta\,,\ N_6 = 4\eta\zeta\,,\\ N_7 &= 4\zeta\xi\,,\ N_8 = 4\xi\kappa\,,\ N_9 = 4\eta\kappa\,,\ N_{10} = 4\zeta\kappa\,, \end{aligned} \tag{17}$$

with $\kappa = 1-\xi-\eta-\zeta$ and a hexahedron with 27 nodes ($n_u = 27$)

$$N_I(\xi\,,\eta\,,\zeta) = N_I(\xi)\,N_I(\eta)\,N_I(\zeta) \tag{18}$$

[1] It is well known that ill-conditioning can occur when a large penalty parameter C_c is selected. Thus in reality the penalty formulation is only able to approximately enforce the constraint condition (8).

[2] In the linear case both conditions, while being different, yield a linear dependence on the components of the displacement gradient. Thus there the choice of using the same ansatz function for the pressure (incompressibility) and the fiber stress (anisotropy) is justified.

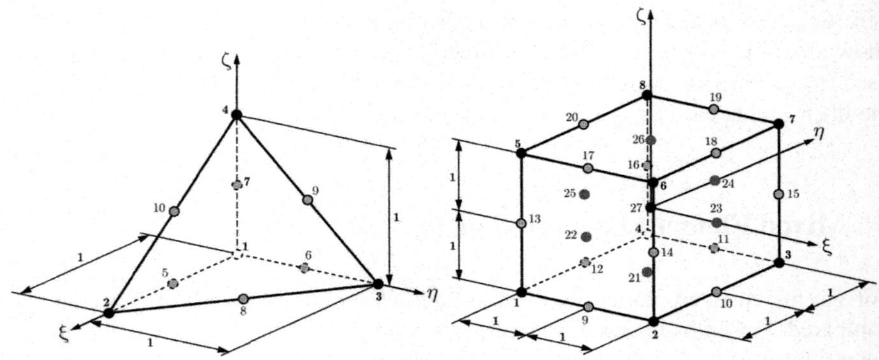

Fig. 1 Nodes of the quadratic tetrahedral and hexahedral element

where $I = 1, \ldots, 27$. $N_I(s)$ is given for the vertex nodes by

$$N_I(s) = \frac{1}{2}(1 - s_I)[s(s-1)] + \frac{1}{2}(1 + s_I)[s(s+1)]$$

for s being either ξ , η or ζ. Here s_I is related to a specific coordinate of a vertex node of the hexahedron in the space of the reference coordinates $(\xi\,,\eta\,,\zeta)$ with $\xi_I = \{-1\,,+1\}$, $\eta_I = \{-1\,,+1\}$ and $\zeta_I = \{-1\,,+1\}$, see Fig. 1. For the mid nodes the shape function $N_I(s)$ is given by

$$N_I(s) = (1 - s^2)$$

where now $\xi_I = 0$, $\eta_I = 0$ and $\zeta_I = 0$. Furthermore, the linear shape functions for the interpolation of the Lagrange multipliers $\sigma_{c\,i}$ are defined for the tetrahedron with respect to the four edge nodes ($n_\sigma = 4$)

$$N_{\sigma\,1} = \xi\,, \quad N_{\sigma\,2} = \eta\,, \quad N_{\sigma\,3} = \zeta\,, \quad N_{\sigma\,4} = \kappa \tag{19}$$

and for the hexahedron with respect to the eight edge nodes ($n_\sigma = 8$ and $K = 1, \ldots, 8$) as

$$N_{\sigma\,K} = \frac{1}{8}(1 + \xi\,\xi_K)(1 + \eta\,\eta_K)(1 + \zeta\,\zeta_K) \tag{20}$$

these will be used to interpolate the Lagrange multipliers (fiber stresses) $\sigma_{c\,i}$ related to the constraint within the element

$$\sigma_{c\,i} = \sum_{K=1}^{n_\sigma} N_{\sigma\,K}(\xi\,,\eta\,,\zeta)\,\sigma_{K\,i}\,. \tag{21}$$

Furthermore we need to define the coordinates within the finite element to formulate the isoparametric mapping. With $\boldsymbol{\zeta} = (\xi\,,\eta\,,\zeta)$ it follows

$$\mathbf{X}_e = \sum_{J=1}^{n_u} N_J(\boldsymbol{\zeta})\,\mathbf{X}_J\,. \tag{22}$$

Based on these ansatz functions the deformation gradient within an element e is computed via

$$\mathbf{F}_e = \mathbf{1} + \text{Grad}\,\mathbf{u}_e = \sum_{I=1}^{n_u} \mathbf{u}_I \otimes \mathbf{J}_e^{-T}\,\nabla_\zeta\,N_I \tag{23}$$

with the Jacobian of the isoparametric map

$$\mathbf{J}_e = \sum_{I=1}^{n_u} \mathbf{X}_I \otimes \nabla_\zeta\,N_I\,.$$

Now the Jacobian J_e of the deformation gradient $\mathbf{F}_e$ is obtained within the element by $J_e = \det\,\mathbf{F}_e$. Furthermore the Cauchy-Green tensor $\mathbf{C}_e$ can be computed at element level from (2). These kinematical relations can then be used to discretize the strain energy function in (3).

Now we have to introduce the discrete version of the different constraint formulations (7), (13) and (15). For that the trace of $\mathbf{C}_e\,\mathbf{M}_{e\,i}$ has to be computed. For the formulation of the mixed finite element we start from Eq. (9). Thus one has to compute the structure tensor $\mathbf{M}_{e\,i}$ that depends on the vector $\mathbf{a}_i$ providing the direction of the anisotropy. Vector $\mathbf{a}_i$ is defined as a unit vector

$$\mathbf{a}_i = \{\,a_{xi}\,,a_{yi}\,,a_{zi}\,\}\,/\,\sqrt{a_{xi}^2 + a_{yi}^2 + a_{zi}^2}\,. \tag{24}$$

With that the complete potential energy for a hyperelastic material is given for the case of the Lagrangian form by

$$\begin{aligned} U(\mathbf{u}\,,\sigma_{c\,i}) = & \int_\Omega \left[\frac{\mu}{2}(\text{tr}\,\mathbf{C}_e - 3 - 2\log J_e) + \frac{\lambda}{4}(J_e^2 - 1 - 2\log J_e)\right] d\Omega \\ & + \sum_{i=1}^{2} \int_\Omega \sigma_{c\,i}\,(\text{tr}[\mathbf{C}_e\mathbf{M}_{e\,i}] - 1)\,d\Omega - \int_{\Gamma_\sigma} \hat{\mathbf{t}}\cdot\mathbf{u}\,d\Gamma\,. \end{aligned} \tag{25}$$

where $\hat{\mathbf{t}}$ is the surface traction that is applied at the boundary Γ_σ of the solid Ω.

In this contribution we will employ the tool *AceGen* to produce the finite element code. Thus (25) can be formulated in terms of the kinematical quantities and interpolations. This is sufficient when AceGen is used to derive the element residual vector and the tangent matrix.

4 Formulation of the Virtual Element Projection

Within the virtual element method (VEM) we have two ingredients that have to be considered. The first is a Galerkin projection of the deformation onto a specific ansatz space. The second ingredient is the stabilization of the formulation. Thus the virtual element method relies on the split of the ansatz space into a part $\Pi \mathbf{u}_h$ (projection) and a remainder (stabilization)

$$\mathbf{u}_h = \Pi \mathbf{u}_h + (\mathbf{u}_h - \Pi \mathbf{u}_h)\,. \tag{26}$$

For the stabilization a new concept was introduced in Wriggers et al. [45] which is used in this paper. The advantage of the method is that a domain Ω can be partitioned into non-overlapping polygonal elements which need not be convex.

In this paper we use a low-order approach with linear ansatz functions where nodes are placed only at the vertices of the polygonal elements. Furthermore, the restriction of the element shape functions to the element boundaries are linear function, see Fig. 2.

4.1 Ansatz Functions for VEM

The first part of a virtual element formulation is related to the computation of the projection $\Pi \mathbf{u}_h$ which is modeled at element level by a linear function

$$\Pi \mathbf{u}_h = \mathbf{H}\,\mathbf{c} = \begin{bmatrix} 1 & 0 & x & 0 & y & 0 \\ 0 & 1 & 0 & x & 0 & y \end{bmatrix} \mathbf{c} \tag{27}$$

where the constants to be determined are $\mathbf{c} = \{\, c_1\,, c_2\,, \ldots\,, c_6\,\}$.

Additionally a linear ansatz for the deformation along the element edge is selected for a boundary segment k of the virtual element, defined by the local nodes (1)–(2) by, see right side of Fig. 3,

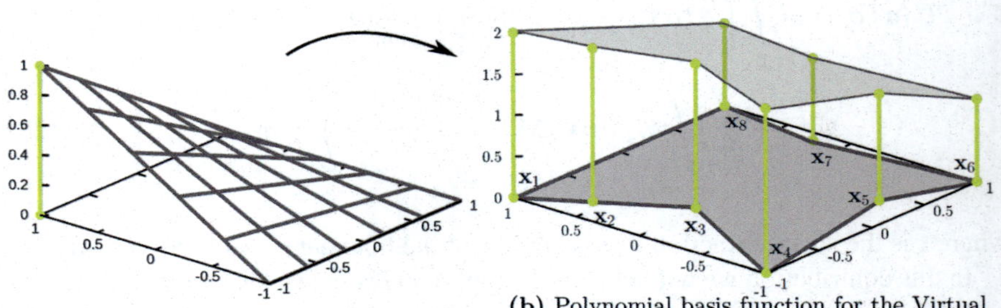

(a) Finite Element ansatz

(b) Polynomial basis function for the Virtual Element ansatz with vertices $\mathbf{x}_I$

Fig. 2 Comparison of the ansatz functions for FEM and VEM formulations

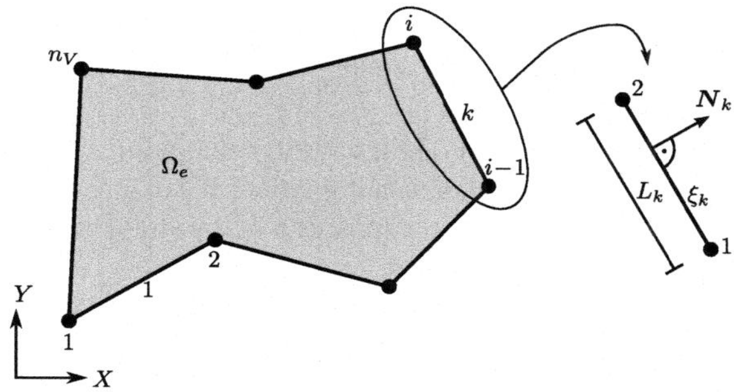

Fig. 3 Virtual element with n_V nodes and local boundary segment of the element

$$(\mathbf{u}_h)_k = (1 - \xi_k)\,\mathbf{u}_1 + \xi_k\,\mathbf{u}_2 = M_{k\,1}\,\mathbf{u}_1 + M_{k\,2}\,\mathbf{u}_2 \quad \text{with } \xi_k = \frac{x_k}{L_k}\,. \tag{28}$$

Here, for example, $M_{k\,1}$ is the ansatz function along a segment k related to node (1), ξ_k is the local dimensionless coordinate and $\mathbf{u}_1$ is the nodal value at that node, see Fig. 3.

The projection $\Pi \mathbf{u}_h$ is defined such that it satisfies, see Beirão Da Veiga et al. [7],

$$\nabla \Pi \mathbf{u}_h|_e = \frac{1}{\Omega_e} \int_{\Omega_e} \operatorname{Grad} \mathbf{u}_h \, d\Omega = \frac{1}{\Omega_e} \int_{\Gamma_e} \mathbf{u}_h \otimes \mathbf{N} \, d\Gamma \tag{29}$$

where $\mathbf{N}$ is the normal at the boundary Γ_e of the domain Ω_e. From (27), the gradient of the projection is constant at element level

$$\nabla \Pi \mathbf{u}_h|_e = \begin{bmatrix} c_3 & c_5 \\ c_4 & c_6 \end{bmatrix}. \tag{30}$$

The right hand side of (29) yields with (28)

$$\frac{1}{\Omega_e} \int\limits_{\Gamma_e} \mathbf{u}_h \otimes \mathbf{N} \, d\Gamma = \frac{1}{\Omega_e} \sum_{k=1}^{n_V} \int\limits_{\Gamma_k} \begin{bmatrix} u_x(\mathbf{x}_k) N_x & u_x(\mathbf{x}_k) N_y \\ u_y(\mathbf{x}_k) N_x & u_y(\mathbf{x}_k) N_y \end{bmatrix} L_k \, d\Gamma \tag{31}$$

where we have used $\mathbf{N} = \{\, N_x\,,\, N_y\,\}^T$ and $\mathbf{u} = \{\, u_x\,,\, u_y\,\}^T$, and n_V are the number of segments of the element. Note that the normal vector $\mathbf{N}$ changes from segment to segment. All quantities are related to the undeformed or initial configuration. Furthermore we have used a form of $\mathbf{N}_k$ that is not normalized since the length L_k cancels out when the integral in (31) is evaluated over the edges. The integral in (31) can be evaluated for the ansatz functions (28) exactly by using the trapezoidal or Gauss-Lobatto rule, see Wriggers et al. [45].

This projection determines the ansatz $\Pi \mathbf{u}_h$ on (27) within an element up to a constant. Thus this result has to be supplemented by a further condition to ensure uniqueness. For this purpose we adopt the condition that the sum of the nodal values of $\mathbf{u}_h$ and of its projection $\Pi \mathbf{u}_h$ are equal. This yields for each element Ω_e

$$\frac{1}{n_V}\sum_{I=1}^{n_V}\Pi\mathbf{u}_h(\mathbf{x}_I) = \frac{1}{n_V}\sum_{I=1}^{n_V}\mathbf{u}_h(\mathbf{x}_I)\,, \tag{32}$$

where $\mathbf{x}_I$ are the coordinates of the nodal point I and the sum includes all boundary nodes. Note that the constant gradient $\nabla\Pi\mathbf{u}_h|_e$ can be computed directly using (30) and (31). Thus for the computation of the strain energy it is not necessary to evaluate Eq. (32).

The Lagrangian parameters are approximated in the virtual element by a constant value $\bar{\sigma}_{c\,i}$, leading to

$$\sigma_{c\,i} = \bar{\sigma}_{c\,i} \tag{33}$$

4.2 Construction of the Virtual Element

A virtual element is based only on the projection $\Pi\,\mathbf{u}_h$ would lead to a rank deficient element once the number of vertices is greater than 3. Thus the formulation has to be stabilized like the classical one-point integrated elements developed by Flanagan and Belytschko [15], Belytschko and Bindeman [8], Reese et al. [32], Reese and Wriggers [33], Reese [31], Mueller-Hoeppe et al. [29], Korelc et al. [23], Krysl [26]. The anisotropic version, discussed here, is based on the formulation developed in Wriggers et al. [45]. In the following we will first discuss the formulation of the element part that stems from the projection, see last section, for the anisotropic case. Furthermore, the stabilization of the virtual element will be performed as developed in Wriggers et al. [45]. Thus the anisotropic behaviour is only introduced in the first term of the potential that is related to the projection.

With this the development of the virtual using the hyperelastic potential function (25) yields

$$U(\mathbf{u}\,,\sigma_{c_i}) = \mathop{\mathsf{A}}_{e=1}^{n_e}\left[U(\Pi\mathbf{u}_h|_e\,,\bar{\sigma}_{c_i}) + U_{stab}(\mathbf{u}_h|_e - \Pi\mathbf{u}_h|_e)\right]. \tag{34}$$

4.2.1 Constant Part Due to Projection

The simplest possible formulation for a finite deformation virtual element is a split into a constant part of the deformation gradient and an associated stabilization term. This was developed for the linear case in Beirão Da Veiga et al. [6] and also in Wriggers et al. [46] where in the latter work the focus was on contact mechanics. The same approach can be found in the work of Beirão Da Veiga et al. [7], Chi et al. [14] and Wriggers et al. [45] for the nonlinear case where different approaches for the stabilization were introduced.

Now the anisotropic part of the potential in (25) will be used to develop the constant part of the virtual element. Since $\bar{\sigma}_{c_i}$, $\mathbf{M}_i$ and $\mathbf{C}(\Pi\mathbf{u}_h|_e)$ are constant within the virtual element Ω_e the integration of the last two terms in (25) is trivial, leading with the perturbed Lagrangian formulation, see (13), to

$$U(\Pi\mathbf{u}_h|_e, \bar{\sigma}_{c_i}) = U^{iso}(\Pi\mathbf{u}_h|_e) + \sum_{i=1}^{2}\left[\bar{\sigma}_{c_i}(\text{tr}[\mathbf{C}(\Pi\mathbf{u}_h|_e)\,\mathbf{M}_i] - 1) - \frac{1}{2C_{ci}}\bar{\sigma}_{c_i}^2\right]\Omega_e\,. \tag{35}$$

where $U^{iso}(\Pi\mathbf{u}_h|_e)$ is the discretized first term in (25) denoting the isotropic matrix behaviour. This formulation introduces two additional variables $\bar{\sigma}_{c_i}$ which can however be eliminated at element level in case of the perturbed Lagrangian formulation. Note that this is not the case for the pure Lagrangian multiplier formulation in (9).

The perturbed Lagrangian formulation in (35) is a regularized method for enforcing the constraint of rigid fibers where the parameters $C_c i$ are penalty parameters that can be interpreted as stiffnesses of the fibers in directions $\mathbf{a}_i$. Thus this formulation can also be used for the numerical simulation of anisotropic materials with a given fiber stiffness that allows for deformation of the fibers.

All derivations with respect to the unknown displacements leading to the residual vector $\mathbf{R}_e^c$ and the tangent matrix $\mathbf{K}_{Te}^c$ were performed with the symbolic tool AceGen, see Korelc and Wriggers [24]. This yields for (35)

$$\mathbf{R}_e = \frac{\partial U(\Pi\mathbf{u}_h|_e, \bar{\sigma}_{c_i})}{\partial \mathbf{p}_e} \quad \text{and} \quad \mathbf{K}_{Te} = \frac{\partial \mathbf{R}_e^c(\mathbf{p}_e)}{\partial \mathbf{p}_e} \tag{36}$$

where $\mathbf{p}_e = \{\mathbf{u}_e\,,\bar{\sigma}_{c_i}\}$ denotes the vector unknowns of the virtual element Ω_e.

4.2.2 Nonlinear Stabilization

In this contibution a novel stabilization approach is applied for virtual elements which was derived in Wriggers et al. [45]. This stabilization is based on a scheme proposed in Krysl [25]. This techniques is based on the introduction of a new, positive definite strain energy $\hat{\Psi}$. Then the stabilization contribution to the strain energy can be defined by

$$U_{stab}(\mathbf{u}_h|_e - \Pi\mathbf{u}_h|_e) = \hat{U}(\mathbf{u}_h|_e) - \hat{U}(\Pi\mathbf{u}_h|_e)\,. \tag{37}$$

The second term on the right side ensures consistency of the total potential energy. This means that once the element size is very small, a constant strain will occur in each virtual element. In that case $\hat{U}(\mathbf{u}_h)$ will approach $\hat{U}(\Pi\mathbf{u}_h)$ and thus will not influence the final result. With (37) the total potential yields

$$U(\mathbf{u}_h) = U(\Pi\mathbf{u}_h) + \hat{U}(\mathbf{u}_h) - \hat{U}(\Pi\mathbf{u}_h)\,. \tag{38}$$

The choice

$$\hat{U}(\mathbf{u}_h) = \sum_{e=1}^{n_e}\int_{\Omega_e}\hat{W}(\mathbf{u}_h|_e)\,d\Omega\,. \tag{39}$$

results in a purely isotropic strain energy analogous to (3). Following the approach advocated in Krysl [25], we propose the stabilization strain energy

$$\hat{U}(\mathbf{u}_h|_e) = \int\limits_{\Omega} \left[\frac{\hat{\lambda}}{2}(J_m - 1)^2 + \frac{\hat{\mu}}{2}(\operatorname{tr}\mathbf{C}_m - 3 - 2\ln J_m) \right] d\Omega\,. \tag{40}$$

The anisotropic part (13) was neglected since the basic physical behaviour has to be covered by (35) which represents the projection onto the virtual element space. The terms involving $\Pi\mathbf{u}_h$ can be integrated in the same way as (35). It remains to develop a simple procedure for the integration of the first term in (37) involving the displacement $\mathbf{u}_h|_e$.

In Wriggers et al. [45] the displacement field $\mathbf{u}_h|_e$ in (26) was approximated by an inscribed triangular finite element mesh, see Fig. 4 which then can be used to compute the stabilization energy. The specific choice of the mesh using n_{int} linear three-noded triangles that are connected to the nodes of the virtual element does not introduce extra degrees of freedom. Note, that triangularization is always possible. Once an ansatz is formulated for the approximation within each triangle Ω^i_m of the inscribed mesh for the displacement field, here denoted by $\mathbf{u}_m$ the deformation measures J_m and $\mathbf{C}_m$ can be easily computed in the standard way, either by using an isoparametric formulation for the three noded triangle or by direct evaluation of the ansatz functions.

The gradient $\operatorname{Grad}\mathbf{u}_m$ is constant over each inscribed element Ω^i_m as well as the deformation gradient $\mathbf{F}^i_m = \mathbf{1} + \operatorname{Grad}\mathbf{u}_m$. The potential (40) can now be evaluated for the internal triangular mesh. All further derivations leading to the residual vector $\mathbf{R}_e$ and the tangent matrix $\mathbf{K}_{Te}$ were performed with the symbolic tool ACEGEN, see Korelc and Wriggers [24]. This yields for (37)

$$\mathbf{R}^s_e = \frac{\partial U_{stab}(\mathbf{u}_e)}{\partial \mathbf{u}_e} \quad \text{and} \quad \mathbf{K}^s_{Te} = \frac{\partial \mathbf{R}^s_e(\mathbf{u}_e)}{\partial \mathbf{u}_e}\,. \tag{41}$$

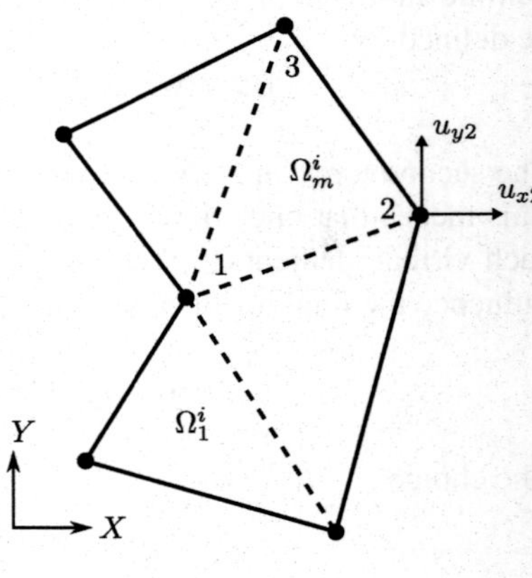

Fig. 4 Internal triangular mesh

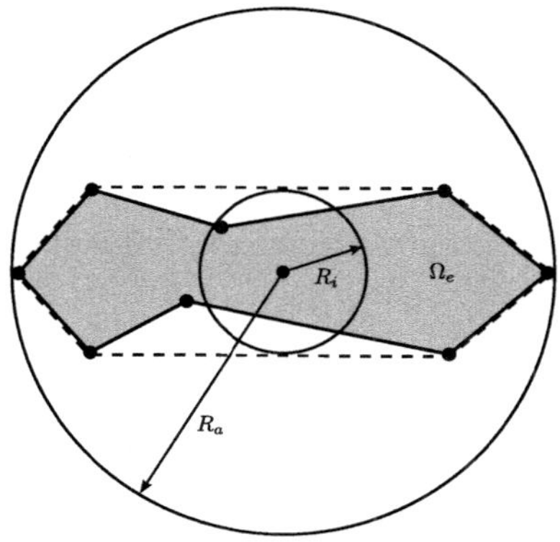

Fig. 5 Inner and outer radius of a virtual element

Thus the total residual and tangent matrix of the virtual element are given by the sum of expressions (36) and (41): $\mathbf{R}_e = \mathbf{R}_e^c + \mathbf{R}_e^s$ and $\mathbf{K}_{Te} = \mathbf{K}_{Te}^c + \mathbf{K}_{Te}^s$.

The values of the Lamé parameters in the strain energy (40) have to be determined in a proper way. Wriggers et al. [45] used a procedure that takes into account the element distortion to compute the Lamé parameters $\hat{\lambda}$ and $\hat{\mu}$, see also Krysl [27]. The initial element distortion can be obtained from the geometry of a virtual element, see Fig. 5. The algorithm to determine the material parameters for the stabilization energy is as follows, see Krysl [27] and Wriggers et al. [45]:

- Convert the Lamé parameters, related to the problem, to Young's modulus and Poisson ratio.
- Compute a geometry parameter by using the inner and outer radii, R_i^2, R_a^2 respectively, see Fig. 5, to obtain

$$\beta = 2\sqrt{2}\,(1+\nu)\frac{R_i^2}{R_a^2 - R_i^2}\,. \tag{42}$$

  The inner radius is computed by using the distance from the geometrical centre to the convex hull of the virtual element while the outer radius is defined by the maximum distance of nodes related to the virtual element, see Fig. 5.
- Compute the correction

$$\hat{E} = E\frac{\beta}{1+\beta} \qquad \hat{\nu} = 0.3 \tag{43}$$

  which is based on a comparison of the bending energy of a rectangular virtual element with that of a beam in order to enhance the bending behaviour of the element, see Krysl [27]. Thus the Young's modulus and the Poisson ratio of the virtual element can be computed. Note that $\bar{\nu}$ is kept constant since the Poisson ratio does not influence the convergence behaviour of the element and avoids locking in the stabilization term for incompressible problems. A similar procedure was developed in the work of Nadler and Rubin [30] who stabilized a Cosserat brick element using 18 deformation modes for bending and torsion and matched the stiffness related to these modes by comparisons with linear elastic solutions. Extensions for non-rectangular elements were provided in Loehnert et al. [28] and Boerner et al. [10].

- The Lamé parameters for the stabilization energy are then obtained from

$$\hat{\lambda} = \frac{\hat{E}\,\hat{\nu}}{(1+\hat{\nu})(1-2\hat{\nu})} \qquad \hat{\mu} = \frac{\hat{E}}{2(1+\hat{\nu})} \tag{44}$$

5 Examples

Several numerical examples are considered to show the performance of the mixed formulation for anisotropic materials using finite and virtual elements. In the examples the following discretization schemes are compared:

- Tetrahedral elements for the constraint formulation (9) and (10) with quadratic ansatz functions (17) for the deformations and linear ansatz, see (19), for the Lagrangian multiplier σ_c. These elements are labeled T2-A1 in the following.
- The same ansatz is also used perturbed Lagrangian formulation (13). These elements are labeled T2-A1-P.
- Hexahedral elements for the constraint formulation (9) and (10) with quadratic ansatz functions (18) for the deformations and linear ansatz, see (20), for the Lagrangian multiplier σ_c. These elements are labeled H2-A1 in the following.
- For the perturbed Lagrangian formulation the same ansatz (18) is used. These hexahedral elements are labeled H2-A1-P.

For comparison reasons standard displacement elements were formulated as well based on the penalty method (14). These elements are denoted by T2 and H2.
For the virtual elements different element types are used.

- A sepcial virtual element with 8 nodes is labeled by VEM-T1.
- Additionally Voronoi meshes are used with elements that have arbitrary number of nodes. This discretization is called VEM-T1-VO.

All examples are subjected to loads that lead to finite deformation strain states.

5.1 *Cook's Membrane Problem*

An example that will show a clear anisotropic response is the Cook's membrane problem of a tapered cantilever beam, clamped at the left end. The structure is loaded at the right end by a constant vertical load, as depicted in Fig. 6.

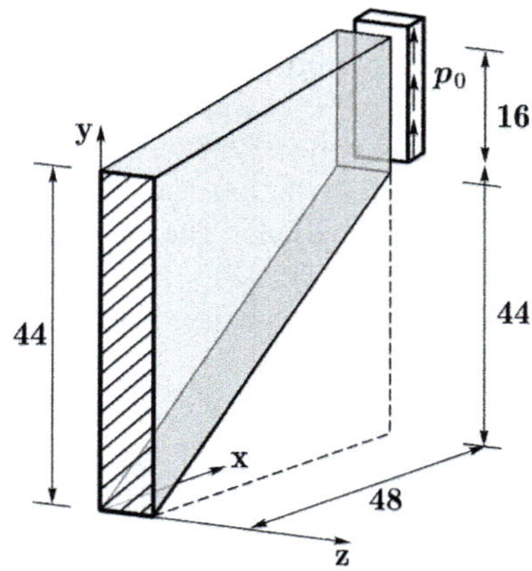

Fig. 6 Initial configuration of the cantilever beam

The selected dates for the Lame constants are $\mu = 500$ and $\lambda = 1000$. The direction of anisotropy is given by $\mathbf{a} = \frac{1}{\sqrt{3}}\{\,1\,,1\,,1\,\}$. The beam is clamped at its left end thus all displacements at $X = 0$ all were set to zero in x-, y- and z-direction. The total load of $p_0 = 250$ was applied in different loading steps.

Different mesh densities where used to compute the solution, see Fig. 7 for the tetrahedra. For hexahedra elements the same mesh sequence is used. The sequence of meshes is selected such that the finer meshes are included in the coarser meshes. This enables convergence studies that will depict differences of the formulations. The number N denotes the mesh division. Since the membrane has a thickness of $H_z = 10$ the number of elements in this direction are reduced which yields a mesh of $N \times N \times N\,/\,2$.

In a first computation a mesh with $N = 16$ was used to obtain the load displacement curve for Cook's membrane problem. The element used for this simulation was

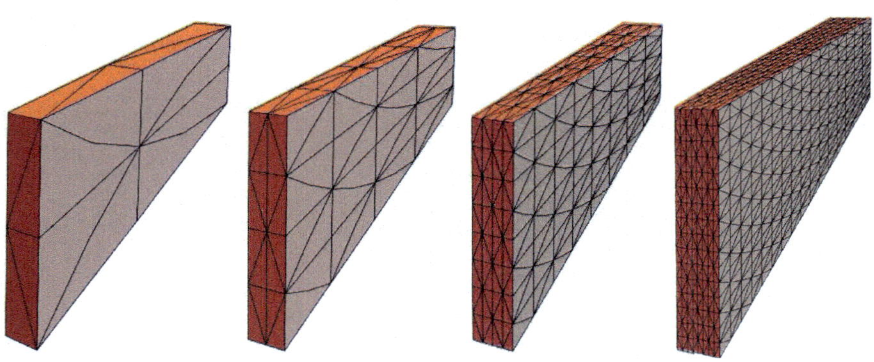

Fig. 7 Tetrahedral meshes of the cantilever beam with $N = 2, 4, 8, 16$

the H2-A1-P formulation. The load was applied in 10 even load increments λ with $\Delta\lambda = 0.25$. The parameter for the perturbed formulation was selected as $C_c = 10^6$.

The deformed mesh on the right in Fig. 8 was computed with a mesh of $16 \times 16 \times 8$ elements which led to a total number of 59058° of freedoms. The deformation at the final configuration clearly depicts the twist in the deformed shape due to the anisotopic constraint at large deformations. The solution was computed using 8 load steps for all discretizations reported in Fig. 9. The convergence behaviour was robust, 6 Newton iterations per load step were needed for all discretizations to obtain quadratic convergence

A convergence study is performed for the fully constraint case, using the Lagrangian multiplier formulation (9). The element formulations H2-A1 and T2-

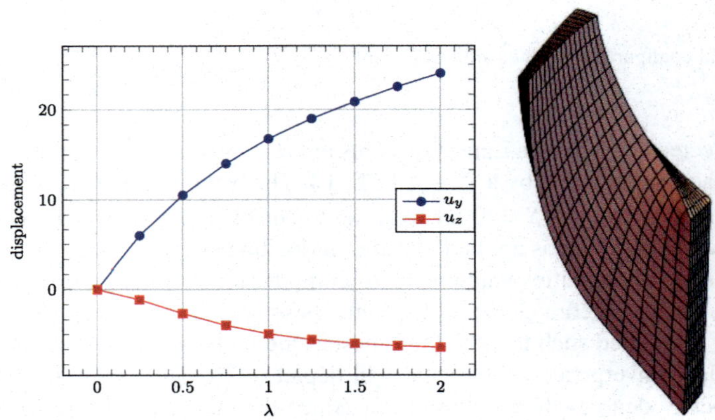

Fig. 8 Load displacement curve: λ versus displacement components in y- and z-direction at point (48, 52, 5) and deformed shape at final configuration

Fig. 9 Convergence Study, constraint case

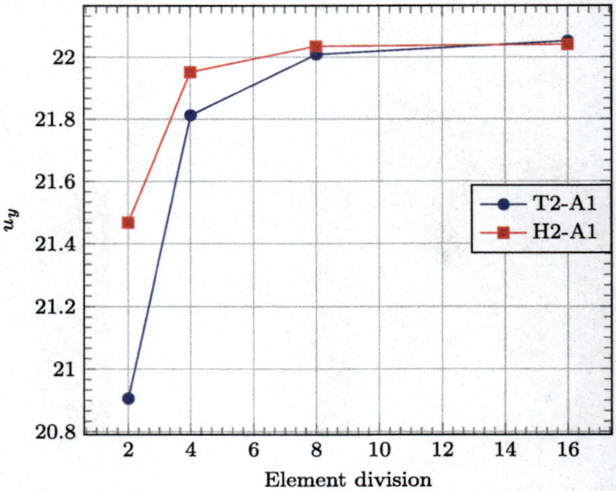

A1 are compared. Figure 9 depicts the convergence of the vertical displacement at point (48, 60, 0). The load displacement curve, see Fig. 8, is computed for the vertical displacement of the mid node $(X, Y, Z) = (48, 52, 5)$ of the plane at the right end of the cantilever beam which is in the direction of the load p_0, see Fig. 6. Furthermore the out-of-plane displacement in z-direction is plotted that shows the out-of-plane deformation of the cantilever beam due to the anisotropic material. It can be observed that the hexahedral element performs slightly better for coarse meshes. Here one has to acknowledge that the coarsest mesh ($N = 2$) of the triangularization for the tetrahedral elements is not symmetric and thus will have a certain bias. Nevertheless the displacement for the coarsest mesh is close to the final result, being approximately only 5% off.

In order to show the dependency of the solution on the penalty or fiber stiffness parameter C_c a series of computations were performed. The perturbed formulation (12) was used and a mesh division of $N = 8$ selected. Note that the anisotropic constraint is not enforced for a penalty parameter $C_c \leq 10$. Decreasing C_c leads to an intermediate stage where the stiffness of the fiber changes the deformation state. This is observed for parameters between $10 \leq C_c \leq 10^5$. Finally from $C_c > 10^5$ on there is no further change, thus the parameter is sufficient to enforce the constraint. Additionally we note, that for $C_c > 10^7$ the result is the same that is obtained with the pure Lagrangian multiplier formulation (9).

A convergence study is now performed for the perturbed Lagrangian formulation, see (12) which is compared with the penalty formulation (14) for a parameter of $C_c = 10^6$. The results can be found in Fig. 11. It can be seen that the penalty formulation does not converge to the same solution as the perturbed Lagrangian formulation. Here a penalty parameter was used that is sufficient to fulfill the constraint, see Fig. 10. Thus it is clear from Fig. 11 that the penalty formulation locks. Furthermore it is interesting to observe that for a penalty parameter of $C_c > 10^7$ the penalty method

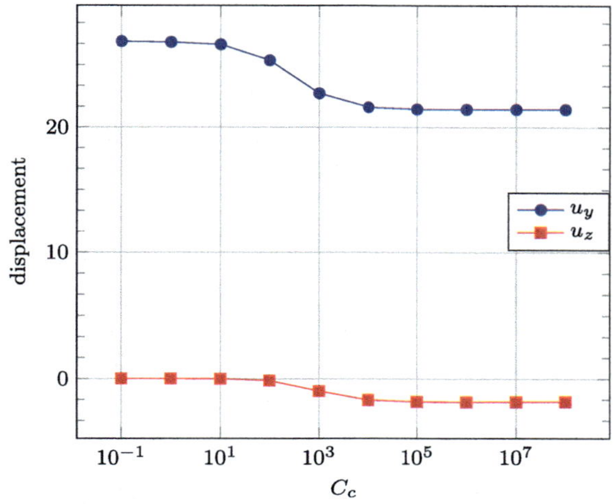

Fig. 10 Influence of the stiffness parameter C_c on the displacement components in y- and z-direction at point (48, 60, 5)

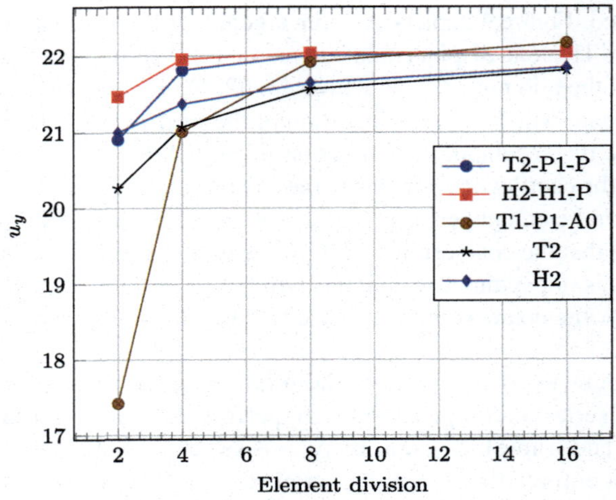

Fig. 11 Convergence Study, penalty versus perturbed Lagrangian, $C_p = 10^6$

for the H2 as well as for the T2 element the Newton iteration diverged while the perturbed Lagrangian formulations H2-A1-P and T2-A1-P still converges. Thus the different ansatz spaces for u and σ are necessary for a robust element formulation.

5.2 *Shear Deformation of a Beam*

When a beam that is clamped at one end is subjected to an end load then the beam will usually bend in the direction of the loading. Here we will investigate the case of a special anisotropic constraint that allows no deformation in horizontal direction. In that case the axial movement is constraint and the beam can only undergo shear deformations. The beam has a length of $L = 2$, a height of $H = 0.5$ and a thickness $T = 1$ (in dimensionless coordinates), see Fig. 12. The constitutive data for the Lame constants are $\mu = 40$ and $\lambda = 100$. The direction of anisotropy is given by $\mathbf{a} = \{1, 0\}$ which enforces the constraint in horizontal x-direction. The beam is clamped at the left end using the boundary conditions: $u_x = u_y = 0$ for all nodes at $x = 0$. It is loaded by a constant load of $p_y = 1$ at the right end.

The loading is such that a moderate strain state occurs for pure shear. This leads to a deformed state that is reported in Fig. 13. In this figure the deformation is scaled by a factor of 5. The displacement at the right side of the beam is $u_y = 0.0531$. This result can be checked using the classical beam theory where the shear deformation at the end of the beam is

$$u_{yB} = \frac{Q\,L}{\mu\,A}$$

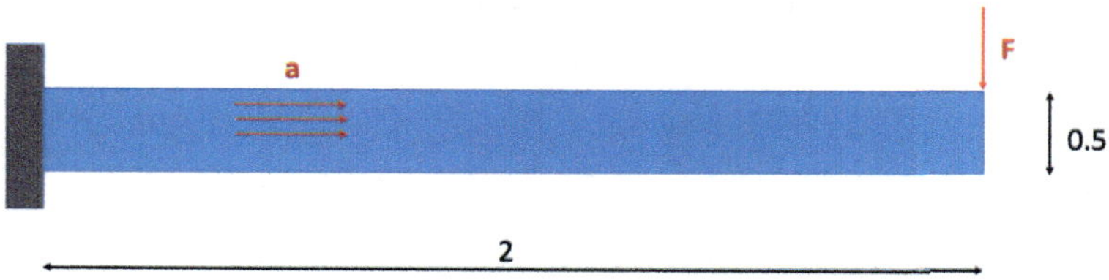

Fig. 12 Beam under end load

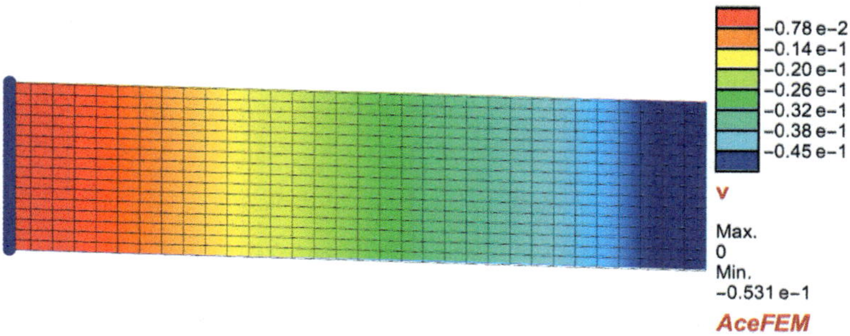

Fig. 13 Pure shear deformation of the beam for a mesh with 32 × 16 elements

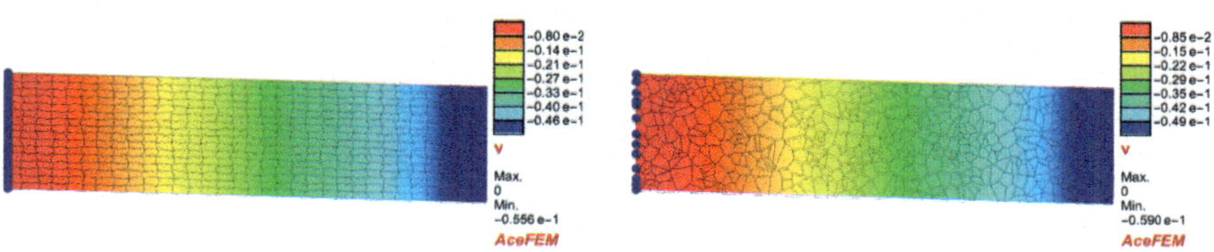

Fig. 14 Pure shear deformation two distorted meshes with 32 × 16 elements

with $A = 0.5$, $Q = p_y\, A = 0.5$ and it follows $u_{yB} = 0, 05$. A convergence study shows that the solution converges to $u_y = 0.05$ for a high density mesh with 512 × 256 virtual elements. Figure 15 depicts the convergence behaviour of different meshes. In Fig. 15 a regular mesh with 8-node virtual elements (VEM-8), a distorted mesh with 8-node elements (VEM-8-D), a Voronoi mesh (VEM-VO) is used as well as the T2-A1 ansatz. The deformed shapes for the distorted meshes can be found in Fig. 14. All meshes converge to the correct solution, however at the beginning the deviation from the converged solution is quite high.

The relatively slow convergence for this simple problem stems from the fact that the anisotropic constraint in the virtual element formulation is fulfilled as mean value for an element Ω_e. This is somehow equivalent to a fulfillment at mid point and actually leads for only one element to pure bending since at a beam axis the constraint is fulfilled exactly in bending.

The same behaviour can also be observed in Fig. 15 when the computation is performed with the tetrahedral element T2-A1 using the Taylor-Hood ansatz for

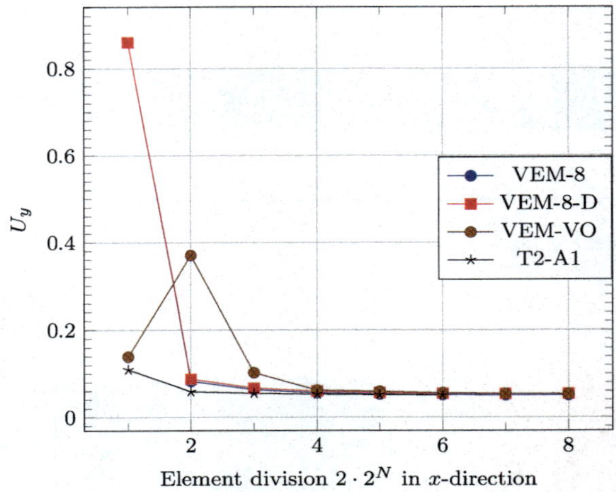

Fig. 15 Convergence study for the shear beam: VEM-T1- and Q2S-element

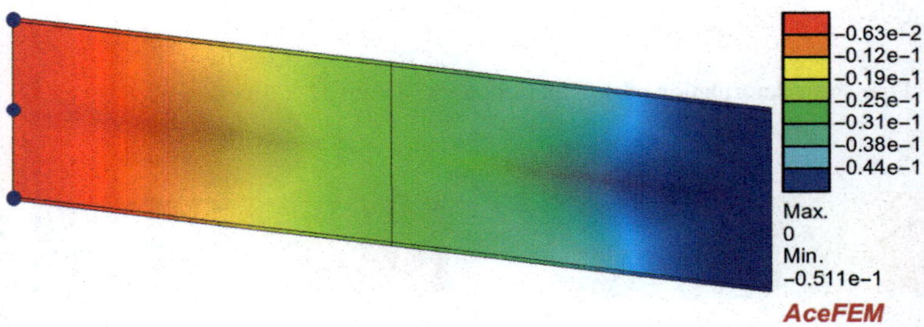

Fig. 16 Pure shear deformation of the beam for a mesh with 2×3 virtual elements, scaled by a factor of 5

displacements and pressure where the displacements in the third directions are constraint to obtain a plane strain state. However the deviations are smaller since this is an element with quadratic ansatz for the displacements and the anisotropic constraint is fulfilled using a linear ansatz for the Lagrangian multipliers.

Thus if we use a special mesh where a very thin upper and lower layer is introduced then the constraint is enforced at the upper and lower side of the beam and the pure shear mode is recovered, see Fig. 16. The solution is with $u_y = 0.0511$ very close to the converged solution despite the very coarse mesh with only six virtual elements. Also note that in this example the very thin upper and lower elements with an element ratio of 1/100 do not lead to locking.

Hence we can conclude that the shear deformation pattern can be computed for different mesh types when using the virtual element method. The results clearly show that the constraint due to anisotropic behaviour can be reproduced which is also true for the T2-A1 tetrahedral element. Finally it should be noted that the bending deformation state related to the given load would lead to large strains and deflections when no anisotropic constraint is present.

5.3 Bias Extension Test

A problem where tension locking can occur is the tensile test where fibers are oriented in ±45° in the initial configuration. This bias-extension test was used in ten Thjie and Akkerman [42] and Hamila and Boisse [16] to investigate behaviour of standard finite element formulations and special interpolation techniques to avoid locking. The test is performed on a rectangular specimen, see Fig. 17 for the finite element mesh in the initial configuration. The length of the specimen is $L = 300$, its width is $H = 100$ and the thickness of the specimen is $T = 10$. The specimen is clamped at both ends and pulled using a constant displacement $\bar{u}_x = 65$. In order to obtain a two-dimensional plane strain state, as it was used in Hamila and Boisse [16], the displacements of all nodes where set to zero in thickness direction at $Z = 0$ for the T2-A1 elements. The material properties of the matrix material are described by the Lame constants $\lambda = 1$ and $\mu = 1$. The fiber stiffness is $C_c = 4000$.

When the specimen is stretched from L to $L + \bar{u}_x$ different in-plane shear zones occur, see Hamila and Boisse [16]. As depicted also in this paper, the computation using a standard element formulation, here a pure displacement formulation using T2 elements, yields a non physical deformation state, see left side of Fig. 18. On the other hand the new T2-A1 element yields even with a relatively coarse mesh a correct deformation pattern, which is depicted on the right side of Fig. 18 and has the same form as described in Hamila and Boisse [16]. The deformation of the finest mesh, see Fig. 19, actually shows also the different shear zones.

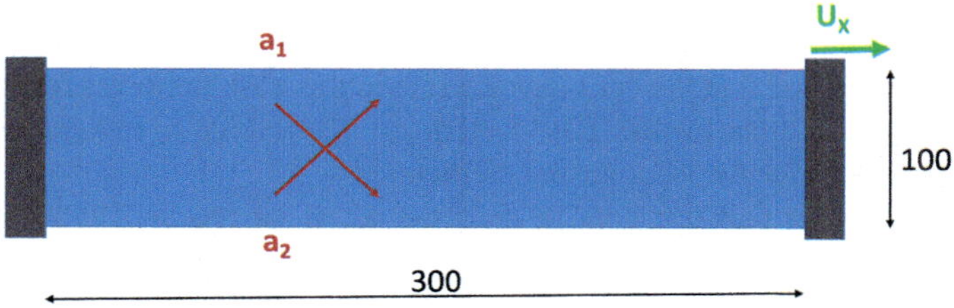

Fig. 17 Bias extension test of a woven composite

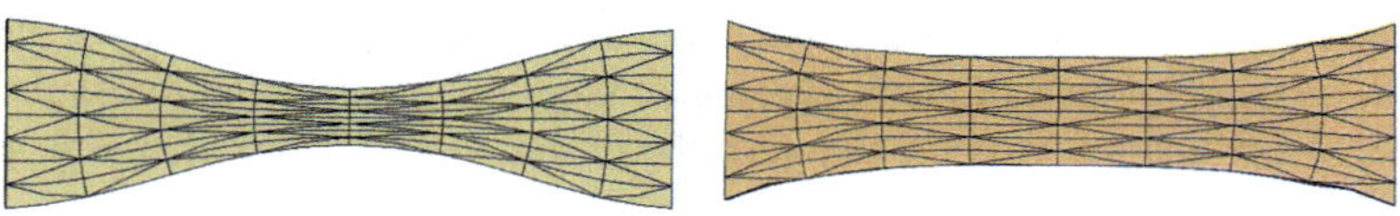

Fig. 18 Deformations states for T2 and T2-A1 element formulations

Fig. 19 Different shear zones obtained with a T2-A1 mesh of 40960 elements

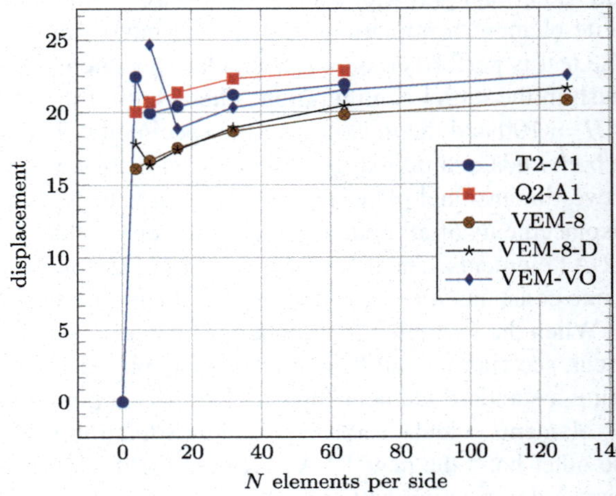

Fig. 20 Mesh convergence of the displacement u_y in the midst of the specimen

The plot in Fig. 20 shows the mesh convergence for the T2-A1 and the virtual element formulations using $N = 4, 8, 16, 32$ and 64 elements per side. As can be seen the result is insensitive with respect to the mesh size. The deviation for $N = 4$ is related to the fact that the mesh cannot model the different shear zones, see Fig. 19. It can also be observed that the convergence rate in this case is slower for the virtual element formulation than for the T2-A1 element. This is due to the approximation which is one order lower for the virtual elements.

It is worth noting that the final displacement can be reached with the T2-A1 element in one single load step for all mesh sizes, while the T2 element needs about 25 load steps to reach the final configuration. Thus the new T2-A1 element is a lot more robust than the T2 element for such applications. The virtual element formulation needs also only one load step.

Figure 21 shows the deformed mesh that was computed with the virtual element formulation using a Voronoi mesh. Again can see the large deformation state of the specimen after loading. We note that also the virtual element formulation is able to predict the correct deformed shape of the specimen.

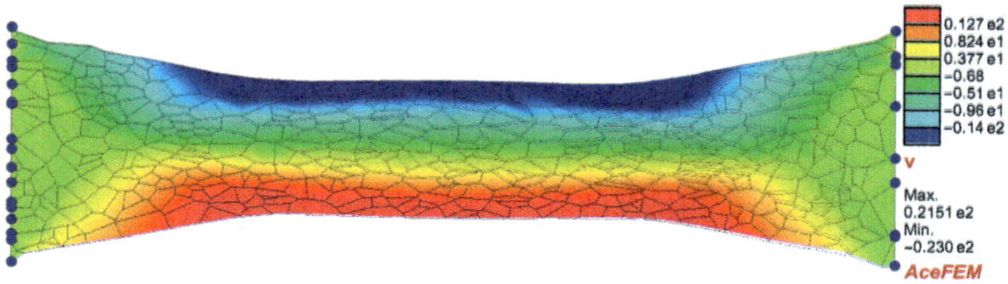

Fig. 21 Bias-extension: deformed configuration computed with a Voronoi mesh

6 Conclusions

Finite elements for large strain anisotropic behaviour were developed in this paper. Special emphasis was put on a formulation that was able to enforce inextensible fiber extensions for anisotropic materials exactly using a constraint formulation. This led to a Lagrange multiplier method with different ansatz spaces for the deformations and the Lagrangian multipliers (fiber stresses). The mixed approach shows a robust convergence behaviour and does not lock. A comparison with standard quadratic elements depicts the locking behaviour of these elements when the constraint was added via a penalty term. Furthermore the mixed approach led to a more robust behaviour in the iterative procedure needed to solve the associated nonlinear problems.

Acknowledgements The first and third author acknowledge the support of the "Deutsche Forschungsgemeinschaft" under contract of the Priority Program 1748 'Reliable simulation techniques in solid mechanics: Development of non-standard discretization methods, mechanical and mathematical analysis' under the project WR 19/50-1 and SCHR 570/23-1.

References

1. Auricchio, F., de Veiga, L.B., Lovadina, C., Reali, A.: A stability study of some mixed finite elements for large deformation elasticity problems. Comput. Methods Appl. Mech. Eng. **194**, 1075–1092 (2005)
2. Auricchio, F., da Velga, L.B., Lovadina, C., Reali, A., Taylor, R.L., Wriggers, P.: Approximation of incompressible large deformation elastic problems: some unresolved issues. Comput. Mech. **52**, 1153–1167 (2013)
3. Babuska, I.: The finite element method with lagrangian multipliers. Numer. Math. **20**(3), 179–192 (1973)
4. Babuska, I., Suri, M.: Locking effects in the finite element approximation of elasticity problems. Numer. Math. **62**(1), 439–463 (1992)
5. Bathe, K.J.: Finite Element Procedures. Prentice Hall (2006)
6. Beirão Da Veiga, L., Brezzi, F., Marini, L.: Virtual elements for linear elasticity problems. SIAM J. Numer. Anal. **51**(2), 794–812 (2013)
7. Beirão Da Veiga, L., Lovadina, C., Mora, D.: A virtual element method for elastic and inelastic problems on polytope meshes. Comput. Methods Appl. Mech. Eng. **295**, 327–346 (2015)

8. Belytschko, T., Bindeman, L.P.: Assumed strain stabilization of the 4-node quadrilateral with 1-point quadrature for nonlinear problems. Comput. Methods Appl. Mech. Eng. **88**(3), 311–340 (1991)
9. Belytschko, T., Ong, J.S.J., Liu, W.K., Kennedy, J.M.: Hourglass control in linear and nonlinear problems. Comput. Methods Appl. Mech. Eng. **43**, 251–276 (1984)
10. Boerner, E., Loehnert, S., Wriggers, P.: A new finite element based on the theory of a cosserat point—extension to initially distorted elements for 2D plane strain. Int. J. Numer. Methods Eng. **71**, 454–472 (2007)
11. Brezzi F. On the existence, uniqueness and approximation of saddle-point problems arising from lagrangian multipliers. Revue francaise d'automatique, informatique, recherche operationnelle. Anal. Numer. **8**(2), 129–151 (1974)
12. Brezzi, F., Fortin, M.: Mixed and Hybrid Finite Element Methods. Springer, Berlin (1991)
13. Chapelle, D., Bathe, K.J.: The inf-sup test. Comput. Struct. **47**, 537–545 (1993)
14. Chi, H., Beirão da Veiga, L., Paulino, G.: Some basic formulations of the virtual element method (VEM) for finite deformations. Comput. Methods Appl. Mech. Eng. (2016). https://doi.org/10.1016/j.cma.2016.12.020
15. Flanagan, D., Belytschko, T.: A uniform strain hexahedron and quadrilateral with orthogonal hour-glass control. Int. J. Num. Methods Eng. **17**, 679–706 (1981)
16. Hamila, N., Boisse, P.: Locking in simulation of composite reinforcement deformations. Analysis and treatment. Compos. Part A, 109–117 (2013)
17. Helfenstein, J., Jabareen, M., Mazza, E., Govindjee, S.: On non-physical response in models for fiber-reinforced hyperelastic materials. Int. J. Solids Struct. **47**(16), 2056–2061 (2010)
18. Holzapfel, G., Gasser, T., Ogden, R.: A new constitutive framework for arterial wall mechanics and a comparative study of material models. J. Elast. Phys. Sci. Solids **61**(1–3), 1–48 (2000)
19. Hughes, T.R.J.: The Finite Element Method. Prentice Hall, Englewood Cliffs, New Jersey (1987)
20. Korelc, J.: Automatic generation of finite-element code by simultaneous optimization of expressions. Theor. Comput. Sci. **187**, 231–248 (1997)
21. Korelc, J.: Automatic generation of numerical codes with introduction to AceGen 4.0 symbolic code generator (2000). http://www.fgg.uni-lj.si/Symech
22. Korelc, J.: Computational Templates (2016). http://www.fgg.uni-lj.si/Symech
23. Korelc, J., Solinc, U., Wriggers, P.: An improved EAS brick element for finite deformation. Comput. Mech. **46**, 641–659 (2010)
24. Korelc, J., Wriggers, P.: Automation of Finite Element Methods. Springer, Berlin (2016)
25. Krysl, P.: Mean-strain eight-node hexahedron with optimized energy-sampling stabilization for large-strain deformation. Int. J. Num. Methods Eng. **103**, 650–670 (2015)
26. Krysl, P.: Mean-strain eight-node hexahedron with stabilization by energy sampling. Int. J. Numer. Methods Eng. **103**, 437–449 (2015)
27. Krysl, P.: Mean-strain 8-node hexahedron with optimized energy-sampling stabilization. Finite Elem. Anal. Des. **108**, 41–53 (2016)
28. Loehnert, S., Boerner, E., Rubin, M., Wriggers, P.: Response of a nonlinear elastic general cosserat brick element in simulations typically exhibiting locking and hourglassing. Comput. Mech. **36**, 255–265 (2005)
29. Mueller-Hoeppe, D.S., Loehnert, S., Wriggers, P.: A finite deformation brick element with inhomogeneous mode enhancement. Int. J. Numer. Methods Eng. **78**, 1164–1187 (2009)
30. Nadler, B., Rubin, M.: A new 3-D finite element for nonlinear elasticity using the theory of a cosserat point. Int. J. Solids Struct. **40**, 4585–4614 (2003)
31. Reese, S.: On a consistent hourglass stabilization technique to treat large inelastic deformations and thermo-mechanical coupling in plane strain problems. Int. J. Numer. Methods Eng. **57**, 1095–1127 (2003)
32. Reese, S., Kuessner, M., Reddy, B.D.: A new stabilization technique to avoid hourglassing in finite elasticity. Int. J. Numer. Methods Eng. **44**, 1617–1652 (1999)
33. Reese, S., Wriggers, P.: A new stabilization concept for finite elements in large deformation problems. Int. J. Numer. Methods Eng. **48**, 79–110 (2000)

34. Sansour, C.: On the physical assumptions underlying the volumetric-isochoric split and the case of anisotropy. Eur. J. Mech. A/Solids **27**(1), 28–39 (2008)
35. Schröder, J.: Anisotropic polyconvex energies. In: Schröder, J. (ed.) Polyconvex Analysis, vol. 62, pp. 1–53. CISM, Springer, Wien (2009)
36. Schröder, J., Neff, P.: Invariant formulation of hyperelastic transverse isotropy based on polyconvex free energy functions. Int. J. Solids Struct. **40**(2), 401–445 (2003)
37. Schröder, J., Viebahn, N., Balzani, D., Wriggers, P.: A novel mixed finite element for finite anisotropic elasticity; the SKA-element simplified kinematics for anisotropy. Comput. Methods Appl. Mech. Eng. **310**, 475–494 (2016)
38. Simo, J.C., Armero, F.: Geometrically non-linear enhanced strain mixed methods and the method of incompatible modes. Int. J. Numer. Methods Eng. **33**, 1413–1449 (1992)
39. Simo, J.C., Armero, F., Taylor, R.L.: Improved versions of assumed enhanced strain tri-linear elements for 3D finite deformation problems. Comput. Methods Appl. Mech. Eng. **110**, 359–386 (1993)
40. Simo, J.C., Rifai, M.S.: A class of assumed strain methods and the method of incompatible modes. Int. J. Numer. Methods Eng. **29**, 1595–1638 (1990)
41. Simo, J.C., Taylor, R.L., Pister, K.S.: Variational and projection methods for the volume constraint in finite deformation elasto-plasticity. Comput. Methods Appl. Mech. Eng. **51**, 177–208 (1985)
42. ten Thjie, R.H.W., Akkerman, R.: Solutions to intra-ply shear locking in finite element analyses of fibre reinforced materials. Compos. Part A, 1167–1176 (2008)
43. Weiss, J.A., Maker, B.N., Govindjee, S.: Finite element implementation of incompressible, transversely isotropic hyperelasticity. Comput. Methods Appl. Mech. Eng. **135**(1), 107–128 (1996)
44. Wriggers, P.: Nonlinear Finite Elements. Springer, Berlin (2008)
45. Wriggers, P., Reddy, B., Rust, W., Hudobivnik, B.: Efficient virtual element formulations for compressible and incompressible finite deformations. Comput. Mech. (2017). https://doi.org/10.1007/s00466-017-1405-4
46. Wriggers, P., Rust, W., Reddy, B.: A virtual element method for contact. Comput. Mech. **58**, 1039–1050 (2016)
47. Wriggers, P., Schröder, J., Auricchio, F.: Finite element formulations for large strain anisotropic materials. Int. J. Adv. Model. Simul. Eng. Sci. **3**(25), 1–18 (2016)
48. Zdunek, A., Rachowicz, W., Eriksson, T.: A novel computational formulation for nearly incompressible and nearly inextensible finite hyperelasticity. Comput. Methods Appl. Mech. Eng. **281**, 220–249 (2014)
49. Zienkiewicz, O.C., Taylor, R.L.: The Finite Element Method, vol. 2, 5th edn. Butterworth-Heinemann, Oxford, UK (2000)
50. Zienkiewicz, O.C., Taylor, R.L., Too, J.M.: Reduced integration technique in general analysis of plates and shells. Int. J. Numer. Methods Eng. **3**, 275–290 (1971)

Part III
Gradient Enhanced Modeling

A Micromorphic Damage-Plasticity Model to Counteract Mesh Dependence in Finite Element Simulations Involving Material Softening

Tim Brepols, Stephan Wulfinghoff and Stefanie Reese

Abstract A gradient-extended damage-plasticity material model is presented which belongs to the class of micromophic media as proposed by Forest (J Eng Mech 135:117–131, 2009) [17]. A 'two-surface' formulation is utilized in which damage and plasticity are treated as independent but strongly coupled dissipative phenomena. To this end, separate yield and damage criteria as well as loading/unloading conditions are introduced. The model is thermodynamically consistent and accounts for both nonlinear kinematic and isotropic hardening as well as damage hardening. Various theoretical and numerical aspects of the formulation are discussed. Emphasis is also put on a procedure to enforce stress constraints at the local integration point level which provides, for instance, the basis for a straightforward integration of 3D gradient-extended material models into beam or shell elements or for their usage in 2D plane stress computations. A structural example problem illustrates the merits of the model and its ability to deliver mesh-independent results in coupled damage-plasticity finite element simulations.

1 Introduction

Studying damage and fracture processes within materials and structures is a subject of major interest in the mechanics community. The availability of today's computing power, the ongoing improvement of sophisticated simulation software as well as the increasing knowledge of the physical backgrounds of the aforementioned phenomena due to better experimental insights have provided the basis for a steadily growing

T. Brepols (✉) · S. Wulfinghoff · S. Reese
Institute of Applied Mechanics, RWTH Aachen University,
52074 Aachen, Germany
e-mail: tim.brepols@rwth-aachen.de

S. Wulfinghoff
e-mail: stephan.wulfinghoff@rwth-aachen.de

S. Reese
e-mail: stefanie.reese@rwth-aachen.de

J. Sorić et al. (eds.), *Multiscale Modeling of Heterogeneous Structures*,
Lecture Notes in Applied and Computational Mechanics 86,
https://doi.org/10.1007/978-3-319-65463-8_12

number of interesting numerical studies on this topic in more recent times (among many others, see e.g. [12, 13, 15, 27, 31, 38, 40, 41, 49]). Naturally, along with these developments has come also a considerable progress in and application of advanced and 'non-standard' material modeling techniques.

A common example is the consideration of 'nonlocal' gradient effects in damage and fracture computations. For long, it has been recognized that finite element simulations involving conventional 'local' continuum damage models may suffer from strongly mesh-dependent results which are caused by an oversimplified approach to model material softening (see e.g. [4, 7, 11, 20, 25]). Gradient-enhanced damage models can provide a way out of this dilemma by incorporating internal material length scales acting as 'localization limiters' in the underlying formulation and thus restoring its physical and mathematical soundness. For this reason and due to the above mentioned technical achievements, gradient-extended material models find an increasing application in more and more practically relevant computational investigations nowadays (e.g. [37, 39, 48, 57], to name only a few).

However, especially with regard to gradient-enhanced models coupling damage and plasticity, a lot of open questions remain and a need for further clarification and research is noticeable within the scientific community. In this regard, the present study aims at making a valuable contribution by presenting and discussing a gradient-extended damage-plasticity model which fits into the rather general framework of micromorphic media as proposed by Forest [17, 18]. The latter may be considered an approach for unifying concepts among the various existing types of gradient-enhancements in the literature that is applied by an increasing number of authors, see e.g. also [5] or [58]. A 'two-surface' formulation is utilized in which damage and plasticity are modeled as separate dissipative mechanisms using independent damage and yield criteria and corresponding sets of loading/unloading conditions. The latter fact makes the model flexibly adaptable to various cases in which the material's behavior is either only (quasi-)brittle-like, ductile-like or possibly anything in between. Nonlinear kinematic and isotropic hardening as well as damage hardening are considered and thermodynamical consistency of the model is ensured via suitable choices for the evolution equations.

An additional point is the discussion of a procedure to enforce stress constraints for 3D gradient-extended material models at the local integration point level. The method makes it possible to use these models without modifications in 2D plane stress computations or to integrate the latter e.g. in beam or shell element formulations. In a variety of different ways, many authors have worked in the past on the subject of incorporating stress constraints into conventional 'local' material models at the integration point level (among many others, see e.g. [1, 10, 14, 16, 24, 32, 33, 51]). The present study, however, is concerned with gradient-extended materials and explains e.g. how to appropriately condense the additional material tangent operators showing up due to a gradient-extension. This makes it possible to retain the quadratic rate of convergence in a Newton-Raphson iteration scheme at the global finite element level. To the best knowledge of the authors, this issue has not been presented or discussed for gradient-extended material models in the literature, yet.

Finally, the application of the just-mentioned procedure and the mesh regularization properties of the presented model are carefully examined by means of a structural 2D plane stress benchmark simulation.

2 Constitutive Theory

This section addresses the theoretical details and constitutive theory of the gradient-extended damage-plasticity model. Starting with a brief explanation of the origin of the considered type of gradient-extension and a presentation of the model's strong and weak form in Sect. 2.1, the fundamental kinematic assumptions and general thermodynamic considerations are dealt with in Sect. 2.2. In Sect. 2.3, the individual parts of the underlying Helmholtz free energy are defined, whereas Sect. 2.4 describes the derivation of the state relations and thermodynamic conjugate forces of the model based on the second law of thermodynamics. Finally, Sects. 2.5 and 2.6 are concerned with a thermodynamically consistent modeling of plasticity and damage, respectively.

2.1 Micromorphic Extension, Strong and Weak Form of the Problem

The gradient-extended model under consideration is derived from the micromorphic theory proposed by Forest [17, 18] which may be regarded as a systematic approach for constructing higher-order gradient material models from already existing 'local' counterparts. In the case of the present study, a scalar 'micromorphic' damage variable is introduced as an additional degree of freedom of the model and a further generalized micromorphic balance equation needs to be solved next to the classical balance of linear momentum. As will be described in Sect. 2.3, a coupling between the micromorphic damage variable and its local counterpart is enforced by introducing a direct dependence of the underlying free energy on the difference between the just-mentioned quantities. Omitting any details of the derivation process for brevity (the interested reader is kindly referred to the above mentioned references), the 'strong form' of the problem consists of the following partial differential equations and boundary conditions which are valid in a pointwise sense:

$$
\begin{array}{llcrl}
\underline{\text{linear momentum balance}}: & & & \underline{\text{micromorphic balance}}: & \\
\operatorname{div}(\boldsymbol{\sigma}) + \mathbf{f} = \mathbf{0} & \text{in } \Omega & & H(D - \bar{D}) + A \operatorname{div}(\nabla \bar{D}) = 0 & \text{in } \Omega \\
\boldsymbol{\sigma}[\mathbf{n}] = \hat{\mathbf{t}} & \text{on } \Gamma_t & & \nabla \bar{D} \cdot \mathbf{n} = 0 & \text{on } \Gamma \\
\mathbf{u} = \hat{\mathbf{u}} & \text{on } \Gamma_u & & &
\end{array}
\qquad (1)
$$

In the above, $\boldsymbol{\sigma}$ denotes the stress tensor, $\mathbf{u}$ is the displacement vector, D is the local damage variable and $\bar{D}$ its corresponding micromorphic counterpart. Conventional body forces acting in the domain of body Ω are represented by $\mathbf{f}$, whereas $\hat{\mathbf{t}}$ and $\hat{\mathbf{u}}$ are prescribed tractions and displacements at the Dirichlet (Γ_t) and Neumann (Γ_u) parts of the boundary, i.e. $\Gamma = \Gamma_u \cup \Gamma_t$, $\Gamma_u \cap \Gamma_t = \emptyset$. Moreover, by $\mathbf{n}$ appropriate unit outward normals at the boundary are denoted and A, H are micromorphic material parameters to be described in more detail in Sect. 2.3. It is noted that the very same set of Eq. (1) may also be used to model size effects in, for example, gradient crystal plasticity, if D is replaced by an equivalent plastic strain measure, see [59] and [60].

With (1) at hand, the 'weak form' of the problem is obtained in the usual way by multiplying the partial differential equations (linear momentum balance and micromorphic balance) by appropriate vectorial and scalar test functions $\delta\mathbf{u}$ and $\delta\bar{D}$, respectively, and integrating over the domain Ω. By further considering Gauss' theorem and taking into account the boundary conditions, the weak form of the problem follows as[1]:

$$g(\mathbf{u}, \bar{D}, \delta\mathbf{u}) := \int_\Omega (\boldsymbol{\sigma} \cdot \nabla^s \delta\mathbf{u} - \mathbf{f} \cdot \delta\mathbf{u})\, \mathrm{d}\Omega - \int_{\Gamma_t} \hat{\mathbf{t}} \cdot \delta\mathbf{u}\, \mathrm{d}\Gamma = 0 \qquad \forall \delta\mathbf{u} \tag{2}$$

$$h(\mathbf{u}, \bar{D}, \delta\bar{D}) := \int_\Omega \left(H\,(D - \bar{D})\, \delta\bar{D} - A\, \nabla\bar{D} \cdot \nabla\delta\bar{D} \right) \mathrm{d}\Omega = 0 \qquad \forall \delta\bar{D} \tag{3}$$

Since g and h are generally nonlinear, a linearization is carried out with respect to the unknowns $\mathbf{u}$ and $\bar{D}$ which leads to a coupled system of linear algebraic equations in the increments $\Delta\mathbf{u}$ and $\Delta\bar{D}$. The latter equation system is then ready for being solved e.g. by means of an iterative finite element analysis (details are omitted here).

2.2 Kinematic Assumptions and Thermodynamic Considerations

A geometrically linear theory is assumed. As such, the total strain $\boldsymbol{\varepsilon} := \nabla^s \mathbf{u}$ is separable into elastic ($\boldsymbol{\varepsilon}_e$) and plastic ($\boldsymbol{\varepsilon}_p$) parts in an additive manner. Furthermore, in order to model nonlinear kinematic hardening, an additional additive split of the plastic part into recoverable ($\boldsymbol{\varepsilon}_{p_e}$) and irrecoverable ($\boldsymbol{\varepsilon}_{p_i}$) parts is considered, leading to the following relationship:

$$\boldsymbol{\varepsilon} = \boldsymbol{\varepsilon}_e + \boldsymbol{\varepsilon}_p = \boldsymbol{\varepsilon}_e + \boldsymbol{\varepsilon}_{p_e} + \boldsymbol{\varepsilon}_{p_i} \tag{4}$$

The additional split of the plastic strain is physically motivated. According to Lion [36], $\boldsymbol{\varepsilon}_{p_e}$ accounts for dislocation-induced lattice rotations and stretches on the microscale, whereas $\boldsymbol{\varepsilon}_{p_i}$ results from local plastic deformations coming from inelastic slip on crystallographic slip systems.

[1] $\nabla^s(\bullet) := \frac{1}{2}\left[(\bullet) + (\bullet)^T\right]$ denotes the symmetric part of a quantity $(\bullet)$.

With regard to the isotropic continuum damage framework considered in this study, the scalar internal damage variable D measures the degree of material degradation at a certain point in an averaged sense. Furthermore, the model makes use of the concept of effective stress, originally introduced by Kachanov [28] and Rabotnov [43], and the hypothesis of strain equivalence, proposed by Lemaitre [34]. The effective stress is denoted and defined as

$$\tilde{\boldsymbol{\sigma}} := \frac{\boldsymbol{\sigma}}{f_{\text{dam}}(D)} \tag{5}$$

where $f_{\text{dam}}(D)$ is a scalar-valued positive and decreasing function which is unity for a virgin material ($D = 0$) and progressively approaches zero for increasing damage, until a predefined critical value is reached ($D = D_{\text{cr}}$) that represents complete material failure.

The model is based on the framework of irreversible thermodynamics (see e.g. [8, 19, 35, 52]). All thermomechanical processes within the material are characterized by means of state couples $(\mathbf{a}_i, \mathbf{A}_i)$ where index i is used to distinguish the various processes available. Quantity $\mathbf{a}_i$ is a strain-like state variable and $\mathbf{A}_i$ a thermodynamically conjugated stress-like state variable, both can be tensors of arbitrary order according to the physical phenomena they represent.

In more detail, the present damage-plasticity model is characterized in every material point by a set of couples $\{(\boldsymbol{\varepsilon}_e, \boldsymbol{\sigma}), (\boldsymbol{\varepsilon}_{p_e}, \mathbf{X}_p), (\xi_p, q_p), (D, Y), (\xi_d, q_d), (\bar{D}, a_\text{i}), (\nabla\bar{D}, \mathbf{b}_\text{i})\}$. The pairs $(\boldsymbol{\varepsilon}_e, \boldsymbol{\sigma})$, $(\boldsymbol{\varepsilon}_{p_e}, \mathbf{X}_p)$ and (ξ_p, q_p) describe the state of plastic flow, kinematic and isotropic plastic hardening, whereas (D, Y) and (ξ_d, q_d) characterize the state of damage and damage hardening, respectively. The two couples $(\bar{D}, a_\text{i})$ and $(\nabla\bar{D}, \mathbf{b}_\text{i})$ are related to the micromorphic damage and its first gradient, respectively.

2.3 State Potential in Terms of the Helmholtz Free Energy

The reversible processes taking place within the material are described by means of a state potential which is expressed in terms of the Helmholtz free energy ψ. The latter energy functional depends on the strain-like variables of the model and is assumed to be additively separable into individual parts. More precisely, there exist energetic contributions related to elasticity (ψ_e), plasticity (ψ_p), damage hardening (ψ_d) and micromorphic damage ($\psi_{\bar{d}}$), leading in total to the following expression and dependencies:

$$\psi = \psi_e(\boldsymbol{\varepsilon}_e, D) + \psi_p(\boldsymbol{\varepsilon}_{p_e}, \xi_p, D) + \psi_d(\xi_d) + \psi_{\bar{d}}(D - \bar{D}, \nabla\bar{D}) \tag{6}$$

The elastic part of the energy is defined as

$$\psi_e(\boldsymbol{\varepsilon}_e, D) := f_{\text{dam}}(D)\,\frac{1}{2}\,\boldsymbol{\varepsilon}_e \cdot \mathscr{C}[\boldsymbol{\varepsilon}_e] \tag{7}$$

The quantity $\mathscr{C} := \lambda \mathbf{I} \otimes \mathbf{I} + 2\mu\, \mathscr{I}$ denotes the isotropic fourth-order elasticity tensor in which λ, μ are the Lamé parameters, $\mathbf{I}$ is the second-order identity tensor and $\mathscr{I}$ is the fourth-order symmetric identity tensor. The plastic part of the energy is also affected by damage through function $f_{\mathrm{dam}}(D)$ and reads

$$\psi_p(\boldsymbol{\varepsilon}_{p_e}, \xi_p, D) := f_{\mathrm{dam}}(D) \left(\frac{1}{2} a \boldsymbol{\varepsilon}_{p_e} \cdot \boldsymbol{\varepsilon}_{p_e} + e \left(\xi_p + \frac{\exp(-f\, \xi_p) - 1}{f} \right) \right) \tag{8}$$

The first term inside the outer parentheses accounts for kinematic hardening, the second term is related to nonlinear isotropic hardening. The constants a, e and f are corresponding plastic material parameters. In simple analogy, the energetic part related to nonlinear damage hardening is defined as

$$\psi_d(\xi_d) := r \left(\xi_d + \frac{\exp(-s\, \xi_d) - 1}{s} \right) \tag{9}$$

with r and s being damage material parameters. Finally, the part of the energy belonging to micromorphic damage is expressed as

$$\psi_{\bar{d}}(D - \bar{D}, \nabla \bar{D}) := \frac{1}{2} H \left(D - \bar{D} \right)^2 + \frac{1}{2} A \nabla \bar{D} \cdot \nabla \bar{D} \tag{10}$$

and consists of two terms. The first term penalizes the difference between the local and the micromorphic damage variable with H being a penalty parameter, thus enforcing a certain degree of coupling between the two quantities. The second term accounts for the influence of the gradient of $\bar{D}$ in the free energy. Parameter A introduces implicitly an 'internal material length' into the formulation.

2.4 State Relations and Thermodynamic Conjugate Forces

The second law of thermodynamics in terms of the Clausius-Duhem inequality is used to derive consistent state relations of the model. The latter reads in its isothermal local form (see [17] for further information):

$$-\dot{\psi} + \boldsymbol{\sigma} \cdot \dot{\boldsymbol{\varepsilon}} + \underbrace{a_{\mathrm{i}}\, \dot{\bar{D}} + \mathbf{b}_{\mathrm{i}} \cdot \nabla \dot{\bar{D}}}_{\text{micromorphic extension}} \overset{!}{\geq} 0 \tag{11}$$

Naturally, two additional terms appear in the Clausius-Duhem inequality due to the micromorphic extension which account for the contributions of the micromorphic damage and its first gradient to the second law.

With the chain rule of differentiation and expressions (4) and (6) in mind, inequality (11) can be rewritten as:

$$\left(\boldsymbol{\sigma} - \frac{\partial \psi}{\partial \boldsymbol{\varepsilon}_e}\right) \cdot \dot{\boldsymbol{\varepsilon}} + \left(a_{\mathrm{i}} - \frac{\partial \psi}{\partial \bar{D}}\right) \dot{\bar{D}} + \left(\mathbf{b}_{\mathrm{i}} - \frac{\partial \psi}{\partial \nabla \bar{D}}\right) \cdot \nabla \dot{\bar{D}} + \\ + \frac{\partial \psi}{\partial \boldsymbol{\varepsilon}_e} \cdot \dot{\boldsymbol{\varepsilon}}_p - \frac{\partial \psi}{\partial D} \dot{D} - \frac{\partial \psi}{\partial \boldsymbol{\varepsilon}_{p_e}} \cdot \dot{\boldsymbol{\varepsilon}}_{p_e} - \frac{\partial \psi}{\partial \xi_p} \dot{\xi}_p - \frac{\partial \psi}{\partial \xi_d} \dot{\xi}_d \geq 0 \tag{12}$$

The latter expression must hold for arbitrary thermomechanical processes. One possibility is now to set the expressions in the parentheses intentionally to zero, such that the state relations of the model follow as:

$$\boldsymbol{\sigma} = \frac{\partial \psi}{\partial \boldsymbol{\varepsilon}_e} = f_{\mathrm{dam}}(D)\, \mathscr{C}[\boldsymbol{\varepsilon}_e]\,, \quad a_{\mathrm{i}} = \frac{\partial \psi}{\partial \bar{D}} = -H\,(D - \bar{D})\,, \quad \mathbf{b}_{\mathrm{i}} = \frac{\partial \psi}{\partial \nabla \bar{D}} = A\, \nabla \bar{D} \tag{13}$$

As shown, using the micromorphic approach one obtains in addition to the stress-strain relation further state relations for the thermodynamic conjugate forces a_{i} and $\mathbf{b}_{\mathrm{i}}$ to $\bar{D}$ and $\nabla \bar{D}$, respectively.

What remains left to be fulfilled is the remaining dissipation inequality which reads

$$\left(\boldsymbol{\sigma} - \mathbf{X}_p\right) \cdot \dot{\boldsymbol{\varepsilon}}_p + Y\, \dot{D} + \mathbf{X}_p \cdot \dot{\boldsymbol{\varepsilon}}_{p_i} - q_p\, \dot{\xi}_p - q_d\, \dot{\xi}_d \geq 0 \tag{14}$$

The abbreviations of the thermodynamic conjugate forces have been introduced for notational convenience and are defined as:

$$Y = -\frac{\partial \psi}{\partial D} = -\frac{\mathrm{d} f_{\mathrm{dam}}(D)}{\mathrm{d}D} \left(\frac{1}{2}\, \boldsymbol{\varepsilon}_e \cdot \mathscr{C}[\boldsymbol{\varepsilon}_e] + \frac{1}{2}\, a\, \boldsymbol{\varepsilon}_{p_e} \cdot \boldsymbol{\varepsilon}_{p_e} + \\ + e\left(\xi_p + \frac{\exp(-f\, \xi_p) - 1}{f}\right)\right) - H\,(D - \bar{D}) \tag{15}$$

$$\mathbf{X}_p = \frac{\partial \psi}{\partial \boldsymbol{\varepsilon}_{p_e}} = f_{\mathrm{dam}}(D)\, a\, \boldsymbol{\varepsilon}_{p_e} \tag{16}$$

$$q_p = \frac{\partial \psi}{\partial \xi_p} = f_{\mathrm{dam}}(D)\, e\, \left(1 - \exp(-f\, \xi_p)\right) \tag{17}$$

$$q_d = \frac{\partial \psi}{\partial \xi_d} = r\,(1 - \exp(-s\, \xi_d)) \tag{18}$$

As can be seen from Eq. (15), Y (sometimes loosely referred to as the 'energy release rate') is directly influenced by the micromorphic damage variable $\bar{D}$ and will in Sect. 2.6 be used in the definition of the damage loading function. According to this fact, a 'nonlocal' character is introduced into the constitutive equations at the local integration point level. Further, Y contains both elastic and plastic contributions which proves to be beneficial since a purely elastic 'energy release rate' would maybe be too restrictive from a physical point of view (for arguments supporting this hypothesis, see e.g. [8, 9, 26, 29, 47]).

2.5 Modeling of Plasticity

Deviatoric plastic behavior of the material is assumed for simplicity which is modeled by means of a von Mises-type yield function[2]:

$$\Phi_p(\tilde{\boldsymbol{\sigma}}, \tilde{\mathbf{X}}_p, \tilde{q}_p) = \sqrt{\frac{3}{2}} \left\| \tilde{\boldsymbol{\sigma}}' - \tilde{\mathbf{X}}_p' \right\| - (\sigma_0 + \tilde{q}_p) \leq 0 \tag{19}$$

The initial yield stress is given by material parameter σ_0 and nonlinear kinematic and isotropic hardening are taken into account via $\tilde{\mathbf{X}}_p$ and $\tilde{q}_p$, respectively. As indicated, a further coupling between plasticity and damage is considered in the formulation since Φ_p is expressed in terms of effective quantities, i.e. in terms of $\tilde{\boldsymbol{\sigma}}$ as defined in (5) and

$$\tilde{\mathbf{X}}_p := \frac{\mathbf{X}_p}{f_{\text{dam}}(D)}, \qquad \tilde{q}_p := \frac{q_p}{f_{\text{dam}}(D)} \tag{20}$$

Furthermore, the following thermodynamically consistent plastic evolution equations for the strain-like plastic internal variables are postulated:

$$\dot{\boldsymbol{\varepsilon}}_p = \dot{\lambda}_p \frac{\partial \Phi_p}{\partial \boldsymbol{\sigma}} = \sqrt{\frac{3}{2}} \frac{\dot{\lambda}_p}{f_{\text{dam}}(D)} \frac{\boldsymbol{\sigma}' - \mathbf{X}_p'}{\|\boldsymbol{\sigma}' - \mathbf{X}_p'\|} \tag{21}$$

$$\dot{\boldsymbol{\varepsilon}}_{p_i} = \frac{\dot{\lambda}_p}{f_{\text{dam}}(D)} \frac{b}{a} \tilde{\mathbf{X}}_p' = \frac{\dot{\lambda}_p}{f_{\text{dam}}(D)} b\, \boldsymbol{\varepsilon}_{p_e} \tag{22}$$

$$\dot{\xi}_p = -\dot{\lambda}_p \frac{\partial \Phi_p}{\partial q_p} = \frac{\dot{\lambda}_p}{f_{\text{dam}}(D)} \tag{23}$$

Constant b in expression (22) is an additional plastic material parameter necessary for modeling nonlinear kinematic hardening. By inspecting the remaining dissipation inequality (14), it can be easily verified that evolution equations (21)–(23) lead to a thermodynamically consistent plastic behavior of the model.

As usual, the plastic loading/unloading conditions of the model have to be taken into account, i.e.

$$\dot{\lambda}_p \geq 0, \qquad \Phi_p \leq 0, \qquad \dot{\lambda}_p \, \Phi_p = 0 \tag{24}$$

2.6 Modeling of Damage

Damage is modeled by means of the damage loading function

$$\Phi_d(Y, q_d) = Y - (Y_0 + q_d) \leq 0 \tag{25}$$

[2] $(\bullet)' := (\bullet) - \frac{1}{3}\,\text{trace}\,(\bullet)\,\mathbf{I}$ denotes the deviatoric part of a second-order tensor $(\bullet)$.

The structural analogy to plasticity is apparent. The onset of damage is represented by material parameter Y_0, whereas q_d accounts for nonlinear damage hardening.

Thermodynamically consistent evolution equations for the strain-like damage variables are postulated as

$$\dot{D} = \dot{\lambda}_d \frac{\partial \Phi_d}{\partial Y} = \dot{\lambda}_d \tag{26}$$

$$\dot{\xi}_d = -\dot{\lambda}_d \frac{\partial \Phi_d}{\partial q_d} = \dot{\lambda}_d \tag{27}$$

and, finally, the damage loading/unloading conditions need to be considered, i.e.

$$\dot{\lambda}_d \geq 0, \qquad \Phi_d \leq 0, \qquad \dot{\lambda}_d \, \Phi_d = 0 \tag{28}$$

As was mentioned in Sect. 2.4, a 'nonlocal' character of the constitutive equations exists due to the dependence of Φ_d on Y which itself is directly influenced by the micromorphic damage variable $\bar{D}$.

To summarize, there exist 12 material parameters of the model: λ, μ, σ_0, a, b, e, f, r, s, Y_0, A and H. In the present study, constant H plays the role of a penalty constant controlling the degree of coupling between D and $\bar{D}$ which is typically desired to be strong. In other words, H is intentionally chosen as a large number. All other parameters should be determined via suitable experiments.

3 Numerical and Algorithmic Aspects

The present section discusses some fundamental numerical and algorithmic aspects which are practically relevant, if the model is to be implemented into, e.g., finite element codes. More concretely speaking, Sect. 3.1 is concerned with the time-discretization of the model's evolution equations and a solution strategy for the resulting algebraic equation system, whereas Sect. 3.2 explains a procedure to enforce stress constraints for 3D gradient-extended material models at the local integration point level in finite elements.

3.1 Discretization of the Evolution Equations and Remarks on the Incremental Problem

The evolution equations (21)–(23) and (26)–(27) of the model are discretized in time by means of a fully implicit backward Euler scheme, leading to a coupled system of nonlinear algebraic equations. Considering a time interval $[t_n, t_{n+1}]$ during the Newton-Raphson iteration at the global finite element level, this equation system needs to be solved iteratively (for given values of ε and $\bar{D}$) in every integration point for the following unknowns at time t_{n+1}: $\varepsilon_p, \varepsilon_{p_i}, \xi_p, D, \xi_d, \Delta\lambda_p, \Delta\lambda_d$, where $\Delta\lambda_p = \Delta t\, \dot{\lambda}_p$ and $\Delta\lambda_d = \Delta t\, \dot{\lambda}_d$ denote the incremental plastic and damage multipliers, respectively.[3] In the actual iteration process, the variables ε_{p_i}, ξ_p, D and ξ_d do not need to be treated as additional unknowns but can be expressed in terms of the quantities ε_p, $\Delta\lambda_p$, $\Delta\lambda_d$, thus reducing the computational effort. The stress-like quantities $\boldsymbol{\sigma}, \mathbf{X}_p, q_p, Y$ and q_d at time t_{n+1} can be computed according to the formulas given in Sect. 2.4.

The resulting equation system is additionally subject to the time-discrete version of the loading/unloading conditions (24) and (28) which leads to four mutually exclusive scenarios: the step can either be (1) elastic, (2) elastic with concurrently evolving damage, (3) elastoplastic or (4) elastoplastic with concurrently evolving damage. In order to determine the correct solution, a suitable algorithmic strategy needs to be adopted. With regard to monolithic approaches in which the plastic and damage variables are computed simultaneously, a classical approach would be e.g. to apply a local active set search strategy similar as in multi-surface plasticity (see e.g. [50]). Another more recent alternative would be to reformulate the two sets of loading/unloading conditions by means of the so-called Fischer-Burmeister functional, thus rigorously eliminating the need for any kind of case differentiation within the algorithm (see e.g. [2, 3, 6, 30, 31]). Both algorithmic strategies possess advantages and disadvantages. Further details about this and other numerical issues (e.g. finite element discretization of the equations, computation of the consistent tangent operators etc.) are omitted here for brevity but are thoroughly discussed by Brepols et al. [6].

The final equations which need to be solved in case of an elastoplastic step with concurrently evolving damage are given below in a residuum format[4]:

$$\begin{aligned} \mathbf{r}_1 &= \boldsymbol{\varepsilon}_p - \boldsymbol{\varepsilon}_{p_n} - \sqrt{\frac{3}{2}} \frac{\Delta\lambda_p}{f_{\text{dam}}(D_n + \Delta\lambda_d)} \frac{\boldsymbol{\sigma}' - \mathbf{X}_p'}{\|\boldsymbol{\sigma}' - \mathbf{X}_p'\|} = \mathbf{0} \\ r_2 &= \Phi_p = 0 \\ r_3 &= \Phi_d = 0 \end{aligned} \tag{29}$$

Taking symmetry into account, the resulting number of scalar equations and unknowns to be solved for in this particular case is eight (six components of ε_p plus two scalars $\Delta\lambda_p$ and $\Delta\lambda_d$).

---

[3] In the following, if not explicitly denoted otherwise, any quantity is referred to time $t_{n+1} = t_n + \Delta t$.

[4] The corresponding residuals in case of a purely elastoplastic step or elastic step with concurrently evolving damage are obtained by deleting either r_3 or $\mathbf{r}_1$ and r_2 in (29), respectively.

3.2 Enforcing Stress Constraints for 3D Gradient-Extended Material Models in Finite Elements

In engineering applications, it is frequently desirable and convenient to work with beam or shell instead of classical solid finite elements. The validity of some of these element formulations fundamentally depends upon special zero-stress conditions prevailing within the respective elements, a fact which needs to be considered by the applied material formulation as well. However, sophisticated material models are often only available for three-dimensional problems and a consideration of such zero-stress conditions would necessitate a reformulation of these models which proves to be complicated. Similar problems can arise, if 2D plane stress computations are to be performed using a complex 3D material model, as is e.g. the case in the present study.

In this section, it is shown how arbitrary nonlinear 3D gradient-extended material models (introducing one additional degree of freedom) can be consistently integrated into beam, shell or 2D plane stress elements of the above mentioned kind which rely upon additional zero-stress conditions. For this, an iterative algorithm will be described which enforces these constraints at the local integration point level. The style of presentation is very similar to that given by Klinkel and Govindjee [33] who explained a corresponding strategy for conventional 'local' 3D material models. In the following, the method is extended to account also for gradient-extended material models of the type presented in this study.

For the development of the procedure, consider the time-discretized version of the weak form constituted by (2)–(3). During a time interval $[t_n, t_{n+1}]$ in a global Newton-Raphson iteration, the stress $\boldsymbol{\sigma}$ and the damage variable D are considered as functions of the strain $\boldsymbol{\varepsilon}$ and the micromorphic damage variable $\bar{D}$ only, i.e.[5]

$$\boldsymbol{\sigma} \equiv \boldsymbol{\sigma}(\boldsymbol{\varepsilon}, \bar{D}), \qquad D \equiv D(\boldsymbol{\varepsilon}, \bar{D}) \tag{30}$$

A consistent linearization of the weak form requires to derive $\boldsymbol{\sigma}$ and D with respect to $\boldsymbol{\varepsilon}$ and $\bar{D}$, respectively, leading to

$$\Delta\boldsymbol{\sigma} = \mathscr{A}[\Delta\boldsymbol{\varepsilon}] + \mathbf{B}\,\Delta\bar{D}, \qquad \Delta D = \mathbf{E}\cdot\Delta\boldsymbol{\varepsilon} + F\,\Delta\bar{D} \tag{31}$$

where $\mathscr{A} := \frac{\partial\boldsymbol{\sigma}}{\partial\boldsymbol{\varepsilon}}\Big|_{\bar{D}=\text{const}}$ (fourth-order tensor), $\mathbf{B} := \frac{\partial\boldsymbol{\sigma}}{\partial\bar{D}}\Big|_{\boldsymbol{\varepsilon}=\text{const}}$, $\mathbf{E} := \frac{\partial D}{\partial\boldsymbol{\varepsilon}}\Big|_{\bar{D}=\text{const}}$ (second-order tensors) and $F := \frac{\partial D}{\partial\bar{D}}\Big|_{\boldsymbol{\varepsilon}=\text{const}}$ (scalar).

[5]Remember that, during the considered time interval, $\boldsymbol{\sigma}$ and D do additionally depend on the history variables at time t_n which are, however, constant within $[t_n, t_{n+1}]$.

In all the element formulations considered in this section, certain components of the stress tensor vanish. Assuming a standard Cartesian frame, these are components $\{\sigma^{22}, \sigma^{33}, \sigma^{23}\}$ in case of beam formulations, whereas σ^{33} and $\{\sigma^{33}, \sigma^{13}, \sigma^{23}\}$ are assumed to be zero in case of shell and 2D plane stress elements, respectively. To describe a single algorithm which is applicable in all cases, a matrix-vector notation of all quantities is adopted[6] and the expressions in (31) are rewritten in the following partitioned matrix format:

$$\begin{bmatrix} \Delta\boldsymbol{\sigma}_m \\ \Delta\boldsymbol{\sigma}_z \end{bmatrix} = \begin{bmatrix} \mathscr{A}_{mm} & \mathscr{A}_{mz} \\ \mathscr{A}_{zm} & \mathscr{A}_{zz} \end{bmatrix} \begin{bmatrix} \Delta\boldsymbol{\varepsilon}_m \\ \Delta\boldsymbol{\varepsilon}_z \end{bmatrix} + \begin{bmatrix} \mathbf{B}_m \\ \mathbf{B}_z \end{bmatrix} \Delta\bar{D} \tag{32}$$

$$\Delta D = \begin{bmatrix} \mathbf{E}_m & \mathbf{E}_z \end{bmatrix} \begin{bmatrix} \Delta\boldsymbol{\varepsilon}_m \\ \Delta\boldsymbol{\varepsilon}_z \end{bmatrix} + F\,\Delta\bar{D} \tag{33}$$

In the respective cases, $\boldsymbol{\sigma}_m$ and $\boldsymbol{\sigma}_z$ are defined as

beam: $\boldsymbol{\sigma}_m = \left[\sigma^{11}, \sigma^{12}, \sigma^{13}\right]^{\mathrm{T}}, \quad \boldsymbol{\sigma}_z = \left[\sigma^{22}, \sigma^{33}, \sigma^{23}\right]^{\mathrm{T}} = \mathbf{0}$ (34)

shell: $\boldsymbol{\sigma}_m = \left[\sigma^{11}, \sigma^{22}, \sigma^{12}, \sigma^{13}, \sigma^{23}\right]^{\mathrm{T}}, \quad \boldsymbol{\sigma}_z = \left[\sigma^{33}\right] = \mathbf{0}$ (35)

2D plane stress: $\boldsymbol{\sigma}_m = \left[\sigma^{11}, \sigma^{22}, \sigma^{12}\right]^{\mathrm{T}}, \quad \boldsymbol{\sigma}_z = \left[\sigma^{33}, \sigma^{13}, \sigma^{23}\right]^{\mathrm{T}} = \mathbf{0}$ (36)

The vectors $\boldsymbol{\sigma}_m$ and $\boldsymbol{\varepsilon}_m$ contain the components of the stress and strain which are actually included in the weak form of the underlying formulation.

An algorithm enforcing the stress constraint(s) at the local integration point level can be obtained by arguing that $\boldsymbol{\sigma}_z$ should vanish. Expanding this condition in a Taylor series in terms of the unknown strain $\boldsymbol{\varepsilon}_z$ delivers

$$\boldsymbol{\sigma}_z^{(i+1)} = \boldsymbol{\sigma}_z^{(i)} + \left.\frac{\partial\boldsymbol{\sigma}_z}{\partial\boldsymbol{\varepsilon}_z}\right|^{(i)} \Delta\boldsymbol{\varepsilon}_z + \cdots \stackrel{!}{=} \mathbf{0} \tag{37}$$

where a superscript denotes the corresponding local iteration number. By neglecting higher-order terms and identifying $\mathscr{A}_{zz}^{(i)} = (\partial\boldsymbol{\sigma}_z/\partial\boldsymbol{\varepsilon}_z)|^{(i)}$, expression (37) can then be rewritten as

$$\Delta\boldsymbol{\varepsilon}_z = -\mathscr{A}_{zz}^{(i)^{-1}}\,\boldsymbol{\sigma}_z^{(i)} \tag{38}$$

which suggests the following iterative procedure for the calculation of $\boldsymbol{\varepsilon}_z$ at time t_{n+1}:

$$\boldsymbol{\varepsilon}_z^{(i+1)} \longleftarrow \boldsymbol{\varepsilon}_z^{(i)} + \Delta\boldsymbol{\varepsilon}_z, \qquad \boldsymbol{\varepsilon}_z^{(0)} := \boldsymbol{\varepsilon}_z \text{ at time } t_n \tag{39}$$

The iteration must be performed until $\|\boldsymbol{\sigma}_z^{(i)}\| \approx 0$; the rate of convergence is quadratic. As indicated, the algorithm requires to additionally store the converged components of $\boldsymbol{\varepsilon}_z$ during the computation.

[6]No extra symbols are introduced to avoid an excessive proliferation of notation. The change in meaning of the quantities is implicitly understood.

A new point discussed here is the computation of the appropriate material tangent moduli in case of gradient-extended material models. The latter are necessary to retain a quadratic rate of convergence in a Newton-Raphson iteration at the global finite element level. For this, one needs to compute the variations of $\boldsymbol{\sigma}_m$ and D with respect to $\boldsymbol{\varepsilon}_m$ and $\bar{D}$, respectively. Since these quantities do implicitly depend on the complete strain state $\boldsymbol{\varepsilon}$, a proper condensation procedure is necessary as follows:

Remember that $\Delta\boldsymbol{\sigma}_z = \mathbf{0}$ and plug this condition into the second equation of (32), leading to

$$\Delta\boldsymbol{\varepsilon}_z = -\mathscr{A}_{zz}^{-1}\mathscr{A}_{zm}\,\Delta\boldsymbol{\varepsilon}_m - \mathscr{A}_{zz}^{-1}\mathbf{B}_z\,\Delta\bar{D} \tag{40}$$

Inserting this result both into the first equation of (32) and Eq. (33) delivers, after simple rearranging, the expressions:

$$\Delta\boldsymbol{\sigma}_m = \widehat{\mathscr{A}}\,\Delta\boldsymbol{\varepsilon}_m + \widehat{\mathbf{B}}\,\Delta\bar{D}, \qquad \Delta D = \widehat{\mathbf{E}}\,\Delta\boldsymbol{\varepsilon}_m + \widehat{F}\,\Delta\bar{D} \tag{41}$$

where $\widehat{\mathscr{A}}$, $\widehat{\mathbf{B}}$, $\widehat{\mathbf{E}}$ and $\widehat{F}$ denote the appropriate material tangent moduli of the stress-constrained problem in matrix format:

$$\widehat{\mathscr{A}} := \left.\frac{\partial\boldsymbol{\sigma}_m}{\partial\boldsymbol{\varepsilon}_m}\right|_{\bar{D}=\text{const}} = \mathscr{A}_{mm} - \mathscr{A}_{mz}\mathscr{A}_{zz}^{-1}\mathscr{A}_{zm} \tag{42}$$

$$\widehat{\mathbf{B}} := \left.\frac{\partial\boldsymbol{\sigma}_m}{\partial\bar{D}}\right|_{\boldsymbol{\varepsilon}=\text{const}} = \mathbf{B}_m - \mathscr{A}_{mz}\mathscr{A}_{zz}^{-1}\mathbf{B}_z \tag{43}$$

$$\widehat{\mathbf{E}} := \left.\frac{\partial D}{\partial\boldsymbol{\varepsilon}_m}\right|_{\bar{D}=\text{const}} = \mathbf{E}_m - \mathbf{E}_z\mathscr{A}_{zz}^{-1}\mathscr{A}_{zm} \tag{44}$$

$$\widehat{F} := \left.\frac{\partial D}{\partial\bar{D}}\right|_{\boldsymbol{\varepsilon}=\text{const}} = F - \mathbf{E}_z\mathscr{A}_{zz}^{-1}\mathbf{B}_z \tag{45}$$

The above procedure can equally be applied in case of a geometrically nonlinear theory. It could furthermore readily be extended to account for more general cases with multiple gradient variables.

4 Numerical Example

In the following, in order to illustrate the merits of the gradient-extended material model and to test the functionality of the procedure presented in Sect. 3.2, a 2D plane stress finite element problem is considered. A tensile specimen with length $\ell =$ 200 mm, width 80 mm and two sharp vertical notches of 16 mm length (zero width) located at the upper and lower edges is subjected to a state of uniaxial tension by

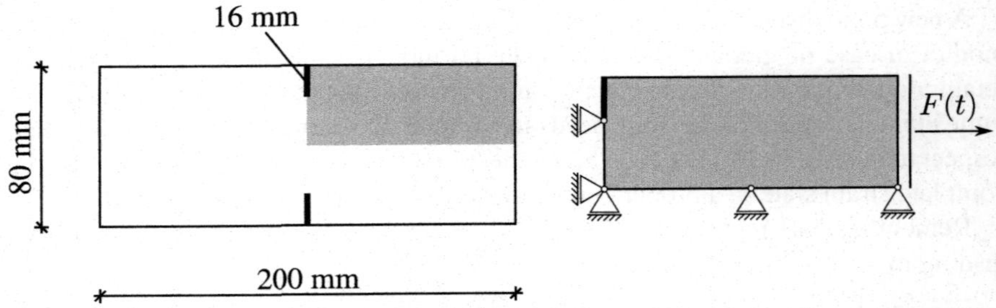

(a) Geometry and dimensions (b) Boundary-value problem

Fig. 1 Double-edge notched specimen

applying a horizontal force $F(t)$ at the left and right ends of the sample, respectively. Note that the applied force $F(t)$ is (pseudo-)time-dependent: the tensile specimen is first loaded gradually up to the point of maximum load $F_{\max}$ and further deformed afterwards at a decreasing load level (due to evolving damage) until the structure may essentially be regarded as 'broken'.

The geometry and its dimensions are shown in Fig. 1a. By exploiting symmetry conditions, the original boundary-value problem can be reduced to a quarter of its original size, see Fig. 1b. The loaded edge on the right is constrained such that it remains straight and perpendicular to the loading direction throughout the analysis. Moreover, note that, due to the assumed elastoplastic isotropy of the model, the plane stress condition implies that components σ_{13}, σ_{23} and ε_{13}, ε_{23} of the stress and strain tensor automatically vanish, meaning that $\boldsymbol{\sigma}_z$ in Eq. (36) as well as $\boldsymbol{\varepsilon}_z$ directly reduce to $\boldsymbol{\sigma}_z = [\sigma_{33}] = 0$ and $\boldsymbol{\varepsilon}_z = [\varepsilon_{33}]$, respectively.

The regularization properties of the model shall be investigated by conducting a convergence study using different mesh-sizes (640, 1715, 4415 and 8330 elements) and the following set of material parameters: $\lambda = 5000$ MPa, $\mu = 7500$ MPa, $a = 1900$ MPa, $b = 8.5$, $e = 400$ MPa, $f = 2.5$, $r = 0.5$ MPa, $s = 0.1$, $\sigma_0 = 20$ MPa, $Y_0 = 0.1$ MPa, $H = 10^5$ MPa and $A = 10$ MPa mm^2. The function in Eqs. (5) and (20) is defined as $f_{\text{dam}}(D) = (1 - D)^2$. As will be illustrated below, this choice allows for very stable computations up to values of $D = D_{\text{cr}} \approx 1$. Linear finite element approximations of the displacement field and the micromorphic damage field are considered. The meshes are strongly refined in the region of the notch tip(s) and the area(s) where damage is expected to develop during the process, see Fig. 2 illustrating exemplarily both the coarsest and finest meshes with 640 and 8330 elements, respectively.[7]

In the process, damage starts to develop at the notch tip(s) and from there on progresses into the interior of the specimen until the latter may be regarded as fully 'broken', meaning that, practically, no further load can be carried by the structure.

[7]Quadrilateral elements are preferred in the meshing process, triangular ones are only used rarely as transitional elements.

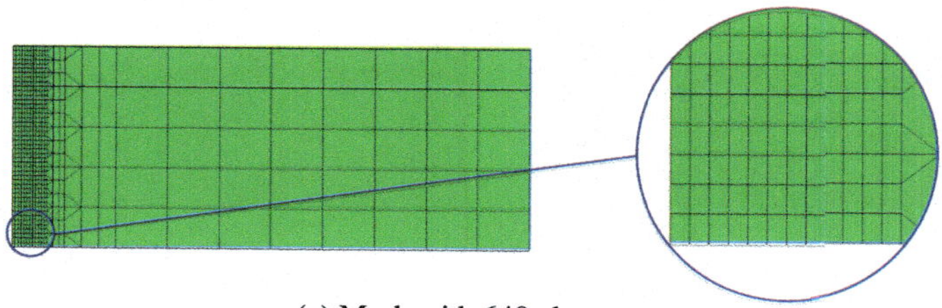

(a) Mesh with 640 elements

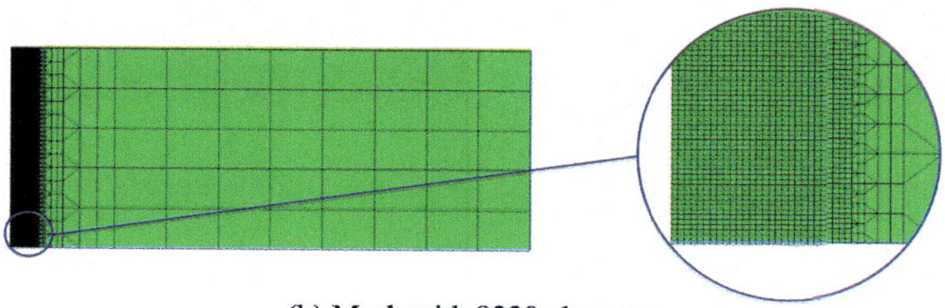

(b) Mesh with 8330 elements

Fig. 2 Mesh refinement

Three different stages during the process are exemplarily illustrated in Fig. 3. The corresponding accumulated plastic strain is shown as well.[8]

Figure 4 shows the global response of the structure, i.e. the normalized load-displacement curves obtained by using the different meshes. Here, u denotes again the horizontal displacements of the loaded ends at the left and right side, respectively. Mesh convergence is recognized, the curves lie very closely together. The points at which the damage contours of Fig. 3 were taken are indicated by small rectangular boxes.

A noticeable snapback of the load-displacement curve is visible shortly after exceeding the point of maximum load which deserves further clarification. It should be kept in mind that a quasi-static analysis is performed of a process which is accompanied by dynamic effects at the point when severe strain localization sets in and the 'crack' starts to run through the specimen. If these effects were taken into account, the load-displacement curve would show a 'sudden' vertical drop when the tangent to the curve becomes infinite. This usually marks the beginning of dynamic crack propagation which cannot be modeled using a purely quasi-static framework (see e.g. [21–23] for more detailed explanations). Instead, an arclength method is used in the present study to compute the equilibrium solution of the process, practically considering the 'crack' to grow quasi-statically.

[8]The top and bottom values in the legends of the contour plots always indicate the corresponding minimum and maximum values attained in the computations.

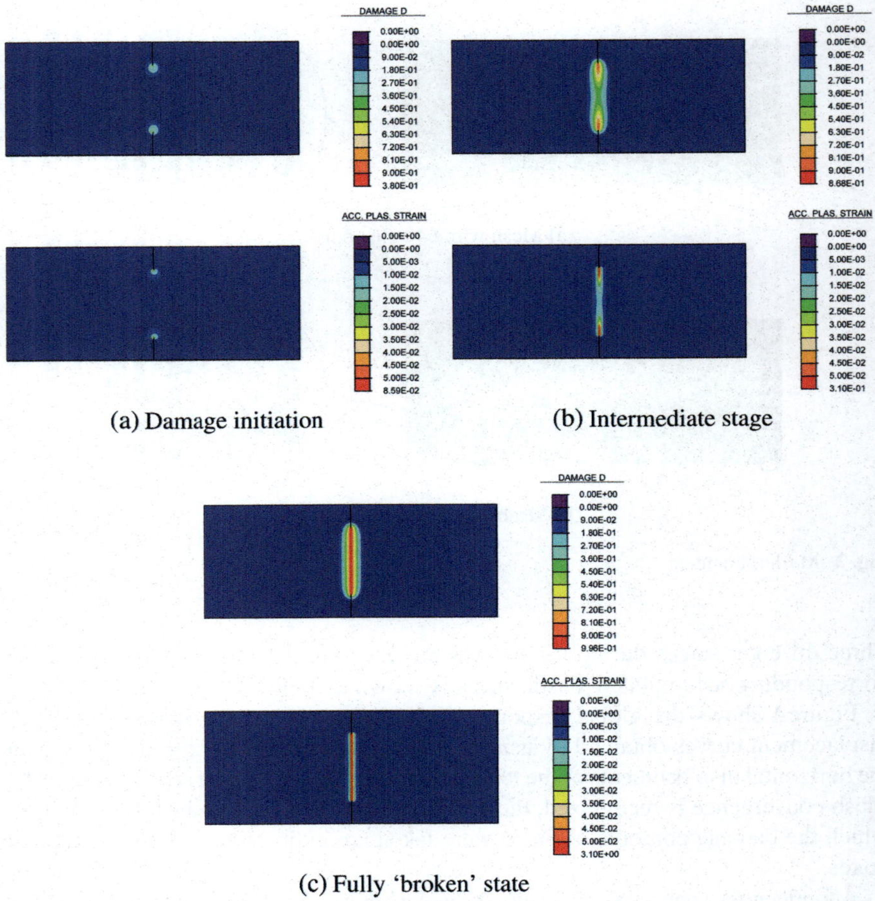

Fig. 3 Damage contours D (top) and accumulated plastic strain (bottom) during three different stages of the process (8330 elements)

Convergence of the local fields is exemplarily demonstrated in Fig. 5 for the case of the damage variable D. For brevity, only the end states of the process for the coarsest and finest meshes are shown, respectively. The highly damaged zone ($D \geq 0.9$, marked in red color) is confined to a relatively thin area (representing a crack in a continuous sense) and slightly wider in case of the coarser mesh; however, the differences between the plots can be considered negligibly small.

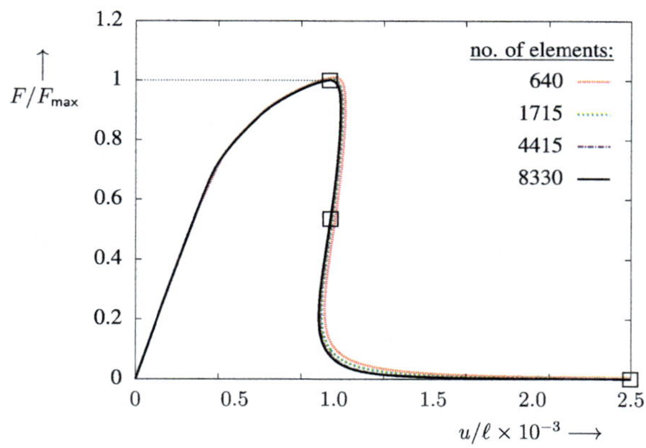

Fig. 4 Convergence study: load-displacement curves

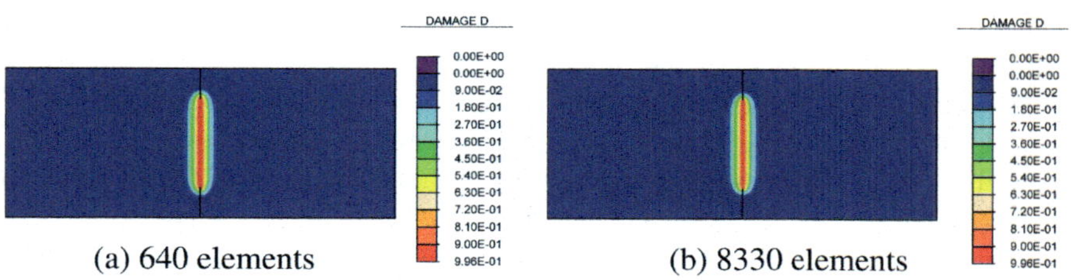

Fig. 5 Convergence study: damage contours D at the end of the simulation

5 Conclusions

A gradient-enhanced damage-plasticity material model was presented and discussed in this study which can be considered as a particular instance of the general class of micromorphic media as proposed by Forest [17, 18]. The gradient-extension is characterized by a Helmholtz-type partial differential equation for a newly introduced 'micromorphic' (i.e. nonlocal) damage variable which needs to be solved in addition to the classical balance of linear momentum. In finite element computations, this micromorphic damage variable shows up as a further degree of freedom at the nodes (similar e.g. to the temperature in thermomechanically coupled problems). Damage and plasticity are regarded as distinct physical mechanisms which was realized by postulating individual damage and yield criteria as well as corresponding sets of loading/unloading conditions. Nonlinear kinematic and isotropic hardening as well as damage hardening are considered in the formulation.

Besides a detailed description of the model's equations and some remarks about its numerical implementation into finite element codes, the present study focused also on a procedure to locally enforce stress constraints in computations involving 3D gradient-extended material models. Special emphasis was put here on a condensation procedure for the four consistent tangent operators which is necessary to retain the rate of quadratic convergence in a Newton-Raphson iteration scheme at the global finite element level. The procedure can e.g. be utilized to easily carry out 2D plane stress computations using 3D gradient-extended material models or to incorporate the latter in a straightforward manner in beam or shell elements without the need of a (possibly) complicated reformulation of the model's constitutive equations.

Finally, a numerical benchmark test was conducted in which the evolution of damage within a 2D double-edge notched specimen was examined. To demonstrate the practical applicability of the above mentioned procedure, the underlying mechanical problem was assumed to be plane stress. The results showed that the micromorphic damage-plasticity model is able to suitably regularize the mesh-dependence which is otherwise present for finite element simulations involving conventional 'local' continuum damage models. Stable computations were performed up to very high values of the damage variable at which the structure could essentially be regarded as 'broken'.

The promising results of the study naturally pave the way for further possible enhancements of the model. Currently, the authors are working on a finite strain extension which relies upon a multiplicative description of plasticity and nonlinear kinematic hardening in the spirit of [54, 55]. Furthermore, the incorporation of suitable finite element technology (see e.g. [44–46]) into the formulation in order to avoid artificial locking effects and concurrently reducing the computational effort seems sensible, especially with regard to large plastic deformations. Another interesting point would be to take account of anisotropic plasticity and/or damage (for related works of the authors on the mentioned subject, see e.g. [42, 53, 56]).

References

1. Aravas, N.: On the numerical integration of a class of pressure-dependent plasticity models. Int. J. Numer. Meth. Eng. **24**(7), 1395–1416 (1987)
2. Auricchio, F., Bonetti, E., Scalet, G., Ubertini, F.: Theoretical and numerical modeling of shape memory alloys accounting for multiple phase transformations and martensite reorientation. Int. J. Plast. **59**, 30–54 (2014)
3. Bartel, T., Hackl, K.: A micromechanical model for martensitic phase-transformations in shape-memory alloys based on energy-relaxation. ZAMM J. Appl. Math. Mech. **89**(10), 792–809 (2009)
4. Bažant, Z.P., Belytschko, T., Chang, T.-P.: Continuum theory for strain-softening. J. Eng. Mech. **110**(12), 1666–1692 (1984)
5. Bayerschen, E., Stricker, M., Wulfinghoff, S., Weygand, D., Böhlke, T.: Equivalent plastic strain gradient plasticity with grain boundary hardening and comparison to discrete dislocation dynamics. Proc. R. Soc. A **471**(2184) (2015)
6. Brepols, T., Wulfinghoff, S., Reese, S.: Gradient-extended two-surface damage-plasticity: micromorphic formulation and numerical aspects. Int. J. Plast. **97**, 64–106 (2017)
7. Cervera, M., Chiumenti, M.: Mesh objective tensile cracking via a local continuum damage model and a crack tracking technique. Comput. Meth. Appl. Mech. Eng. **196**(1–3), 304–320 (2006)

8. Chaboche, J.-L.: Thermodynamics of local state: overall aspects and micromechanics based constitutive relations. Tech. Mech. **23**(2–4), 113–119 (2003)
9. Chow, C.L., Wang, J.: An anisotropic theory of continuum damage mechanics for ductile fracture. Eng. Fract. Mech **27**(5), 547–558 (1987)
10. de Borst, R.: The zero-normal-stress condition in plane-stress and shell elastoplasticity. Commun. Appl. Numer. Meth. **7**(1), 29–33 (1991)
11. de Borst, R., Sluys, L.J., Mühlhaus, H.-B., Pamin, J.: Fundamental issues in finite element analyses of localization of deformation. Eng. Comput. **10**(2), 99–121 (1993)
12. Dimitrijevic, B.J., Hackl, K.: A method for gradient enhancement of continuum damage models. Tech. Mech. **28**(1), 43–52 (2008)
13. Dimitrijevic, B.J., Hackl, K.: A regularization framework for damage-plasticity models via gradient enhancement of the free energy. Int. J. Numer. Method. Biomed. Eng. **27**(8), 1199–1210 (2011)
14. Dodds, R.H.: Numerical techniques for plasticity computations in finite element analysis. Comput. Struct. **26**(5), 767–779 (1987)
15. Duda, F.P., Ciarbonetti, A., Sánchez, P.J., Huespe, A.E.: A phase-field/gradient damage model for brittle fracture in elastic-plastic solids. Int. J. Plast. **65**, 269–296 (2015)
16. Dvorkin, E.N., Pantuso, D., Repetto, E.A.: A formulation of the MITC4 shell element for finite strain elasto-plastic analysis. Comput. Meth. Appl. Mech. Eng. **125**(1), 17–40 (1995)
17. Forest, S.: Micromorphic approach for gradient elasticity, viscoplasticity, and damage. J. Eng. Mech. **135**(3), 117–131 (2009)
18. Forest, S.: Nonlinear regularization operators as derived from the micromorphic approach to gradient elasticity, viscoplasticity and damage. Proc. R. Soc. A **472**(2188), 20150755 (27 pages) (2016)
19. Germain, P., Nguyen, Q.S., Suquet, P.: Continuum thermodynamics. J. Appl. Mech. **50**(4b), 1010–1020 (1983)
20. Grassl, P., Jirásek, M.: On mesh bias of local damage models for concrete. In: Proceedings of FraMCoS-5, pp. 252–262. Vail, USA (2004)
21. Heinrich, C., Aldridge, M., Wineman, A.S., Kieffer, J., Waas, A.M., Shahwan, K.W.: The role of curing stresses in subsequent response, damage and failure of textile polymer composites. J. Mech. Phys. Solid. **61**(5), 1241–1264 (2013)
22. Hütter, G., Linse, T., Mühlich, U., Kuna, M.: Simulation of ductile crack initiation and propagation by means of a non-local gurson-model. Int. J. Solid. Struct. **50**(5), 662–671 (2013)
23. Hütter, G., Roth, T.L.S., Mühlich, U., Kuna, M.: A modeling approach for the complete ductile-brittle transition region: cohesive zone in combination with a non-local gurson-model. Int. J. Fract. **185**(1), 129–153 (2014)
24. Jetteur, P.: Implicit integration algorithm for elastoplasticity in plane stress analysis. Eng. Comput. **3**(3), 251–253 (1986)
25. Jirásek, M., Grassl, P.: Evaluation of directional mesh bias in concrete fracture simulations using continuum damage models. Eng. Fract. Mech **75**(8), 1921–1943 (2008)
26. Ju, J.W.: On energy-based coupled elastoplastic damage theories: constitutive modeling and computational aspects. Int. J. Solid. Struct. **25**(7), 803–833 (1989)
27. Junker, P., Schwarz, S., Makowsk, J., Hackl, K.: A relaxation-based approach to damage modeling. Continuum Mech. Therm. **29**(1), 291–310 (2016)
28. Kachanov, L.M.: 'Time of the rupture process under creep conditions', Izvestiya Akademii Nauk SSSR. Otdelenie Tekhnicheskikh Nauk **8**, 26–31 (1958)
29. Kachanov, M.: Elastic solids with many cracks and related problems. In: Hutchinson, J. W., Wu, T. Y. (eds.) Advances in Applied Mechanics, vol. 30, pp. 259–445. Elsevier (1993)
30. Kiefer, B., Bartel, T., Menzel, A.: Implementation of numerical integration schemes for the simulation of magnetic sma constitutive response. Smart Mater. Struct. **21**(9), 094007 (8 pages) (2012)
31. Kiefer, B., Waffenschmidt, T., Sprave, L., Menzel, A.: A gradient-enhanced damage model coupled to plasticity—multi-surface formulation and algorithmic concepts. Int. J. Damage Mech. 1–43 (2017)

32. Kirchner, E., Reese, S., Wriggers, P.: A finite element method for plane stress problems with large elastic and plastic deformations. Commun. Numer. Meth. Eng. **13**(12), 963–976 (1997)
33. Klinkel, S., Govindjee, S.: Using finite strain 3D-material models in beam and shell elements. Eng. Comput. **19**(8), 902–921 (2002)
34. Lemaitre, J.: Evaluation of dissipation and damage in metals submitted to dynamic loading. In: Proceedings International Conference Mechanical Behavior of Materials, vol. 1. Kyoto, Japan (1971)
35. Lemaitre, J., Chaboche, J.-L.: Mechanics of Solid Materials, 1st edn. Cambridge University Press (1990)
36. Lion, A.: Constitutive modelling in finite thermoviscoplasticity: a physical approach based on nonlinear rheological models. Int. J. Plast. **16**(5), 469–494 (2000)
37. Miehe, C.: Variational gradient plasticity at finite strains. Part I: mixed potentials for the evolution and update problems of gradient-extended dissipative solids. Comput. Meth. Appl. Mech. Eng. **268**, 677–703 (2014)
38. Miehe, C., Aldakheel, F., Raina, A.: Phase field modeling of ductile fracture at finite strains: a variational gradient-extended plasticity-damage theory. Int. J. Plast. **84**, 1–32 (2016)
39. Miehe, C., Welschinger, F., Aldakheel, F.: Variational gradient plasticity at finite strains. Part II: local-global updates and mixed finite elements for additive plasticity in the logarithmic strain space. Comput. Meth. Appl. Mech. Eng. **268**, 704–734 (2014)
40. Mozaffari, N., Voyiadjis, G.Z.: Coupled gradient damage—viscoplasticty model for ductile materials: phase field approach. Int. J. Plast. **83**, 55–73 (2016)
41. Naderi, M., Jung, J., Yang, Q.D.: A three dimensional augmented finite element for modeling arbitrary cracking in solids. Int. J. Fract. **197**(2), 147–168 (2016)
42. Pietryga, M.P., Vladimirov, I.N., Reese, S.: A finite deformation model for evolving flow anisotropy with distortional hardening including experimental validation. Mech. Mater. **44**, 163–173 (2012)
43. Rabotnov, Y. N.: Paper 68: on the equation of state of creep. Proc. Inst. Mech. Eng. (Conf. Proc.) **178**(1), 117–122 (1963)
44. Reese, S.: On the equivalent of mixed element formulations and the concept of reduced integration in large deformation problems. Int. J. Nonlinear Sci. Numer. Simul. **3**(1), 1–34 (2002)
45. Reese, S.: On a consistent hourglass stabilization technique to treat large inelastic deformations and thermo-mechanical coupling in plane strain problems. Int. J. Numer. Meth. Eng. **57**(8), 1095–1127 (2003)
46. Reese, S.: On a physically stabilized one point finite element formulation for three-dimensional finite elasto-plasticity. Comput. Meth. Appl. Mech. Eng. **194**(45), 4685–4715 (2005)
47. Saanouni, K., Forster, C., Ben Hatira, F.: On the anelastic flow with damage. Int. J. Damage Mech. **3**(2), 140–169 (1994)
48. Saanouni, K., Hamed, M.: Micromorphic approach for finite gradient-elastoplasticity fully coupled with ductile damage: Formulation and computational aspects. Int. J. Solid. Struct. **50**(14–15), 2289–2309 (2013)
49. Seabra, M.R.R., Šuštarič, P., Cesar de Sa, J.M.A., Rodič, T.: Damage driven crack initiation and propagation in ductile metals using XFEM. Comput. Mech. **52**(1), 161–179 (2013)
50. Simo, J.C., Kennedy, J.G., Govindjee, S.: Non-smooth multisurface plasticity and viscoplasticity. Loading/unloading conditions and numerical algorithms. Int. J. Numer. Meth. Eng. **26**(10), 2161–2185 (1988)
51. Simo, J.C., Taylor, R.L.: A return mapping algorithm for plane stress elastoplasticity. Int. J. Numer. Meth. Eng. **22**(3), 649–670 (1986)
52. Truesdell, C., Noll, W.: The non-linear field theories of mechanics. In: Encyclopedia of Physics, vol. III/3. Springer, Berlin, Heidelberg (1965)
53. Vladimirov, I.N., Pietryga, M.P., Kiliclar, Y., Tini, V., Reese, S.: Failure modelling in metal forming by means of an anisotropic hyperelastic-plasticity model with damage. Int. J. Damage Mech. **23**(8), 1096–1132 (2014)
54. Vladimirov, I.N., Pietryga, M.P., Reese, S.: On the modelling of non-linear kinematic hardening at finite strains with application to springback—comparison of time integration algorithms. Int. J. Numer. Meth. Eng. **75**(1), 1–28 (2008)

55. Vladimirov, I.N., Pietryga, M.P., Reese, S.: Prediction of springback in sheet forming by a new finite strain model with nonlinear kinematic and isotropic hardening. J. Mater. Process. Tech. **209**(8), 4062–4075 (2009)
56. Vladimirov, I.N., Pietryga, M.P., Reese, S.: Anisotropic finite elastoplasticity with nonlinear kinematic and isotropic hardening and application to sheet metal forming. Int. J. Plast. **26**(5), 659–687 (2010)
57. Waffenschmidt, T., Polindara, C., Menzel, A., Blanco, S.: A gradient-enhanced large-deformation continuum damage model for fibre-reinforced materials. Comput. Meth. Appl. Mech. Eng. **268**, 801–842 (2014)
58. Wulfinghoff, S., Bayerschen, E., Böhlke, T.: A gradient plasticity grain boundary yield theory. Int. J. Plast. **51**, 33–46 (2013)
59. Wulfinghoff, S., Böhlke, T.: Equivalent plastic strain gradient enhancement of single crystal plasticity: theory and numerics. Proc. R. Soc. A **468**(2145), 2682–2703 (2012)
60. Ziemann, M., Chen, Y., Kraft, O., Bayerschen, E., Wulfinghoff, S., Kirchlechner, C., Tamura, N., Böhlke, T., Walter, M., Gruber, P.A.: Deformation patterns in cross-sections of twisted bamboo-structured Au microwires. Acta Mater. **97**, 216–222 (2015)

Modeling of Material Deformation Responses Using Gradient Elasticity Theory

Jurica Sorić, Tomislav Lesičar, Filip Putar and Zdenko Tonković

Abstract Realistic description of material deformation responses demands more accurate modeling at both macroscopic and microscopic scales. Multiscale techniques employing several homogenization schemes are mostly used, in which a transition between nonlocal and local continuum formulations has been performed. Therein the transition of state variables is not defined fully consistently. In the present contribution a novel multiscale approach is proposed, where the same nonlocal theories at both scales are coupled, and discretisation is performed only by means of the C^1 finite element based on the strain gradient theory. The advantage of the new computational procedure is discussed in comparison with the approach using a local concept at microlevel. Employing the strain gradient continuum theory, a damage model for quasi-brittle materials is proposed and embedded into the C^1 continuity triangular finite element. The softening response of homogeneous materials under assumption of isotropic damage law is considered. The regularization superiority over the conventional implicit gradient enhancement procedure is demonstrated.

Keywords Multiscale approach · Nonlocal continuum · Heterogeneous materials · C^1 continuity finite element · Quasi brittle damage · Strain gradient formulation

J. Sorić (✉) · T. Lesičar · F. Putar · Z. Tonković
Faculty of Mechanical Engineering and Naval Architecture, University of Zagreb,
Ivana Lučića 5, 10000 Zagreb, Croatia
e-mail: jurica.soric@fsb.hr

T. Lesičar
e-mail: tomislav.lesicar@fsb.hr

F. Putar
e-mail: filip.putar@fsb.hr

Z. Tonković
e-mail: zdenko.tonkovic@fsb.hr

J. Sorić et al. (eds.), *Multiscale Modeling of Heterogeneous Structures*,
Lecture Notes in Applied and Computational Mechanics 86,
https://doi.org/10.1007/978-3-319-65463-8_13

1 Introduction

High demands on structural integrity in recent years lead to development and application of new materials with complex microstructures giving desired material properties. Numerical analysis of mechanical behavior of this new class of materials emerges necessity for an advanced numerical tools enabling more realistic material description. The geometrical and mechanical properties of the constituents making up the microstructure have a significant impact on the material behavior observed at macroscale [1]. In addition, the external loading applied at macroscale might cause changes in the microstructural morphology e.g., void formation, damage as well as cracking, which can put structural integrity at risk. Therefore, in order to assess structural integrity and to predict structural lifetime, an analysis of the evolving microstructure is necessary.

Special attention has been directed to the investigation of relations between the macroscopic properties of materials and their microstructure. It is well known that classical continuum mechanics does not consider structural effects in the material at microlevel. Therefore, to overcome this problem, multiscale techniques have been developed that model materials at multiple levels using homogenization procedures. In more recent formulations the computational homogenization approach has mostly been used [2]. This approach is based on the solution of two boundary value problems, one at the macroscopic and one at the microscopic scale. The results obtained by the simulation of a statistically representative sample of material, named Representative Volume Element (RVE), at the microscopic scale are used as input data for the model at the macrolevel. To solve the boundary value problems, the finite element method (FEM) is mostly applied [2–4], but relatively new meshless methods may also be employed due to some numerical advantages [5, 6]. Furthermore, small strain and large strain multi-scale formulations are available, see e.g. [2–4, 7–9].

Based on the micro-macro variable dependence, the first-order and the second-order homogenization procedures are available. The multiscale analysis using the first-order computational homogenization scheme allows explicit modeling of the microstructure, but retains the essential assumptions of continuum mechanics, and thus gives satisfactory results only for simple loading cases. It includes only the first gradient of macroscopic displacement field and it is based on the principles of a local continuum. The local assumption adopted in the classical continuum theory, whereby the stresses at a point depend only on the strains (and other state variables) at the same point, is no longer adequate for the problems, where the macroscopic stress–strain behavior also depends on the characteristic size of the microstructure, and higher order effects are present.

Due to the mentioned shortcomings, the first-order computational homogenization scheme has been extended to the second-order framework, where the second-order stress and strain are included. The formulation is based on a nonlocal continuum theory (NL) that takes into account the influence of a surrounding material on the behavior of a considered material point [10, 11]. For a general case of the nonlinear material with softening behavior or fracturing, the information about the higher-order strains have to be transmitted to the microstructure to get meaningful results. Moreover, in case of engineering problems for which the computation of stress and strain distributions at micro level is essential, the second-order approach is necessary when the size of the microstructure is significant. Furthermore, the multiscale analysis using the second-order homogenization approach may describe more complex deformation modes. On the other hand, it requires C^1 continuity at the macrolevel discretization, which implicates the requirement that both displacements and displacement gradients must be continuous functions leading to more complex formulations. The RVE discretization at the microlevel is usually performed by assuming C^0 continuity, where the standard local continuum theory (SL) is employed.

Unfortunately, the NL-SL second-order computational homogenization approach suffers from difficulties in the scale transition methodology due to the coupling between the nonlocal theory at the macroscale and the local concept on the RVE. Namely, the second-order macrolevel gradient of state variables cannot be related to the microlevel higher-order gradient as a true volume average. Therefore, in the micro-to-macro scale transition, after resolving the Hill-Mandel energy condition, the homogenized double stress requires a modified definition at the microstructural level. Furthermore, in case of generalized periodic boundary conditions, an artificial stress concentration appears at the RVE corner nodes, as result of suppressed microfluctuations at the RVE corners [2].

In this contribution a new second-order computational homogenization scheme employing the nonlocal theory at both scales (NL-NL) is considered. Therein the consistency of the transition methodology is ensured, considering conforming continuum theories used at different scales. The computational scheme is derived adopting the gradient elasticity theory and small strain setting. The discretization at both the macro and the microlevel is performed by the C^1 continuity plane strain triangular finite element derived in [12, 13]. The macro-to-micro scale transition methodology is derived using the gradient displacement and gradient generalized periodic boundary conditions. The derived scale transition methodology, as well as homogenization procedure are embedded into the finite element program ABAQUS [14] by means of user subroutines. The performance and accuracy of the proposed approach have been verified by an example considering an elastic plate with a pre-existing crack.

As mentioned above, at microstructural level some voids may appear which can lead to microcracks and damage initiation, which is macroscopically characterized by decrease in elastic material stiffness or so-called softening. It may result in crack development in engineering materials, which can significantly decrease structural load-carrying capacity and lead to a complete loss of mechanical integrity. It is well-known that the strain softening cannot be properly resolved with the application of the classical continuum mechanics, where the local loss of positive definiteness of the material tangent stiffness arises. The mathematical description of the model then becomes ill-posed and numerical solutions do not converge to a physically meaningful result [15]. To overcome these problems, various regularization techniques have been proposed.

Basically, there are two different approaches regarding the implementation of the material nonlocality in computational model, the integral and the gradient approach.

The integral approach, introduced in [16], accounts for the influence of microstructural interactions through the weighted average of a variable driving the damage process, typically strain. This leads to very complicated constitutive relations made of convolution-type integrals, making the numerical implementation very demanding. In case of the gradient approach, either the classical constitutive relation is enhanced with the strain gradients, or both the strain gradients and their stress conjugates are introduced in the model via the higher-order continuum. When only strain-gradients are used as an enhancement of the constitutive relation, the explicit and especially the implicit gradient formulations are usually applied when dealing with softening, either in elasticity context [17], plasticity context [18, 19] or in analysis of the elastic wave propagation [20]. Although the structural responses are mesh objective, the mentioned formulations suffer from the spurious damage growth reported in [21]. This is explained in more detail in [22], where a new model based on the decreasing microstructural interactions is presented, recognizing that the width of the fracture process zone localizes towards a macroscopic crack in the quasi-brittle fracture.

The development of a damage model based on the strain gradient continuum theory is presented in this contribution, which includes both the strain gradients and their stress conjugates. The quasi-brittle damage model proposed in [23] under assumption of homogeneous material is adopted. The isotropic damage law is implemented into the constitutive relations of the strain gradient theory, whereby the constitutive matrices, which describe the intensity of the material nonlocal behavior, are directly decreased by the term involving damage variable. The new proposed constitutive model is embedded into the same C^1 triangular finite element formulation as applied in the aforementioned multiscale procedure. All derived numerical algorithms are implemented into the FE software ABAQUS too. In this way a physically correct structural response standing behind a fracturing process can be captured, unlike in the formulation based on the conventional implicit gradient damage model, where the spurious damage growth can be observed [21, 24]. The accuracy of the damage evolution is demonstrated by an example where the elastic plate is again considered.

2 Formulation of C^1 Continuity Triangular Finite Element

The discretization at macrolevel is performed by using the C^1 continuity plane strain triangular finite element already derived in [25] whose formulation is here briefly summarized. As shown in Fig. 1, the element consists of three nodes and 36 degrees of freedom. The displacement field is approximated by the condensed fifth order polynomial. The nodal degrees of freedom are the two displacements and their first- and second order derivatives with respect to the Cartesian coordinates. As usually, the derivation of the element equations is performed by means of the principle of virtual work, which can be expressed for the strain gradient continuum under assumption of small strain as

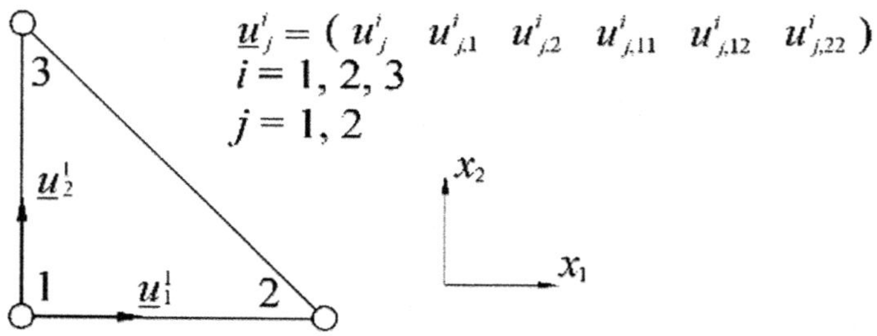

Fig. 1 C^1 triangular finite element [20]

$$\int_A \delta\boldsymbol{\varepsilon}^T \boldsymbol{\sigma} \mathrm{d}A + \int_A \delta\boldsymbol{\eta}^T \boldsymbol{\mu} \mathrm{d}A = \int_s \delta\mathbf{u}^T \mathbf{t}\,\mathrm{d}s + \int_s \delta\left(\mathrm{grad}\,\mathbf{u}^T\right)\mathbf{T}\,\mathrm{d}s. \tag{1}$$

Herein $\boldsymbol{\sigma}$ is the Cauchy stress tensor, and $\boldsymbol{\varepsilon}$ represents the strain tensor. $\boldsymbol{\mu}$ stands for the third-order double-stress tensor, representing an energy conjugate measure to the strain gradient tensor $\boldsymbol{\eta}$. The values $\mathbf{t}$ and $\boldsymbol{\tau}$ denote the traction and the double surface traction vectors, respectively, while $\mathbf{u}$ is the displacement vector. A is the element surface, and s represents the element perimeter. $\mathbf{T}$ is the double traction tensor, $\mathbf{T} = \boldsymbol{\tau}\,\mathbf{n}$, with $\mathbf{n}$ as the unit outward surface normal.

The strain and strain gradient tensors are given by

$$\boldsymbol{\varepsilon} = \begin{pmatrix} \varepsilon_{11} \\ \varepsilon_{22} \\ 2\varepsilon_{12} \end{pmatrix} = \mathbf{B}_\varepsilon \mathbf{v},\ \boldsymbol{\eta} = \begin{pmatrix} \eta_{111} \\ \eta_{222} \\ \eta_{221} \\ \eta_{112} \\ 2\eta_{121} \\ 2\eta_{212} \end{pmatrix} = \mathbf{B}_\eta \mathbf{v}, \tag{2}$$

where $\mathbf{B}_\varepsilon$ and $\mathbf{B}_\eta$ represent the matrices containing first and second derivatives of the element shape functions, while $\mathbf{v}$ is the vector of the nodal degrees of freedom. Since a nonlinear problem should be considered, Eq. (1) is transformed into the incremental form. Therefore, the displacement vector $\mathbf{u}$, the stress tensor $\boldsymbol{\sigma}$ and the double stress $\boldsymbol{\mu}$ are updated according to

$$\begin{aligned} \mathbf{u} &= \mathbf{u}^{i-1} + \Delta\mathbf{u}, \\ \boldsymbol{\sigma} &= \boldsymbol{\sigma}^{i-1} + \Delta\boldsymbol{\sigma}, \\ \boldsymbol{\mu} &= \boldsymbol{\mu}^{i-1} + \Delta\boldsymbol{\mu}, \end{aligned} \tag{3}$$

where the exponent $(i - 1)$ refers to the last converged equilibrium state, and the symbol Δ indicates an incremental change. The incremental constitutive relations are defined as

$$\begin{aligned} \Delta\boldsymbol{\sigma} &= \mathbf{C}_{\sigma\varepsilon}\Delta\boldsymbol{\varepsilon} + \mathbf{C}_{\sigma\eta}\Delta\boldsymbol{\eta}, \\ \Delta\boldsymbol{\mu} &= \mathbf{C}_{\mu\varepsilon}\Delta\boldsymbol{\varepsilon} + \mathbf{C}_{\mu\eta}\Delta\boldsymbol{\eta} \end{aligned} \tag{4}$$

with $\mathbf{C}_{\sigma\varepsilon}$, $\mathbf{C}_{\sigma\eta}$, $\mathbf{C}_{\mu\varepsilon}$ and $\mathbf{C}_{\mu\eta}$ as the constitutive stiffness matrices. By employing Eq. (2), the strain and the second-order strain increments can be expressed in terms of the displacement vector increment $\Delta\mathbf{v}$ as

$$\begin{aligned} \Delta\boldsymbol{\varepsilon} &= \mathbf{B}_{\varepsilon}\Delta\mathbf{v}, \\ \Delta\boldsymbol{\eta} &= \mathbf{B}_{\eta}\Delta\mathbf{v}. \end{aligned} \tag{5}$$

After well-known derivation procedure, the finite element relation is derived and may be expressed in the standard linearized form

$$\mathbf{K}\Delta\mathbf{v} = \mathbf{F}_{\mathrm{e}} - \mathbf{F}_{\mathrm{i}} \tag{6}$$

with $\mathbf{K}$ as the element stiffness matrix which may be decomposed in the following parts

$$\mathbf{K} = \mathbf{K}_{\sigma\varepsilon} + \mathbf{K}_{\sigma\eta} + \mathbf{K}_{\mu\varepsilon} + \mathbf{K}_{\mu\eta}, \tag{7}$$

where the particular matrices are written as

$$\begin{aligned} \mathbf{K}_{\sigma\varepsilon} &= \int_A \left(\mathbf{B}_{\varepsilon}^T \mathbf{C}_{\sigma\varepsilon} \mathbf{B}_{\varepsilon}\right) \mathrm{d}A, \\ \mathbf{K}_{\sigma\eta} &= \int_A \left(\mathbf{B}_{\varepsilon}^T \mathbf{C}_{\sigma\eta} \mathbf{B}_{\eta}\right) \mathrm{d}A, \\ \mathbf{K}_{\mu\varepsilon} &= \int_A \left(\mathbf{B}_{\eta}^T \mathbf{C}_{\mu\varepsilon} \mathbf{B}_{\varepsilon}\right) \mathrm{d}A, \\ \mathbf{K}_{\mu\eta} &= \int_A \left(\mathbf{B}_{\eta}^T \mathbf{C}_{\mu\eta} \mathbf{B}_{\eta}\right) \mathrm{d}A. \end{aligned} \tag{8}$$

Furthermore, $\mathbf{F}_{\mathrm{e}}$ and $\mathbf{F}_{\mathrm{i}}$ are the external and internal nodal force vectors, which are expressed by the relations

$$\begin{aligned} \mathbf{F}_{\mathrm{e}} &= \int_s \left(\mathbf{N}^T \mathbf{t} + \mathrm{grad}\mathbf{N}^T \mathbf{T}\right) \mathrm{d}s, \\ \mathbf{F}_{\mathrm{i}} &= \int_A \left(\mathbf{B}_{\varepsilon}^T \boldsymbol{\sigma}^{i-1} + \mathbf{B}_{\eta}^T \boldsymbol{\mu}^{i-1}\right) \mathrm{d}A. \end{aligned} \tag{9}$$

Discretization at microlevel is also performed using the C^1 continuity plane strain triangular finite elements, but their formulation is based on the Aifantis strain gradient theory, where a microstructural length scale parameter is employed. More can be found in [26].

3 Computational Homogenization and Micro-Macro Scale Transition

The basic of the second-order computational homogenization procedure, which preserves the nonlocal theory is explained. The geometry of the RVE at microscale is assumed to be square shaped with the coordinate system placed in the centroid, as

Fig. 2 Representative volume element

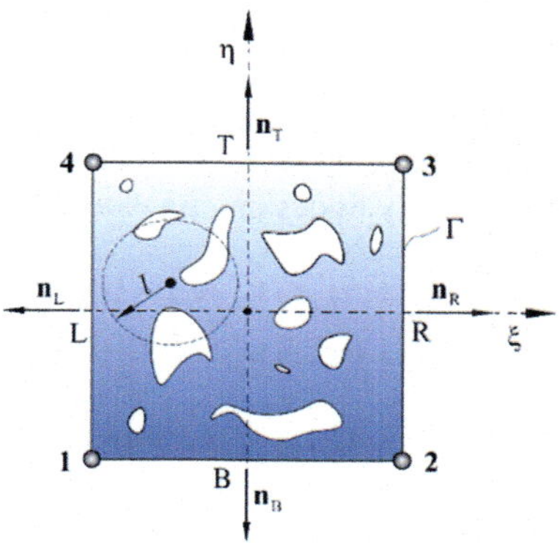

displayed in Fig. 2. All variables appearing at the microstructural level are denoted by the subscript "m", while the macrostructural quantities are referenced by the index "M". The RVE displacement field is represented by a Taylor series expansion depending on the macrolevel strain $\boldsymbol{\varepsilon}_M$ and strain gradient $\nabla\boldsymbol{\varepsilon}_M$. The displacement imposed along the RVE boundaries is defined as

$$\mathbf{u}_m = \mathbf{x}^T \boldsymbol{\varepsilon}_M + \frac{1}{2}\left[\mathbf{x}^T (\nabla\boldsymbol{\varepsilon}_M)\,\mathbf{x}\right] + \mathbf{r}. \tag{10}$$

In Eq. (10), $\mathbf{x}$ is a spatial coordinate on the RVE, $\boldsymbol{\varepsilon}_M$ denotes the strain tensor at the macrolevel and $\mathbf{r}$ represents the microfluctuation field. In the further derivation, the following integral relations arise

$$\frac{1}{V}\int_V (\nabla_m \mathbf{u}_m)\,dV = \frac{1}{V}\int_V \boldsymbol{\varepsilon}_m dV = \boldsymbol{\varepsilon}_M + \frac{1}{V}\int_V (\nabla\boldsymbol{\varepsilon}_M \mathbf{x})\,dV + \frac{1}{V}\int_V (\nabla_m \mathbf{r})\,dV, \tag{11}$$

$$\frac{1}{V}\int_V (\nabla_m \boldsymbol{\varepsilon}_m)\,dV = \nabla\boldsymbol{\varepsilon}_M + \frac{1}{V}\int_V [\nabla_m (\nabla_m \mathbf{r})]\,dV. \tag{12}$$

Using common mathematical procedures in the homogenization approach, the following microfluctuation integrals should be explicitly satisfied from Eqs. (11) and (12) as

$$\frac{1}{V}\int_V (\nabla_m \mathbf{r})\,dV = \frac{1}{V}\int_\Gamma \left(\mathbf{n}^T \mathbf{r}\right) d\Gamma = \mathbf{0}, \tag{13}$$

$$\frac{1}{V}\int_V [\nabla_m (\nabla_m \mathbf{r})]\,dV = \frac{1}{V}\int_\Gamma \left[\mathbf{n}^T (\nabla_m \mathbf{r})\right] d\Gamma = \mathbf{0}, \tag{14}$$

where Γ stands for an outer RVE boundary. Furthermore, by means of Eq. (10) and introducing the coordinate matrices $\mathbf{D}$, $\mathbf{H}_1$ and $\mathbf{H}_2$, the nodal degrees of freedom prescribed on the RVE boundaries are derived from

$$\mathbf{u} = \mathbf{D}^T \boldsymbol{\varepsilon}_\mathrm{M} + \left(\mathbf{H}_1^T\right)(\boldsymbol{\varepsilon},_1)_\mathrm{M} + \left(\mathbf{H}_2^T\right)(\boldsymbol{\varepsilon},_2)_\mathrm{M}\,. \tag{15}$$

The gradient generalized periodic boundary conditions predefine corner nodal degrees of freedom of the RVE using Eq. (15), while remaining degrees of freedom of boundary nodes on opposite sides are related by the periodicity equations as

$$\begin{aligned}\mathbf{u}_\mathrm{R} - \mathbf{u}_\mathrm{L} &= \left(\mathbf{D}_\mathrm{R}^T - \mathbf{D}_\mathrm{L}^T\right)\boldsymbol{\varepsilon}_\mathrm{M} + \left[\left(\mathbf{H}_1^T\right)_\mathrm{R} - \left(\mathbf{H}_1^T\right)_\mathrm{L}\right](\boldsymbol{\varepsilon},_1)_\mathrm{M} + \left[\left(\mathbf{H}_2^T\right)_\mathrm{R} - \left(\mathbf{H}_2^T\right)_\mathrm{L}\right](\boldsymbol{\varepsilon},_2)_\mathrm{M}\,,\\ \mathbf{u}_\mathrm{T} - \mathbf{u}_\mathrm{B} &= \left(\mathbf{D}_\mathrm{T}^T - \mathbf{D}_\mathrm{B}^T\right)\boldsymbol{\varepsilon}_\mathrm{M} + \left[\left(\mathbf{H}_1^T\right)_\mathrm{T} - \left(\mathbf{H}_1^T\right)_\mathrm{B}\right](\boldsymbol{\varepsilon},_1)_\mathrm{M} + \left[\left(\mathbf{H}_2^T\right)_\mathrm{T} - \left(\mathbf{H}_2^T\right)_\mathrm{B}\right](\boldsymbol{\varepsilon},_2)_\mathrm{M}\,.\end{aligned} \tag{16}$$

As may be observed, the above periodicity equations prescribe constraints not only on the displacements, but also on the first and second displacement derivatives available as the nodal degrees of freedom. This gives possibility to prescribe the complete second-order gradient field from the macrolevel on the RVE boundaries without need for the microfluctuation integrals, which is not the case in the NL-SL homogenization, where the transition between the strain gradient nonlocal formulation at the macrolevel and the standard local formulation at the microscale is performed.

In the micro-to-macro scale transition the starting point is the Hill-Mandel energy condition written in the form

$$\frac{1}{V}\int_V \left[\delta\boldsymbol{\varepsilon}_\mathrm{m}^T\boldsymbol{\sigma}_\mathrm{m} + \nabla_\mathrm{m}\delta\boldsymbol{\varepsilon}_\mathrm{m}^T\boldsymbol{\mu}_\mathrm{m}\right]\mathrm{d}V = \delta\boldsymbol{\varepsilon}_\mathrm{M}^T\boldsymbol{\sigma}_\mathrm{M} + \nabla\delta\boldsymbol{\varepsilon}_\mathrm{M}^T\boldsymbol{\mu}_\mathrm{M}. \tag{17}$$

From Eq. (17) the homogenized macroscale stress tensors are extracted and defined as

$$\boldsymbol{\sigma}_\mathrm{M} = \frac{1}{V}\int_V \boldsymbol{\sigma}_\mathrm{m}\mathrm{d}V, \tag{18}$$

$$\boldsymbol{\mu}_\mathrm{M} = \frac{1}{V}\int_V \left(\boldsymbol{\mu}_\mathrm{m} + \mathbf{x}^T\boldsymbol{\sigma}_\mathrm{m}\right)\mathrm{d}V. \tag{19}$$

To account the contributions of a heterogeneous microstructure at the macrolevel, the following generalized constitutive behaviour is assumed

$$\Delta\boldsymbol{\sigma}_\mathrm{M} = \mathbf{C}_{\sigma\varepsilon}\Delta\boldsymbol{\varepsilon}_\mathrm{M} + \mathbf{C}_{\sigma\varepsilon_{x_1}}\Delta(\boldsymbol{\varepsilon},_1)_\mathrm{M} + \mathbf{C}_{\sigma\varepsilon_{x_2}}\Delta(\boldsymbol{\varepsilon},_2)_\mathrm{M}\,, \tag{20}$$

$$\Delta\left(\boldsymbol{\mu}_{x_1}\right)_\mathrm{M} = \mathbf{C}_{\mu_{x_1}\varepsilon}\Delta\boldsymbol{\varepsilon}_\mathrm{M} + \mathbf{C}_{\mu_{x_1}\varepsilon_{x_1}}\Delta(\boldsymbol{\varepsilon},_1)_\mathrm{M} + \mathbf{C}_{\mu_{x_1}\varepsilon_{x_2}}\Delta(\boldsymbol{\varepsilon},_2)_\mathrm{M}\,, \tag{21}$$

$$\Delta\left(\boldsymbol{\mu}_{x_2}\right)_\mathrm{M} = \mathbf{C}_{\mu_{x_2}\varepsilon}\Delta\boldsymbol{\varepsilon}_\mathrm{M} + \mathbf{C}_{\mu_{x_2}\varepsilon_{x_1}}\Delta(\boldsymbol{\varepsilon},_1)_\mathrm{M} + \mathbf{C}_{\mu_{x_2}\varepsilon_{x_2}}\Delta(\boldsymbol{\varepsilon},_2)_\mathrm{M}\,. \tag{22}$$

From Eqs. (20)–(22) it is clear that 9 constitutive material operators are required in the NL homogenization. Using the static condensation procedure, they are expressed in terms of the condensed RVE stiffness $\tilde{\mathbf{K}}_\mathrm{bb}$ and the coordinate matrices as

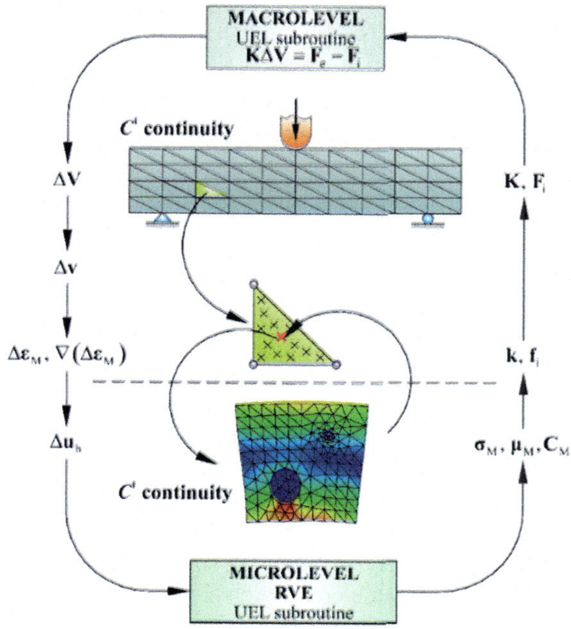

Fig. 3 NL-NL multiscale computational scheme

$$\begin{aligned}
&\mathbf{C}_{\sigma\varepsilon}=\tfrac{1}{V}\mathbf{D}\tilde{\mathbf{K}}_{\mathrm{bb}}\mathbf{D}^T;\ \mathbf{C}_{\sigma\varepsilon_{x_1}}=\tfrac{1}{V}\mathbf{D}\tilde{\mathbf{K}}_{\mathrm{bb}}\mathbf{H}_1^T;\ \mathbf{C}_{\sigma\varepsilon_{x_2}}=\tfrac{1}{V}\mathbf{D}\tilde{\mathbf{K}}_{\mathrm{bb}}\mathbf{H}_2^T;\\
&\mathbf{C}_{\mu_1\varepsilon}=\tfrac{1}{V}\mathbf{H}_1\tilde{\mathbf{K}}_{\mathrm{bb}}\mathbf{D}^T;\ \mathbf{C}_{\mu_1\varepsilon_{x_1}}=\tfrac{1}{V}\mathbf{H}_1\tilde{\mathbf{K}}_{\mathrm{bb}}\mathbf{H}_1^T;\ \mathbf{C}_{\mu_1\varepsilon_{x_2}}=\tfrac{1}{V}\mathbf{H}_1\tilde{\mathbf{K}}_{\mathrm{bb}}\mathbf{H}_2^T;\\
&\mathbf{C}_{\mu_2\varepsilon}=\tfrac{1}{V}\mathbf{H}_2\tilde{\mathbf{K}}_{\mathrm{bb}}\mathbf{D}^T;\ \mathbf{C}_{\mu_2\varepsilon_{x_1}}=\tfrac{1}{V}\mathbf{H}_2\tilde{\mathbf{K}}_{\mathrm{bb}}\mathbf{H}_1^T;\ \mathbf{C}_{\mu_2\varepsilon_{x_2}}=\tfrac{1}{V}\mathbf{H}_2\tilde{\mathbf{K}}_{\mathrm{bb}}\mathbf{H}_2^T.
\end{aligned}\tag{23}$$

More details on the here presented homogenization approach can be found in the authors' previous publication [27].

3.1 *Numerical Implementation of Multiscale Algorithm*

The derived micro-macro scheme has been implemented into the FE program ABAQUS by means of user subroutines, as shown in Fig. 3. In the NL-NL scheme both scales are discretized by the C^1 triangular finite element. Therein the Aifantis elasticity theory is adopted at the microstructural level. Since only the linear elastic behaviour is considered, the homogenization of the constitutive matrices is required only once at the preprocessing step and will be used in the subsequent computations. But, in order to observe the phenomena of evolving microstructure, the RVE analysis is necessary, where the stress homogenization has to be conducted. The RVE size defines some characteristic microstructural size (e.g. the average size of voids) and captures the nonlocal effects at macroscale. The subroutine of the macrolevel finite element serves as a master routine. It prescribes the RVE boundary conditions based on the macrolevel gradients and runs the microlevel analysis as a new boundary value problem. The RVE boundary value problem is solved by the UEL subroutine of the microscale finite element model, where the stress tensors are homogenized and passed to the macrolevel analysis. Therein, the microstructural parameter l^2 must be explicitly defined. Once all necessary data are computed, the macrolevel analysis checks for the convergence and continues.

4 Damage Algorithm Based on Strain Gradient Theory

The reduction of the elastic stiffness properties in an isotropic damage model is expressed by the following well known relation

$$\mathbf{C}^{\text{eff}} = (1 - D)\,\mathbf{C}, \tag{24}$$

where D is a scalar damage variable ranging from zero (undamaged material) to one (fully damaged material), while $\mathbf{C}^{eff}$ and $\mathbf{C}$ are the effective and the elastic stiffness tensors, respectively. The damage state is governed by the monotonically increasing scalar history parameter κ which can be determined as an average local equivalent scalar measure of the strain ε_{eq} through Kuhn-Tucker relations

$$\kappa \geq 0,\, \varepsilon_{\text{eq}} - \kappa \leq 0,\, \dot{\kappa}\left(\varepsilon_{\text{eq}} - \kappa\right) = 0. \tag{25}$$

The two different equivalent elastic strain measures are generally used in the context of the softening behavior of a quasi-brittle material. The first one is defined in [28] as

$$\varepsilon_{\text{eq}} = \sqrt{\sum_{i=1}^{3} \langle \varepsilon_i \rangle^2} \tag{26}$$

with ε_i $(i = 1, 2, 3)$ representing the principal strains. It is clear that, in this case, the equivalent elastic strain measure depends only on the positive principal strains, making it more sensitive to tensile than to compressive strains. On the other hand, the von Mises equivalent strain measure according to [29] and given by

$$\varepsilon_{\text{eq}} = \frac{k-1}{2k\,(1-2\nu)} I_1 + \frac{1}{2k}\sqrt{\frac{(k-1)^2}{(1-2\nu)^2} I_1^2 - \frac{12k}{(1+\nu)^2} J_2}, \tag{27}$$

includes a parameter k which represents the ratio between uniaxial compressive and tensile strength of the material. For the $k = 1$, meaning that both compression and tension influence the equivalent strain measure equally, Eq. (27) results in

$$\varepsilon_{\text{eq}} = \frac{1}{1+\nu}\sqrt{-3J_2}. \tag{28}$$

In above expressions I_1 and J_2 are the first invariant of the strain tensor and the second invariant of the deviatoric strain tensor, respectively.

The damage evolution is usually governed by the linear softening law [30] as

$$D = \begin{cases} \frac{\kappa_u(\kappa-\kappa_0)}{\kappa(\kappa_u-\kappa_0)} \text{ if } \kappa_0 \le \kappa \le \kappa_u, \\ 1 \text{ if} \kappa > \kappa_u, \end{cases} \tag{29}$$

where κ_0 and κ_u are the material parameters representing the threshold strain at which the damage is initiated, and the strain at which material completely loses its stiffness, respectively. The softening in the real materials is usually nonlinear, where the application of the exponential softening law is the most common [30]

$$D = 1 - \frac{\kappa_0}{\kappa}\{1 - \alpha + \alpha \exp[\beta(\kappa_0 - \kappa)]\} \text{ if } \kappa > \kappa_0 \tag{30}$$

with α and β as model parameters. As evident from above, the damage-driving state variable is a local equivalent strain, which differs from most gradient-enhanced formulations, where the damage is governed by the nonlocal state variable. In this contribution the nonlocality is incorporated through the strain gradient continuum theory which is embedded into the finite element formulation.

Applying the isotropic damage model to the incremental constitutive relations described by Eq. (4), the following incremental expressions may be written

$$\begin{aligned} \Delta\boldsymbol{\sigma} &= \Delta\left[(1-D)\,\mathbf{C}_{\sigma\varepsilon}\boldsymbol{\varepsilon} + (1-D)\,\mathbf{C}_{\sigma\eta}\right], \\ \Delta\boldsymbol{\mu} &= \Delta\left[(1-D)\,\mathbf{C}_{\mu\varepsilon}\boldsymbol{\varepsilon} + (1-D)\,\mathbf{C}_{\mu\eta}\right]. \end{aligned} \tag{31}$$

In addition to (3), now the damage variable is to be updated

$$D = D^{i-1} + \Delta D. \tag{32}$$

Since in this contribution the damage evolution is modeled only under assumption of material homogeneity, the tangent stiffness matrices $\mathbf{C}_{\sigma\eta}$ and $\mathbf{C}_{\mu\varepsilon}$ are equal to zero [31]. The remaining two matrices can be computed analytically [31, 32], which may be written symbolically in the form

$$\begin{aligned} \mathbf{C}_{\sigma\varepsilon} &= \mathbf{C}_{\sigma\varepsilon}(E, \nu), \\ \mathbf{C}_{\mu\eta} &= \mathbf{C}_{\mu\eta}(E, \nu, l), \end{aligned} \tag{33}$$

where l denotes the microstructural length scale. On the other hand, the stiffness matrices can be also computed numerically using the procedure presented in [25]. Therein the microstructural parameter is expressed by the relation

$$l^2 = \frac{L^2}{12}, \tag{34}$$

where L is the RVE side length. As displayed above, the material nonlocality is included into the second-gradient continuum theory in terms of the microstructural parameter l through the tangent stiffness matrices. When these matrices are multiplied by the term $(1 - D)$ according to Eq. (31), the nonlocality decreases if the damage rises.

For homogeneous material Eq. (31) can further be written as

$$\begin{aligned} \Delta\boldsymbol{\sigma} &= (1 - D^{i-1})\mathbf{C}_{\sigma\varepsilon}\Delta\boldsymbol{\varepsilon} - \Delta D\mathbf{C}_{\sigma\varepsilon}\boldsymbol{\varepsilon}^{i-1}, \\ \Delta\boldsymbol{\mu} &= (1 - D^{i-1})\mathbf{C}_{\mu\eta}\Delta - \Delta D\ \mathbf{C}_{\mu\eta}{}^{i-1}. \end{aligned} \tag{35}$$

Here the incremental change of the damage variable may be expressed by

$$\Delta D = \left(\frac{\mathrm{d}D}{\mathrm{d}\boldsymbol{\varepsilon}}\right)^{i-1} \Delta\boldsymbol{\varepsilon} \tag{36}$$

since the damage variable is assumed to be a function only of the strain tensor $D = D\,(\boldsymbol{\varepsilon})$.

The substitution of Eqs. (5), (35) and (36) into Eq. (1), after some straightforward calculus, leads to the following finite element relation

$$\left(\mathbf{K}_{\varepsilon\varepsilon} + \mathbf{K}_{\eta\varepsilon} + \mathbf{K}_{\eta\eta}\right)\Delta\mathbf{v} = \mathbf{F}_{\mathrm{e}} - \mathbf{F}_{\mathrm{i}}, \tag{37}$$

where the particular element stiffness matrices are defined as

$$\begin{aligned} \mathbf{K}_{\varepsilon\varepsilon} &= \int_A \mathbf{B}_\varepsilon^T \left[\left(1 - D^{i-1}\right)\mathbf{C}_{\sigma\varepsilon} - \mathbf{C}_{\sigma\varepsilon}\boldsymbol{\varepsilon}^{i-1}\left(\tfrac{\mathrm{d}D}{\mathrm{d}\boldsymbol{\varepsilon}}\right)^{i-1}\right]\mathbf{B}_\varepsilon \mathrm{d}A, \\ \mathbf{K}_{\eta\varepsilon} &= -\int_A \mathbf{B}_\eta^T \mathbf{C}_{\mu\eta}{}^{i-1}\left(\tfrac{\mathrm{d}D}{\mathrm{d}\boldsymbol{\varepsilon}}\right)^{i-1}\mathbf{B}_\varepsilon \mathrm{d}A, \\ \mathbf{K}_{\eta\eta} &= \int_A \mathbf{B}_\eta^T \left(1 - D^{i-1}\right)\mathbf{C}_{\mu\eta}\mathbf{B}_\eta \mathrm{d}A. \end{aligned} \tag{38}$$

The external and internal nodal force vectors $\mathbf{F}_{\mathrm{e}}$ and $\mathbf{F}_{\mathrm{i}}$ are the same as in (9).

5 Numerical Examples

5.1 Multiscale Analysis of Plate Tension with Pre-existing Crack

The proposed multiscale procedure is tested by the computation of the plate with a pre-existing crack under tension, as shown in Fig. 4. The presented length has been taken as $h = 100$ mm. Firstly, the computation is performed under assumption of a homogeneous material and the Aifantis gradient theory employing a microstructural parameter. The results are compared with the solution obtained by the standard

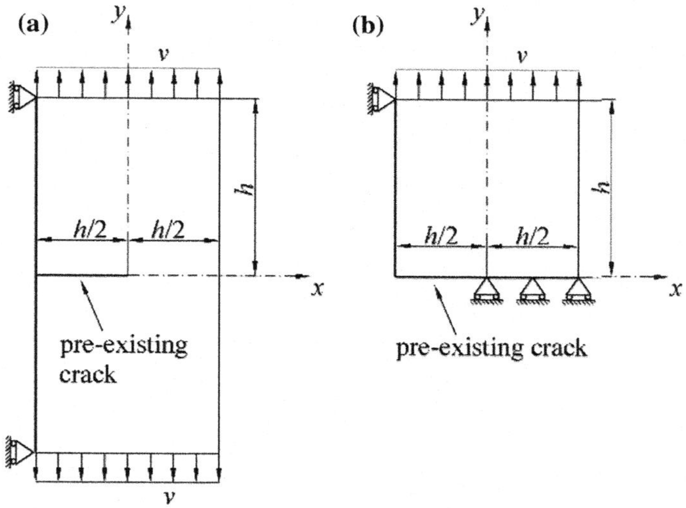

Fig. 4 Plate with a pre-existing crack subjected to tensile load: **a** geometry and loading **b** computational model of half plate

ABAQUS plain strain elements using singular elements at the crack tip. Due to symmetry, only the upper half of the plate is discretized by 169 C^1 finite elements and 324 CPE8 Abaqus finite elements [14] imposing the appropriate boundary conditions, as depicted in Fig. 4b. As displayed in Fig. 4, the upper plate edge is subjected to the vertical displacement of $v = 1$ mm. The material data are the Young's modulus $E = 1000$ MPa and the Poisson's ratio $\nu = 0, 2$. The von Mises equivalent strain measure expressed in (27) is computed, where the parameter k is set to $k = 10$. The results obtained by the C^1continuity strain gradient formulation for two different values of the microstructural parameter l in comparison with the solutions obtained by using singular elements are presented in Fig. 5. As evident, the strain concentration is clearly captured, however the deformation responses vary. Using the strain gradient theory which employs the nonlocality, the maximum of the equivalent strain at the crack tip is significantly decreased, as expected. With the increase of parameter l, the strain concentration at the crack tip drops, which is a direct consequence of an increase of a surrounding continuum taken into account.

Next an academic example of heterogeneous material is considered, which is represented by the RVE of the side length $L = 0.2$ mm as shown in Fig. 6. The material properties of the RVE matrix are the same as in the homogeneous specimen, while the porosities of 13% are randomly distributed with the average radius $r_{\text{ave}} = 0.0086$ mm. For the computation of the microstructural boundary value problem, the gradient generalized periodic boundary conditions have been employed. As mentioned above, the nonlocal effect of the strain gradient theory is employed at the macrolevel through the RVE size L. Besides, due to the Aifantis strain gradient theory employing in the RVE, another microstructural parameter l appears at microlevel, which captures a mechanism below the microstructural size.

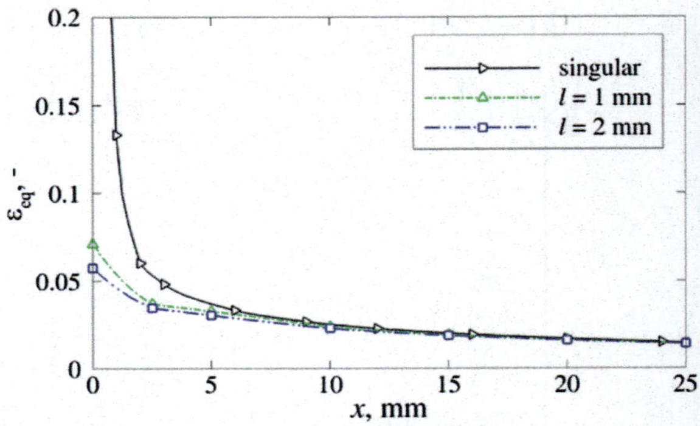

Fig. 5 Distribution of equivalent elastic strain ε_{eq} in front of the crack tip

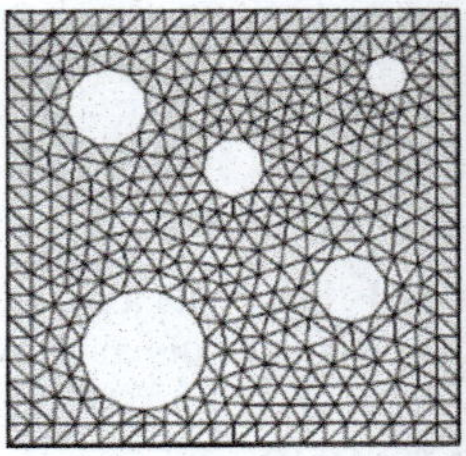

Fig. 6 RVE for heterogeneous material

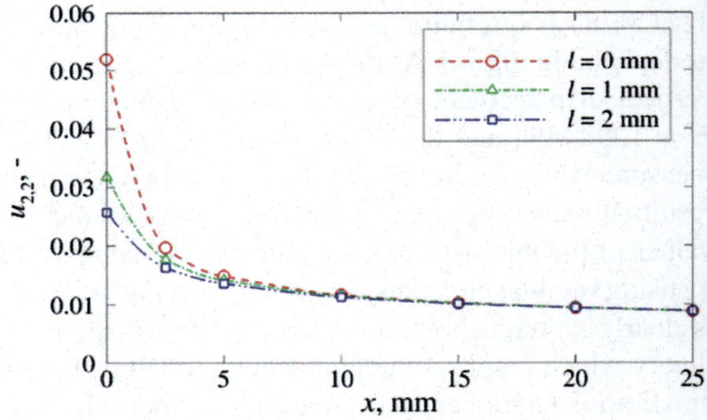

Fig. 7 Distribution of $u_{2,2}$ in front of the crack tip

Therefore, the nonlocal effect is here expressed by the two microstructural values L and l, as described in [27]. Figures 7 and 8 show the distributions of displacement gradients $u_{2,2}$ and $u_{2,12}$ for the different l and using RVE presented in Fig. 6. The regularizing effect expressed by the microstructural parameters is clearly displayed. In the vicinity of the crack tip the displacement gradients are very high, but their maximal values decrease with the increase of l.

The contour plots of the distribution of the displacement gradients $u_{2,2}$ and $u_{2,12}$ over the RVE positioned in front of the crack tip are shown in Figs. 9 and 10.

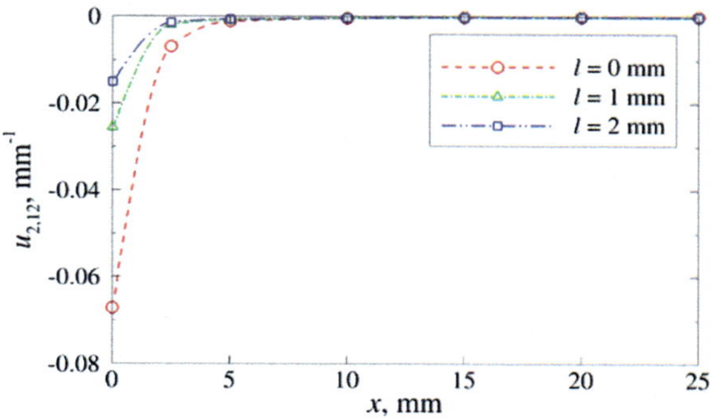

Fig. 8 Distribution of $u_{2,12}$ in front of the crack tip

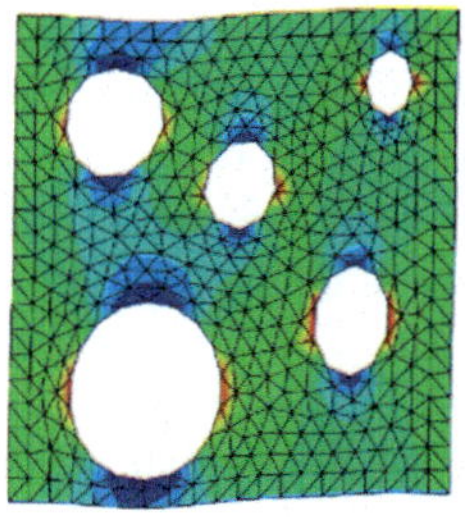

Fig. 9 Distribution of $u_{2,2}$ over the RVE

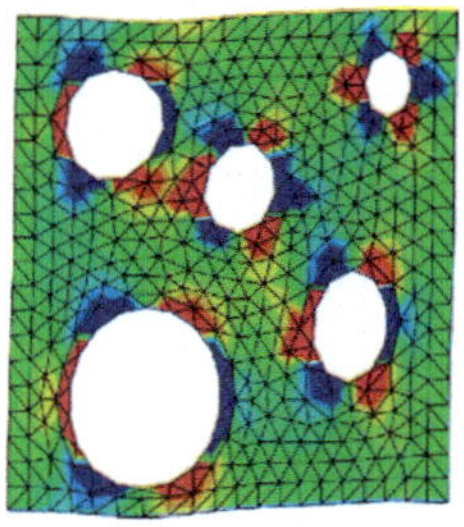

Fig. 10 Distribution of $u_{2,12}$ over the RVE

As evident, the deformed RVE shapes demonstrate the pure stretch in the loading direction, as expected. The contribution of the microfluctuation field is more visible on the top and bottom edge, while the left and right edge remain mostly straight. It is clearly displayed that the strains and the strain gradients start to grow around the porosities, which can yield the damage initiation. Here described nonlocal formulation at the microscale could be very useful in the modeling of damage evolution in the microstructural RVE space, which can lead to the crack development at the macrolevel.

5.2 Damage Analysis of Plate Tension with Pre-existing Crack

In the further consideration of the specimen described above, a damage response is modelled under assumption of the homogeneous material with the unchanged material data. The quasi-brittle damage model described by the exponential softening law (30) with the parameters $\kappa_0 = 0.0001$, $\alpha = 0.99$ and $\beta = 300$ is employed. The modified von Mises equivalent elastic strain measure (27) for the parameter $k = 10$ is used for the computation. The discretization is performed by the previously described C^1 continuity finite elements. The same specimen has already been studied in [21] with the adoption of the damage model based on the conventional implicit gradient enhancement, resulting in a non-physical damage evolution. The damage distributions over the computational model, Fig. 4b, are shown in Fig. 11 for the two different internal length scales, of $l = 0.08\,h$ and $l = 0.2\,h$. For the sake of comparison the length scale values are taken from [22].

As obvious, the damage spreads along the line in front of the crack tip, as expected. It compares well with the experimental observations, of the brittle damage evolution in concrete specimens in [33] as well as with the damage profile obtained in [22], where the spurious damage growth is avoided. In this reference the localizing gradient damage model has been derived in the micromorphic framework. The damage distributions ahead of the crack tip for several loading levels are presented in Fig. 12. As evident, the maximum values of the damage variable are achieved at the crack tip, which is physically realistic. The mesh sensitivity is examined using the two different finite element discretizations. The first of 800 elements is applied for the results presented above, and the second one comprises 3200 elements. As evident in Fig. 13, there are no differences in the damage responses. The damage profile is correctly captured for both discretization sizes.

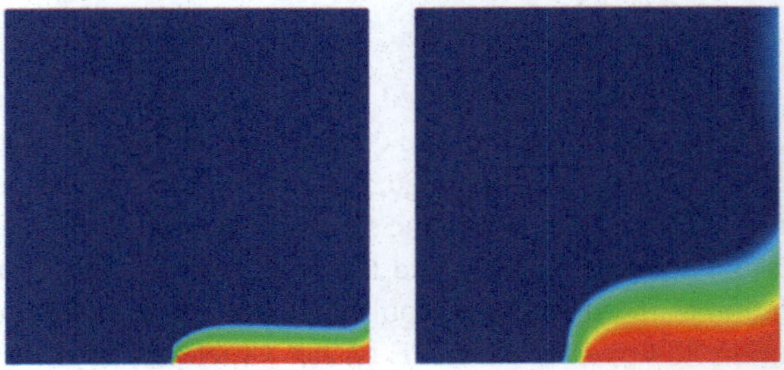

Fig. 11 Damage distributions for length scales l = 8 mm (left) and l = 20 mm (right)

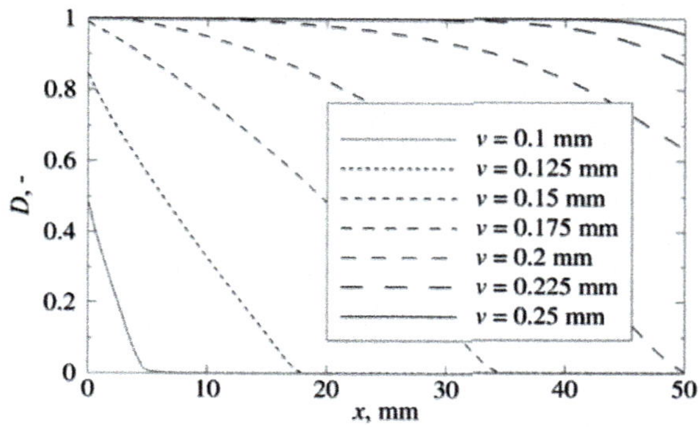

Fig. 12 Damage distributions in front of the crack tip

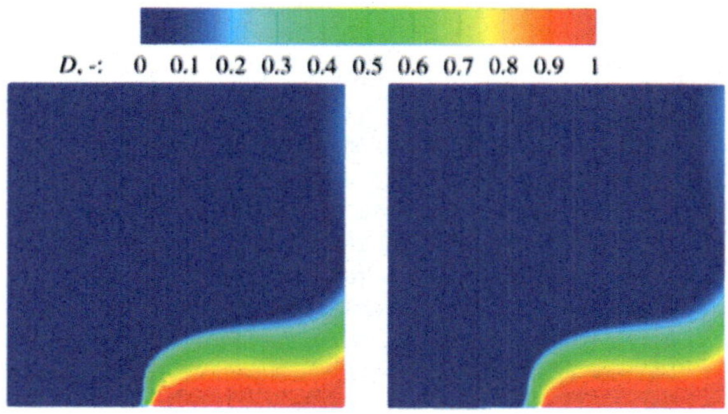

Fig. 13 Comparison of damage distributions for two uniform discretizations consisting of 800 (left) and 3200 (right) C^1 triangular finite elements

6 Conclusions

A two-scale computational approach employing the strain gradient elasticity theory at both macro- and microscale is proposed. The formulation of the nonlocal theory is embedded into the finite element framework using the C^1 three node triangular plane strain finite element. It has been demonstrated that the employment of the nonlocal theory at the microstructural level resolves the inconsistencies arising in the available multiscale procedure, where the transition between the nonlocal and local theory has been performed. In the new proposed approach the same strain gradient variables are defined at both scales, which contributes to the consistency and accuracy in the homogenization procedure. Herein the nonlocality is described by the RVE size at macrolevel and using the Aifantis microstructural parameter at microscale.

Furthermore, a computational approach employing the strain gradient continuum theory for modeling of quasi-brittle damage phenomena in homogeneous materials is presented. The strain gradient constitutive relation comprising the damage variable is implemented into the formulation of the above mentioned C^1 triangular finite element. In contrast to the results obtained in the literature, where the conventional implicit gradient damage formulation is adopted, the proposed damage algorithm yields a physically realistic softening response.

Further research in damage modeling will be concerned with the consideration of heterogeneous material at microlevel using the multiscale procedure. It requires the application of the constitutive relations directly at the microlevel considering all material constituents of the RVE and, after a homogenization procedure, the transfer of the state variables to the macrostructural level.

Acknowledgements This work has been supported by the Alexander von Humboldt Foundation, Germany, as well as by the Croatian Science Foundation.

References

1. Čanžar, P., Tonković, Z., Kodvanj, J.: Microstructure influence on fatigue behaviour of nodular cast iron. Mater. Sci. Eng. A **556**, 88–99 (2012)
2. Kouznetsova, V.G., Geers, M.G.D., Brekelmans, W.A.M.: Multi-scale second-order computational homogenization of multi-phase materials: a nested finite element solution strategy. Comput. Methods Appl. Mech. Eng. **193**, 5525–5550 (2004)
3. Miehe, C., Koch, A.: Computational micro-to-macro transitions of discretized microstructures undergoing small strains. Arch. Appl. Mech. **72**, 300–317 (2002)
4. Kaczmarczyk, L.K., Pearce, C.J., Bicanic, N.: Scale transition and enforcement of RVE boundary conditions in second-order computational homogenization. Int. J. Numer. Methods Eng. **74**, 506–522 (2008)
5. Dang, T.D., Sankar, B.V.: Meshless local Petrov-Galerkin micromechanical analysis of periodic composites including shear loadings. Comput. Model. Eng. Sci. **26**, 169 (2008)
6. Ahmadi, I., Aghdam, M.M.: A truly generalized plane strain meshless method for combined normal and shear loading of fibrous composites. Eng. Anal. Boundary Elem. **35**, 395–403 (2011)
7. Miehe, C.: Computational micro-to-macro transitions for discretized micro-structures of heterogeneous materials at finite strains based on the minimization of averaged incremental energy. Comput. Methods Appl. Mech. Eng. **192**, 559–591 (2003)
8. Hain, M., Wriggers, P.: Numerical homogenization of hardened cement paste. Comput. Mech. **42**, 197–212 (2008)
9. Temizer, Ä., Wriggers, P.: An adaptive multiscale resolution strategy for the finite deformation analysis of microheterogeneous structures. Comput. Meth. Appl. Mech. Eng. **200**, 2639–2661 (2011)
10. Mindlin, R.D.: Second gradient of strain and surface-tension in linear elasticity. Int. J. Solids Struct. **1**, 417–438 (1965)
11. Toupin, R.A.: Theories of elasticity with couple-stress. Arch. Ration. Mech. Anal. **17**, 85–112 (1964)
12. Lesičar, T., Tonković, Z., Sorić, J.: C1 continuity finite element formulation in second-order computational homogenization scheme. J. Multiscale Modell. **04**, 1250013 (2012)
13. Lesičar, T., Tonković, Z., Sorić, J.: Second-order computational homogenization scheme preserving microlevel C1 continuity. Key Eng. Mater. **627**, 381–384 (2015)
14. ABAQUS/Standard. ABAQUS Dassault Systemes, Providence, RI, USA (2014)
15. Peerlings, R.H.J., De Borst, R., Brekelmans, W.A.M., De Vree, J.H.P.: Gradient enhanced damage for quasi-brittle materials. Int. J. Numer. Methods Eng. **39**, 3391–3403 (1996)
16. Bažant, Z.P., Belytschko, T., Chang, T.P.: Continuum theory for strain-softening. J. Eng. Mech. **110**(2), 666–1692 (1984)

17. Peerlings, R.H.J., de Borst, R., Brekelmans, W.A.M., Geers, M.G.D.: Gradient-enhanced damage modelling of concrete fracture. Mech. Cohes. Frict. Mater. **3**, 323–342 (1998)
18. De Borst, R., Mühlhaus, H.-B.: Gradient-dependent plasticity: formulation and algorithmic aspects. Int. J. Numer. Methods. Eng. **35**, 521–539 (1992)
19. Engelen, R.A.B., Geers, M.G.D., Baaijens, F.P.T.: Nonlocal implicit gradient-enhanced elasto-plasticity for the modelling of softening behaviour. Int. J. Plast. **19**, 403–433 (2003)
20. Sluys, L.J., de Borst, R., Muhlhaus, H.B.: Wave propagation, localization and dispersion in a gradient-dependent medium. Int. J. Solids Struct. **30**, 1153–1171 (1993)
21. Simone, A., Askes, H., Sluys, L.J.: Incorrect initiation and propagation of failure in non-local and gradient enhanced media. Int. J. Solids. Struct. **41**, 351–363 (2004)
22. Sun, G., Poh, L.H.: Homogenization of intergranular fracture towards a transient gradient damage model. J. Mech. Phys. Solids **95**, 374–392 (2016)
23. Yang, Y., Misra, A.: Higher-order stress-strain theory for damage modeling implemented in an element-free Galerkin formulation. Comput. Model. Eng. Sci. **64**(1), 1–36 (2010)
24. Geers, M.G.D., de Borst, R., Brekelmans, W.A.M., Peerlings, R.H.J.: Strain-based transient-gradient damage model for failure analyses. Comput. Methods Appl. Mech. Eng. **160**, 133–153 (1998)
25. Lesičar, T., Tonković, Z., Sorić, J.: A second-order two-scale homogenization procedure using C1 macrolevel discretization. Comput. Mech. **54**, 425–441 (2014)
26. Lesičar, T.: Multiscale modeling of heterogeneous materials using second-order homogenization. Faculty of Mechanical Engineering and Naval Architecture, 183 pp. Doctoral Thesis, University of Zagreb, Zagreb
27. Lesičar, T., Tonković, Z., Sorić, J.: Two-scale computational approach using strain gradient theory at microlevel. Int. J. Mech. Sci. **126**, 67–78 (2017)
28. Mazars, J., Pijaudier-Cabot, G.: Continuum damage theory-application to concrete. J. Eng. Mech. **115**(2), 345–365 (1989)
29. De Vree, J.H.P., Brekelmans, W.A.M., van Gils, M.A.J.: Comparison of nonlocal approaches in continuum damage mechanics. Comput. Struct. **55**, 581–588 (1995)
30. Peerlings, R.H.J.: Enhanced damage modeling for fracture and fatigue. Ph.D. Thesis (1999). Eindhoven University of Technology, Netherlands
31. Kaczmarczyk, L., Pearce, C.J., Bićanić, N.: Scale transition and enforcement of RVE boundary conditions in second-order computational homogenization. Int. J. Numer. Methods Eng. **74**, 506–522 (2008)
32. Kouznetsova, V.G., Geers, M.G.D., Brekelmans, W.A.M.: Size of a representative volume element in a second-order computational homogenization framework. Int. J. Multiscale Comput. Eng. **2**(4), 575–598 (2004)
33. Otsuka, K., Date, H.: Fracture process zone in concrete tension specimen. Eng. Fract. Mech. **65**, 111–131 (2000)

3D Dynamic Crack Propagation by the Extended Finite Element Method and a Gradient-Enhanced Damage Model

M. Pezeshki, S. Loehnert, P. Wriggers, P.A. Guidault and E. Baranger

Abstract A combined continuous-discontinuous approach to fracture is presented to model crack propagation under dynamic loading. A gradient-enhanced damage model is used to evaluate degradation of the material ahead of the crack. This type of model avoids mesh dependency and pathological effects of local damage models. Discrete cracks are reflected by means of extended finite elements (XFEM) and level sets. For the transition between damage and discrete fracture a damage based criterion is utilized. A discrete crack propagates if a critical damage value at the crack front is reached. The propagation direction is also determined through the damage field. Finally a dynamic mode II crack propagation example is simulated to show the capabilities and robustness of the employed approach.

1 Introduction

Dynamic failure in forms of shear band or crack propagation may happen in real engineering structures under dynamic loading. Modelling these failure phenomena helps to gain better insight into the involved mechanisms, and to improve the design and increase the safety of the structure. A fracture process includes material degradation, microcrack and macrocrack formation, crack propagation and final rupture. Although traditionally material degradation used to be studied within the framework of continuum mechanics and crack propagation within the framework of fracture mechanics, there is a close relation between these two. Microcrack formation as a result of material softening is the onset of macrocrack formation and advancement.

There is a variety of approaches to model fracture in solids which may generally be divided into discrete and continuum approaches. To recall discrete frac-

M. Pezeshki (✉) · S. Loehnert · P. Wriggers
Institute of Continuum Mechanics, Leibniz Universität Hannover,
Appelstr. 11A, 30167 Hanover, Germany
e-mail: pezeshki@ikm.uni-hannover.de

P.A. Guidault · E. Baranger
LMT/ENS Cachan/CNRS/Université Paris-Saclay, 94235 Cachan, France

J. Sorić et al. (eds.), *Multiscale Modeling of Heterogeneous Structures*,
Lecture Notes in Applied and Computational Mechanics 86,
https://doi.org/10.1007/978-3-319-65463-8_14

ture methods, apart from the standard finite element with remeshing [1, 2], one may indicate the cohesive zone or inter-element method [3–6], the embedded discontinuity approach [7, 8], peridynamics [9, 10], meshfree methods [11, 12], and the extended/generalized finite element method (XFEM/GFEM) [13–18] which is employed in this study. The XFEM incorporates a discontinuity into the displacement field by using additional degrees of freedom and enrichment functions. The mesh does not require to conform to cracks and with a few exceptions generally no remeshing is needed. XFEM implementations require additional algorithms to track the advancement of discontinuities in the domain, and they may face geometrical difficulties in 3D cases for crack branching or coalescence. However, their advantage compared to many other approaches is their generally high accuracy even for rather coarse meshes, their relatively low computational effort and their mesh independence.

To indicate continuum-based approaches to fracture the first candidate is a typical local damage model. Since these basic models exhibit strong spurious mesh dependency, several remedies were proposed in literature including rate dependent continuum models, delayed damage models, non-local [19] and gradient-enhanced damage models [20]. Non-local and gradient-enhanced models use an additional parameter as a localization limiter. More recently phase-field approaches to fracture gained a lot of attention among the continuum-based methods. The phase-field considers cracks as a diffusive type field and avoids the explicit modeling of discontinuities [21, 22]. Although by using phase-field approaches crack branching and coalescence can easily be modeled, they suffer some disadvantages such as requiring a very fine mesh resolution to achieve accurate results, demanding an extensive number of Newton-Raphson iterations or even special solution techniques [23], and needing extreme computational effort for 3D simulations.

For accurate modeling of fracture processes a coupled continuous-discontinuous approach is a good alternative. In this study a combined approach of gradient-enhanced damage and XFEM is used. In the continuum part of the model, gradient-enhanced damage accounts for mesh independent material degradation and softening and the loss of stiffness of the structure due to microcracking. As damage develops within the domain, highly-damaged areas show the potential locations where strong discontinuities may occur in the displacement field.

Discrete crack propagation modeled with the XFEM occurs if the gradient-enhanced damage value along the discrete crack front exceeds a critical value. A method is proposed to govern the length of the crack extension increment by the damage field distribution. Whenever damage in front of the crack is not critical the crack advancement is stopped until the critical state of damage is reached. In this way the damage evolution governs the discrete crack advancement until the final failure of the structure. Although the continuum model is able to predict crack initiation, the focus in this study is on structures with pre-existing macrocracks. To decide into which direction a crack may propagate, a strategy based on the maximum damage value is proposed which is similar to the method used in [24] but extended to the 3D case. The XFEM requires an additional method to represent the crack geometry and to track crack propagation. For this purpose, here we use a simple and fast level set approach where nodal level set values are found without solving an additional equation for the level set field which makes this method computationally efficient. The mesh resolution required to obtain an accurate solution for the damage field is fine enough to ensure an accurate level set representation of the crack.

To obtain a sharp representation of a crack using continuum-based models, highly refined mesh resolutions are necessary leading to extreme computational expenses for 3D crack simulations. The proposed approach combining continuum damage models with discrete crack representations using the XFEM and level sets provides an acceptable accuracy at a significantly lower computational cost.

The paper is structured as follows: The momentum balance and gradient-enhanced damage equations are expressed in Sects. 2 and 3. The implementation details of the XFEM are described in Sect. 4 followed by the solution procedure in Sect. 5 and the damage-based crack propagation criterion and direction in Sect. 6. Finally, a dynamic fracture test is modeled and discussed in Sect. 7.

2 Governing Equations

The strong form of the linear momentum balance of structural dynamics reads:

$$\rho \ddot{\mathbf{u}} = \nabla \cdot \boldsymbol{\sigma} + \rho \mathbf{b} \tag{1}$$

with the Cauchy stress tensor $\boldsymbol{\sigma}$, the acceleration $\ddot{\mathbf{u}}$, the mass density ρ, and the body force $\mathbf{f}^b = \rho \mathbf{b}$. The boundary conditions for Γ_t, Γ_u and Γ_c which are traction, displacement and crack boundaries can be written as:

$$\mathbf{u}(\mathbf{x}, t) = \bar{\mathbf{u}}(\mathbf{x}, t) \text{ on } \Gamma_u, \quad \boldsymbol{\sigma} \cdot \mathbf{n} = \mathbf{t} \text{ on } \Gamma_t, \quad \boldsymbol{\sigma} \cdot \mathbf{n} = \mathbf{0} \text{ on } \Gamma_c \tag{2}$$

Here $\mathbf{t}$ is the external traction vector. The initial conditions are:

$$\mathbf{u}(\mathbf{x}, 0) = \bar{\mathbf{u}}(0), \quad \dot{\mathbf{u}}(\mathbf{x}, 0) = \bar{\dot{\mathbf{u}}}(0) \tag{3}$$

By multiplying with the test function $\mathbf{w}_u$, integrating over the domain, applying the divergence theorem as well as the boundary conditions (Eq. 2) the weak form of Eq. 1 becomes

$$\int_{\Omega} \rho \mathbf{w}_u \cdot \ddot{\mathbf{u}} \, d\Omega = - \int_{\Omega} \nabla \mathbf{w}_u : \boldsymbol{\sigma} \, d\Omega + \int_{\Omega} \mathbf{w}_u \cdot \mathbf{b} \, d\Omega + \int_{\Gamma} \mathbf{w}_u \cdot \mathbf{t} \, d\Gamma \tag{4}$$

3 Gradient-Enhanced Damage Model

Attempts to consider microstructural effects on the deformation and failure of material resulted in damage mechanics theory. This theory introduces the damage variable d which describes the local degradation of the structure:

$$\boldsymbol{\sigma}_{ij} = (1 - d)\mathbf{C}_{ijkl}\boldsymbol{\varepsilon}_{kl} \tag{5}$$

where $d = 0$ and $d = 1$ represent the undamaged and fully damaged states, respectively. Typical microstructural defects are microcracks or microvoids which in most engineering problems are not considered explicitly. Instead, their effect on the material behavior is interpreted as degradation or damage. As a result, if microcracks grow, damage increases. Beyond a certain threshold microcracks coalesce and form a macrocrack which is large enough to be modeled explicitly. This motivates the transition between damage and discrete crack modeling.

It is well known that local damage models lead to mesh dependent finite element solutions. One of the remedies is to replace the local damage model by a non-local model [19]. Even though a non-local model yields mesh independent results, a drawback is the significantly higher numerical effort. In 1996 Peerlings et al. [20] proposed an alternative technique by transforming the integral to evaluate the non-local quantities into a scalar valued partial differential equation. This transformation in general is not exact, but it yields very similar results for the non-local equivalent quantities and its advantage compared to the non-local model in [19] is the reduced numerical effort. The additional partial differential equation for the gradient-enhanced equivalent strain $\bar{\varepsilon}$ is

$$\bar{\varepsilon} - c\nabla^2\bar{\varepsilon} = \tilde{\varepsilon} \tag{6}$$

with the boundary condition proposed in [20]:

$$\nabla\bar{\varepsilon} \cdot \mathbf{n} = 0 \tag{7}$$

where $\mathbf{n}$ is the outward normal vector to the boundary of the problem domain and c is the gradient parameter which acts as a localization limiter. This parameter cannot be calibrated directly from experiments but can be found through an inverse approach. The element size should be adjusted with respect to this parameter to achieve accurate numerical results.

For brittle materials in general the driving force for the gradient-enhanced strain $\bar{\varepsilon}$ is chosen to be a function of the local strain evaluated at each integration point $\tilde{\varepsilon}$. Among multiple options available for this function, we apply [25]:

$$\tilde{\varepsilon} = \sqrt{\sum_{i=1}^{3} < \boldsymbol{\varepsilon}_i >^2} \tag{8}$$

Here, $\boldsymbol{\varepsilon}_i$ are the principal strains and $< . >$ are the McAuley brackets $< x > = \frac{1}{2}(x + |x|)$. In this formulation only positive eigenvalues of the strain tensor have an effect. Hence, compressive strains do not contribute to damage and crack propagation.

After the determination of the gradient-enhanced equivalent strain, the chosen damage model can be evaluated using $\bar{\varepsilon}$ instead of the local equivalent strain quantity $\tilde{\varepsilon}$. The history data κ associated to the gradient-enhanced strain $\bar{\varepsilon}$ takes the largest value of $\bar{\varepsilon}$ ever reached by the material at the considered point to retain the previous loading history and avoid healing effects. In the beginning $\kappa = \kappa_0$, where κ_0 is a threshold value. If $\dot{\kappa} > 0$ damage evolves. These update conditions for κ can be formulated mathematically by means of Kuhn-Tucker conditions:

$$\begin{aligned} f(\bar{\varepsilon}, \kappa) &= \bar{\varepsilon} - \kappa \\ \dot{\kappa} \geq 0, \quad f &\leq 0, \quad \dot{\kappa} f = 0 \end{aligned} \tag{9}$$

where f represents the equivalent strain space (similar to the yield space in elasto-plasticity). Then κ can be mapped to a scalar damage value by considering a semi-monotonic relation between d and κ:

$$d = d(\kappa) \tag{10}$$

Different damage evolution laws that are proposed for local damage models can be used for gradient-enhanced models as well. Here, we apply a model proposed by Mazars and Pijaudier-Cabot [25]:

$$d = 1 - \frac{\kappa_0}{\kappa}(1 - \alpha) - \alpha e^{-\beta(\kappa - \kappa_0)} \tag{11}$$

where α and β are the material parameters.

4 Extended Finite Element Method (XFEM)

The XFEM allows us to handle discontinuities within the displacement field almost independently of the mesh by an explicit construction of a suitable approximation space. The XFEM employs enrichment functions to account for special features of the solution in the entire domain. By multiplying the standard finite element shape functions with a set of chosen enrichment functions a new set of enriched shape functions is created which is able to represent the desired features of the primary variable or its derivatives. The enrichment functions we choose for the displacement field and the gradient-enhanced equivalent strain field are described further in Sect. 4.2. In general enrichment functions depend on the geometry of the desired feature. In case a crack needs to be modeled, to capture the discontinuity within the displacement field the location and geometry of the crack needs to be known as well as the distance of an arbitrary point to the crack surface or the crack front and the angle of the point within a plane perpendicular to the tangential plane at the crack front. One way to model the crack geometry and to assure a fast evaluation of the required geometrical data is to use level sets. The following section describes the level set approach helping us to easily evaluate the enrichment functions and to track the movement of the discontinuity as it advances within the domain.

4.1 The Level Set Method for Crack Front Tracking

The level set method (LSM) was developed by Osher and Sethian to track the motion of interfaces [26]. Since an advancing crack can be considered as a moving open interface, this method is employed widely in the context of the XFEM [16, 27–29]. In this work two level set functions are used, ϕ which describes the smallest signed distance to the crack surface Γ_c, and ψ which describes the smallest signed distance to the surface which is perpendicular to the crack surface and passing through the crack front line $\partial\Gamma_c$:

$$\begin{aligned} \phi(x) &= \| \mathbf{x} - \mathbf{x}_{\Gamma_c} \| \ \mathrm{sign}(\mathbf{n}_{\Gamma_c} \cdot (\mathbf{x} - \mathbf{x}_{\Gamma_c})) \\ \psi(x) &= \| \mathbf{x} - \mathbf{x}_{\partial\Gamma_c} \| \ \mathrm{sign}(\mathbf{n}_{\partial\Gamma_c} \cdot (\mathbf{x} - \mathbf{x}_{\partial\Gamma_c})) \end{aligned} \tag{12}$$

The nearest projection points of the crack surface and the crack front line are $\mathbf{x}_{\Gamma_c}$ and $\mathbf{x}_{\partial\Gamma_c}$, respectively. The corresponding normal vectors to these interfaces are $\mathbf{n}_{\Gamma_c}$ and $\mathbf{n}_{\partial\Gamma_c}$. The crack surface Γ_c is the iso-zero surface of the level set function ϕ. The location of the crack tip/front, $\partial\Gamma_c$ can be expressed by means of the intersection of the zero level sets of ϕ and ψ:

$$\mathbf{x} \in \Gamma_c \setminus \partial\Gamma_c \quad \Leftrightarrow \quad \psi < 0 \ \wedge \ \phi = 0 \tag{13}$$

$$\mathbf{x} \in \partial\Gamma_c \quad \Leftrightarrow \quad \phi = 0 \ \wedge \ \psi = 0 \tag{14}$$

The discontinuity always corresponds to the set expressed in Eq. 13. The gradients of the two level set functions are chosen to be orthogonal:

$$\nabla\phi \cdot \nabla\psi = 0. \tag{15}$$

The level set values can be initialized by using Eq. 12. In this study, nodal-wise level set values are utilized which are interpolated within the domain by using the same shape functions as for the geometry approximation.

Based on the crack propagation criterion which is described in Sect. 6, the advancement of the crack at each time step is known. Then the level set field must be updated to be able to represent the new crack location within the domain. There is a variety of methods available to update the level set values e.g. by solving the Hamilton-Jacobi equation [28] or by applying the Fast Marching Method [30], the global crack tracking approach [31] or the global geometrical approach [27]. In the global geometrical approach [27] an explicit representation of the crack is created by meshing the iso-zero surface of the level set function. Then the new crack surface resulting from the crack advancement is added to the mesh and after that the updated level sets are recomputed from the explicit representation of the updated crack geometry. This may cause some additional cost at each crack propagation step. In the local geometrical approach employed here, by advancement of the crack in one element, the level set update is performed locally in the neighboring elements by considering a small domain within a tube which makes this approach efficient. The radius of this tube is chosen based on the average element length or the norm of the propagation vector $\mathbf{v}$:

$$r_{\text{tube}} = \max(\sqrt{2}l_{\text{elem}}, \parallel \mathbf{v} \parallel) \tag{16}$$

The orthogonal local basis vectors $\mathbf{b}_k$ at the crack front can be used to describe the propagation vector locally at each point along the crack front line. They can easily be calculated using the two level set fields:

$$\mathbf{b}_1 = \frac{\nabla\psi}{\parallel \nabla\psi \parallel} \quad , \quad \mathbf{b}_2 = \frac{\nabla\phi}{\parallel \nabla\phi \parallel} \quad , \quad \mathbf{b}_3 = \mathbf{b}_1 \times \mathbf{b}_2 \tag{17}$$

The determination of the crack propagation vector is discussed in Sect. 6.

4.2 XFEM Approximation

From the level set representation, the location of the crack within the desired domain is known. The approximation space which enables us to model the crack as a discontinuity by means of enrichment functions and additional degrees of freedom is described here. Let $\mathscr{N}$ be the set of all nodes in the domain. Then we consider the following subsets of $\mathscr{N}$:

- $\mathscr{N}^*$: nodes belonging to elements completely intersected by the crack surface
- $\mathscr{N}^{**}$: nodes belonging to elements partially intersected by the crack surface
- $\mathscr{N}^{\mathscr{H}}$: nodes in $\mathscr{N}^* \setminus \mathscr{N}^{**}$
- $\mathscr{N}^{\mathscr{R}}$: nodes belonging to elements which have at least one node in $\mathscr{N}^{**}$, i.e. $\mathscr{N}^{**} \subset \mathscr{N}^{\mathscr{R}}$

Based on these nodal sets, which are depicted in Fig. 1 and the C^0 continuous shape functions N_i, we choose the extended displacement approximation

$$\mathbf{u}^h(\mathbf{x}, t) = \sum_{i \in \mathscr{N}} N_i \, \mathbf{u}_i(t) + \sum_{i \in \mathscr{N}^{\mathscr{H}}} N_i \, H(\phi) \, \mathbf{a}_i(t) + \sum_{i \in \mathscr{N}^{\mathscr{R}}} \sum_{j=1}^{4} N_i \, B_j(\phi, \psi) \, R \, \mathbf{b}_{ij}(t) \tag{18}$$

Here $\mathbf{u}_i$ expresses the standard DOF and $\mathbf{a}_i$ and $\mathbf{b}_{ij}$ are the enriched DOF corresponding to the discontinuity/singularity enrichment functions for the crack surface

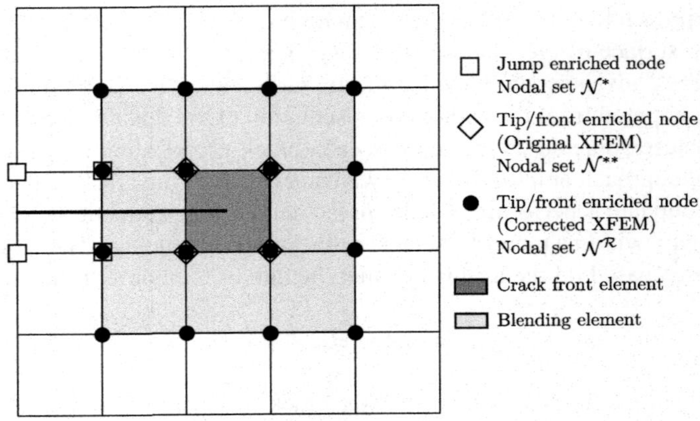

Fig. 1 Nodal sets related to the original and corrected XFEM schemes

and crack front, respectively. The dependency on coordinates on the right-hand side of Eq. 18 is dropped for simplicity. H and B_j are the modified Heaviside and front enrichment functions:

$$H(\mathbf{x}) = H(\phi(\mathbf{x})) = \begin{cases} +1 & \text{if} \quad \phi(\mathbf{x}) \geq 0 \\ -1 & \text{if} \quad \phi(\mathbf{x}) < 0 \end{cases} \tag{19}$$

$$B_j = [\sqrt{r}\sin(\theta/2), \sqrt{r}\cos(\theta/2), \sqrt{r}\sin(\theta/2)\sin(\theta), \sqrt{r}\cos(\theta/2)\sin(\theta)] \tag{20}$$

These functions can be evaluated using the level sets and the definitions of $r(\phi, \psi)$, the distance of the considered point to the crack front, and $\theta(\phi, \psi)$, the angle of the point to the tangent of the crack surface at the crack front:

$$r = \sqrt{\phi^2 + \psi^2}\ , \ \theta = \tan^{-1}(\frac{\phi}{\psi}) \tag{21}$$

As it can be seen from Eq. 18, the modified Heaviside function is applied to all the nodes of the completely cracked elements excluding those nodes belonging to elements partially intersected by the crack. This function allows for a jump in the displacement field at crack faces. The front enrichment functions are chosen such that they span a basis for the LEFM-based analytic asymptotic solution and produce a $1/\sqrt{r}$ singularity in the stress field near the crack tip/front [28, 32]. In 3D cases these functions can represent the analytic asymptotic solution only for a planar crack with a straight crack front. Even though the four front enrichment functions may still significantly improve the solution error, using all of them leads to a significantly higher condition number of the resulting equation system and to a possibly significantly higher numerical effort. For this reason we only use the first of the front enrichment functions which is required to reflect the discontinuity ending within the crack front element.

In the XFEM context, elements which are not completely enriched by the front enrichment function are often called blending elements. Without any special treatment, the partition of unity is not fulfilled in these elements. One remedy is to define the ramp function $R(x)$,

$$R(x) = \sum_{i \in \mathcal{N}^{**}} N_i(x) \tag{22}$$

extend the front enriched domain by one layer of elements, and multiply the ramp function with all front enrichment functions as it is done in Eq. 18. This modification is called corrected XFEM [13, 29]. The ramp function is equal to 1 in the crack front elements, resulting in the same approximation as in the original XFEM within the crack front elements, and it decreases to zero along the boundary of the front enriched elements. By applying the ramp function, the partition of unity is fulfilled in all elements of the domain which in most cases improves the solution error significantly.

Similar to the displacement field, it is necessary to enrich the gradient-enhanced strain field with the modified Heaviside enrichment to account for possible jumps of the gradient-enhanced strains across the crack surface. To obtain a more accurate gradient-enhanced strain field in the vicinity of the crack front, the first crack front enrichment is applied as well. The enriched approximation for the gradient-enhanced strain field reads:

$$\bar{\varepsilon}(\mathbf{x}) = \sum_{i \in \mathcal{N}} N_i\, \bar{\varepsilon}_i + \sum_{i \in \mathcal{N}^{\mathcal{H}}} N_i\, H(\phi)\, \bar{\alpha}_i + \sum_{i \in \mathcal{N}^{\mathcal{R}}} N_i\, B_1(\phi, \psi)\, R\, \bar{\beta}_i \tag{23}$$

Here, $\bar{\alpha}_i$ and $\bar{\beta}_i$ are the enriched DOF for the gradient-enhanced strain field. The finite elements containing non-smooth enrichment functions require a special treatment of the numerical integration since the standard Gauss integration methods are not able to produce sufficiently accurate results [13]. To resolve this problem, here we employ sub-element partitioning [14, 29] with a low order quadrature rule for the sub-elements away from the crack front and a higher order quadrature rule in the vicinity of the crack front.

4.3 Explicit Time Integration and Mass Lumping for XFEM

Due to usually rather high crack propagation velocities the modeling of a dynamically propagating crack requires very small time increments in order to accurately resolve the physics of the involved phenomena. This results in a costly computation and motivates using an explicit time integrator such as the explicit Newmark scheme which is only conditionally stable but does not need iterations in non-linear cases [33]. This leads to a drastic reduction of computational effort per time step. For time-steps Δt larger than the critical time-step Δt_c explicit time integration may lead to numerical instabilities of the solution. For standard finite elements the well known Courant-Friedrichs-Lewy condition can be used to estimate the critical time step size Δt_c. However, due to the enrichments the critical time step size within the XFEM may be reduced. This is discussed in more detail in [34, 35]. For an efficient explicit method the mass matrix should be lumped. Here we apply a lumping strategy proposed in [36]:

$$m_{diag} = \frac{1}{N_{node}} \int_{\Omega_e} \rho[\sum_{i=1}^{N_{node}} N_i(\mathbf{x}) F_i(\mathbf{x})]^2 d\Omega_e \tag{24}$$

where F_i is the applied enrichment function at the respective node. This relationship is the equivalent to the one proposed in [34] where only the Heaviside function is used to enrich the displacement field and it results in a reasonable non-zero critical time step with $0 < \Delta t_c^{XFEM} \leq \Delta t_c^{FEM}$.

5 Solution Procedure

The discretized weak form of the momentum balance (Eq. 4) reads:

$$\int_{\Omega} \mathbf{N_u^T} \rho \ddot{\mathbf{u}} N_u \, d\Omega = -\int_{\Omega} \mathbf{B_u^T} \sigma \, d\Omega + \int_{\Omega} \mathbf{N_u^T f^b} \, d\Omega + \int_{\Gamma_t} \mathbf{N_u^T t} \, d\Gamma \tag{25}$$

Here $\mathbf{N_u}$ and $\mathbf{B_u}$ are the corresponding shape function and gradient operator for the displacement field. The mass matrix that needs to be lumped according to Eq. 24 to obtain an efficient explicit solution scheme is

$$\mathbf{M} = \int_{\Omega} \mathbf{N_u^T} \rho \mathbf{N_u} \, d\Omega \tag{26}$$

For the explicit Newmark scheme to integrate the linear momentum balance we use the Newmark parameters $\gamma = 1/2$ and $\beta = 0$ [33].

The discretized weak form of Eq. 6 to obtain the gradient-enhanced equivalent strain field is

$$\int_{\Omega} (\mathbf{N_\varepsilon^T N_\varepsilon} \bar{\varepsilon} + \mathbf{B_\varepsilon^T} c \mathbf{B_\varepsilon} \bar{\varepsilon}) \, d\Omega = \int_{\Omega} \mathbf{N_\varepsilon^T} \tilde{\varepsilon} \, d\Omega \tag{27}$$

Here, the corresponding shape function and gradient operators for the gradient-enhanced equivalent strain are $\mathbf{N_\varepsilon}$ and $\mathbf{B_\varepsilon}$. The linear momentum balance equation (Eq. 1) together with the gradient-enhanced equation (Eq. 6) constitute the equation system for the combined approach to fracture. Their discretized forms are expressed in Eqs. 25 and 27. Since $\bar{\varepsilon}$ depends non-linearly on the strain state, and the damage variable depends non-linearly on $\bar{\varepsilon}$ their weak forms are strongly coupled. Instead of a fully coupled approach, this system of equations can be divided into two boundary value problems and solved by using a fixed point (staggered) approach. In an equilibrium step of the momentum balance with fixed damage value ($\Delta\bar{\varepsilon} = 0, d = d^{(k)}$), the new displacements and local equivalent strain $\tilde{\varepsilon}$ can be obtained. $(\cdot)^k$ denotes the converged value from the previous time step. Then the discretized form for $\bar{\varepsilon}$ with a converged displacement configuration ($\Delta\mathbf{u} = 0$) can be solved to update $\bar{\varepsilon}$ and consequently d by considering the conditions in Eq. 9 and the evolution law for damage (Eq. 11). Since time steps are really small in the explicit Newmark scheme, the solution of the implicit Helmholtz-type equation (Eq. 27) at each time step can be costly. If it can be assumed that the damage state does not change too much in every time step, this equation can be solved only every n time steps. By doing so, the solution of $\mathbf{u}$ and $\bar{\varepsilon}$ at $t + \Delta t$ are slightly inconsistent but quite acceptable for small time increments. When the damage state is known in the model it can be used to evaluate the advancement of the crack in the body. This issue is discussed in the next section.

6 Crack Propagation Criterion and Direction Based on Gradient-Enhanced Damage

Once the solution for the displacement and gradient-enhanced equivalent strain field are known, the important questions are whether the crack propagates, which parts of the crack propagate, in which direction does the crack propagate and how far does it propagate. These questions are addressed in the following sections.

6.1 Using Damage as the Crack Propagation Criterion

Based on the employed approach, the XFEM is used to introduce a discontinuity into the body as it naturally and physically happens during the fracture process after the creation of microcracks and degradation of the material. When the nodal solution of Eq. 6 is known by using the described procedure in Sect. 3 damage values are available at integration points. Therefore, the damage state at any crack front point can be estimated by using an inverse weighted distance method which considers the surrounding integration points located within the radius of an average element length for each crack front point. Now, it should be decided at which front points and to what extent the crack will propagate. Based on the the solution of the gradient-enhanced equation two simple options are available to serve as a crack propagation criterion: gradient-enhanced equivalent strain values and the damage state.

A gradient-enhanced strain based criterion is used in [37, 38] with the assumption that the largest gradient-enhanced strain values always appear along the crack front. This criterion is valid in case of linear elastic fracture without damage which is not our goal here. The second possibility is a damage-based criterion along the crack front. When a critical damage state is detected ahead of the current crack front, the discontinuity can be advanced. The critical value of damage for this purpose is d_c:

$$\begin{aligned} d \geq d_c &\quad \textit{crack propagation} \\ d < d_c &\quad \textit{no crack propagation} \end{aligned} \tag{28}$$

The fully damaged material theoretically can be expressed by $d_c = 1$ but based on Lemaitre [39] in real structures a material can be considered as degenerated for damage values between 0.2 and 0.5. The critical damage value for the transition from a non-local damage to discrete response is reported to be dependent on the length-scale and band-width of the damage for a combined damage-cohesive XFEM model [40]. With the combination of XFEM and a gradient-enhanced damage model, the critical values between 0.2 and 0.5 are used in [24, 41]. This parameter may depend on the softening curve as well as the damage evolution law, and can be found through an inverse analysis. To avoid energy conservation issues, here we use a higher value which does not lead to a significant difference in crack propagation behavior. The propagation criterion used here is based on a rather simple and pragmatic choice. One can use more complicated and possibly more accurate criteria such as the calculation of the strain tensor eigenvalues and the acoustic tensor [42], the determination of the loss of stability or loss of strong ellipticity in the quasi-static cases as in [43] or loss of hyperbolicity in the dynamic cases [3]. These indicators may be employed for switching to a different dissipation mechanism which is the topic of further development of the current work.

6.2 Crack Propagation Increment Based on the Damage Field

Once the points along the crack front at which the crack will propagate are determined, it is necessary to calculate the crack propagation vector. The determination of the length of this vector as well as its direction is discussed in the following.

In most cases the length of the crack increment in the context of the XFEM is unknown. Often it is set as a user-defined parameter which should be consistent with the crack propagation criterion. In fracture mechanics theory the crack velocity can be derived such that during the propagation the stress intensity factor is equal to the dynamic crack growth toughness [44, 45]. Therefore, with a predefined time step the increment of growth is known. In [3] the crack velocity is derived based on the loss of hyperbolicity. In contrast to LEFM-based approaches, in damage-based models usually the velocity of the crack growth is not controlled [44].

Since cracking is considered irreversible in the current study, the crack growth increments should be small enough to be able to follow the correct path. However, a compromise is needed since a limitation to very small increments can lead to a costly computation. If the propagation criterion is met at some of the considered crack front points but not at all of them, the new crack surface might become rough. This will require additional increments with smaller segment lengths to be able to follow the crack path. To overcome this difficulty, Fries and Baydoun proposed to advance the crack at all the front points whenever the criterion at a front point is fulfilled [27]. This strategy results in a rather smooth crack path, less propagation increments and does not affect the final crack path significantly. Fries and Baydoun used different propagation increments proportional to the element dimensions and reported nearly analogous crack paths. One may advance the crack at each crack front point with respect to the energy release rate value at that point [37, 38]:

$$\exists \mathbf{x} \in \partial\Gamma_c \; : \; \widetilde{G}(\mathbf{x}) = \frac{G(\mathbf{x})}{G_c} \geq 1, \quad \Delta a = k\, l_{\text{elem}} \widetilde{G} \tag{29}$$

where l_{elem} is the average edge length of elements in the domain $\mathcal{N}^{\mathcal{R}}$ and $0 < k \leq 1$ is a factor. In this study, the following equation based on damage values at the front points is used:

$$\exists \mathbf{x} \in \partial\Gamma_c \; : \; \widetilde{d}(\mathbf{x}) = \frac{d(\mathbf{x})}{d_c} \geq 1, \quad \Delta a = k\, l_{\text{elem}} \widetilde{d} \tag{30}$$

Regarding k, values of $0.3 \leq k \leq 0.5$ are found to be optimal since they allow for multiple propagation steps within an element. This is important since the crack can flexibly change its direction within one finite element. This is discussed further in Sect. 6.3. Allowing for multiple propagation steps per time increment until the damage value falls below the propagation criteria in front of the crack ensures a fair approximation for the crack propagation velocity in the simulations.

6.3 Finding the Propagation Direction from the Damage Field

After determining the crack propagation increment for each considered crack front point, it is necessary to calculate the propagation direction. There exist several well known techniques with different theoretical background such as the maximum energy release rate, the minimum strain energy density or the maximum hoop stress. Since in this contribution we persue a transition between damage and discrete fracture we directly use the damage field to determine the direction of crack propagation.

A number of n_{DEP} Damage Evaluation Points (DEP) can be considered along an arc with a user defined radius l_p and a central angle of 2α (See Fig. 2). The location of these points can be described by the local polar coordinate pair $(r = l_p, \theta)$ with its reference located on the crack front point. At each DEP a damage value d_{DEP} can be estimated.

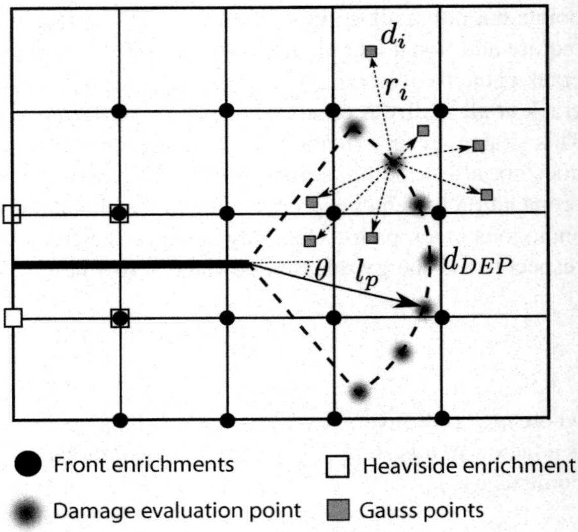

Fig. 2 Finding crack propagation direction based on damage at damage evaluation points

The crack will propagate in the direction θ_c of the point with the highest damage value. This strategy is resembling to the method used by Fries and Baydoun [27] to find the maximum hoop stress, but the evaluated value at each point is damage.

The radius of the arc, l_p is found to be an important parameter as reported in [24]. It should not be too big in a way that the evaluation points be located outside damaged domain and not too small which leads to a zigzag crack path and additional number of iterations. The optimum value found to be $l_p = 1.2l_c$, where l_c is the length scale of the damage model. This is close to the factor 1.5 proposed by Broumand and Khoei for a 2D visco-plastic analysis [24]. It is found that higher or smaller values can lead to very similar crack path but may cause difficulties by generating a rough crack surface. The number of the damage evolution points does not affect the results strongly but too few evaluation points will result in a non-smooth crack surface. Based on the experimental findings for a pure mode II loaded crack, the maximum propagation angle can be about $\pm 71°$ [27]. In this study, the central angle of the arc assumed to be $2\alpha = 90°$ which allows the flexibility of ± 45 at each propagation increment. This assumption does not affect the crack path since within each element a few crack increments are possible as mentioned in Sect. 6.2. This means that within a finite element with at least two crack increments an angle of $\pm 71°$ can easily be achieved and a sharp crack angle can be modeled if needed.

As the crack advances from $\partial\Gamma_{c(k)}$ at time t_k to $\partial\Gamma_{c(k+1)}$ at time t_{k+1}, the crack front points on $\partial\Gamma_{c(k+1)}$ can be located arbitrarily at any location within the element. In general, these positions do not coincide with the location of the integration points. Thus, a strategy is required to evaluate damage values at the crack front points. For this purpose damage values d_i from the surrounding integration points at the distance r_i can be used which provide an accurate and smooth estimation. An inverse weighted distance method is employed to evaluate the damage at any DEP. This weighting approach increases the effects of the closer integration points on the estimated damage value at each front point [24]:

$$d_{\mathrm{DEP}_m} = \frac{\sum_i d_i / r_i^2}{\sum_i 1/r_i^2} \tag{31}$$

where d_{DEP_m} is the estimated damage value at point $\mathbf{x}_{\mathrm{DEP}_m}$.

After finding the propagation direction, the propagation vector, $\mathbf{v}_{k+1}$, is known at each front point $\mathbf{x}_{\partial\Gamma_{c(k)}}$. The location of the new front points $\mathbf{x}_{\partial\Gamma_{c(k+1)}}$ become:

$$\begin{aligned} \mathbf{x}_{\partial\Gamma_{c(k+1)}} &= \mathbf{x}_{\partial\Gamma_{c(k)}} + \mathbf{v_{k+1}} = \\ \mathbf{x}_{\partial\Gamma_{c(k)}} &+ \Delta a \cos(\theta_c)\mathbf{b_1} + \Delta a \sin(\theta_c)\mathbf{b_2} \end{aligned} \tag{32}$$

where Δa and θ_c are found from Eq. 30 and Sect. 6.3, respectively. At each increment of growth, the orthogonal local basis vectors $\mathbf{b}_k$ for each crack point are known based on the the level set fields from Eq. 17. The new crack geometry is known from Eq. 32 and the enrichment scheme can be updated for nodes located within the level set tube expressed in Eq. 16. After the level set update, the new configuration of the crack is set. Since the position of the integration points after the propagation may be different to the position of the integration points before, the state variables (stress, damage, etc.) should be mapped from the old to the new integration points. For this purpose again an inverse weighted distance method [46] with a factor of $1/r^2$ is used. The procedure is similar to the evaluation of the damage values along the crack front (See Eq. 31).

7 Numerical Example

In this section, the performance and robustness of the proposed method is investigated by modeling a plate subjected to impulse loading. This experiment was performed by Kalthoff and is therefore often called the Kalthoff test. [47, 48]. The plate consists of two edge cracks impacted by a projectile at different initial velocities, V_0 (See Fig. 3). Initially this test represents a pure mode II fracture. It could be found that there is a transition from brittle to ductile fracture for increasing V_0. At lower strain rates, brittle failure can be observed with the propagation of two cracks initiated from the notches at an angle of about 70° with respect to the ligament (See path 1 in Fig. 3). At higher impact speeds, a transition between brittle fracture and shear band propagation with a propagation angle of −10° can be observed which is schematically shown as path 2 in Fig. 3. This change of failure mode with impact speed is called dynamic failure mode transition. The failure mode transition in Kalthoff's experiment contradicts the traditional belief that at higher strain rates failure transition is from ductile to brittle fracture. This mode transition attracted attention of many researchers to simulate

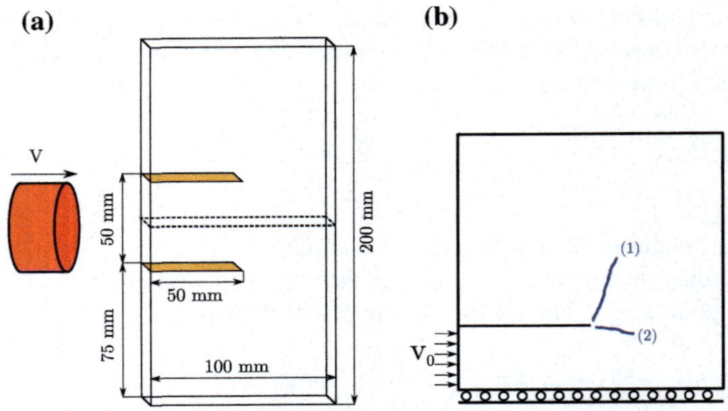

Fig. 3 **a** Geometry and, **b** the boundary conditions and two schematic crack paths of Kalthoff's experiment

these experiments by using different numerical methods such as XFEM [3, 49], cohesive zone approach [3, 5], embedded strong discontinuity approach E-FEM [8], thick level set method [50], smoothed particle hydrodynamics [12], peridynamics [9], and phase-field methods [21, 51].

7.1 Specimen Geometry, Boundary Conditions and Material Properties

The geometry of Kalthoff's test [47] is depicted in Fig. 3a. For the numerical simulation we discretize only a symmetric part of the 3D structure and apply the corresponding symmetry boundary conditions (see Fig. 3b). To avoid frictional contact between the two surfaces of the crack, notches with a finite gap between the two crack edges are considered in experiments as well as in our simulations. A sharp pre-crack is introduced at the end of the notch using the XFEM and level sets.

The projectile is not modeled explicitly. Instead, the impact is modeled by prescribing a constant velocity of $V_0 = 16.5\,\text{m/s}$ as shown in Fig. 3c for a time period of $100\,\mu\text{s}$. In the experiment described in [47] the specimen is not mounted anywhere; therefore, in the simulation traction free boundaries are considered for the remaining surfaces.

The specimen was made of maraging steel with the material constants shown in Table 1. The damage parameters are taken from [3]. Although the damage model used by Belytschko et al. was a local damage model, the constants are related to the damage evolution function which essentially can be used to map an equivalent strain value to a damage value. The same mapping is needed for the gradient-enhanced strains $\bar{\varepsilon}$ in our model. The critical damage value serving as a crack propagation

Table 1 Material parameters for Kalthoff's experiment

Parameter	Value	Parameter	Value
Young' modulus E	190 Gpa	Damage threshold strain	3.0×10^{-3}
Poisson's ratio ν	0.3	Damage law parameter α	1.0
Density ρ	$8000\,\mathrm{Kg/m^3}$	Damage law parameter β	200.0
Characteristic length l_c	0.9 mm		

criterion is considered to be $d_c = 0.9$. Since the damage evolution law used here is an exponential function it saturates quite fast and the stiffness will be degraded to nearly zero. In addition, the stresses within elements ahead of the crack front will be small compared to the stresses in the much less damaged parts of the domain. These two facts allow for a rather smooth transition between damage and fracture. For the damage model we applied in this article, we could not find a significant dependence of the crack propagation behavior for critical damage values beyond $d_c = 0.7$. The characteristic length for the damage model is set to $l_c = 0.9\,\mathrm{mm}$ which is bigger than $2l_{\mathrm{elem}}$.

In those parts of the domain where the crack may propagate and an accurate damage solution is needed we use uniform hexahedral elements of the size $0.4 \times 0.4 \times 0.5\,\mathrm{mm}^3$ (See region 1 in Fig. 4a). Larger elements are used for other parts (region 2) as depicted in Fig. 4a. Before the crack propagates we choose a time increment of $\Delta t = 4 \times 10^{-8}\,\mathrm{s}$. Once the crack propagates we decrease the time step size to $\Delta t = 2 \times 10^{-8}\,\mathrm{s}$. This is consistent with the Courant-Friedrichs-Lewy condition considering that the shear and Rayleigh wave speeds in this medium are $c_S = 3022\,\mathrm{m/s}$ and $c_R = 2799\,\mathrm{m/s}$.

7.2 *Kalthoff's Test Simulation Results*

For an initial velocity of $V_0 = 8\,\mathrm{m/s}$ the crack does not propagate. By increasing the velocity to $V_0 = 16.5$ m/s we observe a brittle fracture mode with a crack propagation

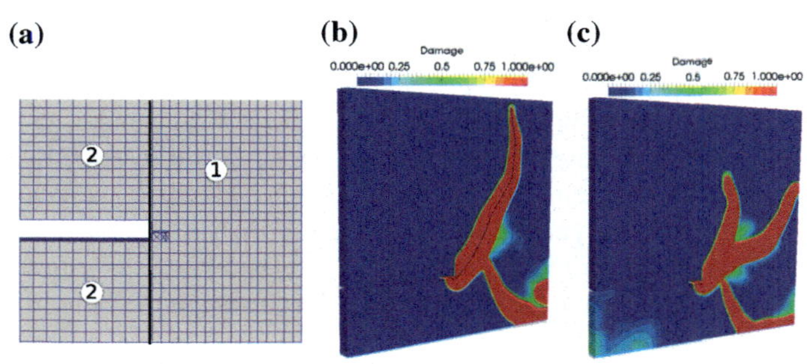

Fig. 4 **a** Mesh resolution for differnet regions of Kalthoff's test, 3d simulation of Kalthoff's test for two projectile velocities of: **b** $V_0 = 16.5$ m/s, **b** $V_0 = 20.0$ m/s

pattern similar to what Kalthoff has reported (See Fig. 4b). The damage field for $V_0 = 20$ m/s is shown in Fig. 4c. Due to the higher impact velocity, higher stress values are imposed and two damage branches appear advancing towards the upper and right edges. For a shear band propagation the impact velocity should be higher than $V_0 = 30\,\mathrm{m/s}$, and temperature dependent plastic deformation should be considered. At higher velocities the plastic deformation and thermal softening suppress brittle fracture response and a shear band propagates through the specimen. This behavior can be classified as a ductile failure mode which involves a temperature rise of about 900–1200 K [52]. This is not within the scope of the current study.

For the case of brittle failure with an impact velocity of $V_0 = 16.5\,\mathrm{m/s}$ the crack pattern is shown in Fig. 5 for four different time steps. The crack starts to propagate at $t = 27\,\mu\mathrm{s}$ at an angle of about 37° with respect to the initial pre-crack. The overall propagation angle, from the ligament to the point at which the crack touches the upper edge is about 69°. This agrees quite well with the value of 70° reported by Kalthoff [47]. The crack path is very similar to the prediction of the phase-field solution (with the same mesh resolution) [51] and the solution using the thick level set method in combination with a non-local damage model [50]. The discrete crack can follow the damage field successfully which confirms the predicted capabilities of the employed approach in Sect. 6.3. Due to the formation of a damage band we obtain similar crack increment lengths in neighboring parts of the crack front line leading to a smooth crack surface during the simulation. This avoids additional iterations to follow the crack path.

Once the compressive waves resulting from the impact arrive on the traction free surfaces of the specimen, they are reflected as tensile waves. Moreover, the diffracted waves from the bottom surface, which accounts for symmetry, can contribute to damage. Because of these reflected waves, the model predicts damage on the bottom-right side of the specimen which develops towards the notch as can be seen in Fig. 5a–d. Since in this contribution we do not consider crack initiation and also we do not introduce an initial crack at the bottom-right side of the specimen, we do not see any discrete crack propagation modeled with the XFEM in that part of the domain.

A similar damage field was reported by using other damage-based models such as local damage [3], phase-field [51] and the thick level set model [50] as well as the embedded strong discontinuities method (E-FEM) [8]. This additional damaged area can be regarded as a mode I crack although no evidence of its existence is reported in the original experiments. Borden reported that for a 2D model with a highly-refined mesh and element edges of about 0.1 mm which is four times finer than our mesh, can alleviate the damage values in this area [21]. Evidence of the similar mode I crack was reported by Ravi-Chandar et al. for a failure mode transition experiment. Their experiment was similar to Kalthoff's but the specimen had only one notch instead of two and it was made of Polymethylmethacrylate (PMMA) [53]. Since the original material for Kalthoff's experiment is steel, using an elasto-plastic constitutive model might give a better representation of the material response in the bottom-right side of the specimen.

The crack propagation speed which is driven by the speed of the development of the damage field is shown in Fig. 6. The crack starts to propagate at $27\,\mu\mathrm{s}$, a bit

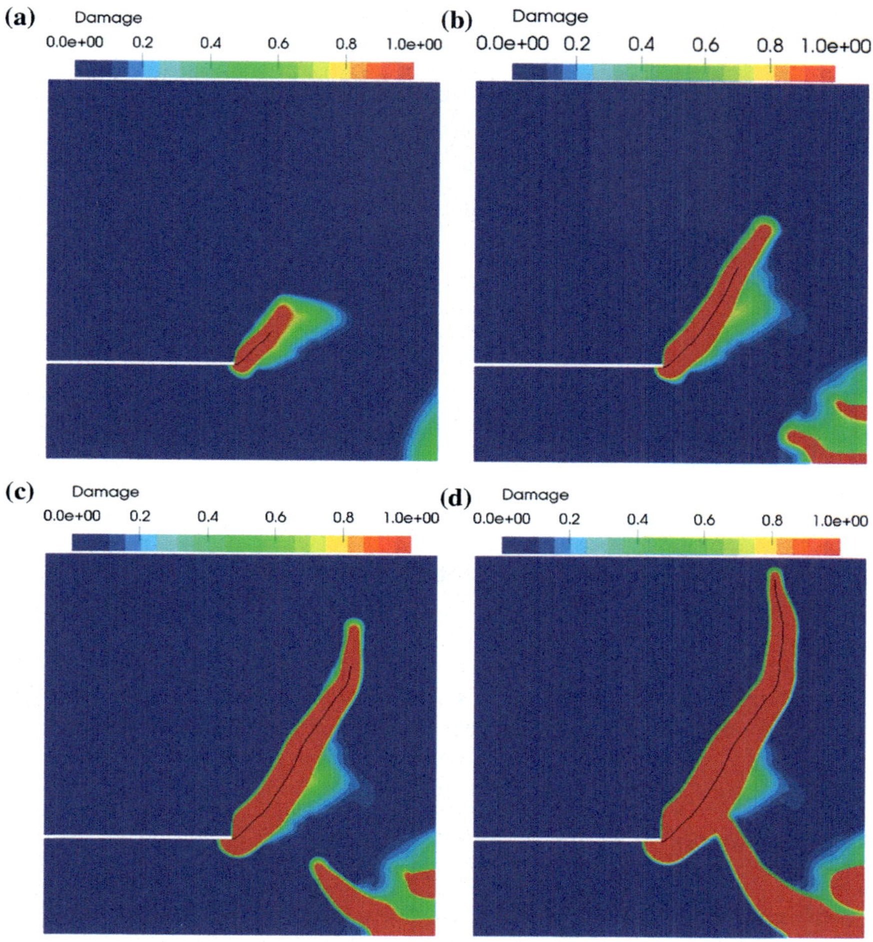

Fig. 5 Crack propagation for Kalthoff's double notch experiment using combined XFEM-gradient-enhanced damage model at **a** $t = 40\,\mu s$, **b** $t = 60\,\mu s$, **c** $t = 80\,\mu s$, **d** $t = 100\,\mu s$

later than in 2D models' predictions and the crack speed is not as high as predicted in 2D plane stress simulations. This speed is bounded by approximately half of the Rayleigh wave speed. The crack decelerates after 80 μ and its speed falls below $0.35c_R$ until $110\,\mu s$ at which time the simulation finishes. In 3D simulations of the test at impact velocities higher than $V_0 = 15\,m/s$ with mesh-free [12] and peridynamics [9] approaches, a simulation time beyond $100\,\mu s$, and an average crack speed of about 900 m/s was reported. The average crack speed, the current crack length devided by the propagation time, is 980 m/s in this study which is slightly lower than $0.4c_R$. Batra and Ravinsankar simulated the same test in 3D and reported a delayed crack propagation in the 3D case compared to 2D plane strain condition simulations [54].

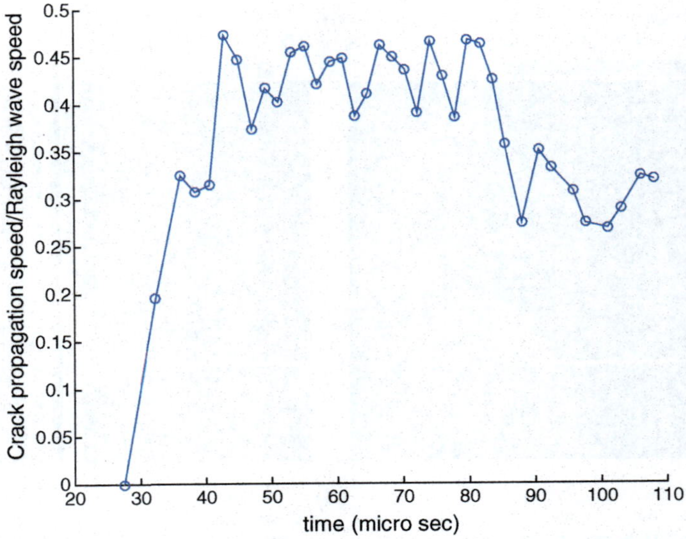

Fig. 6 Velocity of the propagating crack in Kalthoff's test

They claimed that the stress triaxiality and the deformations are more severe in the 2D plane strain case than in the 3D case, and neither the plane stress nor the plane strain assumptions were able to represent the 3D simulation results of the test.

8 Conclusion

A combined continuous-discrete approach is used to model crack propagation under dynamic loading. A damage-based criterion determines the propagation of a discrete crack modeled using the XFEM and level sets. In highly damaged crack front elements the discrete crack is advanced in the direction of the maximum damage value. For this purpose, a geometrical approach based on the damage distribution is successfully used to find the crack propagation direction. The proposed strategy results in a smooth crack surface. To track the discrete crack in the finite element mesh, a level set technique is used. The geometrical level set approach does not require the solution of an additional equation system, and it is found to be computationally efficient. The combined approach benefits from sharp crack representation by the XFEM which does not require special treatment for the mesh behind the crack front. Finally, the robustness and the capabilities of the numerical approach are verified by a numerical example. It is found that the discontinuous crack can successfully follow the damage field and the prediction for the propagation path agrees well with experimental results.

Acknowledgements The German Research Association (DFG) is gratefully acknowledged for the support of the IRTG 1627 program.

References

1. Martha, L.F., Wawrzynek, P.A., Ingraffea, A.R.: Arbitrary crack representation using solid modeling. Eng. Comput. **9**(2), 63–82 (1993)
2. Potyondy, D.O., Wawrzynek, P.A., Ingraffea, A.R.: An algorithm to generate quadrilateral or triangular element surface meshes in arbitrary domains with applications to crack propagation. Int. J. Numer. Methods Eng. **38**(16), 2677–2701 (1995)
3. Belytschko, T., Chen, H., Xu, J., Zi, G.: Dynamic crack propagation based on loss of hyperbolicity and a new discontinuous enrichment. Int. J. Numer. Methods Eng. **58**(12), 1873–1905 (2003)
4. Camacho, G., Ortiz, M.: Computational modelling of impact damage in brittle materials. Int. J. Solids Struct. **33**(20–22), 2899–2938 (1996)
5. Song, J.H., Areias, P.M.A., Belytschko, T.: A method for dynamic crack and shear band propagation with phantom nodes. Int. J. Numer. Methods Eng. **67**(6), 868–893 (2006)
6. Xu, X.P., Needleman, A.: Numerical simulations of fast crack growth in brittle solids. J. Mech. Phys. Solids **42**(9), 1397–1434 (1994)
7. Belytschko, T., Fish, J., Engelmann, B.E.: A finite element with embedded localization zones. Comput. Methods Appl. Mech. Eng. **70**(1), 59–89 (1988)
8. Lloberas-Valls, O., Huespe, A., Oliver, J., Dias, I.: Strain injection techniques in dynamic fracture modeling. Comput. Methods Appl. Mech. Eng. **308**, 499–534 (2016)
9. Silling, S.: Dynamic fracture modeling with a meshfree peridynamic code. Comput. Fluid Solid Mech. **2003**, 641–644 (2003)
10. Silling, S., Askari, E.: A meshfree method based on the peridynamic model of solid mechanics. Comput. Struct. **83**(17), 1526–1535 (2005)
11. Belytschko, T., Organ, D., Gerlach, C.: Element-free galerkin methods for dynamic fracture in concrete. Comput. Methods Appl. Mech. Eng. **187**(3–4), 385–399 (2000)
12. Raymond, S., Lemiale, V., Ibrahim, R., Lau, R.: A meshfree study of the Kalthoffwinkler experiment in 3D at room and low temperatures under dynamic loading using viscoplastic modelling. Eng. Anal. Boundary Elem. **42**, 20–25 (2014)
13. Fries, T.: A corrected XFEM approximation without problems in blending elements. Int. J. Numer. Methods Eng. **75**(5), 503–532 (2008)
14. Moes, N., Dolbow, J., Belytschko, T.: A finite element method for crack growth without remeshing. Int. J. Numer. Methods Eng. **46**(1), 131–150 (1999)
15. Moës, N., Gravouil, A., Belytschko, T.: Non-planar 3D crack growth by the extended finite element and level sets-part I: mechanical model. Int. J. Numer. Methods Eng. **53**(11), 2549–2568 (2002)
16. Stolarska, M., Chopp, D.L., Moës, N., Belytschko, T.: Modelling crack growth by level sets in the extended finite element method. Int. J. Numer. Methods Eng. **51**(8), 943–960 (2001)
17. Strouboulis, T., Babuška, I., Copps, K.: The design and analysis of the generalized finite element method. Comput. Methods Appl. Mech. Eng. **181**, 43–69 (2000)
18. Strouboulis, T., Copps, K., Babuška, I.: The generalized finite element method. Comput. Methods Appl. Mech. Eng. **190**, 4081–4193 (2001)
19. Pijaudier-Cabot, G., Bažant, Z.P.: Nonlocal damage theory. J. Eng. Mech. **113**(0), 1512–1533 (1987)
20. Peerlings, R.H.J., De Borst, R., Brekelmans, W.A.M., De Vree, J.H.P.: Gradient enhanced damage for quasi-brittle materials. Int. J. Numer. Methods Eng. **39**(19), 3391–3403 (1996)
21. Borden, M.J., Verhoosel, C.V., Scott, M.A., Hughes, T.J., Landis, C.M.: A phase-field description of dynamic brittle fracture. Comput. Methods Appl. Mech. Eng. **217**, 77–95 (2012)

22. Miehe, C., Hofacker, M., Welschinger, F.: A phase field model for rate-independent crack propagation: robust algorithmic implementation based on operator splits. Comput. Methods Appl. Mech. Eng. **199**(45), 2765–2778 (2010)
23. Gerasimov, T., De Lorenzis, L.: A line search assisted monolithic approach for phase-field computing of brittle fracture. Comput. Methods Appl. Mech. Eng. **312**, 276–303 (2016)
24. Broumand, P., Khoei, A.: X-FEM modeling of dynamic ductile fracture problems with a non-local damage-viscoplasticity model. Finite Elem. Anal. Des. **99**, 49–67 (2015)
25. Mazars, J.: PijaudierCabot, G.: Continuum damage theory application to concrete. J. Eng. Mech. **115**(2), 345–365 (1989)
26. Osher, S., Sethian, J.A.: Fronts propagating with curvature-dependent speed: algorithms based on Hamilton-Jacobi formulations. J. Comput. Phys. **79**(1), 12–49 (1988)
27. Fries, T.P., Baydoun, M.: Crack propagation with the extended finite element method and a hybrid explicit-implicit crack description. Int. J. Numer. Methods Eng. **89**(12), 1527–1558 (2012)
28. Gravouil, A., Moës, N., Belytschko, T.: Non-planar 3D crack growth by the extended finite element and level sets-part II: level set update. Int. J. Numer. Methods Eng. **53**(11), 2569–2586 (2002)
29. Loehnert, S., Mueller-Hoeppe, D.S., Wriggers, P.: 3D corrected XFEM approach and extension to finite deformation theory. Int. J. Numer. Methods Eng. **86**(4–5), 431–452 (2011)
30. Sukumar, N., Chopp, D., Moran, B.: Extended finite element method and fast marching method for three-dimensional fatigue crack propagation. Eng. Fract. Mech. **70**(1), 29–48 (2003)
31. Oliver, J., Huespe, A.: Continuum approach to material failure in strong discontinuity settings. Comput. Methods Appl. Mech. Eng. **193**(30), 3195–3220 (2004)
32. Sukumar, N., Mos, N., Moran, B., Belytschko, T.: Extended finite element method for three-dimensional crack modelling. Int. J. Numer. Methods Eng. **48**(11), 1549–1570 (2000)
33. Newmark, N.M., Asce, F.: A method of computation for structural dynamics. J. Eng. Mech. Div. **85**, 67–94 (1959)
34. Menouillard, T., Réthoré, J., Combescure, A., Bung, H.: Efficient explicit time stepping for the eXtended Finite Element Method (X-FEM). Inter. J. Numer. Methods Eng. **68**(9), 911–939 (2006)
35. Menouillard, T., Réthoré, J., Moës, N., Combescure, A., Bung, H.: Mass lumping strategies for X-FEM explicit dynamics: application to crack propagation. Int. J. Numer. Methods Eng. **74**(3), 447–474 (2008)
36. Shahbeyk, S., Yaghoobi, M., Vafai, A.: Explicit dynamics X-FEM simulation of heterogeneous materials. Finite Elem. Anal. Des. **56**, 52–79 (2012)
37. Holl, M.: Multiscale crack-propagation and crack coalescence using the XFEM. Ph.D. thesis, Leibniz Universität Hannover (2014)
38. Holl, M., Rogge, T., Loehnert, S., Wriggers, P., Rolfes, R.: 3D multiscale crack propagation using the XFEM applied to a gas turbine blade. Comput. Mech. **53**(1), 173–188 (2014)
39. Lemaitre, J.: A Course on Damage Mechanics. Springer, Berlin (1996)
40. Comi, C., Mariani, S., Perego, U.: An extended FE strategy for transition from continuum damage to mode I cohesive crack propagation. Int. J. Numer. Anal. Methods Geomech. (2007)
41. Seabra, M.R.R., Šuštarič, P., Cesar de Sa, J.M.A., Rodič, T.: Damage driven crack initiation and propagation in ductile metals using XFEM. Comput. Mech. **52**(1), 161–179 (2013)
42. Pijaudier-Cabot, G., Benallal, A.: Strain localization and bifurcation in a nonlocal continuum. Int. J. Solids Struct. **30**(13), 1761–1775 (1993)
43. Areias, P.M.A., Belytschko, T.: Analysis of three-dimensional crack initiation and propagation using the extended finite element method. Int. J. Numer. Meth. Eng. **63**(5), 760–788 (2005)
44. Combescure, A., Gravouil, A., Grégoire, D., Réthoré, J.: X-FEM a good candidate for energy conservation in simulation of brittle dynamic crack propagation. Comput. Methods Appl. Mech. Eng. **197**(5), 309–318 (2008)
45. Grégoire, D., Maigre, H., Réthoré, J., Combescure, A.: Dynamic crack propagation under mixed-mode loading comparison between experiments and X-FEM simulations. Int. J. Solids Struct. **44**(20), 6517–6534 (2007)

46. Wells, G.N., Sluys, L.J., de Borst, R.: Simulating the propagation of displacement discontinuities in a regularized strain-softening medium. Int. J. Numer. Methods Eng. **53**(5), 1235–1256 (2002)
47. Kalthoff, J.F.: Shadow optical analysis of dynamic shear fracture. Opt. Eng. **27**(10), 271035–271035 (1988)
48. Kalthoff, J.F.: Modes of dynamic shear failure in solids. Int. J. Fract. **101**(1/2), 1–31 (2000)
49. Elguedj, T., Gravouil, A., Maigre, H.: An explicit dynamics extended finite element method. Part 1: mass lumping for arbitrary enrichment functions. Comput. Methods Appl. Mech. Eng. **198**(30), 2297–2317 (2009)
50. Moreau, K., Moës, N., Picart, D., Stainier, L.: Explicit dynamics with a non-local damage model using the thick level set approach. Int. J. Numer. Methods Eng. **102**(3–4), 808–838 (2015)
51. Borden, M.J.: Isogeometric analysis of phase-field models for dynamic brittle and ductile fracture. Ph.D. thesis (2012)
52. Zhou, M., Rosakis, A., Ravichandran, G.: On the growth of shear bands and failure-mode transition in prenotched plates: a comparison of singly and doubly notched specimens. Int. J. Plast. **14**(4), 435–451 (1998)
53. Ravi-Chandar, K., Lu, J., Yang, B., Zhu, Z.: Failure mode transitions in polymers under high strain rate loading. Int. J. Fract. **101**(1/2), 33–72 (2000)
54. Batra, R., Ravinsankar, M.: Three-dimensional numerical simulation of the Kalthoff experiment. Int. J. Fract. **105**(2), 161–186 (2000)

Part IV
Multiphysics

A 3D Magnetostrictive Preisach Model for the Simulation of Magneto-Electric Composites on Multiple Scales

J. Schröder and M. Labusch

Abstract In this contribution we derive a three dimensional ferroelectric Preisach model based on an orientation distribution function. Therefore, the classical scalar Preisach model is modified and applied on the individual orientations, which are uniformly distributed in the three dimensional space. This model is used to simulate the behavior of magneto-electric (ME) composites. Such effective multiferroic materials combine two or more ferroic characteristics and can exhibit a coupling between electric polarization and magnetization. Since most of the single-phase ME materials exhibit a weak magneto-electric coupling at low temperatures, two-phase ME composites produce an ME coupling at room temperature. The basic idea for the manufacturing of ME composites is to use the interaction of the ferroelectric and magnetostrictive phases in order to generate strain-induced ME properties. However, in contrast to single-phase multiferroics, the ME coefficient of composites significantly depends on the microscopic morphology and the electro- as well as magneto-mechanical properties of the individual constituents. Therefore, we implemented the 3D Preisach model into the FE^2-method in order to depict the realistic ferroelectric behavior and directly incorporate the microstructure by the consideration of underlying representative volume elements.

1 Introduction

Ferroic materials are used in many devices in sensor and actuator technology. They show interactions between electric fields and mechanical deformations or magnetic fields and deformations, respectively. A special phenomenon, which combines

J. Schröder (✉) · M. Labusch
Faculty of Engineering, Department of Civil Engineering, Institute of Mechanics, University of Duisburg-Essen, Universitätsstraße 15,
45141 Essen, Germany
e-mail: j.schroeder@uni-due.de

M. Labusch
e-mail: matthias.labusch@uni-due.de

J. Sorić et al. (eds.), *Multiscale Modeling of Heterogeneous Structures*,
Lecture Notes in Applied and Computational Mechanics 86,
https://doi.org/10.1007/978-3-319-65463-8_15

both properties, is the magneto-electric (ME) coupling. Materials exhibiting a coupling between their magnetization and electric polarization are known as magneto-electric multiferroics and allow for applications in various fields, such as for instance Magneto-Electric Random Access Memory (MERAM) devices, see [4, 9, 56]. ME composites have been investigated intensively, in e.g. [6, 13, 23, 30, 42, 47, 66, 67]. However, due to physical reasons, only very few natural materials with magneto-electric properties exist and often arise at temperatures far below room temperature, see for instance [1, 7, 43, 44, 63]. Exceptions are made, for example, by the synthetic material chromium(III)-oxide (Cr_2O_3), which also shows an ME coefficient at room temperature (293.15 K), but it quickly decreases for slightly higher temperatures and finally vanishes at a Néel temperature of about $T_n = 307$ K, see [1]. The still relatively low coupling coefficient of most of the synthetic single-phase materials, like bismuth-ferrite ($BiFeO_3$), chromium(III)-oxide (Cr_2O_3) or lithium manganese phosphate ($LiCoPO_4$) results in a limitation of technical exploitation, see also [13]. A further constraint in single-phase materials is the upper bound of the magneto-electric coupling due to their material properties, see [5]. The disadvantage of the very low coupling temperature has been circumvented successfully by manufacturing magneto-electric composites which consist of magneto-mechanically and electro-mechanically coupled constituents, see [10] or [54] for manufacturing processes and [2] or [53] for numerical simulations. The basic idea for the development of such composites is to generate the desired ME effect as a strain-induced product property, see [8, 37, 38, 41, 46, 57, 64]. In general, a product property of a composite is defined as an effective property which is not present in its individual phases, but only appears effectively through their interaction, see [62] for a general treatment on possible product properties. In the case of ME composites, we distinguish between the direct and converse ME effect. The converse effect is characterized by an electrically induced magnetization, where an applied electric field yields deformations of the electric material, which are transferred to the magnetic phase. The magnetostrictive material shows a correlation between strains and magnetic fields, such that the composite exhibits a magnetization. Alternatively, a magnetically induced electric polarization is denoted as the direct magneto-electric effect. Figure 1a shows characteristic dielectric and butterfly hysteresis curves of the ferroelectric phase and Fig. 1c magnetization and magnetostriction curves of the ferromagnetic material. Due to the fact that the effective magneto-electric coupling in composite materials is a strain-induced property, it is obvious that it strongly depends on the characteristics of the individual phases and the morphology of the microstructure.

In order to simulate the effective magneto-electric behavior taking account of the interaction between the different constituents as well as the microscopic morphology of the composite, we make use of a two-scale finite element homogenization scheme, the so called FE^2-method. This approach allows for the direct incorporation of representative volume elements ($\mathscr{RVE}$s) which reflect the microscopic morphology and simultaneously the overall macroscopic properties. Due to a homogenization process over the $\mathscr{RVE}$s a transition between the micro- and the macroscale is performed. Here, we refer to [53] for magneto-electro-mechanically coupled problems and an overview of further homogenization methods.

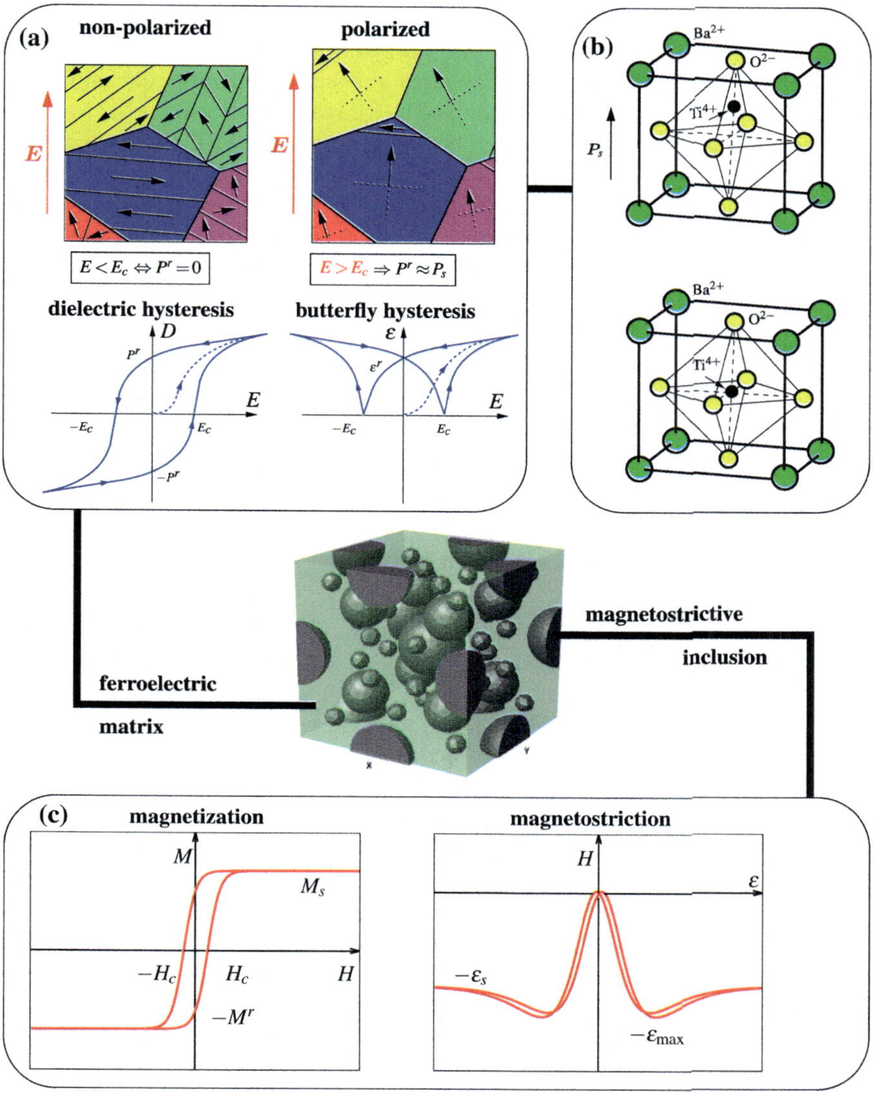

Fig. 1 Design of two-phase magneto-electric composites: **a** Typical macroscopic dielectric and butterfly hysteresis curves of $BaTiO_3$, **b** Unit cell structure on nanoscopic level and **c** typical macroscopic magnetization and magnetostriction of $CoFe_2O_4$

In this contribution we implemented two different material models into the FE^2-method for the simulation of the material behavior of both phases on microscopic level. For the magnetostrictive phase we restrict ourselves to a purely linear model. The ferroelectric matrix material is described by a three dimensional Preisach model, which depicts the typical dielectric and butterfly hysteresis loops. The main idea of this approach is to apply the classical model in each direction of a three dimensional orientation distribution function (ODF).

In Sect. 2 we start with a brief overview of the theoretical procedure of the two-scale homogenization scheme within the framework of magneto-electro-mechanical solids and describe the general treatment of the consistent linearization of the macroscopic field equations. The implemented piezomagnetic and ferroelectric material models are described in Sect. 3. Afterwards, we applied the method to the simulation of magneto-electrically coupled solids, which are discussed in Sect. 4. Finally, we will give a short summary of the present contribution and propose some improvements of the implemented models.

2 A Two-Scale Homogenization Scheme

In the FE^2-approach the microscopic quantities are homogenized to obtain macroscopic constitutive equations instead of defining a macroscopic material model. Therefore, in the case of magneto-electro-mechanical boundary value problems, the macroscopic strains, electric and magnetic fields at each macroscopic integration point are localized to an underlying microstructure, fulfilling suitable boundary conditions. This microstructure should represent the overall material behavior in a sufficient manner and is therefore denoted as a representative volume element ($\mathscr{RVE}$). In order to obtain the microscopic quantities such as the stresses, the electric displacement and the magnetic induction, the weak forms of the balance equations have to be solved on the microscale. Afterwards, a homogenization step is performed, in which average values of the microscopic quantities are determined. These homogenized variables have to be transferred to the associated points on the macroscale. Finally, the macroscopic boundary value problem has to be solved, whereby the entire procedure has to be repeated until an equilibrium state on both scales is reached, see [27, 51–53]. In the sense of multilevel finite element methods and computational homogenization frameworks we refer to [11, 21, 22, 24, 33–36, 39, 48, 49, 59–61]. To ease the readability of the equations, we summarize the magneto-electro-mechanical quantities in Table 1.

2.1 Boundary Value Problem and Scale Transition

The macroscopic body $\mathscr{B} \subset \mathbb{R}^3$ is parameterized in the Cartesian coordinates $\overline{\boldsymbol{x}}$. The macroscopic fundamental balance laws are given by the balance of linear momentum as well as Gauß's laws of electro- and magneto-statics as

Table 1 Overview of the magneto-electro-mechanical quantities and their units

Macro-symbol	Micro-symbol	Continuum mechanical description	Unit
$\overline{\boldsymbol{u}}$	$\boldsymbol{u}$	Displacement vector	m
$\overline{\boldsymbol{\varepsilon}}$	$\boldsymbol{\varepsilon}$	Linear strain tensor	1
$\overline{\boldsymbol{\sigma}}$	$\boldsymbol{\sigma}$	Cauchy stress tensor	N/m^2
$\overline{\boldsymbol{t}}$	$\boldsymbol{t}$	Traction vector	N/m^2
$\overline{\boldsymbol{f}}$	$\boldsymbol{f}$	Mechanical body force vector	N/m^3
$\overline{\phi}$	ϕ	Electric potential	V
$\overline{\boldsymbol{E}}$	$\boldsymbol{E}$	Electric field vector	V/m
$\overline{\boldsymbol{D}}$	$\boldsymbol{D}$	Electric displacement vector	C/m^2
$\overline{Q}$	Q	Electric surface flux density	C/m^2
$\overline{q}$	q	Density of free charge carriers	C/m^3
$\overline{\varphi}$	φ	Magnetic potential	A
$\overline{\boldsymbol{H}}$	$\boldsymbol{H}$	Magnetic field vector	A/m
$\overline{\boldsymbol{B}}$	$\boldsymbol{B}$	Magnetic flux density	T
$\overline{\zeta}$	ζ	Magnetic surface flux density	Vs/m^2

$$\mathrm{div}_{\overline{\boldsymbol{x}}}[\overline{\boldsymbol{\sigma}}] + \overline{\boldsymbol{f}} = \boldsymbol{0}, \quad \mathrm{div}_{\overline{\boldsymbol{x}}}[\overline{\boldsymbol{D}}] = \overline{q} \quad \text{and} \quad \mathrm{div}_{\overline{\boldsymbol{x}}}[\overline{\boldsymbol{B}}] = 0 \quad \text{in} \quad \mathscr{B}\,, \tag{1}$$

with the divergence operator $\mathrm{div}_{\overline{\boldsymbol{x}}}$ with respect to $\overline{\boldsymbol{x}}$. In this contribution we neglect the free charge carriers $\overline{q}$. The macroscopic gradient fields, which are the strain as well as the electric and magnetic field are defined as

$$\overline{\boldsymbol{\varepsilon}} = \mathrm{sym}[\mathrm{grad}_{\overline{\boldsymbol{x}}}\overline{\boldsymbol{u}}]\,, \quad \overline{\boldsymbol{E}} = -\mathrm{grad}_{\overline{\boldsymbol{x}}}\,\overline{\phi} \quad \text{and} \quad \overline{\boldsymbol{H}} = -\mathrm{grad}_{\overline{\boldsymbol{x}}}\,\overline{\varphi}\,, \tag{2}$$

with the macroscopic gradient operator $\mathrm{grad}_{\overline{\boldsymbol{x}}}$ defined with respect to $\overline{\boldsymbol{x}}$. The macroscopic boundary conditions are prescribed through the displacement and the surface traction

$$\overline{\boldsymbol{u}} = \overline{\boldsymbol{u}}_b \quad \text{on } \partial\mathscr{B}_{\overline{u}} \quad \text{and} \quad \overline{\boldsymbol{t}}_b = \overline{\boldsymbol{\sigma}} \cdot \overline{\boldsymbol{n}} \quad \text{on } \partial\mathscr{B}_{\overline{\sigma}} \tag{3}$$

with the relations $\partial\mathscr{B}_{\overline{u}} \cup \partial\mathscr{B}_{\overline{\sigma}} = \partial\mathscr{B}$ and $\partial\mathscr{B}_{\overline{u}} \cap \partial\mathscr{B}_{\overline{\sigma}} = \emptyset$, through the electric potential and the electric surface flux density

$$\overline{\phi} = \overline{\phi}_b \quad \text{on } \partial\mathscr{B}_{\overline{\phi}} \quad \text{and} \quad -\overline{Q}_b = \overline{\boldsymbol{D}} \cdot \overline{\boldsymbol{n}} \quad \text{on } \partial\mathscr{B}_{\overline{D}} \tag{4}$$

with the relations $\partial\mathscr{B}_{\overline{\phi}} \cup \partial\mathscr{B}_{\overline{D}} = \partial\mathscr{B}$ and $\partial\mathscr{B}_{\overline{\phi}} \cap \partial\mathscr{B}_{\overline{D}} = \emptyset$, as well as through the magnetic potential and the magnetic surface flux density

$$\overline{\varphi} = \overline{\varphi}_b \quad \text{on } \partial\mathscr{B}_{\overline{\varphi}} \quad \text{and} \quad -\overline{\zeta}_b = \overline{\boldsymbol{B}} \cdot \overline{\boldsymbol{n}} \quad \text{on } \partial\mathscr{B}_{\overline{B}} \tag{5}$$

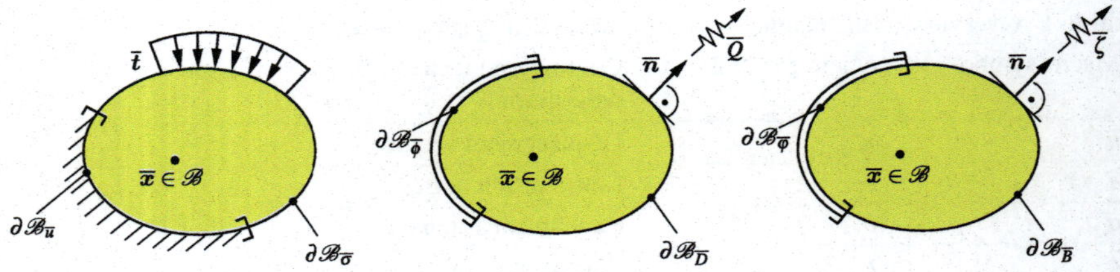

Fig. 2 Boundary conditions on the macroscale: **a** mechanical part, **b** electrical part and **c** magnetical part

with the relations $\partial\mathscr{B}_{\overline{\varphi}} \cup \partial\mathscr{B}_{\overline{B}} = \partial\mathscr{B}$ and $\partial\mathscr{B}_{\overline{\varphi}} \cap \partial\mathscr{B}_{\overline{B}} = \emptyset$, where $\overline{\boldsymbol{n}}$ denotes the outward unit normal vector of the surfaces $\partial\mathscr{B}_{\overline{B}}$, see Fig. 2. To solve the macroscopic boundary value problem within a Newton-Raphson iteration scheme the overall material tangent modulus is required. In the sense of the FE^2-method the effective properties are obtained by a homogenization step, requiring the solution of the microscopic boundary value problem. The representative volume element on the microscale $\mathscr{RVE} \subset \mathbb{R}^3$ is parameterized in $\boldsymbol{x}$. The balance of momentum as well as the Gauß's law of electro- and magneto-statics on microscopic level are given by

$$\mathrm{div}_{\boldsymbol{x}}[\boldsymbol{\sigma}] = \boldsymbol{0}\,, \quad \mathrm{div}_{\boldsymbol{x}}[\boldsymbol{D}] = q \quad \text{and} \quad \mathrm{div}_{\boldsymbol{x}}[\boldsymbol{B}] = 0 \quad \text{in} \quad \mathscr{RVE}. \tag{6}$$

Analogously to Eq. $(1)_2$ we neglect in (6) on microscopic level the free charge carriers q. The microscopic strains as well as the electric and magnetic fields are defined by

$$\varepsilon = \mathrm{sym}[\mathrm{grad}_{\boldsymbol{x}}\boldsymbol{u}]\,, \quad \boldsymbol{E} = -\mathrm{grad}_{\boldsymbol{x}}\,\phi \quad \text{and} \quad \boldsymbol{H} = -\mathrm{grad}_{\boldsymbol{x}}\,\varphi\,. \tag{7}$$

To solve the microscopic boundary value problem, suitable energy functions for the different phases are defined, which describe the corresponding material moduli. The considered $\mathscr{RVE}$ on the microscale should represent the overall material behavior and especially, for the purpose of ME composites, the magneto-electric coupling coefficient. As a consequence, all of the above quantities are defined through an averaging process over the volume of the $\mathscr{RVE}$. Assuming continuity of the displacements, the electric and magnetic potential, we can express the macroscopic variables in terms of simple volume integrals as

$$\overline{\boldsymbol{\lambda}} = \langle\boldsymbol{\lambda}\rangle_V := \frac{1}{V_{\mathscr{RVE}}}\int\limits_{\mathscr{RVE}} \boldsymbol{\lambda}\,dv \quad \text{with} \quad \boldsymbol{\lambda} := \{\varepsilon, \boldsymbol{\sigma}, \boldsymbol{E}, \boldsymbol{D}, \boldsymbol{H}, \boldsymbol{B}\}\,, \tag{8}$$

see for example [53]. In order to determine the microscopic quantities by solving the boundary value problem of the $\mathscr{RVE}$, we have to define boundary conditions on microscopic level. Starting from the fundamental works of [14] and [29] we assume that the individual parts of a generalized magneto-electro-mechanical Hill-Mandel condition of the form

$$\underbrace{\overline{\boldsymbol{\sigma}} : \dot{\overline{\varepsilon}} - \langle \boldsymbol{\sigma} : \dot{\varepsilon} \rangle_V}_{\mathscr{P}_{mech}} = 0, \quad \underbrace{\overline{\boldsymbol{D}} \cdot \dot{\overline{\boldsymbol{E}}} - \langle \boldsymbol{D} \cdot \dot{\boldsymbol{E}} \rangle_V}_{\mathscr{P}_{elec}} = 0 \quad \text{and} \quad \underbrace{\overline{\boldsymbol{B}} \cdot \dot{\overline{\boldsymbol{H}}} - \langle \boldsymbol{B} \cdot \dot{\boldsymbol{H}} \rangle_V}_{\mathscr{P}_{magn}} = 0\,, \tag{9}$$

have to be fulfilled independently, see [48] for the electro-mechanical case and [53] for the magneto-electro-mechanical case. Using the local balance law of linear momentum as well as the Cauchy theorem $\boldsymbol{t} = \boldsymbol{\sigma} \cdot \boldsymbol{n}$ we reformulate $\mathscr{P}_{mech}$ to

$$\mathscr{P}_{mech} = \langle (\boldsymbol{t} - \overline{\boldsymbol{\sigma}} \cdot \boldsymbol{n})(\dot{\boldsymbol{u}} - \dot{\overline{\varepsilon}} \cdot \boldsymbol{x}) \rangle_\Gamma \tag{10}$$

where $\langle \bullet \rangle_\Gamma$ denotes the surface average over the representative volume element, i.e. $\langle \bullet \rangle_\Gamma := \frac{1}{V_{\mathscr{RVE}}} \int_{\partial \mathscr{RVE}} (\bullet)\, da$, and $\boldsymbol{n}$ the outward unit normal on $\partial\mathscr{RVE}$. An evaluation of $\mathscr{P}_{mech} = 0$ leads to the Reuss- and Voigt bounds (constraint conditions)

$$\boldsymbol{\sigma} = \overline{\boldsymbol{\sigma}} = \text{const.} \qquad \text{or} \qquad \dot{\varepsilon} = \dot{\overline{\varepsilon}} = \text{const.} \qquad \forall \boldsymbol{x} \in \mathscr{RVE}\,. \tag{11}$$

Possible periodic boundary conditions, which fulfill $\mathscr{P}_{mech} = 0$, can be derived effectively by assuming a decomposition of the microscopic strains in an affine part $\overline{\varepsilon}$ and a fluctuation field $\widetilde{\varepsilon}$, satisfying $\langle \varepsilon \rangle_V = \overline{\varepsilon}$ and $\langle \widetilde{\varepsilon} \rangle_V = 0$, holding the relation

$$\dot{\varepsilon} = \dot{\overline{\varepsilon}} + \dot{\widetilde{\varepsilon}}\,. \tag{12}$$

The associated periodic boundary conditions for the mechanical part are given by

$$\widetilde{\boldsymbol{u}}(\boldsymbol{x}^+) = \widetilde{\boldsymbol{u}}(\boldsymbol{x}^-) \quad \text{and} \quad \boldsymbol{t}(\boldsymbol{x}^+) = -\boldsymbol{t}(\boldsymbol{x}^-) \quad \text{on} \quad \boldsymbol{x}^\pm \in \Gamma^\pm \tag{13}$$

respectively, see Fig. 3 for a visualization of the mechanical part. Here, $\boldsymbol{x}^+$ and $\boldsymbol{x}^-$ define associated points at Γ^+ and Γ^-, respectively, satisfying $\boldsymbol{n}^+ = -\boldsymbol{n}^-$. Analogous to the mechanical part, we can derive the boundary conditions for the electric and magnetic fields by reformulating Eq. $(9)_2$ and $(9)_3$. Using the Maxwell equations of electro- and magnetostatics and the definitions of the surface tractions Q and ζ yield

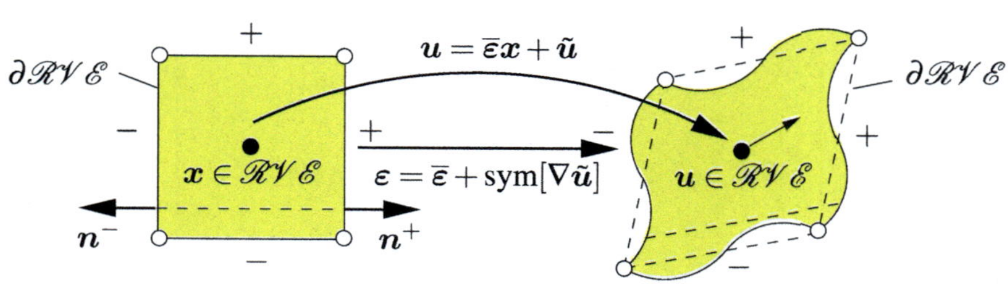

Fig. 3 Possible mechanical periodic boundary conditions on the $\mathscr{RVE}$

$$\mathscr{P}_{elec} = \langle (Q + \overline{\boldsymbol{D}} \cdot \boldsymbol{n})(\dot{\phi} + \dot{\overline{\boldsymbol{E}}} \cdot \boldsymbol{x}) \rangle_{\Gamma} \quad \text{and} \quad \mathscr{P}_{magn} = \langle (\zeta + \overline{\boldsymbol{B}} \cdot \boldsymbol{n})(\dot{\varphi} + \dot{\overline{\boldsymbol{H}}} \cdot \boldsymbol{x}) \rangle_{\Gamma} \,. \tag{14}$$

The evaluation of $\mathscr{P}_{elec} = 0$ and $\mathscr{P}_{magn} = 0$ leads to the constraint conditions for the electric and magnetic fields

$$\begin{aligned} \boldsymbol{D} = \overline{\boldsymbol{D}} = \text{const. or } \dot{\boldsymbol{E}} = \dot{\overline{\boldsymbol{E}}} = \text{const. } \forall \boldsymbol{x} \in \mathscr{RVE} \,, \\ \boldsymbol{B} = \overline{\boldsymbol{B}} = \text{const. or } \dot{\boldsymbol{H}} = \dot{\overline{\boldsymbol{H}}} = \text{const. } \forall \boldsymbol{x} \in \mathscr{RVE} \,. \end{aligned} \tag{15}$$

For the periodic boundary conditions of the electric and magnetic fields we assume analogous to the mechanic fields the following decompositions

$$\dot{\boldsymbol{E}} = \dot{\overline{\boldsymbol{E}}} + \dot{\widetilde{\boldsymbol{E}}} \quad \text{and} \quad \dot{\boldsymbol{H}} = \dot{\overline{\boldsymbol{H}}} + \dot{\widetilde{\boldsymbol{H}}} \,, \tag{16}$$

and the associated periodic conditions are given by

$$\begin{aligned} \widetilde{\phi}(\boldsymbol{x}^+) = \widetilde{\phi}(\boldsymbol{x}^-) \text{ and } Q(\boldsymbol{x}^+) = -Q(\boldsymbol{x}^-) \text{ on } \boldsymbol{x}^\pm \in \Gamma^\pm \,, \\ \widetilde{\varphi}(\boldsymbol{x}^+) = \widetilde{\varphi}(\boldsymbol{x}^-) \text{ and } \zeta(\boldsymbol{x}^+) = -\zeta(\boldsymbol{x}^-) \text{ on } \boldsymbol{x}^\pm \in \Gamma^\pm \,, \end{aligned} \tag{17}$$

respectively. In the following we introduce the discretizations of the microscopic boundary value problems. Furthermore, the determination of the macroscopic material tangent, obtained by a homogenization over the $\mathscr{RVE}$ involving a consistent linearization of the macroscopic constitutive quantities, is shown.

2.2 *Discretizations of the Boundary Value Problems*

For the discretizations of the boundary value problem on the microscale, we apply for actual, virtual and incremental deformations, electric as well as magnetic potential the following discretizations

$$\begin{aligned} \{\underline{\widetilde{u}}, \delta\underline{\widetilde{u}}, \Delta\underline{\widetilde{u}}\} &= \sum_{I=1}^{\mathrm{n}_{node}} \mathbb{N}_u^I \{\underline{\widetilde{d}}_u^I, \delta\underline{\widetilde{d}}_u^I, \Delta\underline{\widetilde{d}}_u^I\} \\ \{\widetilde{\phi}, \delta\widetilde{\phi}, \Delta\widetilde{\phi}\} &= \sum_{I=1}^{\mathrm{n}_{node}} \mathbb{N}_\phi^I \{\widetilde{d}_\phi^I, \delta\widetilde{d}_\phi^I, \Delta\widetilde{d}_\phi^I\} \\ \{\widetilde{\varphi}, \delta\widetilde{\varphi}, \Delta\widetilde{\varphi}\} &= \sum_{I=1}^{\mathrm{n}_{node}} \mathbb{N}_\varphi^I \{\widetilde{d}_\varphi^I, \delta\widetilde{d}_\varphi^I, \Delta\widetilde{d}_\varphi^I\} \,, \end{aligned} \tag{18}$$

where $\mathbb{N}^I$ contains the classical shape-functions associated with node I and the expressions $\{\underline{\widetilde{d}}_u, \widetilde{d}_\phi, \widetilde{d}_\varphi\}$ denote the nodal displacement, the electric and magnetic potential. Due to the constant distribution of the macroscopic part we only have to discretize the fluctuation fields. We approximate the actual, virtual and incremental fluctuation fields of the deformation, electric as well as magnetic field with

$$
\begin{aligned}
\underline{\widetilde{\varepsilon}} = \mathbb{B}^e_u \underline{\widetilde{d}}^e_u\,, \quad \delta\underline{\widetilde{\varepsilon}} = \mathbb{B}^e_u \delta\underline{\widetilde{d}}^e_u\,, \quad \Delta\underline{\widetilde{\varepsilon}} = \mathbb{B}^e_u \Delta\underline{\widetilde{d}}^e_u \\
\underline{\widetilde{E}} = \mathbb{B}^e_\phi \underline{\widetilde{d}}^e_\phi\,, \quad \delta\underline{\widetilde{E}} = \mathbb{B}^e_\phi \delta\underline{\widetilde{d}}^e_\phi\,, \quad \Delta\underline{\widetilde{E}} = \mathbb{B}^e_\phi \Delta\underline{\widetilde{d}}^e_\phi \\
\underline{\widetilde{H}} = \mathbb{B}^e_\varphi \underline{\widetilde{d}}^e_\varphi\,, \quad \delta\underline{\widetilde{H}} = \mathbb{B}^e_\varphi \delta\underline{\widetilde{d}}^e_\varphi\,, \quad \Delta\underline{\widetilde{H}} = \mathbb{B}^e_\varphi \Delta\underline{\widetilde{d}}^e_\varphi\,,
\end{aligned}
\tag{19}
$$

with the $\mathbb{B}^e$-matrices containing the partial derivatives of the shape-functions with respect to the reference coordinates. The following expressions

$$
\underline{\widetilde{\mathbb{X}}} = \mathbb{B}^e_\xi \underline{\widetilde{d}}^e_\xi\,, \quad \delta\underline{\widetilde{\mathbb{X}}} = \mathbb{B}^e_\xi \delta\underline{\widetilde{d}}^e_\xi\,, \quad \Delta\underline{\widetilde{\mathbb{X}}} = \mathbb{B}^e_\xi \Delta\underline{\widetilde{d}}^e_\xi\,,
\tag{20}
$$

give a general form of the discretizations in order to simplify the following derivation of the algorithmic consistent tangent moduli. Here, we introduce the abbreviations $\underline{\mathbb{X}} = \{\varepsilon,\ \boldsymbol{E},\ \boldsymbol{H}\}$ and $\boldsymbol{\xi} = \{\boldsymbol{u},\ \phi,\ \varphi\}$. In analogy to the discretizations on the microscale we discretize the macroscopic boundary value problem.

2.3 *Consistent Linearization of Macroscopic Field Equations*

The boundary value problems on the macroscale as well as on the microscale are solved using the Finite Element Method and involve the derivation of the weak forms on both scales. To obtain the solution of the problems the Newton-Raphson iteration scheme is used, where we want to achieve quadratic convergence on both scales. Therefore, a consistent linearization of the macroscopic constitutive quantities is required. The incremental macroscopic constitutive equations are given by

$$
\underbrace{\begin{bmatrix} \Delta\overline{\boldsymbol{\sigma}} \\ -\Delta\overline{\boldsymbol{D}} \\ -\Delta\overline{\boldsymbol{B}} \end{bmatrix}}_{\Delta\underline{\overline{\Sigma}}} = \underbrace{\begin{bmatrix} \overline{\mathbb{C}} & -\overline{\boldsymbol{e}}^T & -\overline{\boldsymbol{q}}^T \\ -\overline{\boldsymbol{e}} & -\overline{\boldsymbol{\epsilon}} & -\overline{\boldsymbol{\alpha}}^T \\ -\overline{\boldsymbol{q}} & -\overline{\boldsymbol{\alpha}} & -\overline{\boldsymbol{\mu}} \end{bmatrix}}_{\underline{\overline{\mathbb{Z}}}} \underbrace{\begin{bmatrix} \Delta\overline{\boldsymbol{\varepsilon}} \\ \Delta\overline{\boldsymbol{E}} \\ \Delta\overline{\boldsymbol{H}} \end{bmatrix}}_{\Delta\underline{\overline{\mathbb{X}}}}\,,
\tag{21}
$$

with the macroscopic and microscopic tangent modulus $\overline{\mathbb{C}}$ and $\mathbb{C}$, the piezoelectric tangent moduli $\overline{\boldsymbol{e}}$ and $\boldsymbol{e}$, the piezomagnetic tangent moduli $\overline{\boldsymbol{q}}$ and $\boldsymbol{q}$, the electric permittivities $\overline{\boldsymbol{\epsilon}}$ and $\boldsymbol{\epsilon}$, the magnetic permeabilities $\overline{\boldsymbol{\mu}}$ and $\boldsymbol{\mu}$, as well as the magneto-electric tangent moduli $\overline{\boldsymbol{\alpha}}$. In the following we use a matrix-notation denoted by an underline. For the determination of the macroscopic tangent moduli the increments of the macroscopic stresses, electric displacement and magnetic induction are expressed by the volume averages of the corresponding microscopic variables

$$\overline{\underline{\mathbb{Z}}} = \frac{\partial \langle \underline{\boldsymbol{\Sigma}} \rangle_V}{\partial \overline{\underline{\mathbb{X}}}}. \tag{22}$$

The computation of the macroscopic moduli is obviously not straight-forward. The macroscopic stresses, electric displacements and magnetic inductions are defined through volume integrals over the microscopic counterparts. With the application of the chain rule and the additive decomposition of the microscopic quantities into a macroscopic and a fluctuation part $\underline{\mathbb{X}} = \overline{\underline{\mathbb{X}}} + \widetilde{\underline{\mathbb{X}}}$, we obtain the overall material tangent

$$\overline{\underline{\mathbb{Z}}} = \frac{\partial \langle \underline{\boldsymbol{\Sigma}}(\underline{\mathbb{X}}) \rangle_V}{\partial \overline{\underline{\mathbb{X}}}} = \left\langle \frac{\partial \underline{\boldsymbol{\Sigma}}(\underline{\mathbb{X}})}{\partial \underline{\mathbb{X}}} \, \frac{\partial (\overline{\underline{\mathbb{X}}} + \widetilde{\underline{\mathbb{X}}})}{\partial \overline{\underline{\mathbb{X}}}} \right\rangle_V = \underbrace{\langle \underline{\mathbb{Z}} \rangle_V}_{\underline{\mathbb{Z}}^{\text{Voigt}}} + \underbrace{\left\langle \underline{\mathbb{Z}} \, \frac{\partial \widetilde{\underline{\mathbb{X}}}}{\partial \overline{\underline{\mathbb{X}}}} \right\rangle_V}_{\underline{\mathbb{Z}}^{\text{Soft}}} . \tag{23}$$

The first part of the latter equation $\underline{\mathbb{Z}}^{\text{Voigt}}$ is denoted as the Voigt upper bound of the material tangent and is defined as a simple volume average. The second part $\underline{\mathbb{Z}}^{\text{Soft}}$ is called the softening term of the overall tangent and includes the magneto-electric coefficient. In order to derive the softening part, we linearize the microscopic weak forms at an equilibrium state. In the following we define a general weak form for the balance of momentum, Gauß's law of electrostatics and Gauß's law of magnetostatics on microscopic level as

$$\mathbb{G}_\xi = -\int_{\mathscr{RVE}} \delta \underline{\boldsymbol{\xi}}^T \operatorname{div}[\underline{\boldsymbol{\Sigma}}] \, dv = \int_{\mathscr{RVE}} \delta \widetilde{\underline{\mathbb{X}}}^T \, \underline{\boldsymbol{\Sigma}} \, dv - \int_{\partial \mathscr{RVE}} \delta \widetilde{\underline{\boldsymbol{\xi}}}^T \, (\underline{\boldsymbol{\Sigma}} \, \boldsymbol{n}) \, da \, . \tag{24}$$

The linearization of the above general weak form with respect to the microscopic quantities yields

$$\Delta \mathbb{G}_\xi = \int_{\mathscr{RVE}} \delta \widetilde{\underline{\mathbb{X}}}^T \, \underline{\mathbb{Z}} \, \Delta \underline{\mathbb{X}} \, dv = \int_{\mathscr{RVE}} \delta \widetilde{\underline{\mathbb{X}}}^T \, \underline{\mathbb{Z}} \, (\Delta \overline{\underline{\mathbb{X}}} + \Delta \widetilde{\underline{\mathbb{X}}}) \, dv = 0 \, , \tag{25}$$

where we decomposed the microscopic quantities into a macroscopic and a fluctuation part. After a reformulation and the insertion of the discretizations we obtain

$$\begin{aligned} \Delta \mathbb{G}_\xi &= \int_{\mathscr{RVE}} \delta \widetilde{\underline{\mathbb{X}}}^T \, \underline{\mathbb{Z}} \, dv \, \Delta \overline{\underline{\mathbb{X}}} + \int_{\mathscr{RVE}} \delta \widetilde{\underline{\mathbb{X}}}^T \, \underline{\mathbb{Z}} \, \Delta \widetilde{\underline{\mathbb{X}}} \, dv \\ &= \sum_{e=1}^{num_{ele}} \delta \widetilde{\underline{\boldsymbol{d}}}_\xi^{eT} \Big\{ \underbrace{\int_{\mathscr{RVE}} \mathbb{B}_\xi^{eT} \, \underline{\mathbb{Z}} \, dv}_{\underline{\boldsymbol{l}}^e} \, \Delta \overline{\underline{\mathbb{X}}} + \underbrace{\int_{\mathscr{RVE}} \mathbb{B}_\xi^{eT} \, \underline{\mathbb{Z}} \, \mathbb{B}_\xi^e \, dv}_{\underline{\boldsymbol{k}}^e} \, \Delta \widetilde{\underline{\boldsymbol{d}}}_\xi^e \Big\} = 0 \end{aligned} \tag{26}$$

with the number of microscopic finite elements num_{ele}, the element stiffness matrices $\underline{\boldsymbol{k}}^e$ and the sensitivity of the moduli of the finite elements $\underline{\boldsymbol{l}}^e$. The contracted matrix notation of the latter expression can be reformulated with the application of a standard assembling procedure as

$$\sum_{e=1}^{num_{ele}} \delta\underline{\widetilde{\boldsymbol{d}}}_\xi^{eT} \{\underline{\boldsymbol{l}}^e \, \Delta\overline{\mathbb{X}} + \underline{\boldsymbol{k}}^e \, \Delta\underline{\widetilde{\boldsymbol{d}}}_\xi^e\} = \delta\underline{\widetilde{\boldsymbol{D}}}_\xi^T (\underline{\boldsymbol{L}} \, \Delta\overline{\mathbb{X}} + \underline{\boldsymbol{K}} \, \Delta\underline{\widetilde{\boldsymbol{D}}}_\xi) = 0 \,. \tag{27}$$

From this we achieve the microscopic fluctuations depending on the incremental macroscopic process variables and its partial derivative with respect to $\overline{\mathbb{X}}$ as

$$\Delta\underline{\widetilde{\boldsymbol{D}}}_\xi = -\underline{\boldsymbol{K}}^{-1}\underline{\boldsymbol{L}} \, \Delta\overline{\mathbb{X}} \quad \text{and} \quad \frac{\partial \Delta\underline{\widetilde{\boldsymbol{D}}}_\xi}{\partial\overline{\mathbb{X}}} = -\underline{\boldsymbol{K}}^{-1}\underline{\boldsymbol{L}} \, \frac{\partial\Delta\overline{\mathbb{X}}}{\partial\overline{\mathbb{X}}} \tag{28}$$

Inserting the discretizations $(20)_3$ in (23), with the update $\widetilde{\mathbb{X}} \Leftarrow \widetilde{\mathbb{X}} + \Delta\widetilde{\mathbb{X}}$ in consideration of the equilibrium state in the current time step, and the partial derivative $(28)_2$ yield the final expression of the effective magneto-electro-mechanical moduli as

$$\begin{aligned} \overline{\mathbb{Z}} &= \langle\mathbb{Z}\rangle_V + \left\langle \sum_{e=1}^{num_{ele}} \mathbb{Z} \, \frac{\partial\mathbb{B}_\xi^e \, \Delta\underline{\widetilde{\boldsymbol{d}}}_\xi^e}{\partial\overline{\mathbb{X}}} \right\rangle_V = \langle\mathbb{Z}\rangle_V + \frac{1}{V}\underline{\boldsymbol{L}}^T \, \frac{\partial\Delta\underline{\widetilde{\boldsymbol{D}}}_\xi}{\partial\overline{\mathbb{X}}} \\ &= \overline{\mathbb{Z}}^{\text{Voigt}} - \frac{1}{V}\underline{\boldsymbol{L}}^T \, \underline{\boldsymbol{K}}^{-1} \, \underline{\boldsymbol{L}} \end{aligned} \tag{29}$$

For a detailed derivative of the effective tangent moduli for magneto-electro-mechanically coupled material response we refer to [53].

3 Material Modeling

In the following section we briefly discuss the material models for both phases, which are implemented into the FE^2-method. We start with a description of the piezomagnetic model, where we restrict the simulated behavior to the purely linear case. Afterwards, we describe the nonlinear Preisach model for the ferroelectric phase, which is extended to the three dimensional space.

3.1 *Linear Piezomagnetic Model*

For the description of the piezomagnetic material behavior, we use a coordinate-invariant formulation of a magneto-electro-mechanical enthalpy function ψ_1 for transversely isotropic solids adopted from [50] as

$$\psi_1 = \underbrace{\frac{1}{2}\lambda I_1^2 + \mu I_2 + \omega_1 I_5 + \omega_2 I_4^2 + \omega_3 I_1 I_4}_{mechanical} + \underbrace{\kappa_1 I_1 J_2^m + \kappa_2 I_4 J_2^m + \kappa_3 K_1^m}_{piezomagnetic} + \underbrace{\gamma_1 J_1^e + \gamma_2 (J_2^e)^2}_{dielectric} + \underbrace{\xi_1 J_1^m + \xi_2 (J_2^m)^2}_{magnetic} \tag{30}$$

where the individual parts represent functions for the purely mechanical, the piezomagnetic, the purely electric and the purely magnetic behavior, respectively. The individual functions are formulated in terms of the following invariants

$$\begin{aligned} &I_1 := \mathrm{tr}[\varepsilon], \quad I_2 := \mathrm{tr}[\varepsilon^2], \quad I_4 := \mathrm{tr}[\varepsilon \boldsymbol{m}], \quad I_5 := \mathrm{tr}[\varepsilon^2 \boldsymbol{m}], \\ &J_1^m := \mathrm{tr}[\boldsymbol{H} \otimes \boldsymbol{H}], \quad J_2^m := \mathrm{tr}[\boldsymbol{H} \otimes \boldsymbol{a}], \quad K_1^m := \mathrm{tr}[\varepsilon(\boldsymbol{H} \otimes \boldsymbol{a})], \\ &J_1^e := \mathrm{tr}[\boldsymbol{E} \otimes \boldsymbol{E}], \quad J_2^e := \mathrm{tr}[\boldsymbol{E} \otimes \boldsymbol{a}], \end{aligned} \tag{31}$$

in which $\boldsymbol{a}$ describes the preferred direction of the respective phase and $\boldsymbol{m} := \boldsymbol{a} \otimes \boldsymbol{a}$ is the associated structural tensor. In this model, the preferred direction is assumed to coincide with a defined direction of remanent electric polarization. The constitutive magneto-electro-mechanical moduli follow consequently as

$$\begin{aligned} \mathbb{C} &= \lambda \mathbf{1} \otimes \mathbf{1} + 2\mu \mathbb{I} + \omega_3 \left[\mathbf{1} \otimes \boldsymbol{m} + \boldsymbol{m} \otimes \mathbf{1}\right] + 2\omega_2\, \boldsymbol{m} \otimes \boldsymbol{m} + \omega_1\, \boldsymbol{\Xi}, \\ \boldsymbol{q} &= -\kappa_1 \boldsymbol{a} \otimes \mathbf{1} - \kappa_2 \boldsymbol{a} \otimes \boldsymbol{m} - \kappa_3 \boldsymbol{\theta}, \\ \boldsymbol{\epsilon} &= -2\gamma_1 \mathbf{1} - 2\gamma_2 \boldsymbol{m}, \\ \boldsymbol{\mu} &= -2\xi_1 \mathbf{1} - 2\xi_2 \boldsymbol{m}, \end{aligned} \tag{32}$$

with the second-order identity tensor $\mathbf{1}$ and the third- and fourth-order tensors

$$\begin{aligned} &\theta_{ijk} := \tfrac{1}{2}(a_j \delta_{ik} + a_k \delta_{ij}), \quad \mathbb{I}_{ijkl} := \tfrac{1}{2}(\delta_{ik}\delta_{jl} + \delta_{il}\delta_{jk}), \\ &\text{and} \quad \Xi_{ijkl} := a_i \delta_{jk} a_l + a_k \delta_{il} a_j. \end{aligned} \tag{33}$$

Note that in this phase the piezoelectric moduli and the ME-coefficient is zero. In order to determine the overall ME-coefficient we use the homogenization approach of the FE^2-method. The parameters for the coordinate-invariant formulation can be related to the experimental parameters, see [53].

3.2 *Ferroelectric Preisach Model*

Several models for the description of ferroelectric materials have been developed. In previous works we used a ferroelectric switching model, where a homogenization of multiple reoriented microscopic remanent polarizations yield the overall hysteresis loops. See [16, 20] for ferroelectric switching models. In this contribution we use a Preisach model to depict the ferroelectric characteristics. Therefore, we additively decompose the electric polarization and strains into an elastic part ($\bullet_e$) and a remanent part ($\bullet_r$). The linear part is described by a transversely isotropic linear material law, see e.g. [26, 53]. The behavior is characterized by a preferred direction $\boldsymbol{a}$, describing the polarization direction. The nonlinear remanent polarization is described by the Preisach operator. In the following we give a brief description of the classical Preisach model and extend this approach to the three dimensional case.

3.2.1 Classical Preisach Model

The idea of the classical Preisach model goes back to the work of Preisach [40] in the year 1935 and further approaches using a Preisach model are e.g. given in [3, 12, 15, 17, 31, 32, 45, 55, 58, 65]. Based on this approach the remanent magnetization $\boldsymbol{M}_r$ or the remanent electric polarization $\boldsymbol{P}_r$, respectively, can be computed by a scalar hysteresis operator $\mathfrak{p}$, which depends on a finite number of so called hysterons. Each hysteron is characterized by a square hysteresis cycle with the output value γ, which is either $+1$ or -1, depending on the *up* and *down* switching thresholds α and β, see Fig. 4. A superposition of multiple hysterons with different switching thresholds yield the hystersis loop depending on the input variable, in this case the current magnetic field $H(t)$, as

$$\mathfrak{p}(H) = \int_\beta \int_\alpha \omega(\alpha, \beta)\, \gamma(\alpha, \beta)\, H(t) \mathrm{d}\alpha\, \mathrm{d}\beta \,, \tag{34}$$

or with $\mathfrak{p}(E)$ depending on the current electric field in the case of electric polarization. The single hysterons are furthermore multiplied with a weighting function $\omega(\alpha, \beta)$ in order to conincide with measured hysteresis loops. Methods to identify the weighting function from experimental measurements are for example given in [17]. Here, it has to be mentioned, that the classical approach is a scalar valued model, such that the input field H is also considered to be a scalar value. Due to the fact, that we consider closed hysteresis loops, we can visualize the set of possible *up* and *down* switching thresholds to a Preisach plane $S = \{ (\alpha, \beta) \in \mathbb{R}^2 \mid \beta \geq \alpha \}$, see Fig. 4. With a suitable distribution of switching thresholds it is also possible to depict minor hysteresis loops.

3.2.2 Application of the Preisach Model to an ODF

Analogously to the linear piezomagnetic material model, we use a generalized coordinate-invariant formulation of a magneto-electro-mechanical enthalpy function ψ_2 for transversely isotropic solids adopted from Schröder and Gross [50]

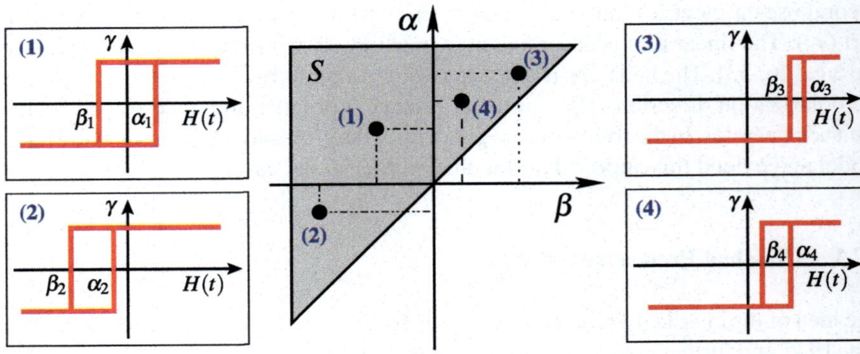

Fig. 4 Visualization of the Preisach plane S with $S = \{ (\alpha, \beta) \in \mathbb{R}^2 \mid \beta \geq \alpha \}$ and possible hysterons with individual switching thresholds $\alpha_{(1-4)}$ and $\beta_{(1-4)}$, respectively

$$\begin{aligned}\psi_2 = &\underbrace{\frac{1}{2}\lambda \hat{I}_1^2 + \mu \hat{I}_2 + \omega_1 \hat{I}_5 + \omega_2 \hat{I}_4^2 + \omega_3 \hat{I}_1 \hat{I}_4}_{mechanical} + \underbrace{\boldsymbol{\kappa}_1 \hat{I}_1 \hat{J}_2^e + \boldsymbol{\kappa}_2 \hat{I}_4 \hat{J}_2^e + \boldsymbol{\kappa}_3 \hat{K}_1^e}_{piezoelectric} \\ &+ \underbrace{\xi_1 \hat{J}_1^m + \xi_2 (\hat{J}_2^m)^2}_{magnetic} + \underbrace{\gamma_1 \hat{J}_1^e + \gamma_2 (\hat{J}_2^e)^2 - (\hat{J}_2^e)\hat{P}_1^e}_{dielectric}\end{aligned} \tag{35}$$

where now the invariants, listed in Eq. (31), depent on the elastic part of the strains ε_e. Here, we introduce the electro-mechanical invariant $\hat{K}_1^e := \text{tr}[\varepsilon_e\,(\boldsymbol{E} \otimes \hat{\boldsymbol{a}})]$ and the electric invariant $\hat{P}_1^e := \text{tr}[\hat{\boldsymbol{P}}_r \otimes \hat{\boldsymbol{a}}]$ with the vector of remanent polarization $\hat{\boldsymbol{P}}_r$. At this point we introduce an additional length scale, the sub-microscale, which is denoted as $(\hat{\bullet})$. It is assumed that the individual hysterons switch to their *up* and *down* positions on sub-microscopic level. For the resulting dielectric displacement as well as the strains we assume an additive decomposition into an elastic part $(\bullet_e)$ and a remanent part $(\bullet_r)$, according to Kamlah and Tsakmakis [19], as

$$\hat{\boldsymbol{D}} = \hat{\boldsymbol{D}}_e + \hat{\boldsymbol{P}}_r \qquad \text{and} \qquad \hat{\varepsilon} = \hat{\varepsilon}_e + \hat{\varepsilon}_r\,, \tag{36}$$

where the remanent polarization vector $\hat{\boldsymbol{P}}_r$ is determined by a Preisach operator $\mathfrak{p}$ and the remanent strains $\hat{\varepsilon}_r$ depend on the current polarization state as

$$\hat{\boldsymbol{P}}_r = \mathfrak{p}(E)\,\hat{\boldsymbol{a}} \qquad \text{and} \qquad \hat{\varepsilon}_r = \frac{3}{2}\varepsilon_s \frac{1}{P_s^2}\text{dev}(\hat{\boldsymbol{P}}_r \otimes \hat{\boldsymbol{P}}_r)\,, \tag{37}$$

with the saturation polarization P_s and saturation strains ε_s. The butterfly hysteresis curve can also be modeled by an enhanced hysteresis operator, however, we assume that the strains result proportional to the squared remanent polarization vector $\hat{\boldsymbol{P}}_r$. Here the Preisach operator depend on the scalar product of the current electric field and the preferred orientation $\hat{\boldsymbol{a}}$ as an input variable and is given by

$$\mathfrak{p}(\boldsymbol{E}, \hat{\boldsymbol{a}}) = \int_{\beta} \int_{\alpha} \omega(\alpha, \beta)\, \gamma(\alpha, \beta)\, \boldsymbol{E}(t) \cdot \hat{\boldsymbol{a}}\, \mathrm{d}\alpha\, \mathrm{d}\beta\,. \tag{38}$$

For a simplification the Preisach weighting function $\omega(\alpha, \beta)$ as well as the switching thresholds α and β are in this case adjusted to the hyperbolic tangent to obtain the typical ferroelectric hysteresis loops. A fitting of the switching thresholds to experimental measurements is for example described in Kaltenbacher et al. [17] and will be done in future works. However, this Preisach operator is defined on a fixed preferred direction. In order to extend the classical Preisach model to the three dimensional space, we apply the Preisach operator to multiple orientations on sub-microscopic level, which are distributed in the three dimensional space. The distribution of the individual directions is based on the construction of a geodesic dome, see [25]. After the determination of the individual Preisach operators, which depend on the current local electric field in the corresponding direction, the constitutive quantities on microscopic level are calculated by a homogenization step over all sub-microscopic orientations n_{orient} as

$$\boldsymbol{\Sigma} = \frac{1}{n_{orient}} \sum_{i=1}^{n_{orient}} \hat{\boldsymbol{\Sigma}}_i \;\text{ with }\; \hat{\boldsymbol{\Sigma}}_i = \{\hat{\boldsymbol{\sigma}}_i, \hat{\boldsymbol{D}}_i, \hat{\boldsymbol{B}}_i, \hat{\mathbb{C}}_i, \hat{\boldsymbol{e}}_i, \hat{\boldsymbol{\epsilon}}_i, \hat{\boldsymbol{\mu}}_i, \hat{\varepsilon}_{r,i}, \hat{\boldsymbol{P}}_{r,i}\}. \tag{39}$$

Then the microscopic constitutive equations of the ferroelectric phase are computed via (39) with the constitutive equations related to the individual orientations of the ODF. Thus for each orientation we compute

$$\begin{aligned} \hat{\boldsymbol{\sigma}}_i &= \hat{\mathbb{C}}_i : (\varepsilon - \hat{\varepsilon}_{r,i}) - \hat{\boldsymbol{e}}_i^T \cdot \boldsymbol{E} \\ \hat{\boldsymbol{D}}_i &= \hat{\boldsymbol{e}}_i : (\varepsilon - \hat{\varepsilon}_{r,i}) + \hat{\boldsymbol{\epsilon}}_i \cdot \boldsymbol{E} + \hat{\boldsymbol{P}}_{r,i} \\ \hat{\boldsymbol{B}}_i &= \hat{\boldsymbol{\mu}}_i \cdot \boldsymbol{H} \end{aligned} \tag{40}$$

In order to regard a dependence of the electromechanical coupling on the polarization, the piezoelectric tensor $\boldsymbol{e}$ is multiplied by $\|\boldsymbol{P}_r\|/P_s$, see e.g. [18]. Although, a large number of relays is necessary to depict a smooth hysteresis loop, it requires large memory space, especially in the case of using an orientation distribution function, and reduces strongly the calculation speed.

4 Numerical Examples

The following chapter is divided into three parts, where we consider the response of the Preisach model for different cases. First, we demonstrate the evolution of the hystersis curves for an increasing number of relays. For simplicity, a homogeneous ferroelectric material with only one orientation is considered. The second part focuses on the application of the Preisach model to an orientation distribution function and the influence on the overall hysteresis curves. Analogously to the first example, a

Table 2 Material parameters of polycrystalline barium titanate ($BaTiO_3$) and cobalt ferrite ($CoFe_2O_4$)

Parameter	Unit	$BaTiO_3$	$CoFe_2O_4$
$\mathbb{C}_{1111}$	N/mm^2	16.6×10^4	21.21×10^4
$\mathbb{C}_{1122}$	N/mm^2	7.7×10^4	7.45×10^4
$\mathbb{C}_{1133}$	N/mm^2	7.8×10^4	7.45×10^4
$\mathbb{C}_{3333}$	N/mm^2	16.2×10^4	21.21×10^4
$\mathbb{C}_{1212}$	N/mm^2	4.3×10^4	6.88×10^4
ε_{11} $(\varepsilon^r_{11})^*$	mC/kVm (1)	0.0112 (1264.9)	8×10^{-5} (9.04)
ε_{33} (ε^r_{33})	mC/kVm (1)	0.0126 (1423.1)	9.3×10^{-5} (10.5)
μ_{11} $(\mu^r_{11})^*$	N/kA^2 (1)	1.26 (1)	157.0 (124.9)
μ_{33} (μ^r_{33})	N/kA^2 (1)	1.26 (1)	157.0 (124.9)
e_{311}	C/m^2	−4.4	0.0
e_{333}	C/m^2	18.6	0.0
e_{113}	C/m^2	11.6	0.0
q_{311}	N/Am	0.0	580.3
q_{333}	N/Am	0.0	−699.7
q_{113}	N/Am	0.0	550.0
$\hat{E}_c$	kV/mm	1.0	0.0
$\hat{\sigma}_c$	N/mm^2	100.0	0.0
$\hat{\varepsilon}_s$	1	0.00834	0.0
$\hat{P}_s$	C/m^2	0.26	0.0

*Depending on the permittivity as well as the permeability of free space $\varepsilon_0 \approx 8.854 \cdot 10^{-12}$ As/Vm and $\mu_0 = 4\pi \cdot 10^{-7}$ N/A^2 we determined the relative electric permittivity $\varepsilon^r = \varepsilon/\varepsilon_0$ and relative magnetic permeability $\mu^r = \mu/\mu_0$, respectively. For a detailed explanation of the chosen parameters we refer to [53]

homogeneous microstructure is taken into account, such that we can investigate the influence of the orientation distribution function on the overall hysteresis curves. Finally, the 3D Preisach model is used for the simulation of magneto-electric composites. Thereby, the ME coefficient is investigated for different types of composites. In the following Table 2, the parameters for the used materials are listed.

4.1 Preisach Model Applied on One Orientation

In the first example, we consider a purely ferroelectric macroscopic cubic body which is loaded with an alternating electric field in vertical direction E_3, see Fig. 5. Here, only one orientation is attached in each integration point. We increase in each simulation the number of relays, which approximate the overall hysteresis loops. Starting with one relay an overall square dielectric hysteresis cycle is obtained, see Fig. 6a. It can clearly be seen that the switchings to the *up* and *down* positions take

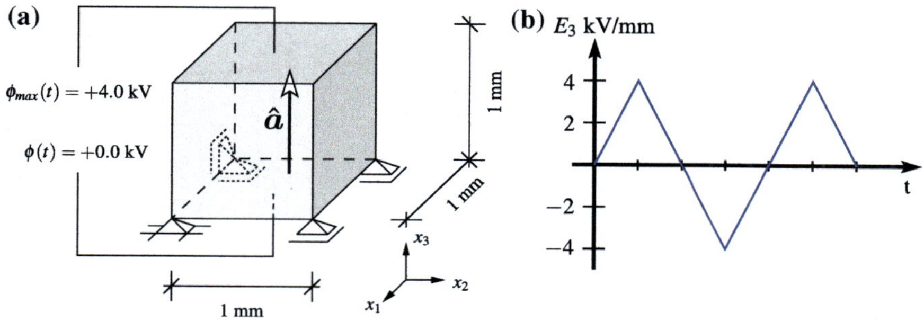

Fig. 5 **a** Macroscopic boundary value problem. The ferroelectric phase is loaded with an electric field E_3 in vertical direction and the preferred direction $\hat{\boldsymbol{a}}$ is also oriented in vertical direction. **b** Loading path of the alternating electric field

place when the applied field reaches the coercivity field strength. Furthermore, we observe a linear behavior of the deformations in vertical direction, see Fig. 6b. In the initial state the relay is defined to start in neutral position, which is neither *up* nor *down*, such that we observe no remanent deformation and polarization. After reaching the coercivity field, the relay switches into the *up* position and a further increase of the applied field result into an elongation of the material. However, a switching of the polarization direction of only one relay does not influence the magnitude of the remanent strains.

Figure 6c, d depict the dielectric and butterfly hysteresis loops for 10 relays. Due to the different switching thresholds of the individual relays, a cascade increase of the dielectric displacement is observed, which slowly approximates to the hyperbolic tangent. Depending on the current polarization state, the deformations slowly increase beyond the coercivity field. In contrast to the example with one relay, the body exhibits no displacement in the initial state, since the remanent deformations are canceled out due to the *up* and *down* positions of the relays. For an increasing electric field we can furthermore observe, that the slope of the linear piezoelectric effect between the switching processes of the single relays increases for a larger number of switched relays in the same direction. This is caused by the assumption that the piezoelectric coefficient $\boldsymbol{e}$ is multiplied by the factor $\|\boldsymbol{P}_r\|/P_s$, which increases for higher polarization states.

In the last case, see Fig. 6e, f 1000 relays were used. Here, it can be clearly seen, that now the shape of the dielectric hysteresis loop coincides with the hyperbolic tangent. Certainly, also 100 relays are sufficient to depict the hyperbolic tangent.

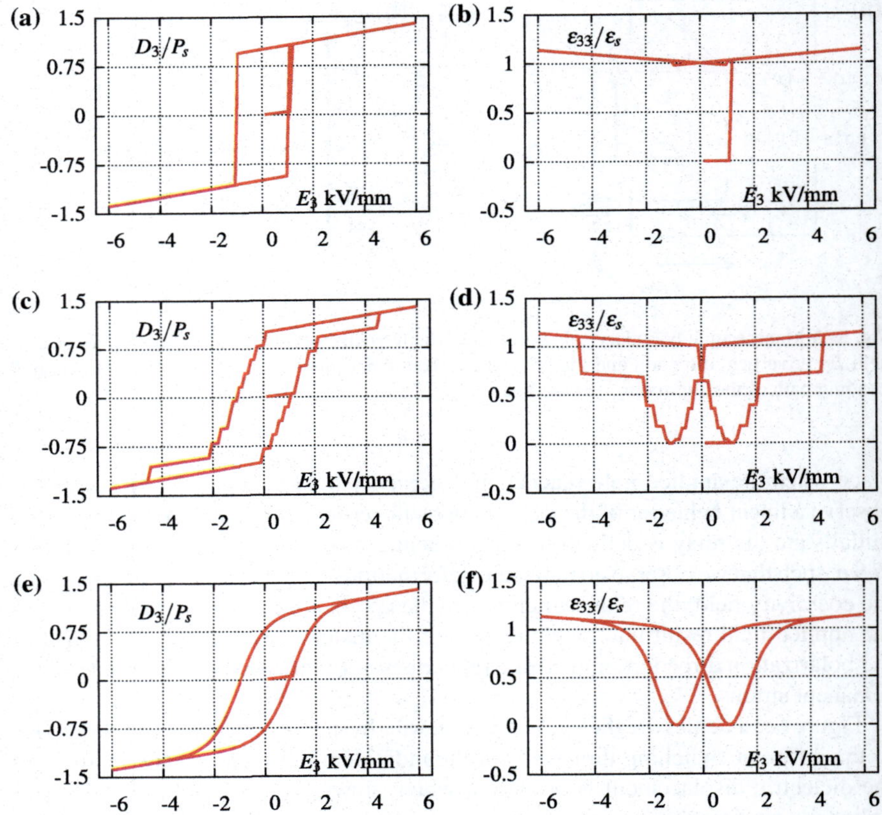

Fig. 6 Dielectric/butterfly hysteresis loop for **a/b** 1 relay, **c/d** 10 relays and **e/f** 1000 relays

4.2 Preisach Model Applied on an ODF

In this example, we consider again a purely ferroelectric body which is loaded with an alternating electric field, see Fig. 7a, b. Now, we attach in each integration point an orientation distribution function. On the individual directions the Preisach operator is applied and the microscopic constitutive quantities are determined through the homogenization over the number of orientations. Taking a closer look on the response of the dielectric and butterfly hysteresis curve, shown in Fig. 7c, d, we can observe a reduction of the maximum polarization and deformation values. Due to the homogenization over multiple orientations, pointing in different directions in the three dimensional space, a decrease of the saturation values can be explained, see for example [25] or [28]. Figures 7c, d show the comparison of the hysteresis loops for 1 and 42 orientations. We can furthermore observe, that in contrast to a single

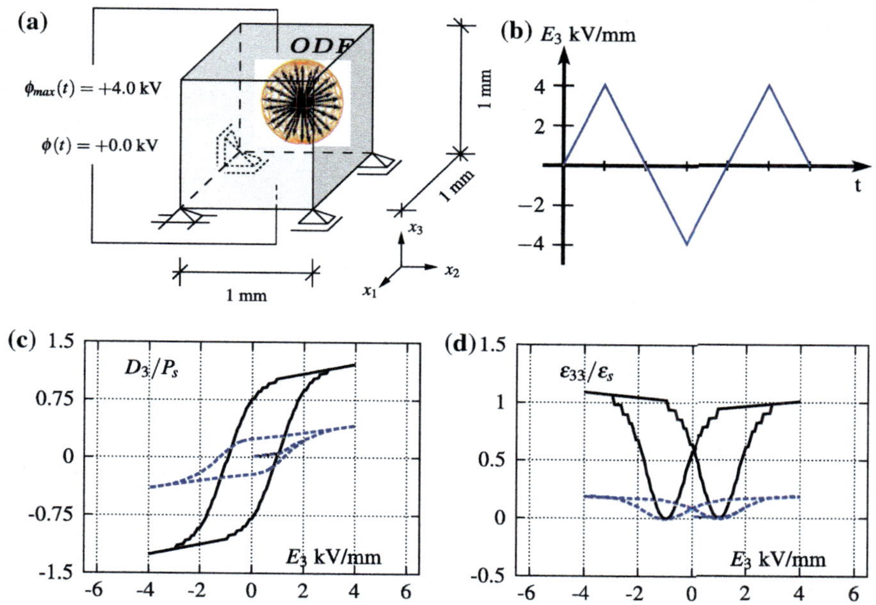

Fig. 7 **a** Macroscopic boundary value problem. The ferroelectric material is loaded with an alternating electric field in vertical direction. **b** Loading path of the electric field. Comparison of the **c** dielectric and **d** butterfly hysteresis curve for 1 (solid black line) and 42 (dashed blue line) orientations

orientation, 50 relays on each orientation of the ODF are sufficient to depict smooth hysteresis curves.

4.3 Simulation of Two-Phase Magneto-Electric Composites

In the last example, we consider two different magneto-electric composites. The first composite is characterized by a ferroelectric matrix with spherical inclusions, whereas the second composite has a cylindrical inclusion aligned in vertical direction. The behavior of the magnetic inclusions is assumed to be piezomagnetic and is simulated with the model described in Sect. 3.1. For the ferroelectric matrix material we use the 3D Preisach model based on an orientation distribution function. In order to investigate the response of the Preisach model on the overall properties, we consider a macroscopic body and apply an alternating vertical electric field. In each macroscopic integration point we attached representative volume elements, where we furthermore attach an orientation distribution function in each microscopic integration point of the electric material, see Fig. 8. Figure 8c, d show the resulting hysteresis curves for the composite with the spherical inclusions. A reduction of the

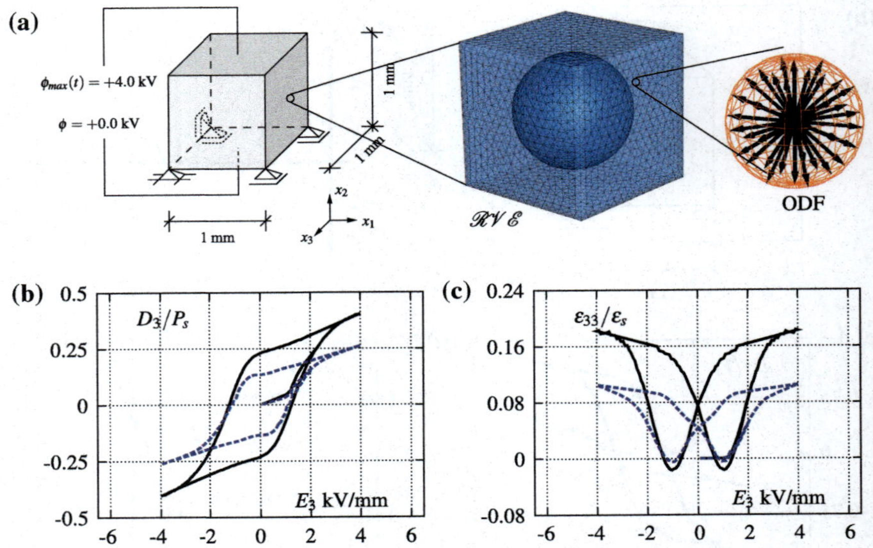

Fig. 8 **a** Macroscopic bvp with attached $\mathscr{RVE}$ and further attached ODF. Comparison of the **b** dielectric and **c** butterfly hysteresis loop for a single phase ferroelectric material (solid black line) and a two-phase composite (dashed blue line)

maximum saturation values compared to a single ferroelectric phase, i.e. replacing the magnetic inclusion by the ferroelectric material, is observable. The reason for this decrease is given by the reduced volume fraction of the electric phase. Furthermore, it can be noted that the slope of the butterfly hysteresis curve after the polarization process ($4 \geq E_3(t) \geq 2$) is flatter than the piezoelectric slope for a single phase ferroelectric body in the same field region. Due to the inhomogeneous microscopic morphology, some areas in the microstructure with local electric field minima are not completely saturated, which explains the reduced piezoelectric slope.

In the following we investigate the magneto-electric coupling behaviors of the different composites, which are depicted in Fig. 9. Here we increased the applied electric field to reach the saturation polarization in a larger area of the microstructure. Due to the strain-induced interaction of both phases the magneto-electric coupling depends on the transferred deformations and therefore including the piezoelectric coupling modulus $\boldsymbol{e}$. Depending on the current polarization state, the coupling modulus is factorized by $\|\boldsymbol{P}_r\|/P_s$. This dependency can clearly be seen in Fig. 9c, where the ME response is depicted for a composite with a cylindrical piezomagnetic inclusion. The shape of the ME curve conform to the shape of the remanent polarization. When the polarization reaches its saturation, the piezoelectric modulus simultaneously reaches its highest value, resulting in a saturation of the magneto-electric coefficient. A subsequent unloading in the field region of approximately $1.5 \leq E_3(t) \leq 6$ kV/mm does not change the ME response, since the piezoelectric response continues due to the remanent polarization. A similar response for an ME two-phase composite

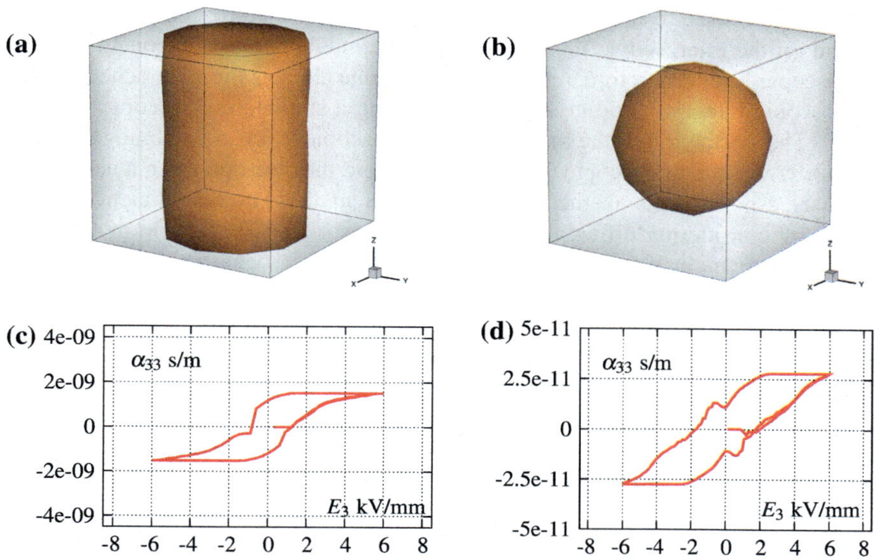

Fig. 9 **a** Atteched $\mathscr{RVE}$ with a cylindrical magnetic inclusion and **c** its ME response. **b** Attached $\mathscr{RVE}$ with spherical magnetic inclusion and **d** its corresponding ME response

with spherical piezomagnetic inclusions is also obtained in [53], where a detailed discussion of the resulting ME coefficient can be found.

Figure 9d shows the ME response of a composite with spherical piezomagnetic inclusions. In comparison to the response of a composite with cylindrical inclusions, we can directly see two main differences. First, the ME response does not reach its saturation point in the same electric field region. Due to the inhomogeneous electric field distribution on microscopic level, the piezoelectric modulus is saturated in far less regions in the ferroelectric material. Second, the maximum value of the ME coefficient is much lower. However, if we compare the obtained dielectric displacement, taking a decrease of the value caused by the homogenization over all orientations into account, with the values given in [53] the also decreased ME coefficient matches very well with the results obtained in the mentioned work. The homogenization of the microscopic constitutive quantities over the orientation distribution function is the reason for the lower maximum values of the hysteresis loops as well as the magneto-electric coupling coefficient.

4.4 Conclusion

In the present contribution, we presented a two-scale homogenization scheme for magneto-electro-mechancially coupled boundary value problems. This so called

FE^2-method is used to simulate the response of magneto-electric composites, which depend on the microscopic morphology and the interaction of the constituents on microscopic level. Therefore, representative volume elements were attached at each macroscopic integration point instead of deriving a suitable macroscopic material model. The interaction of the materials on microscopic level and a homogenization approach over the microscopic quantities then yield the corresponding macroscopic variables, especially the magneto-electric coefficient. The latter coefficient is investigated by considering different composites consist of a ferroelectric matrix with piezomagnetic inclusions. In order to simulate the realistic behavior of the ferroelectric material and to depict the typical dielectric and butterfly hysteresis curves, we implemented a three dimensional Preisach model into the FE^2-scheme. Therefore, the cassical scalar model is extended to the three dimensional space through the application of the model on an orientation distribution function. Numerical examples, demonstrated the performance of the 3D Preisach model, which is capable of showing the typical hysteresis curves. Additionally, we investigated the magneto-electric response of two different composites, one with spherical and the other with cylindrical magnetic inclusions. The obtained ME curves were similar to the results of previous works and the typical behavior could be captured. However, due to the application of the Preisach model on an orientation distribution function and the homogenization over these orientations, the resulting values of the hysteresis loops and the magneto-electric coefficient were slightly reduced. Additionally, the large number of necessary relays to depict smooth hysteresis curves requires large memory space and reduces strongly the calculation speed. Future works will show one solution of this problem, by assuming only one orientation, which can change its direction due to an applied field.

Acknowledgements We gratefully acknowledge the financial support of the German Research Foundation (DFG) in the framework of the research unit 1509 "Ferroic Functional Materials—Multiscale Modeling and Experimental Characterization" (SCHR 570/12-2).

References

1. Astrov, D.N.: Magnetoelectric effect in chromium oxide. J. Exp. Theor. Phys. **40**, 1035–1041 (1961)
2. Avakian, A., Gellmann, R., Ricoeur, A.: Nonlinear modeling and finite element simulation of magnetoelectric coupling and residual stress in multiferroic composites. Acta Mechanica **226**, 2789–2806 (2015)
3. Bermúdez, A., Dupré, L., Gómez, D., Venegas, P.: Electromagnetic computations with Preisach hysteresis model. Finite Elem. Anal. Des. **126**, 65–74 (2017)
4. Bibes, M., Barthélémy, A.: Multiferroics: towards a magnetoelectric memory. Nature Mater. **7**(6), 425–426 (2008). ISSN 1476-1122
5. Brown, W.F., Hornreich, R.M., Shtrikman, S.: Upper bound on the magnetoelectric susceptibility. Phys. Rev. **168**(2), 574–577 (1968)
6. Cheong, S. -W., Mostovoy, M.: Multiferroics: a magnetic twist for ferroelectricity. Nature Mater. **6**(1):13–20 (2007). ISSN 1476-1122

7. Crottaz, O., Rivera, J. -P., Revaz, B., Schmid, H.: Magnetoelectric effect of $Mn_3B_7O_{13}I$ boracite. Ferroelectrics **204**, 125–133 (1997)
8. Eerenstein, W., Mathur, N.D., Scott, J.F.: Multiferroic and magnetoelectric materials. Nature **442**(7104), 759–765 (2006)
9. Eerenstein, W., Wiora, M., Prieto, J.L., Scott, J.F., Mathur, N.D.: Giant sharp and persistent converse magnetoelectric effects in multiferroic epitaxial heterostructures. Nature Mater. **6**(5), 348–351 (2007)
10. Etier, M., Shvartsman, V.V., Gao, Y., Landers, J., Wende, H., Lupascu, D.C.: Magnetoelectric effect in (0–3) CoFe2O4-BaTiO3 (20/80) composite ceramics prepared by the organosol route. Ferroelectrics **448**, 77–85 (2013)
11. Feyel, F., Chaboche, J. -L.: FE^2 multiscale approach for modelling the elastoviscoplastic behavior of long fibre SiC/Ti composite materials. Comput. Methods Appl. Mech. Eng. **183**, 309–330 (2000)
12. Hegewald, T., Kaltenbacher, B., Kaltenbacher, M., Lerch, R.: Efficient modeling of ferroelectric behavior for the analysis of piezoceramic actuators. J. Intell. Mater. Syst. Struct. (2008)
13. Hill, N.A.: Why are there so few magnetic ferroelectrics? J. Phys. Chem. B **104**, 6694–6709 (2000)
14. Hill, R.: Elastic properties of reinforced solids—some theoretical principles. J. Mech. Phys. Solids **11**, 357–372 (1963)
15. Hughes, D., Wen, J.T.: Preisach modeling of piezoceramic and shape memory alloy hysteresis. Smart Mater. Struct. **6**, 287–300 (1997)
16. Hwang, S.C., Lynch, C.S., McMeeking, R.M.: Ferroelectric/ferroelastic interaction and a polarization switching model. Acta Metallurgica et Materialia **43**, 2073–2084 (1995)
17. Kaltenbacher, M., Kaltenbacher B., Hegewald, T., Lerch, R.: Finite element formulation for ferroelectric hysteresis of piezoelectric materials. J. Intell. Mater. Syst. Struct. (2010)
18. Kamlah, M.: Ferroelectric and ferroelastic piezoceramics—modeling of electromechanical hysteresis phenomena. Continuum Mech. Thermodyn. **13**, 219–268 (2001)
19. Kamlah, M., Tsakmakis, C.: Phenomenological modeling of the non-linear electro-mechanical coupling in ferroelectrics. Int. J. Solids Struct. **36**, 669–695 (1999)
20. Kamlah, M., Liskowsky, A.C., McMeeking, R.M., Balke, H.: Finite element simulation of a polycrystalline ferroelectric based on a multidomain single crystal switching model. Int. J. Solids Struct. **42**, 2949–2964 (2005)
21. Keip, M. -A., Rambausek, M.: A multiscale approach to the computational characterization of magnetotheological elastomers. Int. J. Numerical Methods Eng. **107**, 338–360 (2015)
22. Keip, M. -A., Schrade, D., Thai, H.N.M., Schröder, J., Svendsen, B., Müller, R., Gross, D.: Coordinate-invariant phase field modeling of ferroelectrics, Part II: application to composites and polycrystals. GAMM-Mitteilungen **38**(1), 115–131 (2015)
23. Khomskii, D.: Classifying multiferroics: mechanisms and effects. Physics **2**, 20 (2009)
24. Kouznetsova, V., Brekelmans, W.A.M., Baaijens, F.P.T.: An approach to micro-macro modeling of heterogeneous materials. Comput. Mech. **27**, 37–48 (2001)
25. Kurzhöfer, I.: Mehrskalen-Modellierung polykristalliner Ferroelektrika basierend auf diskreten Orientierungsverteilungsfunktionen. Ph.D. thesis, University of Duisburg-Essen (2007)
26. Labusch, M., Etier, M., Lupascu, D.C., Schröder, J., Keip, M. -A.: Product properties of a two-phase magneto-electric composite: synthesis and numerical modeling. Comput. Mech. **54**, 71–83 (2014)
27. Labusch, M., Keip, M. -A., Shvartsman, V.V., Lupascu, D.C., Schröder, J.: On the influence of ferroelectric polarization states on the magneto-electric coupling in two-phase composites. Technische Mechanik **36**, 73–87 (2016)
28. Landis, C.M., Wang, J., Sheng, J.: Micro-electromechanical determination of the possible remanent strain and polarization states in polycrystalline ferroelectrics and the implications for phenomenological constitutive theories. J. Intell. Mater. Syst. Struct. **15**, 513–525 (2004)
29. Mandel, J., Dantu, P.: Conribution à l'étude théorique et expérimentale du coefficient d'élasticité d'un milieu hétérogène mais statistiquement homogène. *Annales des Ponts et Chaussées* (1963)

30. Martin, L.W., Chu, Y. -H., Ramesh, R.: Advances in the growth and characterization of magnetic, ferroelectric, and multiferroic oxide thin films. Mater. Sci. Eng.: R: Rep. **68**(46), 89–133 (2010). ISSN 0927-796X
31. Mayergoyz, I.D.: Mathematical models of hysteresis. IEEE Trans. Magn. **22**, 603–608 (1986)
32. Mayergoyz, I.D., Friedman, G.: Generalized Preisach model of hysteresis. IEEE Trans. Magn. **24**, 212–217 (1988)
33. Miehe, C.: Computational micro-to-macro transitions for discretized micro-structures of heterogeneous materials at finite strains based on the minimization of averaged incremental energy. Comput. Methods Appl. Mech. Eng. **192**, 559–591 (2003)
34. Miehe, C., Bayreuther, C.G.: On mutiscale FE analyses of heterogeneous structures: from homogenization to multigrid solvers. Int. J. Numer. Methods Eng. **71**, 1135–1180 (2007)
35. Miehe, C., Schotte, J., Schröder, J.: Computational micro-macro transitions and overall moduli in the analysis of polycrystals at large strains. Comput. Mater. Sci. **16**(1–4), 372–382 (1999)
36. Miehe, C., Vallicotti, D., Teichtmeister, S.: Homogenization and multiscale stability analysis in finite magneto-electro-elasticity. Application to soft matter EE, ME and MEE composites. Comput. Methods Appl. Mech. Eng. **300**, 294–346 (2016)
37. Nan, C. -W.: Magnetoelectric effect in composites of piezoelectric and piezomagnetic phases. Phys. Rev. B **50**, 6082–6088 (1994)
38. Nan, C. -W., Bichurin, M.I., Shuxiang Dong, D. Viehland, Srinivasan, G.: Multiferroic magnetoelectric composites: historical perspective, status, and future directions. J. Appl. Phys. **103**(3), 031101 (2008)
39. Perić, D., de Souza Neto, E.A., Feijóo, R.A., Partovi, M., Carneiro Molina A.J.: On micro-to-macro transitions for multi-scale analysis of non-linear heterogeneous materials: unified variational basis and finite element implementation. Int. J. Numer. Methods Eng. **87**, 149–170 (2011)
40. Preisach, F.: über die magnetische Nachwirkung. Zeitschrift für Physik **94**, 277–302 (1935)
41. Priya, S., Islam, R., Dong, S.X., Viehland, D.: Recent advancements in magnetoelectric particulate and laminate composites. J. Electroceramics **19**(1), 147–164 (2007)
42. Ramesh, R., Spaldin, N.A.: Multiferroics: progress and prospects in thin films. Nature Mater. 6(1):21–29 (2007). ISSN 1476-1122
43. Rivera, J. -P.: On definitions, units, measurements, tensor forms of the linear magnetoelectric effect and on a new dynamic method applied to Cr-Cl boracite. Ferroelectrics **161**, 165–180 (1994)
44. Rivera, J. -P., Schmid, H.: On the birefringence of magnetoelectric $BiFeO_3$. Ferroelectrics **204**, 23–33 (1997)
45. Robert, G., Damjanovic, D., Setter, N., Turik, A.V.: Preisach modeling of piezoelectric non-linearity in ferroelectric ceramics. J. Appl. Phys. **89**, 5067–5074 (2001)
46. Ryu, J., Priya, S., Uchino, K., Kim, H.E.: Magnetoelectric effect in composites of magnetostrictive and piezoelectric materials. J. Electroceramics **8**, 107–119 (2002)
47. Schmid, H.: Multi-ferroic magnetoelectrics. Ferroelectrics **162**, 317–338 (1994)
48. Schröder, J.: Derivation of the localization and homogenization conditions for electro-mechanically coupled problems. Comput. Mater. Sci. **46**(3), 595–599 (2009)
49. Schröder, J.: A numerical two-scale homogenization scheme: the FE^2–method. In: Schröder, J., Hackl, K. (eds.) Plasticity and Beyond. CISM Courses and Lectures, vol. 55, pp. 1–64. Springer (2014)
50. Schröder, J., Gross, D.: Invariant formulation of the electromechanical enthalpy function of transversely isotropic piezoelectric materials. Archive Appl. Mech. **73**, 533–552 (2004)
51. Schröder, J., Keip, M. -A.: Two-scale homogenization of electromechanically coupled boundary value problems. Comput. Mech. **50**, 229–244 (2012)
52. Schröder, J., Labusch, M., Keip, M. -A., Kiefer, B., Brands, D., Lupascu, D.C.: Computation of non-linear magneto-electric product properties of 0-3 composites. GAMM-Mitteilungen (2015)
53. Schröder, J., Labusch, M., Keip, M. -A.: Algorithmic two-scale transition for magneto-electro-mechanically coupled Problems: FE^2-Sheme: Localization and homogenization. Comput. Methods Appl. Mech. Eng. **302**, 253–280 (2016)

54. Shvartsman, V.V., Alawneh, F., Borisov, P., Kozodaev, D., Lupascu, D.C.: Converse magnetoelectric effect in cofe$_2$o$_4$-batio$_3$ composites with a core-shell structure. Smart Mater. Struct. **20** (2011)
55. Song, G., Zhao, J., Zhou, X., De Abreu-García, J.A.: Tracking control of a piezoceramic actuator with hysteresis compensation using inverse Preisach model. IEEE/ASME Trans. Mechatron. **10**, 198–209 (2005)
56. Spaldin, N.A., Fiebig, M.: The renaissance of magnetoelectric multiferroics. Science **309**, 391–392 (2005)
57. Srinivasan, G.: Magnetoelectric composites. Annu. Rev. Mater. Res. **40**, 1–26 (2010)
58. Stancu, A., Bissell, P.R., Chantrell, R.W.: Interparticle interactions in magnetic recording maedia as obtained from high-order measurements by a Preisach model. J. Appl. Phys. **87**, 8645–8652 (2000)
59. Temizer, I., Wriggers, P.: On the computation of the macroscopic tangent for multiscale volumetric homogenization problems. Comput. Methods Appl. Mech. Eng. **198**, 495–510 (2008)
60. Terada, K., Hori, M., Kyoya, T., Kikuchi, N.: Simulation of the multi-scale convergence in computational homogenization approach. Int. J. Solids Struct. **37**, 2285–2311 (2000)
61. van der Sluis, O., Schreurs, P.J.G., Brekelmans, W.A.M., Meijer, H.E.H.: Overall behavior of heterogeneous elastoviscoplastic materials: effect of microstructural modelling. Mech. Mater. **32**, 449–462 (2000)
62. van Suchtelen, J.: Product properties: a new application of composite materials. Philips Res. Rep. **27**, 28–37 (1972)
63. Vaz, C.A.F., Hoffman, J., Ahn, C.H., Ramesh, R.: Magnetoelectric coupling effects in multiferroic complex oxide composite structures. Adv. Mater. **22**, 2900–2918 (2010)
64. Wang, Y., Hu, J., Lin, Y., Nan, C.-W.: Multiferroic magnetoelectric composite nanostructures. Nature Asia-Pacific, Asia Mater. **2**, 61–68 (2010)
65. Yu, Y., Xiao, Z., Naganathan, N.G., Dukkipati, R.V.: Dynamic Preisach modelling of hysteresis for the piezoceramic actuator system. Mech. Mach. Theory **37**, 75–89 (2002)
66. Zohdi, T.I.: Simulation of coupled microscale multiphysical-fields in particulate-doped dielectrics with staggered adaptive fdtd. Comput. Methods Appl. Mech. Eng. **199**, 3250–3269 (2010)
67. Zohdi, T.I.: Electromagnetic Properties of Multiphase Dielectrics. Springer (2012)

A Multiscale Framework for Thermoplasticity

Marko Čanađija and Neven Munjas

Abstract The chapter describes a homogenization procedure for thermoplasticity problems. The proposed model is suitable for the finite strain regime and supports a very wide class of plasticity models. The methodology starts from the thermodynamically consistent thermoelastic framework already described in the literature. The latter framework is now extended to account for inelastic deformations. The problem is separated by means of the isothermal split into a mechanical and a thermal step, both at the macroscale and the microscale. As demonstrated in an example, the method does provide a way to successfully homogenize microscale variables as well as tangent operators. Finally, limitations of the approach are pointed out.

1 Introduction

Thermoplasticity usually involves nonisothermal inelastic problems that are either dominated by the mechanical effects or the thermal effects. The former is the case in the metal forming processes that take place at the room temperature. In these processes, the primary heat source is the internal dissipation due to inelastic deformation. As an example of the latter class of problems, metal forming can be pointed out once more. In this case, forming takes place at elevated temperature. As known for a long time, the elevated temperature decreases the yield point of a metal and such material softening makes it more suitable for shaping operations.

It is well-known that the temperature increase is accompanied by elongation of a structure. The inverse effect, temperature decrease due to elastic tensile mechanical loading is frequently disregarded as the heat source since in most cases corresponding temperature change is very small. Nevertheless, the effect can be significant

M. Čanađija (✉) · N. Munjas
Faculty of Engineering, University of Rijeka,
Vukovarska 58, 51000 Rijeka, Croatia
e-mail: marko.canadija@riteh.hr

N. Munjas
e-mail: neven.munjas@riteh.hr

J. Sorić et al. (eds.), *Multiscale Modeling of Heterogeneous Structures*,
Lecture Notes in Applied and Computational Mechanics 86,
https://doi.org/10.1007/978-3-319-65463-8_16

if the mechanically triggered thermoplasticity does not involve large strains, especially if the cyclic loading is concerned [1]. The thermoelastic phenomenon is usually accounted for in the classical single-scale thermoplasticity [2, 3]. However, as the physics of the plastic deformation is tightly connected to the microscopic level, usually in the form of dislocations and other defects, it would be advantageous to have a procedure that could address these issues as well. Analyses of this sort are known as multiscale analyses and are performed at two scales. It is assumed that at the smaller scale—microscale—all constitutive behaviour is defined. Variables of interest like stresses and heat flux are calculated at the microscale in a representative volume element by some kind of averaging/homogenizations technique. These results are transferred to the larger scale, i.e. macroscale, where displacements and temperatures are determined.

The research on the multiscale thermoplasticity is still rather scarce. Initial contribution was given in [4]. The paper considered adiabatic thermoplasticity in the case of polycrystalline materials. Under adiabatic conditions, the analysis simplifies considerably since the microscale temperature calculations can be avoided. Instead, homogenization of the internal dissipation was carried out in the mechanical step and supplied to the macroscale as a heat source. The thermoelastic effect was neglected. In the similar manner [5] uses the homogenized internal dissipation as the heat source at the macroscale and specializes the computational procedure to axisymmetrical problems. An approach more suitable for the thermally triggered thermoplasticity was presented in [6]. The mechanical dissipation was not considered in the formulation and thus is not able to capture heat generated by the plastic deformation. Differently to [4, 5], research [7] includes the microscale thermal step as a steady-state problem. The internal dissipation was estimated at the microscale and treated as the external heat source at the macroscale in the manner of [4, 5]. At the end, a paper on thermoelasticity [8] should be pointed out. The paper presents a very strict and consistent thermodynamical framework that should be preferred in multiscale thermomechanical analyses.

The present chapter starts from the mentioned thermoelastic paper [8] and extends it to the thermoplastic framework. Both the thermoelastic effect and the internal dissipation generated by plastic deformation are accounted for. The mechanical part of the microscale analysis is based on the variationally consistent formulation [1, 9–12], while the thermal part is identical to the one used in [8]. Verification of the homogenization procedure is provided on an example.

2 Macroscale Continuum Mechanics Framework

When the multiscale thermomechanical analysis is concerned, there are several issues regarding consistency between macro- and microscale procedures. These are carefully elaborated within the thermoelastic framework in [8]. The present manuscript extends this framework to thermoplasticity. The basics of macroscale continuum mechanics are described first.

2.1 Deformation and Temperature Gradients at the Macro- and Microscale

In text that follows, the overline will denote a macroscale quantity, i.e. $(\bar{\bullet})$. Absence of the overline will indicate a microscale quantity. With this notation at hand, a link between the macroscopic and the microscopic deformation gradient is provided by [6, 13]:

$$\boldsymbol{F} = \overline{\boldsymbol{F}} + \text{GRAD}\ \boldsymbol{w}_F, \tag{1}$$

where $\boldsymbol{w}_F$ denotes the microscale fluctuation field of the configuration. In the same manner, temperature gradients are defined as:

$$\nabla\Theta = \overline{\nabla}\overline{\Theta} + \text{GRAD}\ w_\Theta, \tag{2}$$

where w_Θ denotes the temperature fluctuation field. At the microscale, the configuration in a point $\boldsymbol{X}$ of the representative volume element (RVE) $\mathcal{B}_0$ can be now defined as:

$$\varphi = \overline{\boldsymbol{F}} \cdot \boldsymbol{X} + \boldsymbol{w}_F, \tag{3}$$

while temperature at the same position $\Theta(\boldsymbol{X})$ is:

$$\Theta = \overline{\nabla}\overline{\Theta} \cdot \boldsymbol{X} + w_\Theta. \tag{4}$$

At this point it should be noted that the macroscopic gradients should be equal to the volume averages of the microscopic ones [6, 13]:

$$\begin{aligned} \overline{\boldsymbol{F}} &= \tfrac{1}{|\mathcal{B}_0|} \textstyle\int_{\mathcal{B}_0} \boldsymbol{F} \mathrm{d}V, \\ \overline{\nabla}\overline{\Theta} &= \tfrac{1}{|\mathcal{B}_0|} \textstyle\int_{\mathcal{B}_0} \nabla\Theta \mathrm{d}V. \end{aligned} \tag{5}$$

Introducing Eqs. (1) and (2) into Eq. (5) leads toward notion that both fluctuation fields have to vanish at the RVE boundary, giving:

$$\int_{\partial\mathcal{B}_0} \boldsymbol{w}_F \otimes \boldsymbol{N}\ \mathrm{d}S = 0, \tag{6}$$

$$\int_{\partial\mathcal{B}_0} w_\Theta \boldsymbol{N}\ \mathrm{d}S = 0. \tag{7}$$

These constraints are enforced by a suitable set of boundary conditions. Although several options are available at this point, hereby only periodic boundary conditions will be considered. The application details are pretty much standard nowadays and will be left out from the present discussion, see [6, 8] for an overview. Note that volume averages in Eq. (5) introduce notation that will be used for the volume average of any quantity in text that follows, $\langle\bullet\rangle = 1/|\mathcal{B}_0| \int_{\mathcal{B}_0}(\bullet)\mathrm{d}V$.

2.2 Balance Equations at the Macroscale

Standard balance equations should be now rearranged to account for the homogenization procedure. To this end, the balance of momentum at the macroscale is now:

$$\overline{\mathrm{DIV}}\,\overline{\boldsymbol{P}} + \overline{\rho_0}\boldsymbol{B}_0 = \boldsymbol{0}, \tag{8}$$

where $\overline{\boldsymbol{P}}$ denotes the macroscopic (homogenized) first Piola-Kirchhoff stress tensor, $\overline{\rho_0}$ is the macroscopic specific mass in the reference configuration and $\boldsymbol{B}_0$ are the macroscopic body forces in the reference configuration. It is assumed that the body forces do not depend on the intrinsic material response and thus are not homogenized. In the present case, inertial terms are deemed not to be significant and are disregarded.

In the finite strain thermomechanics, special attention should be given to the first and the second law of thermodynamics. The balance of energy is stated as:

$$\dot{\overline{E}} + \overline{\mathrm{DIV}}\,\overline{\boldsymbol{H}} = \overline{\boldsymbol{P}} : \dot{\overline{\boldsymbol{F}}} + \overline{\rho_0} Q, \tag{9}$$

where the heat source per unit mass Q is also assumed to be independent of the intrinsic material response and consequently is not homogenized. The second law of thermodynamics is introduced as:

$$\overline{\mathcal{D}} = \overline{\Theta}\dot{\overline{N}} - \dot{\overline{E}} + \overline{\boldsymbol{P}} : \dot{\overline{\boldsymbol{F}}} + \overline{\boldsymbol{H}} \cdot \overline{\boldsymbol{G}} \geq 0. \tag{10}$$

In Eqs. (9) and (10) the homogenized quantities are denoted as follows. $\overline{E}$ is the internally stored energy per unit undeformed volume, $\overline{\Theta}$ is the absolute temperature, $\overline{\boldsymbol{H}}$ is the outward heat flux vector, $\overline{\boldsymbol{G}} = -(\overline{\mathrm{GRAD}}\,\overline{\Theta})/\overline{\Theta}$ is the normalized temperature gradient, $\overline{\mathcal{D}}$ is the dissipation and $\overline{N}$ denotes entropy. A superimposed dot implies the time rate.

The second law of thermodynamics places familiar restrictions on every admissible thermodynamical process. Traditionally, the dissipation (10) is additively separated into the part arising from heat conduction $\overline{\mathcal{D}}_{\mathrm{cond}}$ and internal dissipation $\overline{\mathcal{D}}_{\mathrm{int}}$ as:

$$\overline{\mathcal{D}}_{\mathrm{cond}} = \overline{\boldsymbol{H}} \cdot \overline{\boldsymbol{G}} \geq 0, \qquad \overline{\mathcal{D}}_{\mathrm{int}} = \overline{\Theta}\dot{\overline{N}} - \dot{\overline{E}} + \overline{\boldsymbol{P}} : \dot{\overline{\boldsymbol{F}}} \geq 0. \tag{11}$$

The first constraint regarding heat conduction restrains physically allowable processes so that heat can spontaneously flow only from the warmer to the colder environment and not vice-versa. Obviously, the constraint is set on the specific form of the heat flux $\boldsymbol{H}$.

In the presence of inelastic deformation, it is assumed that the internal energy is a function $\overline{E} = \overline{E}(\overline{\boldsymbol{F}}, \overline{N}, \overline{\alpha})$, where $\overline{\alpha}$ represents a suitable set of internal variables governing inelastic behaviour of material. It should be emphasized that neither the nature of internal variables nor the constitutive law are not known at the macroscale, but merely the existence of these is admitted. The constitutive behaviour for internal variables are fully specified at the microscale. By means of Legrende transformation

$$\overline{E} = \overline{\Psi} + \overline{\Theta}\overline{N}, \tag{12}$$

the homogenized Helmholtz free energy $\overline{\Psi} = \overline{\Psi}(\overline{\boldsymbol{F}}, \overline{\Theta}, \overline{\boldsymbol{\alpha}})$ is introduced. Standard arguments [14] now lead toward the following set of relationships:

$$\overline{\boldsymbol{P}} = \partial_{\overline{\boldsymbol{F}}}\overline{\Psi}, \quad \overline{N} = -\partial_{\overline{\Theta}}\overline{\Psi}, \quad \bar{c} = -\overline{\Theta}\partial^2_{\overline{\Theta}}\overline{\Psi}, \quad \overline{\boldsymbol{Q}} = -\partial_{\overline{\boldsymbol{\alpha}}}\overline{\Psi}, \tag{13}$$

where $\overline{\boldsymbol{Q}}$ is a set of variables conjugated to $\overline{\boldsymbol{\alpha}}$ and $\bar{c}$ is the macroscale specific heat capacity. Internal dissipation Eq. (11) can be rephrased with the aid of Eq. (12) as:

$$\overline{\mathcal{D}}_{\text{int}} = \overline{\boldsymbol{P}} : \dot{\overline{\boldsymbol{F}}} - \dot{\overline{\Psi}} - \dot{\overline{\Theta}}N = \overline{\boldsymbol{Q}} \cdot \dot{\overline{\boldsymbol{\alpha}}} \geq 0. \tag{14}$$

Thus, a thermodynamically consistent multiscale model should fulfil inequalities Eqs. (11) and (14) in addition to relationships Eqs. (12) and (13). As it will be demonstrated in the next section, fulfilment of such requirements is far from being straightforward.

The standard procedure now yields the temperature evolution equation at the macroscale:

$$\bar{c}\dot{\overline{\Theta}} = -\overline{\text{DIV}}\overline{\boldsymbol{H}} + \bar{\rho}_0 Q + \overline{\mathcal{D}}_{\text{int}} + \overline{\Theta}\overline{\mathcal{H}}^{\text{ep}}, \tag{15}$$

where $\overline{\mathcal{H}}^{\text{ep}} = (\partial_{\overline{\Theta}}\overline{\boldsymbol{P}} : \dot{\overline{\boldsymbol{F}}} - \partial_{\overline{\Theta}}\overline{\boldsymbol{Q}}\dot{\overline{\boldsymbol{\alpha}}})$ is the structural heating term [3]. The structural heating describes heating due to the thermoelastic effect and latent plastic structural changes caused by temperature dependency of hardening parameters. The latter equation is accompanied by standard boundary conditions and possible heat conduction and radiation terms.

3 Enforcement of Thermodynamical Consistency

To ensure thermodynamical consistency between scale transitions, fundamental macroscopic variables should be equal to the equivalent averaged variables at the microscale. In particular, by extension of the thermoelastic formulation in [8] to inelastic processes there must be:

$$\begin{gathered} \overline{E} = \langle E \rangle, \quad \overline{N} = \langle N \rangle, \quad \overline{\rho_0} = \langle \rho_0 \rangle, \\ \overline{\mathcal{D}}_{\text{int}} = \langle \mathcal{D}_{\text{int}} \rangle, \quad \overline{\mathcal{D}}_{\text{cond}} = \langle \mathcal{D}_{\text{cond}} \rangle, \quad \overline{\mathcal{H}}^{\text{ep}} = \langle \mathcal{H}^{\text{ep}} \rangle. \end{gathered} \tag{16}$$

In the text that follows, some constraints arising from these equalities will be discussed.

3.1 Energy and Entropy in Micro-Macro Transition

As it will be shown in this section, equalities of energy and entropy between both scales set rather strict constraints on the solution procedure [8]. To illustrate these issues, a simple constitutive model is considered with the following assumptions:

- The model is purely thermal. In this case, the microscale Helmholtz free energy takes the form: $\Psi(\Theta) = c_0 [(\Theta - \Theta_0 - \Theta \ln \Theta / \Theta_0)]$. Such selection of the free energy function at microscale, together with Eq. $(13)_3$ gives $c = -\Theta \partial^2_{\overline{\Theta}} \Psi = c_0$.
- The initial temperature field Θ_0 is homogeneous. As a consequence, $\overline{\Theta}_0 = \Theta_0$ in every point.
- The specific heat capacity is assumed to be constant and independent of deformation. This ensures $\bar{c} = \langle c \rangle = c_0$.

For the selected form of the microscale Helmholtz energy, the entropy is:

$$N = -\partial_\Theta \Psi(\Theta) = c_0 \ln \frac{\Theta}{\Theta_0}. \tag{17}$$

With this result at hand, the internal energy can be redefined in terms of temperature:

$$E(N(\Theta)) = \Psi(\Theta) + N(\Theta)\Theta = \Psi(\Theta) + c_0 \Theta \ln \frac{\Theta}{\Theta_0}. \tag{18}$$

Now, note that Eq. $(13)_3$ can be also written in the form:

$$\bar{c} = \partial_{\overline{\Theta}} \overline{E} = \overline{\Theta} \partial_{\overline{\Theta}} \overline{N}, \tag{19}$$

with the analogous counterpart at the microscale:

$$c_0 = \partial_\Theta E = \Theta \partial_\Theta N. \tag{20}$$

Consistent scale transitions of energy and entropy requires that Eq. $(16)_{1,2}$ are simultaneously fulfilled. Bearing this in mind, integration of Eqs. (19) and (20) gives

$$\begin{aligned} \overline{E}_0 + \int_{\overline{\Theta}_0}^{\overline{\Theta}} \bar{c}_0 \mathrm{d}\overline{\Theta}' &= \langle E_0 \rangle + \left\langle \int_{\Theta_0}^{\Theta} \partial_\Theta E \mathrm{d}\Theta' \right\rangle = E_0 + \left\langle \int_{\overline{\Theta}_0}^{\Theta} c_0 \mathrm{d}\Theta' \right\rangle \\ \overline{N}_0 + \int_{\overline{\Theta}_0}^{\overline{\Theta}} \overline{c_0} \frac{1}{\overline{\Theta}'} \mathrm{d}\overline{\Theta}' &= \langle N_0 \rangle + \left\langle \int_{\Theta_0}^{\Theta} \partial_\Theta N \mathrm{d}\Theta' \right\rangle = N_0 + \left\langle \int_{\overline{\Theta}_0}^{\Theta} c_0 \frac{1}{\Theta'} \mathrm{d}\Theta' \right\rangle \end{aligned} \tag{21}$$

or

$$\begin{aligned} \overline{E}_0 + c_0(\overline{\Theta} - \overline{\Theta}_0) &= E_0 + c_0(\langle \Theta \rangle - \overline{\Theta}_0) \\ \overline{N}_0 + c_0 \ln \overline{\Theta}/\overline{\Theta}_0 &= N_0 + c_0(\langle \ln \Theta \rangle - \ln \overline{\Theta}_0). \end{aligned} \tag{22}$$

The only possible solution that does not violate both scale transition constraints is that the microscale temperature is constant and consequently equal to the macroscopic one. This implies that e, η, Ψ, $\boldsymbol{P}$, c, $\boldsymbol{H}$, $\mathcal{D}_{\text{cond}}$, $\mathcal{D}_{\text{int}}$, $\mathcal{H}^{\text{ep}}$ must be calculated with respect to the macroscopic temperature $\overline{\Theta}$. So, at the microscale the problem naturally decomposes into two steps. The mechanical step is an isothermal one in which the temperature increase in each RVE point is equal to the macroscopic temperature increase. The structural heating, as well as the internal dissipation are nevertheless homogenized at the microscale, but are accounted for at the macroscale. The second step—a thermal one—is still needed in order to calculate the homogenized heat flux vector. For a more through discussion on the topic, but within the thermoelastic context, the reader is again refereed to the work [8].

3.2 *Internal Dissipation and the Microscale Mechanical Step*

As described in Sect. 3.1, the microscale solution involves the mechanical step. This part of the procedure is based on variationally consistent updates as originally developed in [15] for the general nonisothermal setting and an adaptation to isothermal setting in [10–12]. The latter research was also extended to cyclic thermoplasticity in [9, 16] and later coupled to nonisothermal damage [1]. The method was already applied to multiscale isothermal plasticity in [13] and relies on postulating a governing microscale potential $\mathcal{E}$:

$$\dot{\mathcal{E}}(\dot{\varphi}, \dot{\boldsymbol{\alpha}}, \Theta) = \int_{\mathcal{B}_0} \left(\dot{\Psi}(\dot{\varphi}, \dot{\boldsymbol{\alpha}}, \Theta) + \mathcal{D}_{\text{int}}(\dot{\boldsymbol{\alpha}})\right) \mathrm{d}V \tag{23}$$

where $\boldsymbol{\alpha}$ is a set of internal variables governing isotropic and kinematic hardening at the microscale level. Now, since $\overline{\boldsymbol{F}}$ is prescribed in the form of boundary conditions at the microlevel, from Eq. (1) follows that the microscale configuration is fully defined by the fluctuation field $\boldsymbol{w}_F$. Internal variables then follow from the minimization of the governing potential in each integration point of the microscale RVE (local level):

$$\dot{\boldsymbol{\alpha}} = \arg \inf_{\dot{\boldsymbol{\alpha}}} \dot{\mathcal{E}}(\dot{\boldsymbol{w}}_F, \dot{\boldsymbol{\alpha}}, \Theta) \mathrm{d}V \Big|_{\dot{\boldsymbol{w}}_F = \text{const}, \overline{\Theta} = \Theta = \text{const}} . \tag{24}$$

Above, temperature is kept constant in each microscopic integration point and is equal to the macroscopic temperature $\Theta = \overline{\Theta}$. The fluctuation field $\boldsymbol{w}_F$ is calculated at the global level of the microscale problem. With internal variables calculated at the local level, the fluctuation field $\boldsymbol{w}_F$ follows from minimization of the reduced potential $\dot{\mathcal{E}}_{\text{red}}$ at the global level:

$$\dot{\boldsymbol{w}}_F = \arg \inf_{\dot{\boldsymbol{w}}_F} \dot{\mathcal{E}}_{\text{red}} \tag{25}$$

where the reduced potential $\mathcal{E}_{\text{red}}$ is:

$$\dot{\mathcal{E}}_{\text{red}} = \inf_{\dot{\alpha}} \dot{\mathcal{E}}(\dot{\boldsymbol{w}}_F, \dot{\boldsymbol{\alpha}}, \Theta)\bigg|_{\dot{\boldsymbol{w}}_F=\text{const}, \overline{\Theta}=\Theta=\text{const}}. \tag{26}$$

The reduced form of the internal dissipation used in Eq. (23) at the microscale can be obtained from the microscale counterpart of Eq. (14), by applying Eq. (13):

$$\mathcal{D}_{\text{int}} = -\frac{\partial \Psi}{\partial \boldsymbol{\alpha}} \dot{\boldsymbol{\alpha}} \geq 0. \tag{27}$$

The set of internal variables $\boldsymbol{\alpha}$ can be now defined more specifically. Let $\boldsymbol{\alpha} = \{\boldsymbol{F}^{\text{p}}, \alpha^{\text{i}}\}$, where $\boldsymbol{F}^{\text{p}}$ is the plastic part of the deformation gradient that follows from the multiplicative decomposition of the deformation gradient $\boldsymbol{F} = \boldsymbol{F}^{\text{e}}\boldsymbol{F}^{\text{p}}$. Symbol α^{i} denotes a scalar variable defining isotropic hardening at the microscale level. With this choice of internal variables, and with Eq. $(13)_4$, the reduced dissipation Eq. (27) takes form:

$$\mathcal{D}_{\text{int}} = \boldsymbol{\Sigma} : \boldsymbol{L}^{\text{p}} + Q^{\text{i}}\dot{\alpha}^{\text{i}} \geq 0, \tag{28}$$

where $\boldsymbol{\Sigma} = 2\boldsymbol{C}^{\text{e}} \cdot \partial_{\boldsymbol{C}^{\text{e}}}\Psi$ is the Mandel stress, $\boldsymbol{C}^{\text{e}} = \boldsymbol{F}^{\text{e}\text{T}} \cdot \boldsymbol{F}^{\text{e}}$ is the elastic right Cauchy-Green strain tensor and $\boldsymbol{L}^{\text{p}} = \dot{\boldsymbol{F}}^{\text{p}} \cdot (\boldsymbol{F}^{\text{p}})^{-1}$ the plastic velocity gradient. Now, existence of the convex yield function ϕ is postulated as

$$\phi = \Sigma^{\text{eq}}(\boldsymbol{\Sigma}) - Q^{\text{i}} - Q_0^{\text{eq}} = 0, \tag{29}$$

where $\Sigma^{\text{eq}}(\boldsymbol{\Sigma})$ is the equivalent stress measure, $Q^{\text{i}} = -\partial_{\alpha^{\text{i}}}\Psi$ and $Q_0^{\text{eq}} = Q_0^{\text{eq}}(\Theta)$ is the initial yield stress which is influenced by the softening induced by temperature increase. It is also assumed that the equivalent stress is a positively homogeneous function of degree one:

$$\Sigma^{\text{eq}}(n\,|\boldsymbol{\Sigma}|) = n\Sigma^{\text{eq}}(\boldsymbol{\Sigma}) \tag{30}$$

for any positive real number n. This assumption is fulfilled by most yield functions, von Mises for example. Evolution laws for the internal variables are obtained as usual by invoking the principle of maximal plastic dissipation:

$$\boldsymbol{L}^{p} = \lambda\,\partial_{\boldsymbol{\Sigma}}\phi, \quad \dot{\alpha}^{\text{i}} = \lambda\,\partial_{Q^{\text{i}}}\phi, \tag{31}$$

where λ is the Lagrange multiplier. The dissipation, Eq. (28) is now:

$$\mathcal{D}_{\text{int}} = \lambda(\boldsymbol{\Sigma} : \partial_{\boldsymbol{\Sigma}}\phi + Q^{\text{i}}\partial_{Q^{\text{i}}}\phi). \tag{32}$$

Finally, the assumption of the positively homogeneous yield function of degree one Eq. (30) and the prototype yield function Eq. (29) give by the application of Euler's homogeneous function theorem very simple result:

$$\mathcal{D}_{\text{int}} \overset{\phi=0}{=} \lambda Q_0^{\text{eq}}. \tag{33}$$

Note that extension to kinematic hardening is straightforward, see [1, 10]. The thermodynamic consistency with respect to the internal dissipation is now simply enforced as:

$$\overline{\mathcal{D}}_{\text{int}} = \langle \mathcal{D}_{\text{int}} \rangle = Q_0^{\text{eq}}(\overline{\Theta}) \langle \lambda \rangle \,, \tag{34}$$

since the microscale temperature field is homogeneous and equal to the macroscopic one as described in Sect. 3.1. Thus, to determine the macroscale dissipation, homogenization of internal variables $\overline{\alpha}$ and conjugated quantities $\overline{Q}^{\text{i}}$ is not needed.

The remaining details about the stress averaging follow closely [8], so only main points will be mentioned. Due to small RVE size assumption, body forces and inertial terms at the microscale can be neglected, resulting in

$$\text{DIV}\boldsymbol{P} = 0 \tag{35}$$

what leads to:

$$\langle \boldsymbol{P} \cdot \boldsymbol{F} \rangle = \langle \boldsymbol{P} \rangle \cdot \langle \boldsymbol{F} \rangle \,, \quad \overline{\boldsymbol{P}} = \langle \boldsymbol{P} \rangle \,, \quad \overline{\boldsymbol{F}} = \langle \boldsymbol{F} \rangle \,, \tag{36}$$

which provides the needed link between the microscopic and the macroscopic stresses.

At the end, the consistency of the specific mass should be addressed. As shown in [8, 17], the condition:

$$\det \overline{\boldsymbol{F}} = \frac{1}{|\mathcal{B}_0|} \int_{\mathcal{B}_0} \det \boldsymbol{F} \mathrm{d}V = \langle \det \boldsymbol{F} \rangle_{\mathcal{B}_0}, \tag{37}$$

leads to:

$$\bar{\rho} = \langle \rho \rangle_{\mathcal{B}}, \quad \bar{\rho}_0 = \langle \rho_0 \rangle_{\mathcal{B}_0}, \quad \bar{\rho} = \langle \det \boldsymbol{F} \rangle_{\mathcal{B}_0} \bar{\rho}_0. \tag{38}$$

Temizer and Wriggers show that the mass criterion is fulfilled for the linear deformation and periodic boundary conditions [17]. For uniform traction the criterion does not hold in general, but order of violation decreases as the sample size increases.

3.3 Dissipation due to Heat Conduction and the Microscale Thermal Step

In order to assure that the model is thermodynamically consistent with respect to the dissipation due to heat conduction, according to Eqs. $(11)_1$ and $(16)_5$, it must be:

$$\overline{\boldsymbol{H}} \cdot \overline{\boldsymbol{G}} = \langle \boldsymbol{H} \cdot \boldsymbol{G} \rangle \,. \tag{39}$$

However, to determine evolution of temperature at the macroscale Eq. (15), the divergence of the heat flux vector $\boldsymbol{H}$ is needed. The heat flux vector can be conveniently obtained if $\langle \boldsymbol{H} \cdot \boldsymbol{G} \rangle = \langle \boldsymbol{H} \rangle \cdot \langle \boldsymbol{G} \rangle$ could be proved. The latter equation is valid if divergence of the heat flux vector at the microscale vanishes, $\mathrm{DIV}\boldsymbol{H} = 0$, see the discussion in [8]. This turns out to be a potentially difficult constraint to fulfil for certain classes of thermoplasticity problems.

To demonstrate these issues, consider the temperature evolution in a point at the microscale:

$$c\dot{\Theta} = -\mathrm{DIV}\boldsymbol{H} + \rho_0 Q + \mathcal{D}_{\mathrm{int}} + \Theta \mathcal{H}^{\mathrm{ep}}, \tag{40}$$

where

$$\mathcal{H}^{\mathrm{ep}} = \left(\partial_\Theta \boldsymbol{P} : \dot{\boldsymbol{F}} - \partial_\Theta Q^{\mathrm{i}} \dot{\alpha}^{\mathrm{i}} \right). \tag{41}$$

This is a standard equation used in single-scale thermoplasticity and augmented with usual boundary conditions. Heat convection and radiation usually do not play any role at the microscale level so they are dropped out. Furthermore, it can be also assumed:

- The heat source at microscale is absent, $Q = 0$.
- The term $\partial_\Theta Q^{\mathrm{i}} \dot{\alpha}^{\mathrm{i}}$ is responsible for possible temperature dependency of the plastic hardening potential and is frequently disregarded in single-scale plasticity. Its contribution at the microscale can be also safely disregarded in most cases.
- The transient term $c\dot{\Theta}$ is also very small in most circumstances. The microscale temperature changes can be considered as instantaneous compared to the macroscale transient problem.
- The thermoelastic term $\partial_\Theta \boldsymbol{P} : \dot{\boldsymbol{F}}$ is frequently neglected in the single-scale thermoplasticity. However, in some situations, like cyclic thermoplasticity [1], this term can be somewhat important. If this is the case, acceptability of the error introduced by neglecting this term at the microscale should be carefully studied. In thermoplasticity problems that are triggered by the external heat supply or by means of convection or radiation at the macroscale, these issues can be disregarded.
- The internal dissipation term $\mathcal{D}_{\mathrm{int}}$ is usually more critical than the thermoelastic term. Same guidance as for the thermoelastic term is recommended here as well.

It should be emphasized that although the thermoelastic, thermoplastic and internal dissipation terms have been neglected at the microscale thermal step, they are accounted for at the macroscale. In [7] these are regarded as external sources of heat. Hence the thermodynamical consistency Eq. $(16)_{4,6}$ requires:

$$\begin{aligned} \overline{\mathcal{D}}_{\mathrm{int}} &= \langle \mathcal{D}_{\mathrm{int}} \rangle = \frac{1}{|\mathcal{B}_0|} \int_{\mathcal{B}_0} \lambda Q_0 \mathrm{d}V \\ \overline{\mathcal{H}}^{\mathrm{ep}} &= \langle \mathcal{H}^{\mathrm{ep}} \rangle = \frac{1}{|\mathcal{B}_0|} \int_{\mathcal{B}_0} \left(\partial_\Theta \boldsymbol{P} : \dot{\boldsymbol{F}} - \partial_\Theta Q^{\mathrm{i}} \dot{\alpha}^{\mathrm{i}} \right) \mathrm{d}V, \end{aligned} \tag{42}$$

what follows from Eqs. (33) and (41). Finally, with above assumptions introduced, the temperature equation Eq. (40) collapses to the steady-state problem $\mathrm{DIV}\boldsymbol{H} = 0$, fulfilling the criteria for separation $\langle \boldsymbol{H} \cdot \boldsymbol{G} \rangle = \langle \boldsymbol{H} \rangle \cdot \langle \boldsymbol{G} \rangle$. Consequently, the heat flux vector $\boldsymbol{H}$ can be conveniently obtained by the averaging procedure $\overline{\boldsymbol{H}} = \langle \boldsymbol{H} \rangle$.

The microscale temperature fluctuation field obtained in this manner is not relevant. Its solely purpose is to introduce the temperature gradient resulting in $\boldsymbol{H}(\boldsymbol{F}, \boldsymbol{G}, \overline{\Theta}) = -\boldsymbol{k} \cdot \text{GRAD}\, \Theta$. Possible temperature dependency of conductivity tensor should involve the macroscopic temperature $\boldsymbol{k} = \boldsymbol{k}(\overline{\Theta})$, according to the discussion in Sect. 3.1.

A final note regarding determination of the macroscopic specific heat capacity $\bar{c}$. Although this term is commonly calculated by simple averaging of the microscopic counterpart $\bar{c} = \langle c \rangle$, such a procedure is not thermodynamically justified. Starting from the analysis [8] and here extended for the plasticity effects, the correct procedure should be:

$$\begin{aligned}\bar{c}(\overline{\boldsymbol{F}}, \overline{\Theta}) &= \langle \partial_{\boldsymbol{F}} E \partial_{\overline{\Theta}} \boldsymbol{F} + \partial_{\boldsymbol{\alpha}} E \partial_{\overline{\Theta}} \boldsymbol{\alpha} + \partial_{\overline{\Theta}} E \rangle \\ &= \langle (\partial_{\boldsymbol{F}} E_0 + \textstyle\int_{\overline{\Theta}_0}^{\overline{\Theta}} \partial_{\boldsymbol{F}} c \, \mathrm{d}\overline{\Theta}') \partial_{\overline{\Theta}} \boldsymbol{F} + (\partial_{\boldsymbol{\alpha}} E_0 + \textstyle\int_{\overline{\Theta}_0}^{\overline{\Theta}} \partial_{\boldsymbol{\alpha}} c \, \mathrm{d}\overline{\Theta}') \partial_{\overline{\Theta}} \boldsymbol{\alpha} + c(\boldsymbol{F}, \overline{\Theta}).\end{aligned} \tag{43}$$

Remark 1 Since the heat flux depends on the macroscopic deformation gradient and the macroscopic temperature gradient, two sets of boundary conditions should be specified in the microscale thermal step. In this case, periodic boundary conditions were selected for both deformation and temperature gradients. For an overview and performance of other possible combinations, see [8].

4 Numerical Aspects

Apart from the above presented multiscale framework, numerical details of analysis steps involve techniques already described in the literature, so only a short overview of these will be presented. The whole framework was implemented into finite element software Abaqus by means of newly developed `UMAT` and `UMATHT` subroutines. In brief, the solution process in each increment is separated into two steps, the macroscale step and the microscale step.

At the macroscale, the procedure is decoupled by the standard isothermal split into a mechanical and a thermal phase in all integration points. In the mechanical phase the `UMAT` is invoked. Based on the supplied macroscale deformation gradient and the macroscale temperature, another instance of Abaqus solves the isothermal microscale problem. Upon solution of the microscale problem, the homogenized stresses, the dissipation and the structural heating are obtained. The solution procedure requires macroscopic tangent operators as well. These are obtained by means of numerical differentiation as described in [18].

The thermal phase utilizes the dissipation and the structural heating obtained in the mechanical step and is coded in a `UMATHT` subroutine. Again, to obtain the homogenized heat flux vector, the steady state heat analysis at the microscale is invoked. The homogenized heat flux vector is obtained. As far as the thermal tangent operator is concerned, the identical procedure used in the thermoelastic case as described in [8] was applied here as well.

Thus, two microscale problems are solved—the mechanical and the steady-state heat transfer. The numerical treatment of the mechanical step described in Sect. 3.2 closely follows numerical procedures described in [10–13]. To obtain the set of internal variables $\boldsymbol{\alpha}_{\mathrm{n+1}} = \left\{\boldsymbol{F}^{\mathrm{p}}_{\mathrm{n+1}}, \alpha^{\mathrm{i}}_{\mathrm{n+1}}\right\}$ at some particular time instant $t_{\mathrm{n+1}}$, the set is rephrased as $\boldsymbol{\alpha}_{\mathrm{n+1}} = \left\{\Delta\lambda_{\mathrm{n+1}}, \tilde{\boldsymbol{\Sigma}}_{\mathrm{n+1}}\right\}$. Here $\Delta\lambda_{\mathrm{n+1}} = \int_{t_{\mathrm{n}}}^{t_{\mathrm{n}}+1} \lambda \mathrm{d}t$, while $\tilde{\boldsymbol{\Sigma}}_{\mathrm{n+1}}$ are so-called pseudo-stresses. The in-depth presentation of the concept of pseudo stresses can be found in [10–12]. In short, main property of the pseudo stresses is that they result in the same flow direction as the usual stress tensor. The advantage of such approach is that the plastic flow constraints are fulfilled in advance. Time discretization of the potential in Eq. (23) now leads towards an incremental potential:

$$I_{\mathrm{inc}}(\boldsymbol{w}_{F,\mathrm{n+1}}, \Delta\lambda_{\mathrm{n+1}}, \tilde{\boldsymbol{\Sigma}}_{\mathrm{n+1}}) = \int_{t_{\mathrm{n}}}^{t_{\mathrm{n}}+1} \int_{\mathcal{B}_0} \dot{\mathcal{E}} \mathrm{d}V \mathrm{d}t \approx \int_{t_{\mathrm{n}}}^{t_{\mathrm{n}}+1} \int_{\mathcal{B}_0} (\Psi_{\mathrm{n+1}} - \Psi_{\mathrm{n}} + \mathcal{D}_{\mathrm{int,n+1}}) \mathrm{d}V \mathrm{d}t. \tag{44}$$

Minimization of the incremental potential at the microscale local level will provide a nonlinear system at fixed configuration $\boldsymbol{w}_{F,\mathrm{n+1}} = \mathrm{const}$:

$$\delta_{\Delta\lambda_{\mathrm{n+1}}} I_{\mathrm{inc}} = 0, \quad \delta_{\tilde{\boldsymbol{\Sigma}}_{\mathrm{n+1}}} I_{\mathrm{inc}} = 0. \tag{45}$$

When the internal variables are determined from the above system, the microscale fluctuation field can be determined by minimization of the discrete counterpart $I_{\mathrm{red,inc}}$ of the reduced potential $\mathcal{E}_{\mathrm{red}}$, Eq. (26):

$$\boldsymbol{w}_{F,\mathrm{n+1}} = \arg \inf_{\boldsymbol{w}_{F,\mathrm{n+1}}} I_{\mathrm{red,inc}} \tag{46}$$

A second UMAT subroutine is developed to implement this step. Periodic boundary conditions were used; these were prepared at the macroscale and then supplied to the microscale analysis.

The microscale thermal step is a steady-state heat transfer analysis. As already mentioned, periodic boundary conditions were considered here for both deformation gradient and temperature. They were also prepared in the macroscale thermal step and supplied to the microscale analysis.

5 Example

The present example considers bending of a cantilever plate subjected to the enforced rotation at the free end. Thus, the problem at hand is mechanically driven and generated heat is solely due to elastic and plastic deformation. Heat exchange with the environment, by means of convection or radiation is neglected, resulting in a thermally insulated system.

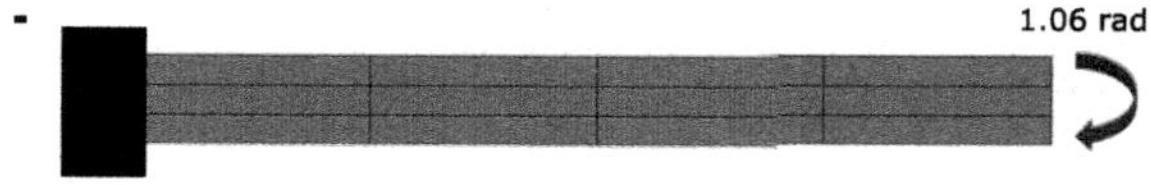

Fig. 1 The macroscale finite element mesh of the cantilever plate

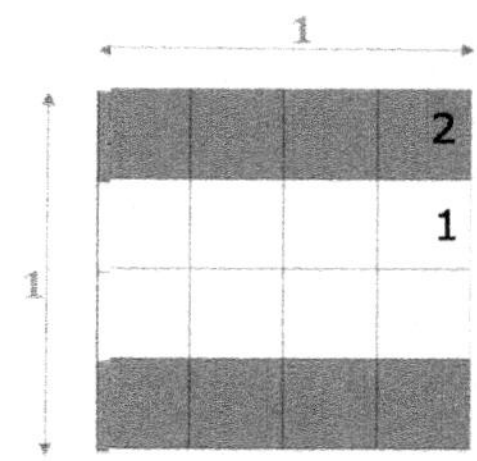

Fig. 2 Microscale finite element mesh. Numbers 1 and 2 indicate component number

The maximum rotation is prescribed as 1.06 rad and assumed to be linearly increasing in time. The total duration of the deformation process is 7.05 s. The plate is 1 m long and 0.1 m thick. Plane strain conditions are assumed. The problem is taken to be both geometrically and materially nonlinear.

The finite element mesh at the macroscale consists of 12 `CPE4RT` finite elements, Fig. 1. These finite elements account for thermomechanical coupling, plane strain conditions, bilinear displacements and temperature and reduced integration with hourglass control. Reduced integration has a beneficial effect on known numerical issues in plasticity problems [19]. Additionally, since the microscale problem is solved at each macroscale integration point, the procedure also reduces the computational burden due to lower number of integration points involved. At the microscale level, the mesh (Fig. 2) consisted of 16 `CPE4H` elements, suitable for plane strain conditions with 4 integration points in the mechanical phase and `DC2D4` elements in the thermal phase. At the RVE boundaries, periodic boundary conditions were enforced as described in Sect. 4. The microscale problem is solved in 5 equal time increments. Material properties of each microscale component is given in Table 1.

In the problem at hand, the Helmholtz free energy is considered to be of the form:

$$\Psi = W\left(\overline{\boldsymbol{C}}^{\mathrm{e}}\right) + U\left(J\right) + M\left(J, \Theta\right) + \Psi^{\mathrm{p,i}}\left(\alpha^{\mathrm{i}}\right) + T\left(\Theta\right), \tag{47}$$

$$W\left(\overline{\boldsymbol{C}}^{\mathrm{e}}\right) = \tfrac{1}{2}\mu\left[\mathrm{tr}\left(\overline{\boldsymbol{C}}^{\mathrm{e}}\right) - 3\right], \tag{48}$$

$$U\left(J^{\mathrm{e}}\right) = \tfrac{1}{2}\kappa\left[\tfrac{1}{2}\left(\left(J^{\mathrm{e}}\right)^2 - 1\right) - \ln J^{\mathrm{e}}\right], \tag{49}$$

$$M\left(J, \Theta\right) = \left(\Theta - \Theta_0\right)\left[-3\,\alpha\, U'\left(J\right)\right], \tag{50}$$

$$\Psi^{\mathrm{p,i}} = \tfrac{1}{2}H(\alpha^{\mathrm{i}})^2, \tag{51}$$

$$T\left(\Theta\right) = c_0\left[\left(\Theta - \Theta_0\right) - \Theta \ln \tfrac{\Theta}{\Theta_0}\right], \tag{52}$$

where $\overline{\boldsymbol{C}}^{\mathrm{e}} = (\det \boldsymbol{F}^{\mathrm{e}})^{-2/3}\boldsymbol{C}^{\mathrm{e}} = (J^{\mathrm{e}})^{-2/3}\boldsymbol{C}^{\mathrm{e}}$. The yield function is assumed to be of standard von Mises type:

Table 1 Material properties of microscale components

Property	Component 1	Component 2
Young's modulus E	200 GPa	220 GPa
Posson's ratio ν	0.3	0.3
Yield stress y_0	380 MPa	400 MPa
Linear hardening coefficient H	700 MPa	1000 MPa
Coefficient of thermal expansion α	1.3×10^{-5} 1/°C	1.6×10^{-5} 1/°C
Conductivity k	30 W/Km	55 W/Km
Specific heat c	400 J/kgK	600 J/kgK
Mass density ρ	7800 kg/m^3	8000 kg/m^3

$$\phi = \left\| \left(\boldsymbol{\Sigma}^{\text{dev}} \right) \right\| - Q_{\text{i}} - Q_0^{\text{eq}} = 0, \tag{53}$$

where the initial yield stress is a linearly softening function:

$$Q_0^{\text{eq}}(\Theta) = y_0\left(\Theta_0\right)\left[1 - \omega_0\left(\Theta - \Theta_0\right)\right]. \tag{54}$$

Heat conduction of each component is governed by the Fourier's law of heat conduction:

$$\boldsymbol{H} = -k \text{ GRAD } \Theta. \tag{55}$$

Evolution of temperature is given in Fig. 3. In the first step, strains are elastic so almost a constant temperature is observed (barely visible temperature drop takes place at the top of the plate and increase at the bottom). The drop is caused by the tensile stresses and accompanying expansion of material. Similarly, in the central

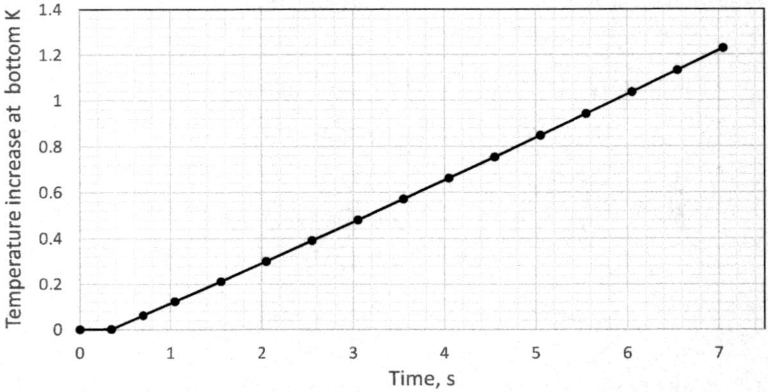

Fig. 3 Temperature versus time at the midpoint of the plate's bottom

point at the bottom of the plate, thermoelastic temperature increase takes place due to compressive character of the normal stress. In subsequent steps plastic deformation evolves and heat release due to mechanical dissipation dominates temperature evolution (in the linear manner).

Distribution of stresses at the macroscale, Fig. 4 shows that stresses are higher near the centre of the plate. Von Mises stresses seem to be rather equal on the top and the bottom surface. At the microscale level, Fig. 5, distribution is pretty much regular and follows the distribution of each component. Throughout all analysis, the

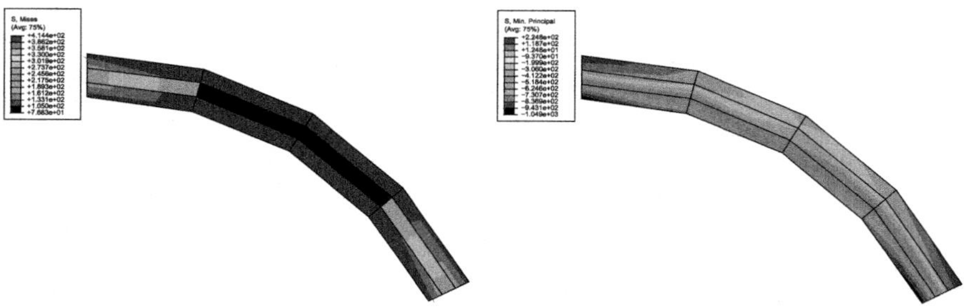

Fig. 4 Distribution of stresses at the macroscale. Left: von Mises stress (max. 414 MPa); Right: minimal principal stresses (min. −1049 MPa)

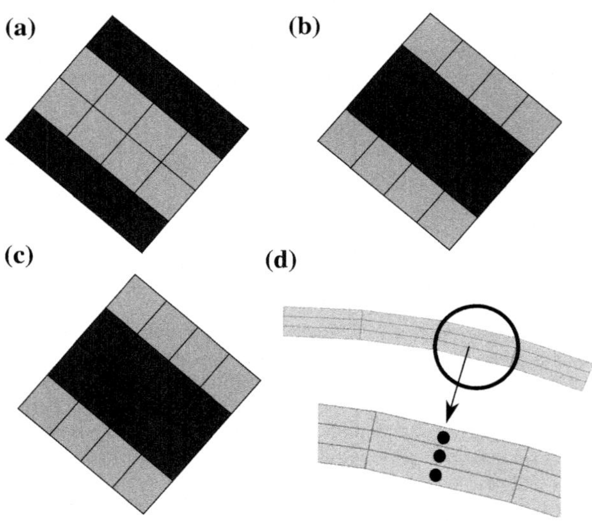

Fig. 5 Microscale stresses. **a** Maximal principal stresses at the top row of finite elements of the plate (dark colour: 203, light colour: 271 MPa, plastic stress state); **b** von Mises stress in the middle of the plate (dark: 78.1, light: 78.5 MPa, elastic stress state), **c** Minimal principal stresses at the bottom row of the plate (dark: −510, light: −436 MPa, plastic stress state), **d** Positions of RVEs in (a–c) at the plate

centre of the plate remains in the elastic state, while representative volume elements positioned at the top and the bottom of the plate exhibit plastic deformation. Also, if absolute values of maximal and minimal principal stresses are considered, the higher stresses occur in the compressive zone, i.e. at the bottom surface of the plate.

6 Conclusions

The chapter presented a procedure suitable for the homogenization in thermoplasticity. The starting point for the current methodology is an earlier work on multiscale thermoelasticity [8]. Such a selection is motivated by very careful elaborations ensuring thermodynamical consistency in the latter paper. The transition between scales takes special care to avoid possible pitfalls frequently met in other papers on the homogenization in thermomechanics. The present research extended this approach to the thermoplasticity regime. It is placed into the finite strain setting and consequently suitable for both geometrically and materially nonlinear problems. Nevertheless, it should be emphasised that some limitations do arise and that the cases where the dissipation term strongly dominates heat production the caution is advocated. The performance of the proposed scheme is verified on an example that was implemented into finite element software Abaqus by means of user subroutines. Obtained solutions support anticipated behaviour.

Acknowledgements This work has been partially supported by Croatian Science Foundation under the project no. 6876—Assessment of structural behaviour in limit state operating conditions. This support is gratefully acknowledged.

References

1. Canadija, M., Mosler, J.: A variational formulation for thermomechanically coupled low cycle fatigue at finite strains. Int. J. Solids Struct. **100**, 388–398 (2016)
2. Canadija, M., Brnic, J.: Associative coupled thermoplasticity at finite strain with temperature-dependent material parameters. Int. J. Plast. **20**, 1851–1874 (2004)
3. Simo, J., Miehe, C.: Associative coupled thermoplasticity at finite strains: formulation, numerical analysis and implementation. Comput. Methods Appl. Mech. Eng. **98**, 41–104 (1992)
4. Miehe, C., Schröder, J., Schotte, J.: Computational homogenization analysis in finite plasticity simulation of texture development in polycrystalline materials. Comput. Methods Appl. Mech. Eng. **171**(3), 387–418 (1999)
5. Celigoj, C.: Thermomechanical homogenization analysis of axisymmetric inelastic solids at finite strains based on an incremental minimization principle. Int. J. Numer. Methods Eng. **71**(1), 102–126 (2007)
6. Özdemir, I., Brekelmans, W., Geers, M.G.: FE2 computational homogenization for the thermo-mechanical analysis of heterogeneous solids. Comput. Methods. Appl. Mech. Eng. **198**(3), 602–613 (2008)
7. Sengupta, A., Papadopoulos, P., Taylor, R.L.: A multiscale finite element method for modeling fully coupled thermomechanical problems in solids. Int. J. Numer. Methods Eng. **91**(13), 1386–1405 (2012)

8. Temizer, I., Wriggers, P.: Homogenization in finite thermoelasticity. J. Mech. Phys. Solids **59**(2), 344–372 (2011)
9. Canadija, M., Mosler, J.: On the thermomechnaical coupling in finite strain plasticity theory with non-linear kinematic hardening by means of incremental energy minimization. Int. J. Solids Struct. **48**(78), 1120–1129 (2011)
10. Mosler, J.: Variationally consistent modeling of finite strain plasticity theory with non-linear kinematic hardening. Comput. Methods Appl. Mech. Eng. **199**, 2753–2764 (2010)
11. Mosler, J., Bruhns, O.: Towards variational constitutive updates for non-associative plasticity models at finite strain: models based on a volumetric-deviatoric split. Int. J. Solids Struct. **46**, 1676–1685 (2009)
12. Mosler, J., Bruhns, O.: On the implementation of rate-independent standard dissaptive solids at finite strain—variational constitutive updates. Comput. Methods Appl. Mech. Eng. **199**, 417–429 (2010)
13. Bleier, N., Mosler, J.: Efficient variational constitutive updates by means of a novel parameterization of the flow rule. Int. J. Numer. Methods Eng. **89**(9), 1120–1143 (2012)
14. Lemaitre, J., Chaboche, J.L.: Mechanics of Solid Materials. Cambridge University Press, Cambridge (2000)
15. Yang, Q., Stainier, L., Ortiz, M.: A variational formulation of the coupled thermo-mechanical boundary-value problem for general dissipative solids. J. Mech. Phys. Solids. **54**, 401–424 (2006)
16. Bartels, A., Bartel, T., Canadija, M., Mosler, J.: On the thermomechanical coupling in dissipative materials: a variational approach for generalized standard materials. J. Mech. Phys. Solids **82**, 218–234 (2015)
17. Temizer, I., Wriggers, P.: On a mass conservation criterion in micro-to-macro transitions. J. Appl. Mech. **75**(5), 054503 (2008)
18. Miehe, C.: Numerical computation of algorithmic (consistent) tangent moduli in large-strain computational inelasticity. Comput. Methods Appl. Mech. Eng. **134**(3), 223–240 (1996)
19. Simo, J., Hughes, T.: Computational Inelasticity. Springer, New York (1998)

A Method of Numerical Viscosity Measurement for Solid-Liquid Mixture

Reika Nomura, Kenjiro Terada, Shinsuke Takase and Shuji Moriguchi

Abstract We present a space-time homogenization procedure for multiscale modeling of solid-liquid mixture. The derived mathematical model enables us to set up two separate governing equations at both macro- and micro-scales. The fluid in the macroscopic governing equation is teated as an equivalent homogeneous medium with average or homogenized viscosity and is regarded as an incompressible Newtonian fluid, whose motion is assumed to be governed by the Navier-Stokes equations. The microscopic equations of motion governing the coupling phenomenon of the fluid and solid particles in a certain local domain and are solved to determine the microscopic flow fields under adequate boundary and loading conditions. Then the macrosopic viscosity is determined as the quantity averaged over the microscopic domain and within a certain time interval. The numerical viscosity measurement (NVM) can be realized by this space-time homogenization procedure. A set of NVMs is presented to demonstrate that the solid-liquid mixture considered in this study possibly exhibits a macroscopic flow characteristics of a special type of non-Newtonian fluids.

R. Nomura (✉) · S. Takase
Department of Civil Engineering, Tohoku University, 468-1 Aza-Aoba, Aramaki, Aoba, Sendai 980-0845, Japan
e-mail: reika.nomura.q4@dc.tohoku.ac.jp

S. Takase
e-mail: takase@irides.tohoku.ac.jp

K. Terada · S. Moriguchi
International Research Institute of Disaster Science (IRIDeS), Tohoku University, 468-1 Aza-Aoba, Aramaki, Aoba, Sendai 980-0845, Japan
e-mail: tei@irides.tohoku.ac.jp

S. Moriguchi
e-mail: s_mori@irides.tohoku.ac.jp

J. Sorić et al. (eds.), *Multiscale Modeling of Heterogeneous Structures*, Lecture Notes in Applied and Computational Mechanics 86, https://doi.org/10.1007/978-3-319-65463-8_17

1 Introduction

Huge number of lives were lost in the tsunami induced by the Great East Japan Earthquake occurred on 11 March 2011. This tsunami, which is expressed and known as "unexpected scale", was an L2 level that indicates an extremely low frequency event to be realized once in several thousand years. Although the predictions of hitting time and wave height of next coming tsunamis cannot be ignored, the characterization of overall flow properties is also a matter of importance. It is indeed pointed out that the damage and delay of recovery in this disaster were caused by the transport of various types of suspended solids or sediments [18], whose effects on tsunami's flow characteristics have gained recognition accordingly. As the flow involving such transport phenomena exhibits a fully three-dimensional two-phase flow characteristics, the standard two-dimensional single-phase flow simulations are inadequate. However, most of tsunami simulations so far have treated tsunami as a single-phase flow and focused on the maximum wave height or tsunami run-up areas. It is, therefore, essential to comprehensively characterize the mechanical behavior of a solid-liquid mixture, when we are concerned with damage prediction by tsunamis.

To take into account the behavior of suspended solids or solid particles, a variety of numerical methods have been developed in the field of computational engineering and science and can be classified into two types according to the spatial scales to describe physical problems under consideration. One of them is a coarse or macroscopic approach to simulate the overall behavior of a single phase flow with experimental or semi-experimental formula describing the effects of motions of suspended solids [9, 10]. Although this approach is capable of dealing with real scale simulations for tsunami disasters over large areas, the underlying mechanism of the actual motions of numerous solid particles can merely be reflected in the characterization of flow properties. In contrast, the other is a direct treatment of motions of solid particles at fine or microscopic scale and often referred to as a direct numerical simulation (DNS). Although the effects of moving particles on the overall fluid motion can be directly taken into account in numerical simulations, DNS requires high computational costs. In fact, DNS has a limitation in the size of analysis regions due to the necessity of high resolution of meshes or grids and, therefore, is not suitable for disaster simulations in practical levels.

To reflect the microscopic coupled mechanisms between a fluid and solid particles in the macroscopic flow simulations for damage prediction by tsunamis, this study presents our first trial for a space-time homogenization procedure for multiscale modeling of solid-liquid mixture. By introducing two separate spatial scales to represent the motions of a solid-liquid mixture, we provide two mathematical models at both macro- and micro-scales. The fluid in the macroscopic governing equation is teated as an incompressible homogeneous Newtonian fluid with average or homogenized viscosity. Also, the microscopic problem describes the motion governing the coupling phenomenon of the fluid and suspended solid particles in a certain local domain and its solution under adequate boundary and loading conditions is averaged over over the microscopic domain and within a certain time interval to evaluate the macrosopic viscosity. This averaged procedure, which is called space-time homogenization, is originally introduced in this study and enables us to realize the numerical viscosity measurement (NVM) for evaluating the macroscopic viscosity. A set of NVMs on an ideal solid-liquid mixture is presented to demonstrate that a macroscopic flow characteristics of a special type of non-Newtonian fluids can be reproduced. In this particular study, we employ the stabilized finite cover method (FCM) [14, 15] to simulate the microscopic flow coupled with motions of solid particles whose equations are solved by the distinct element method (DEM) [4, 5].

2 Multiscale Modeling of Solid-Liquid Mixture

2.1 Separation of Spatial and Temporal Scales

We consider a coupled problem of a solid-liquid mixture as shown in Fig. 1a, which contains numerous solid particles at micro-scale as depicted in Fig. 1b. Denoting l and L by representative lengths for the microscopic or macroscopic flow fields, respectively, we introduce a parameter $\epsilon = l/L$ to express the dependency of the overall domain on the microscopic domain as Ω^ϵ, into which the macroscopic coordinate system $\boldsymbol{x} = (x_1, x_2, x_3)$ is introduced. The microscopic domain is regarded as a representative volume element (RVE) and is denoted by $Y = Y_\text{f} \cup Y_\text{s}$ where Y_f and $Y_\text{s} = \sum Y_\text{s}^i$ $(i = 1, 2, \ldots)$ are the domains of the fluid and solid particles, respectively, along with their boundary surfaces denoted by ∂Y_f and ∂Y_s. To describe the motions of the fluid and solid particles at micro-scale, the microscopic coordinate system $\boldsymbol{y} = (y_1, y_2, y_3)$ is introduced in Y.

In this study, the liquid is assumed to be an incompressible Newtonian fluid, while each solid particle is assumed to be a rigid body that has a smooth external surface. Then, we assume that the overall flow can be represented by the flow of a homogeneous liquid equivalent to the original liquid in Ω^ϵ. Each material point $\boldsymbol{x}$ in the macroscopic domain of this equivalent homogeneous fluid, which is denoted by Ω, can be identified with the corresponding microscopic domain Y; see Fig. 1c. Therefore, the macroscopic flow properties are expected to be characterized by averaging the microscopic motions of the fluid and solid particles in a way similar to that of mathematical homogenization theory [1, 7, 12]. However, it should be noted that the time variation of the microscopic flow field is more rapid than that of the macroscopic one. Also, the non-stationary state at micro-scale cannot be averaged within a certain interval of time so as to become a stationary states. In this regard, to distinguish between the non-stationary and rapidly varying microscopic flow field and the macroscopically stationary one, we introduce separate time variables, t and τ, both of which have the same time scale, but are assumed to have different resolutions. That is, t is a coarse time scale to measure the macroscopic flow, while τ is the fine time scale to measure the microscopic one.

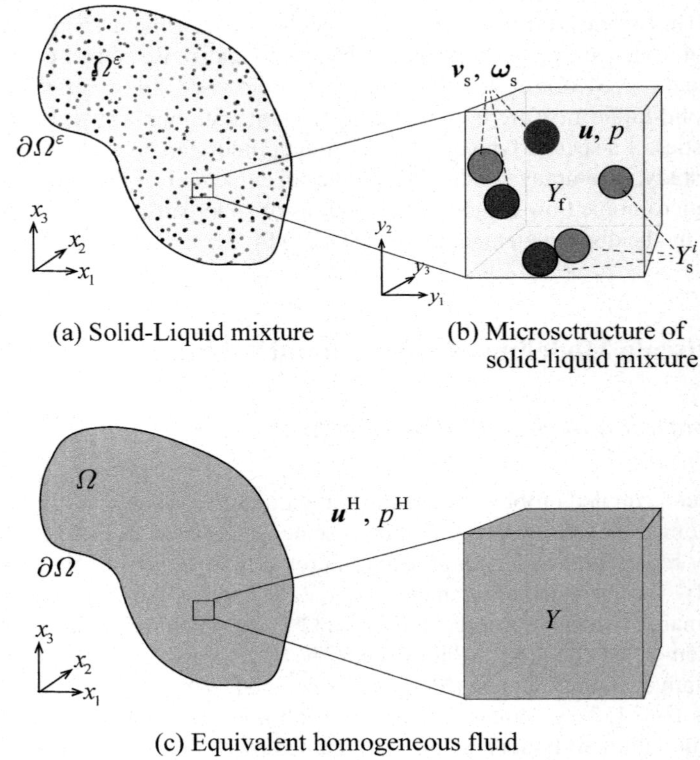

Fig. 1 Solid-fluid mixture and its equivalent homogeneous fluid

In the subsequent sections, with the setting described above, we present a two-scale initial-boundary value problem that is composed of micro- and macroscopic problems.

2.2 *Microscopic Problem*

We describe the liquid and solid motions in the microscopic domain with the Eulerian and Lagrangian descriptions, respectively. Since each microscopic domain corresponds to a macroscopic material point $\boldsymbol{x}$, all the microscopic variables in Y depend on not only the microscopic coordinate $\boldsymbol{y}$ and microscopic time τ, but also the coordinate macroscopic $\boldsymbol{x}$ and macroscopic time t. For example, the liquid velocity and pressure are expressed as $\boldsymbol{u}(\boldsymbol{x}, t; \boldsymbol{y}, \tau)$ and $p(\boldsymbol{x}, t; \boldsymbol{y}, \tau)$, respectively, to clearly specify such dependencies.

The liquid motion is described by the following Navier-Stokes and continuity equations as usual:

$$\left.\begin{aligned} &\rho_{\mathrm{f}}\left(\frac{\partial \boldsymbol{u}}{\partial \tau} + \boldsymbol{u}\cdot\nabla_y \boldsymbol{u}\right) - \nabla_y \cdot \boldsymbol{\sigma}(\boldsymbol{u}, p) = \boldsymbol{0} \\ &\nabla_y \cdot \boldsymbol{u} = 0 \end{aligned}\right\} \quad \text{in} \quad Y_{\mathrm{f}} \tag{1}$$

where ∇_y is the gradient operator with respect to the microscopic coordinate $\boldsymbol{y}$, ρ_{f} and $\boldsymbol{\sigma}(\boldsymbol{x}, t; \boldsymbol{y}, \tau)$ represent the fluid's mass density and the stress acting on the fluids, respectively. Here, the body force is missing, as it is negligibly small at micro-scale. Assuming a Newtonian-fluid at micro-scale, we have the following constitutive equation:

$$\boldsymbol{\sigma} = -p\boldsymbol{I} + 2\mu\boldsymbol{D} \tag{2}$$

along with the shear flow velocity $\boldsymbol{D}$ defined as the symmetric and deviatoric component of the velocity gradient. Here, μ is the viscosity, which is assumed be constant at microscale, $\boldsymbol{I}$ is the 2nd order identigy tensor. Together with appropriate boundary and initial conditions, the microscopic problem can be well-posed to be solved without any difficulty.

On the other hand, the motion of a solid particle at micro-scale is identified with the current position $\boldsymbol{y}_{\mathrm{s}}(\boldsymbol{Y}_{\mathrm{s}}, \tau)$ associated with its initial position $\boldsymbol{Y}_{\mathrm{s}}$ so that the displacement and velocity can be defined as $\boldsymbol{d}_{\mathrm{s}} = \boldsymbol{y}_{\mathrm{s}} - \boldsymbol{Y}_{\mathrm{s}}$ and $\dot{\boldsymbol{d}}_{\mathrm{s}}$. Dividing the velocity into the translational and rotational components, $\boldsymbol{v}_{\mathrm{s}}$ and $\boldsymbol{\omega}_{\mathrm{s}}$, we can write the equations of motions as

$$\frac{d(m_s \boldsymbol{v}_s)}{dt} = \boldsymbol{F}_{\mathrm{s}} \tag{3}$$

$$\frac{d(\boldsymbol{J}\boldsymbol{\omega}_{\mathrm{s}})}{dt} = \boldsymbol{T} \tag{4}$$

where m_s is the particle's mass density, $\boldsymbol{J}$ is the inertia moment tensor and $\boldsymbol{\omega}$ is the angular velocity vector. Here, $\boldsymbol{F}_{\mathrm{s}}$ and $\boldsymbol{T}$ are respectively the force and torque acting on the particle with respect to the centroid. In addition, $\boldsymbol{F}_{\mathrm{s}}$ in Eq. (3) is expressed as the summation of the contact force $\boldsymbol{F}_{\mathrm{c}}$ and the fluid force $\boldsymbol{F}_{\mathrm{f}}$ such that

$$\boldsymbol{F}_{\mathrm{s}} = \sum \boldsymbol{F}_{\mathrm{c}} + \boldsymbol{F}_{\mathrm{f}} \tag{5}$$

Here, $\boldsymbol{F}_{\mathrm{f}}$ is the external force applied by the surrounding fluid and is computed as

$$\boldsymbol{F}_{\mathrm{f}} = \int_{\partial Y_{\mathrm{s}}} \boldsymbol{\sigma} \cdot \boldsymbol{n} dS \tag{6}$$

where ∂Y_s represents the interface between the solid particle and the fluid and $\boldsymbol{n}$ is the unit normal vector on it. Endowed with appropriate boundary, initial and frictional-contact conditions, the microscopic problem for solid particles reaches completion by DEM.

2.3 Macroscopic Problem

The macroscopic motion of the equivalent homogeneous liquid is assumed to be described by the following set of governing equations:

$$\left.\begin{aligned} &\rho^{\mathrm{H}}\left(\frac{\partial \boldsymbol{u}^{\mathrm{H}}}{\partial t}+\boldsymbol{u}^{\mathrm{H}}\cdot\nabla_x\boldsymbol{u}^{\mathrm{H}}\right)-\nabla_x\cdot\boldsymbol{\sigma}^{\mathrm{H}}\left(\boldsymbol{u}^{\mathrm{H}},\ p^{\mathrm{H}}\right)=0\\ &\nabla_x\cdot\boldsymbol{u}^{\mathrm{H}}=0\\ &\boldsymbol{\sigma}^{\mathrm{H}}=-p^{\mathrm{H}}\mathbf{1}+2\mu^{\mathrm{H}}\boldsymbol{D}^{\mathrm{H}}\\ &\boldsymbol{D}^{\mathrm{H}}=\frac{1}{2}\left(\nabla_x\boldsymbol{u}^{\mathrm{H}}+(\nabla_x\boldsymbol{u}^{\mathrm{H}})^{\mathrm{T}}\right) \end{aligned}\right\}\quad \text{in}\quad \Omega \tag{7}$$

along with appropriate boundary and initial conditions, where ∇_x is the gradient operator with respect to the macroscopic coordinate $\boldsymbol{x}$ and superscript H indicates the macroscopic, averaged or homogenized quantities. Here, the macroscopic mass density can easily be computed by the rule of mixture as $\rho^{\mathrm{H}}=\rho_{\mathrm{s}}|Y_{\mathrm{s}}|/|Y|+\rho_{\mathrm{f}}|Y_{\mathrm{f}}|/|Y|$ where $|Y_{\mathrm{f}}|$ and $|Y_{\mathrm{s}}|$ are the volumes of the fluid and solid particles, respectively, and $|Y|$ is the total volume of Y. On the contrary, the macroscopic viscosity μ^{H} depends not only the microscopic configuration, but also the microscopic quantities, such as velocities of the liquid and solid particles, as

$$\mu^{\mathrm{H}}=f(Y;\ \rho_{\mathrm{f}},\ \mu,\ \boldsymbol{u},\ p,\ \rho_{\mathrm{s}},\ \boldsymbol{v}_{\mathrm{s}}) \tag{8}$$

Thus, the macroscopic problem is nonlinear, since the macroscopic viscosity is a function of the microscopic velocity and pressure that must be determined according to the macroscopic flow conditions [3]. To evaluated the macroscopic viscosity, the present study concerned with a method of numerical viscosity measurement (NVM) that will be introduced in the next section.

3 Numerical Viscosity Measurement

This section presents a method of NVM for a solid-liquid mixture to calculate the macroscopic viscosity of an equivalent homogeneous liquid.

3.1 Viscosity Measurement with Hagen-Poiseuille Equation

Since the macroscopic viscosity is regarded as a resistance against the macroscopic shear velocity, the NVM is realized by evaluating the relationship between the averaged pressure loss and the macroscopic flow rate or velocity. Therefore, the microscopic analysis must be carried out in a certain microscopic domain to determine the microscopic velocity and pressure fields so that they are averaged over the domain. This microscopic domain is referred to as as a representative volume element (RVE) in the context of homogenization or averaging. In this study, we employ a cylindrical pipe channel as a RVE which is often used in a standard gravimetric capillary viscometer, and follow the corresponding measurement principle. That is, the so-called Hagen-Poiseuille's law is utilized to calculate the macroscopic viscosity by using the results of microscopic flow simulations in the RVE.

When a steady state flow is given in a pipe channel with a radius of r, the viscosity μ is determined as

$$\mu = \frac{\pi r^4 P_x}{8Q} \tag{9}$$

where Q is the flow rate and P_x is the pressure gradient expressed as $\Delta p/\Delta l$. Here, Δl is the length and Δp is the pressure loss between the test surfaces. It is known that Ostward viscometer, which is depicted in Fig. 2, was originated based on the measurement principle represented by this equation. This type of viscometer is also called a gravimetric viscometer, implying that the pressure gradient P_x is always constant as $-\rho g h/\Delta l$ where h is the so-called effective height.

Although the driving force cannot be controlled in actual gravimetric viscometers, the NVM enables us to apply arbitrary values of macroscopic pressure gradient P_x in the NVM to produce various macroscopic flow rates, as will be seen later. Then,

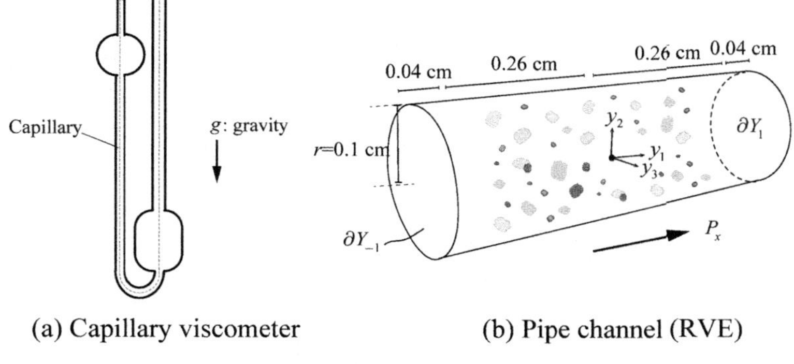

Fig. 2 Viscosity measurement: **a** Standard gravimetric capillary viscometer (Ostward type); **b** Cylindrical RVE used in numerical viscosity measurement

the viscosity μ and the flow rate Q in Hagen-Poiseuille equation (9) are replaced by the macroscopic viscosity $\mu^{\mathrm{H}}$and the macroscopic flow rate Q^{H} that can be obtained in NVMs.

3.2 Microscopic Problem for NVM

Following the idea of actual measurement procedure with a capillary type viscometer, we present a method of NVM using a cylindrical pipe channel as a RVE. The RVE employed in this study is illustrated in Fig. 2b, where $\partial Y_{\pm 1}$ is the positive and negative boundary surfaces perpendicular to the y_1-axis. Both the liquid and solid particles are assumed to move periodically at the end surfaces of the pipe so that

$$\boldsymbol{u}|_{\partial Y_{-1}} = \boldsymbol{u}|_{\partial Y_1} \tag{10}$$

Also, the following non-slip condition is imposed on the circumferential wall as

$$\boldsymbol{u}|_{\partial Y_r} = \boldsymbol{0} \tag{11}$$

With reference to the homogenization for heterogeneous solid materials [1, 12], it is assumed that the microscopic pressure gradient appearing in (1) with (2) can be represented as

$$\nabla_y p(\boldsymbol{x}, t; \boldsymbol{y}, \tau) = \nabla_x p^{\mathrm{H}}(\boldsymbol{x}, t) + \nabla_y p^*(\boldsymbol{x}, t; \boldsymbol{y}, \tau) \tag{12}$$

where $\nabla_x p^{\mathrm{H}}(\boldsymbol{x}, t)$ is the macroscopic pressure gradient, whose x_1-component corresponds to P_x in (9). Also, $\nabla_y p^*(\boldsymbol{x}, t; \boldsymbol{y}, \tau)$ is the fluctuation of the microscopic pressure gradient and assumed to satisfy the periodic boundary condition as

$$p^*\big|_{\partial Y_{-1}} = p^*\big|_{\partial Y_1} \tag{13}$$

as well as the condition $\int_Y p^*(\boldsymbol{x}, \boldsymbol{y}, t, \tau) dY = 0$. When the dominant macroscopic flow direction is parallel to the y_1-axis, Eq. (12) can be rewritten as

$$\frac{\partial p}{\partial y_1}(\boldsymbol{x}, t; \boldsymbol{y}, \tau) = P_x(\boldsymbol{x}, t) + \frac{\partial p^*}{\partial y_1}(\boldsymbol{x}, t; \boldsymbol{y}, \tau) \tag{14}$$

In above equation (12), the space-mean pressure gradient was defined as macro pressure gradient P_x. Then, we introduce the following special type of space-time averaging operations:

$$\begin{aligned}\frac{1}{|Y|}\int_Y \frac{1}{T}\int_t^{t+T} \frac{\partial p}{\partial y_1}(\boldsymbol{x},t;\,\boldsymbol{y},\tau)\,d\tau dY \\ &= P_x(\boldsymbol{x},t) + \frac{1}{|Y|}\int_Y \frac{1}{T}\int_t^{t+T} \frac{\partial p^*}{\partial y_1}(\boldsymbol{x},t;\,\boldsymbol{y},\tau)\,d\tau dY \\ &= P_x(\boldsymbol{x},t)\end{aligned} \tag{15}$$

where Eq. (13) has been utilized. Here, $\frac{1}{|Y|}\int_Y \bullet dY$ indicates the spatial average of quantity $\bullet$, while $\frac{1}{T}\int_t^{t+T} \bullet d\tau$ is the temporal average within time interval T that must be properly set in actual computation.

With the relationship (12), the set of microscopic governing equations, (1) along with (2), can be rewritten as follows:

$$\left.\begin{aligned}&\rho\left(\frac{\partial \boldsymbol{u}}{\partial \tau} + \boldsymbol{u}\cdot\nabla_y \boldsymbol{u}\right) - \nabla_y\cdot\boldsymbol{\sigma}(\boldsymbol{u},\ p^*) = \boldsymbol{f}\\ &\boldsymbol{f} = -\nabla_x p^{\mathrm{H}}\\ &\nabla_y\cdot\boldsymbol{u} = 0\\ &\boldsymbol{\sigma} = -p^*\boldsymbol{I} + 2\mu\boldsymbol{D}\\ &\boldsymbol{D} = \frac{1}{2}\left(\nabla_y\boldsymbol{u} + (\nabla_y\boldsymbol{u})^{\mathrm{T}}\right)\end{aligned}\right\}\quad \text{in} \quad Y_{\mathrm{f}} \tag{16}$$

Note here that $\boldsymbol{f} = -\nabla_x p^{\mathrm{H}} = [-P_x, 0, 0]^{\mathrm{T}}$, which plays a role of a body force, has only the y_1-component, as we are using the pipe channel as shown in Fig. 2b.

3.3 *Space-Time Homogenization*

By using the solution of (16) along with the motions of solid particles, we defined the following mass-density-weighted spatial average of the microscopic velocity $u_1(\boldsymbol{x},t;\,\boldsymbol{y},\tau)$ to calculate the y_1-component of the macroscopic non-stationary velocity, u_1^{H}:

$$\bar{u}_1^{\mathrm{H}}(\boldsymbol{x},t;\tau) = \frac{1}{\rho_{\mathrm{f}}|Y_{\mathrm{f}}| + \rho_{\mathrm{s}}|Y_{\mathrm{s}}|}\cdot\left[\int_{Y_{\mathrm{f}}}\rho_{\mathrm{f}} u_1(\boldsymbol{x},\boldsymbol{y},t,\tau)dY + \sum_{i=1}^{N}\int_{Y_{\mathrm{s}}^i}\rho_{\mathrm{s}} v_{\mathrm{f},1}^i(\boldsymbol{x},\boldsymbol{y},t,\tau)dY\right] \tag{17}$$

Thus, the space averaged flow rate $\bar{Q}^{\mathrm{H}}$ can be computed as

$$\bar{Q}^{\mathrm{H}}(\boldsymbol{x},t;\tau) = |\partial Y_1|\cdot\bar{u}_1^{\mathrm{H}}(\boldsymbol{x},t;\tau) \tag{18}$$

However, both the macroscopic velocity (17) and flow rate (18) depend on the microscopic time τ, they might be rapidly oscillating within a certain time interval. Therefore, assuming that the macroscopic flow rate is stationary within $[t, t+T]$ when measured by the coarse or macroscopic time scale t, we define the following time-space averaged flow rate:

$$Q^{\mathrm{H}}(\boldsymbol{x},\ t) = \frac{1}{T}\int_{t}^{t+T} \bar{Q}^{\mathrm{H}}(\boldsymbol{x},\ t,\ \tau)d\tau \tag{19}$$

Here, T is the same as the one used in Eq. (15) and can be identified with the representative time interval to properly evaluate the temporal average.

In summary, by prvoiding the macroscopic pressure gradient as a driving force for the microscopic problem (16) in the cylindrical RVE, we solve for the microscopic velocity as well as the microscopic pressure and then apply the space-time averaging operations, (17)–(19), to compute the macroscopic flow rate $Q^{\mathrm{H}}(\boldsymbol{x},\ t)$. Then, with the Hagen-Poiseuille equation (9) is used to determine the macroscopic viscosity μ^{H}.

4 Results of NVM

The proposed method of NVM is applied to a specific solid-liquid mixture that contains a number of spherical rigid bodies in the cylindrical RVE introduced in the previous section. After introducing the macroscopic Reynolds number and the macroscopic shear strain rate as evaluation indices, we carry our a set of NVMs on the cylindrical RVE to calculate the macroscopic viscosities with different macroscopic velocities and different numbers of solid particles involved and then investigate the macro- and microscopic flow properties and their correlation.

4.1 Numerical Methods and Fixed Conditions in NVM

The motions of rigid particles are calculated by the distinct element method (DEM) [4, 5], while the fluid motion and the interaction between liquid and solid particles are simulated with the help of the stabilized finite cover method (FCM) [2, 14–16]. Also, the level set method is applied to capture the time evolution of moving interfaces; see [13] for the details of the employed schemes.

In all the calculation cases presented in this section, the liquid is assumed to be water at a temperature of 20 °C with mass density $\rho_{\mathrm{f}} = 1.0\ \mathrm{g/cm^3}$ and viscosity $\mu_{\mathrm{f}} = 0.01$ poise. Also, the rotations are suppressed so that $\boldsymbol{\omega} = \boldsymbol{0}$ all the time in this particular study, in order to represent the friction effects with perfectly spherical rigid body. Generally, rolling resistance model [8] is employed to controll particle rotations but we forced to be zero in this study.

4.2 Evaluation Indices for Numerical Investigations

Let us define two indices, the macroscopic Reynolds number Re^{H} and the macroscopic shear strain rate γ^{H} in preparation for the investigation of the macroscopic viscosity μ^{H}.

It is known that the Hagen-Poiseuille equation is valid, only if the laminar flow condition is satisfied. Therefore, the following definition of the macroscopic Reynolds number is essential for the validation:

$$Re^{\mathrm{H}} = \frac{\rho^{\mathrm{H}} u_1^{\mathrm{H}} d}{\mu^{\mathrm{H}}} \tag{20}$$

where d is the diameter of the cylindrical RVE and equal to $2r$. Note that the critical Reynolds number Re_c is generally 2300 to 4000 in case of a circular pipe flow [6].

Although the component of the macroscopic shear flow velocity $\boldsymbol{D}^{\mathrm{H}}$ can directly be used for an evaluation index, we introduce the following macroscopic shear strain rate for the sake of simplicity:

$$\gamma^{\mathrm{H}}(\boldsymbol{x}, t) = \frac{1}{|\partial Y_r|} \int_{\partial Y_r} \int_t^{t+T} \gamma(\boldsymbol{x}, t, \boldsymbol{y}, \tau) d A d\tau \tag{21}$$

Here, $\gamma(\boldsymbol{x}, t; \boldsymbol{y}, \tau)$ is the microscopic shear strain rate defined as

$$\gamma(\boldsymbol{x}, t; \boldsymbol{y}, \tau) = \sqrt{2\mathrm{tr}\boldsymbol{D}^2} \tag{22}$$

4.3 Effect of Macroscopic Strain Rate on Macroscopic Viscosity

Several NVMs are carried out for various levels of macroscopic pressure gradient P_x. Since the variation of P_x implies the variation of the macroscopic strain rate γ^{H} in the proposed NVM, the relationship between macroscopic strain rate γ^{H} and macroscopic viscosity μ^{H} can easily be investigated.

The number of solid particles in the cylindrical RVE is fixed at $N = 116$ and all the particles have the same diameter of 0.01 [cm]. Then, $P_x = 0.75,\ 1.00,\ 1.50$ [b/cm] are the prepared values of macroscopic pressure gradients.

Figure 3 shows the stream line along with particle configurations in the cylindrical RVE at the final step obtained in some cases. Although slight disturbance is observed around the ends where the periodic boundary condition is introduced, vortices and strong turbulence are not seen in the simulated results. Also, the marcoscopic Reynolds numbers calculated with Eq. (20) are provided in Table 1. As can be seen from the figure and the table, it is clear that the laminar flow condition is satisfied so that the Hagen-Poiseuille formula is valid. The time variation of the spa-

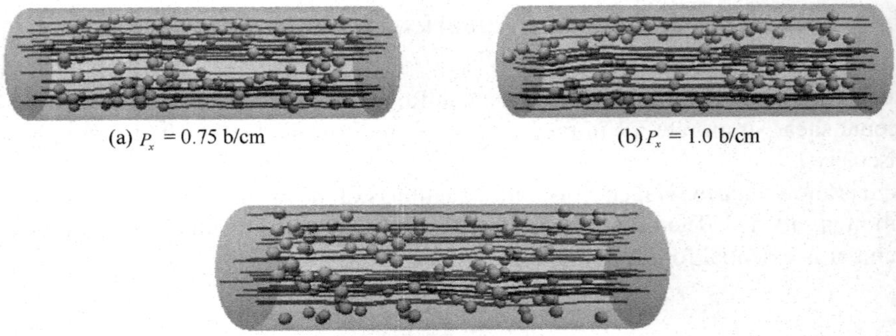

(a) $P_x = 0.75$ b/cm (b) $P_x = 1.0$ b/cm (c) $P_x = 2.0$ b/cm

Fig. 3 Stream line with particle configurations ($N = 116$) in the cylindrical RVE at the final step

Table 1 NVM results with different macroscopic pressure gradients[a]

P_x (b/cm)	$Q^{\rm H}$ (cm^3/s)	$u_1^{\rm H}$ (cm/s)	Re$^{\rm H}$
0.75	0.0027	0.086	1.58
1.00	0.0034	0.108	1.88
2.00	0.0060	0.189	7.95

[a] $N = 116$, $\rho_s = 1.5$ [g/cm^3], particle diameter $\phi_s = 0.01$ [cm]

tial average of flow rate $\bar{Q}^{\rm H}(\boldsymbol{x}, t; \tau)$ calculated with Eq. (18) is shown in Fig. 4a. The macroscopic flow rate attains the maximum value immediately after the beginning of NVM, gradually decreases and tends to become a steady state.

The time t and time length T used for the temporal averaging are set in accordance with 5000th step and 2000 steps, respectively. The macroscopic flow rate $Q^{\rm H}(\boldsymbol{x}, t)$ evaluated by Eq. (19) along with (18) and the corresponding acroscopic strain rate $\gamma^{\rm H}$ calculated by Eq. (22) along with (21) are summarized in Table 1. With the macro-flow rate $Q^{\rm H}$, we calculatedthe macroscopic viscosity $\mu^{\rm H}$ by using (8) and show the relationship between the macroscopic strain rate and $\mu^{\rm H}$ in Fig. 4b. As can be seen from the figure, the viscosity tends to decrease, as the strain rate rises. This is a typical non-Newtonian behavior, as the viscosity depends on the flow rate.

It is, therefore, safe to conclude that the solid-liquid mixture behaves as a non-Newtonian fluid at macro-scale due to the motions of solid particles at micro-scale. However, the tendency that the higher the macroscopic strain rate the larger the macroscopic viscosity is different from the fact reported for liquefied soils [3]. We may not be able to deny that certain sorts of microscopic mechanisms have not be reflected in our NVMs. In fact, since the macroscopic viscosity exceeds the viscosity $\mu_{\rm f}$ used for the liquid phase in the RVE, the suppression of the rotational motion possibly caused much friction effect on the microscopic flow than expected.

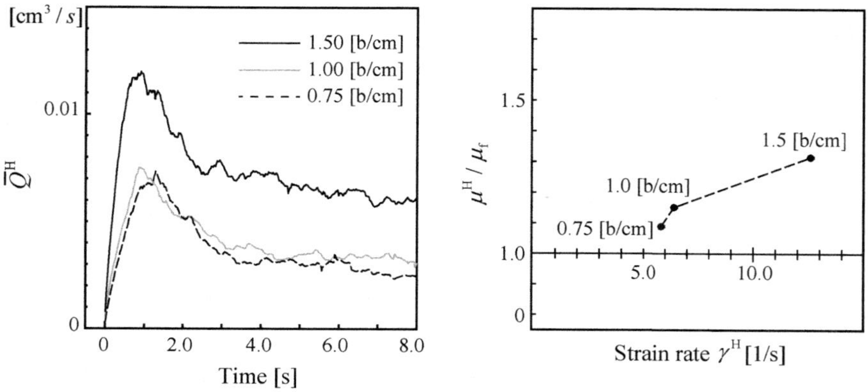

(a) Time history of spatially averaged flow rate

(b) Relation between normalized viscosity and strain rate

Fig. 4 Effect of macroscopic strain rate on macroscopic viscosity

4.4 Effect of Solid's Volume Fraction on Macrosopic Viscosity

It can be generally considered that the fluidity of solid-liquid mixture becomes low with increase in volume occupancy of a solid phase, and the turbulence induced by the drastic motion of particles may also affect the fluidity. In this section, a series of NVMs are performed with different number of solid particles to investigate the effect of the volume fraction of solid particles on the macroscopic viscosity.

By changing the number of solid particles, we carry out the NVMs and calculate the macroscopic flow velocity $\boldsymbol{u}^{\mathrm{H}}(\boldsymbol{x}, t)$ and the corresponding flow rate $\boldsymbol{Q}^{\mathrm{H}}(\boldsymbol{x}, t)$ by (19) along with (18). The results are provided in Table 2. With the values in the table, the relationship between the macroscopic viscosity μ^{H} and the number of solid particles is obtained as shown in Fig. 5a. The vertical axis is the macroscopic viscosity μ^{H} normalized by the liquid viscosity μ_{f}, while the horizontal axis is the volume fraction of solid particles, which is defined as the ratio of $|Y_s|$ to the entire volume of RVE $|Y|$. In the cases of $N \leq 100$ ($N = 10, 20, 67$), the macroscopic viscosity μ^{H} tends to decrease as the volume fraction of solid particles increases. However, when $N \geq 100$ ($N = 116, 243$), the tendency becomes opposite.

In order to study this tendency, let us investigate the effect of contacting particles on the macroscopic viscosity. Figure 5b shows the time history of the number of contacting particles with the circumferential wall. As can be seen from the figure, about 80–90 particles always contact with the wall in the case with the largest number of solid particles ($N = 243$). Then, the number of contacting particles decreases, as the volume fraction of solid particles decreases. Since the rotational motion of

Table 2 NVM results with different numbers of solid particles[a]

N	Q^H (mm^3/s)	u_1^H (cm/s)	Re^H
10	3.47	0.110	1.95
20	3.63	0.115	2.08
67	4.01	0.128	2.61
116	3.35	0.107	1.82
243	2.55	0.081	1.06

[a] $P_x = 1.0$ [b/cm], $\rho_s = 1.5$ [g/cm^3] and particle diameter $\phi_s = 0.01$ [cm]

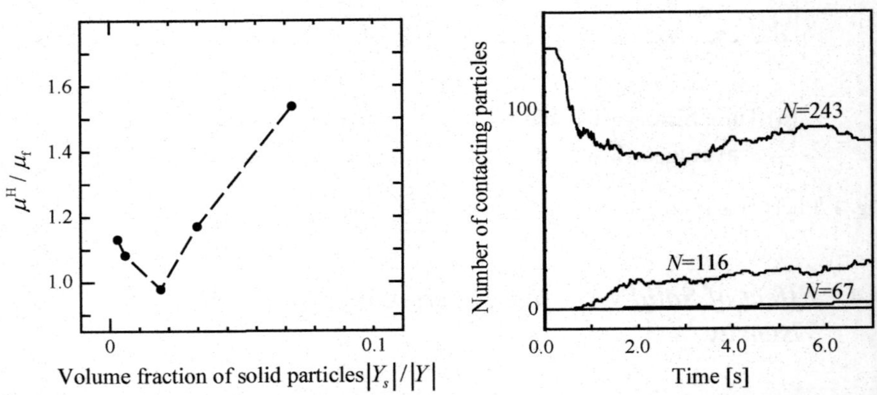

(a) Relation between normalized viscosity and volume fraction of solid particles

(b) Time history of spatially averaged flow rate

Fig. 5 Effect of number of solid particles on macroscopic viscosity

particles are suppressed in the calculations, the resistance against the flow around the contacting particles may become large and accordingly increases the macroscopic viscosity. Also, almost no particles contact with the wall in the cases of $N \leq 100$ ($N = 10, 20, 67$).

The influence of solid particles on the microscopic flow characteristics is also investigated. Figure 6 shows the distributions of the microscopic flow velocity components, u_{y_2} and u_{y_3}, at a certain time in the cases of $N = 243$ and $N = 10$. It can be seen from the figure that the flow regime is fluctuating around the particles, though these flow velocity components are lower than that of u_1^H by two orders of magnitude, Since the disturbance and the resulting energy dissipation are caused by the viscosity that produces the interaction between the particles and the surrounding fluid at micro-scale, its degree depends on both the number of particles. In fact, the disturbance of the microscopic velocity components in the case of $N = 243$ is more severe than that of $N = 10$.

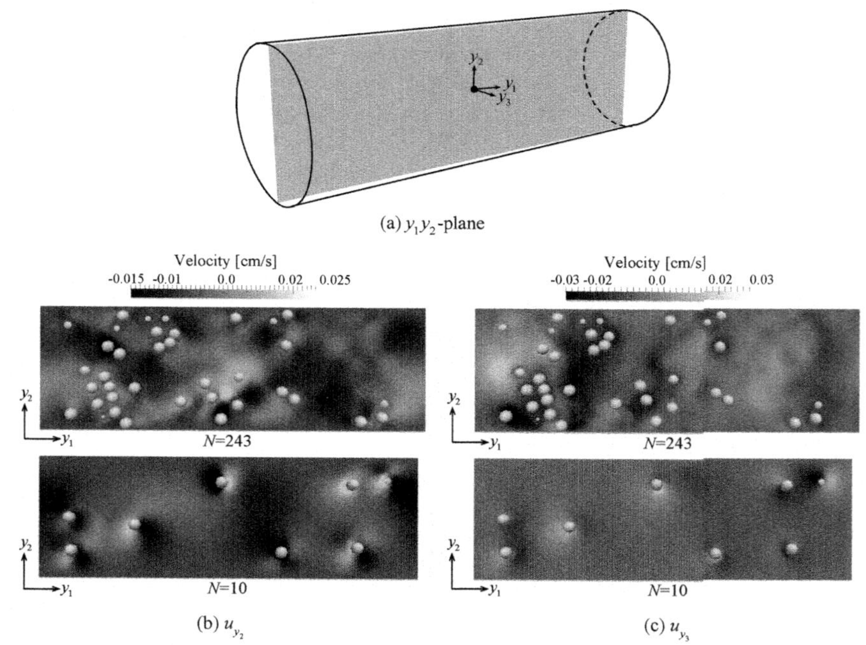

Fig. 6 Spatial distributions of microscopic tangential velocities u_{y2} and u_{y3} in y_1-y_2-plane

5 Conclusion

We have developed a method of numerical viscosity measurement (NVM) for solid-liquid mixed mixture that can be realized by the space-time homogenization procedure originally proposed in this study.

The governing equations at both macro and microscales were first set up and the relationships between macro- and microscopic variables are introduced. Here, the macroscopic fluid was teated as an equivalent homogeneous medium with macroscopic or homogenized viscosity, whose motion was assumed to be governed by the Navier-Stokes equations, while the coupling phenomenon of the fluid with suspended solid particles was described in a microscopic domain. The corresponding microscopic equations governed the microscopic flow fields under adequate boundary and loading conditions and then the macrosopic viscosity was evaluated by the space-time homogenization.

The space-time homogenization was nothing but the averaging procedure for microscopic quantities, which ware supposed to be obtained in NVMs, over the microscopic domain and within a certain time interval. The microscopic domain for NVM was regarded as representative volume element (RVE) and was a cylindrical pipe in this particular study so that the Hagen-Poiseuille's law can be utilized to calculate the macroscopic viscosity. Also, the microscopic or fine time scale was used to determine the stationary values of macroscopic quantities so that they can be measured by the macroscopic or coarse time scale.

Case studies of NVMs were conducted with different macroscopic flow rate and with different volume fractions of solid particles. The obtained macroscopic viscosity exhibited typical non-Newtonian behavior. Then, we studied the effect of microscopic flow properties on the macroscopic viscosity and clarified the underlying microscopic mechanisms. In particular, we reached the conclusion that, since the microscopic motion of suspended solid particles in the RVE caused the energy dissipations due to the fluctuations of microscopic flow regimes and the frictional-contact with the wall, the increase of the number of suspended particles resulted in the increase of the macroscopic viscosity.

References

1. Benssousan, A., Lions, J.L., Papanicoulau, G.: Asymptotic Analysis for Periodic Structures. North-Holland, Amsterdam (1978)
2. Brooks, A.N., Hughes, T.J.R.: Streamline-upwind/Petrov-Galerkin formulations for convection dominated flows with particular emphasis on the incompressible Navier-Stokes equations. Comput. Meth. Appl. M. **32**, 199–259 (1982)
3. Coussot, P., Piau, J.M.: On the behaviour of fine mud suspentions. Rheol. Acta. **33**, 175–184 (1994)
4. Cundall, P.A., Strack, O.D.L.: A discrete numerical model for granular assemblies. Geotechnique. **29**(1), 47–65 (1979)
5. Cundall,P. A., Strack O.D.L.: The distinct element method as a tool for research in granular media. Department of Civil and Mineral Engineering Report, Part 2, University of Minnesota, Minnesota (1979)
6. Holman, J.P.: Heat Transfer, vol. 207. McGraw-Hill (2002)
7. Hornung, U.: Homogenization and Porous Media. Springer, New York (1991)
8. Iwashita, K., Oda, M.: Rolling resistance at contacts in simulation of shear band development by DEM. J. Eng. Mech. **124**(3), 285–292 (1998)
9. Nakamura, T., Yim, S.C.: A non linear three-dimensional coupled fluid-sediment interaction model for large seabed deformation. J. Offshore. Mech. Arct. Eng. **133**, 031103 (2011)
10. Ontowirjo, B., Mano, A.: A turbulent and suspended sediment transport model for plunging breakers. Coast. Eng. J. **50**, 349–367 (2008)
11. Rouse, H.: Modern conceptions of the mechanics of turbulence. T. Am. Soc. Civil. Eng. **102**, 463–543 (1937)
12. Sanchez-Palencia, E.: Non-homogeneous media and vibration theory. In: Lecture Note in Physics, vol. 127. Springer, Berlin (1980)
13. Sussman, M., Smereka, P., Osher, S.: A level set approach for computing solutions to incompressible two-phase flow. J. Comput. Phys. 146–159 (1994)
14. Terada, K., Asai, M., Yamagishi, M.: Finite cover method for linear and nonlinear analyses of heterogeneous solids. Int. J. Numer. Meth. Eng. **58**(9), 1321–1346 (2003)
15. Terada, K., Kurumatani, M.: Performance assessment of generalized elements in the finite cover method. Finite Elem. Anal. Des. **41**, 111–132 (2004)
16. Tezduyar, T.E.: Stabilized finite element formulations for incompressible flow computations: Adv. Appl. Mech. **28**, 1–44 (1991)
17. Ushijima, S., Kuroda, N.: Multiphase modeling to predict finite deformations of elastic objects in free surface. In: Fluid Structure Interaction *V*, vol. 105, pp. 34–45. WIT Press (2009)

18. Yamashita, K., Sugawara, D., Takahashi, T., Imamura, F., Saito, Y., Imato, Y., Kai, T., Uehara, H., Kato, T., Nakata, K., Saka, R., Nishikawa, A.: Numerical simulations of large-scale sediment transport caused by the 2011 Tohoku earthquake tsunami in Hirota Bay. South. Sanriku Coast. Coastal Eng. J. **58**(4), 1640015, 1–28 (2016)

Numerical Simulation of Hydrogen Embrittlement at the Example of a Cracked Pipeline

Milena Möhle, Udo Nackenhorst and Olivier Allix

Abstract A continuum model for numerical simulation of hydrogen induced embrittlement of pipeline material is discussed within this work. For that, a transient hydrogen model considering trapping is coupled with an elasto-plastic material model considering von Mises yielding. The hydrogen enhanced plasticity (HELP) mechanism is assumed to be active within this problem statement and is realized by a hydrogen dependent reduction of the yield strength. An iterative numerical solution scheme is applied to solve the coupled problem. At the example of a pipeline with a blunted crack, the influence of hydrogen is investigated. A localized plastic zone is observed for high hydrogen concentrations, in line with the inherent phenomena of the HELP mechanism. However, when applying hydrogen boundary conditions which are considered to be realistic for an existing natural gas pipeline, no pronounced effect of hydrogen based on reducing the yield strength could be observed. Nevertheless, this numerical results do not imply a judgment if the HELP mechanism in general could be the prevalent mechanism for failure.

1 Introduction

The detrimental effect of hydrogen on steel structures has been investigated for over one hundred years. As early as in 1875 Johnson [1] showed that hydrogen reduces the toughness of specimens and that this process is reversible [2, 3]. However, his

M. Möhle (✉) · U. Nackenhorst
Institute of Mechanics and Computational Mechanics,
Appelstr. 9A, Hannover, Germany
e-mail: milena.moehle@ibnm.uni-hannover.de

U. Nackenhorst
e-mail: nackenhorst@ibnm.uni-hannover.de

O. Allix
L.M.T. Cachan, 61 Avenue du Président Wilson,
94235 Cachan Cedex, France
e-mail: allix@lmt.ens-cachan.fr

J. Sorić et al. (eds.), *Multiscale Modeling of Heterogeneous Structures*,
Lecture Notes in Applied and Computational Mechanics 86,
https://doi.org/10.1007/978-3-319-65463-8_18

findings did not get high attention until in World War II several of the Liberty ships broke without warning. Gerberich [2] stated that most text books were incomplete due to excluding the effect of hydrogen induced during welding on the brittle crack growth in the trunk of the Liberty ships and referred to a study by Zapffe and Sims [4]. From that time on, the interest in hydrogen embrittlement evoke [2]. However, the actual underlying mechanism of hydrogen embrittlement still remains mainly unexplored. A new field of interest for hydrogen embrittlement was raised within the last years by discussing how the irregularly produced energy from renewable sources could effectively be stored. Thus, one idea is to form hydrogen gas from the redundant energy and transport it through the existing pipeline system for natural gas. Here, also a mixing of the two gases is feasible [5]. However, when doing so, problems can arise as hydrogen can diffuse inside the material and subsequently embrittle the pipeline. This is of special interest if the pipeline is damaged during its service time by corrosion for example.

The process of hydrogen induced embrittlement starts with the entrance of hydrogen into the system. Therefore, the hydrogen has to be debonded from molecular hydrogen to atomic one to enter the lattice of the material [6]. Within the lattice, the stress driven diffusion process leads to an accumulation of hydrogen at places of maximal hydrostatic stresses [7].

Several possible mechanisms of how hydrogen causes material embrittlement are discussed in literature. In the following, a brief overview of the main mechanisms shall be given. For a more detailed description it is referred to [8]:

- The mechanism of "hydride induced embrittlement" includes a nucleation of the lattice material and the accumulated hydrogen at the crack tip into a brittle hydride. This mechanism was shown for hydridforming materials within several experiments (see [9, 10]).
- The "hydrogen-enhanced decohesion" (HEDE) mechanism describes the decohesion of atomic bonds in the presence of hydrogen and therefore leads to brittle failure. Unfortunately, there are no suitable tests to prove the influence of this mechanism. This is due to the fact that brittleness and plasticity coexist simultaneously within experimental surroundings.
- According to the "hydrogen enhanced localized plasticity" (HELP) mechanism, the hydrogen either shields dislocations from each other leading to dislocation motion at lower stresses or decreases the interaction between the dislocations. This enhancement of plastic energy leads to a localized softening of the material which on macro scale appears like brittle failure.
- The "adsorption-induced dislocation-emission" (AIDE) mechanism adapts the fundamentals of the HEDE mechanism in a more complex framework. It is assumed that hydrogen is adsorbed at the first atomic layers on the surface of the material and thus lowers the atomic bounds in these regions. This leads to facilitated dislocation nucleation by shearing processes over several atomic distances. At sufficiently high stresses, it is assumed that the dislocation cores emit from the surface into the material and form voids. This leads to a more brittle failure on macro scale.

It is widely agreed upon that not a single one of the mechanisms described above suffices to explain hydrogen embrittlement. Still, usually one of the mechanisms is dominant in different systems under various conditions [11].

In literature different kinds of models for hydrogen embrittlement are described. One way of modeling the HEDE mechanism is to introduce cohesive elements at a crack face and modify the traction separation law of these elements in dependency of hydrogen [12]. However, in this case, the crack surface has to be defined beforehand.

Dadfarnia et al. [13] investigated hydrogen embrittlement at a cracked pipeline in a continuum sense. The authors used plane strain conditions to transfer the three dimensional problem to a two dimensional one. Here, the effect of hydrogen on the mechanical properties was accounted for by introducing a pure dilatation straining, dependent on the hydrogen concentration. This numerical realization is difficult to link directly to one of the discussed mechanisms above. The authors' goal was to study in a first instance the effect of the applied boundary conditions on the steady state hydrogen concentration.

Takayama et al. [14] investigated the influence of the varying hydrogen gas pressure within the pipeline on the hydrogen distribution within the material without considering an effect of hydrogen on the mechanical properties.

Within this study, one possible way of how hydrogen affects the material properties in a continuum sense is investigated using the example of a pipeline with a radial crack. It is assumed that within the investigated low carbon steel, local plastic deformations in front of a blunted crack tip under static loading conditions establishes the conditions for failure, and thus the HELP mechanism dominates the hydrogen embrittlement process. The HELP mechanism is applied by reducing the materials yield strength dependent on the present hydrogen concentration.

In a first step, the physical representation of hydrogen embrittlement is briefly summarized. After the coupled elasto-plastic material model and the transient hydrogen model are presented, the theoretical embrittlement of a perfect pipeline is discussed. The analytical observations on a perfect pipeline are then transferred to the numerical investigation on a cracked pipeline. Here, the numerically determined stress states and hydrogen concentrations at a blunted crack tip is shown and discussed.

2 Physical Representation of Hydrogen Embrittlement

In this work, a stress driven hydrogen model accounting for trapping effects is used [13, 15]. This is coupled with an elasto-plastic mechanical model considering the previously discussed HELP mechanism. In the following, these two coupled models are summarized.

2.1 Hydrogen Transport

In the stress-driven hydrogen model accounting for trapping effects it is assumed that the hydrogen can either be trapped at defects like dislocations or grain boundaries, or remain in the lattice sites of the material. According to Oriani's theory these two hydrogen populations are always in equilibrium [16]

$$\frac{\theta_T}{1-\theta_T} = \frac{\theta_L}{1-\theta_L} exp\left(\frac{W_B}{RT}\right), \tag{1}$$

where θ_T represents the ratio of occupied trapping sites to the total available ones while θ_L denotes the ratio of occupied interstitial lattice sites to the total available ones. The equilibrium constant $K_T = exp\left(\frac{W_B}{RT}\right)$ accounts for the trap binding energy W_B, the universal gas constant $R = 8.314\,\mathrm{J/molK}$ and the absolute temperature T. The hydrogen concentration in lattice sites can be defined by

$$C_L = \theta_L \beta N_L \tag{2}$$

with β defining the number of normal interstitial lattice sites per host atom and $N_L = N_A/V_M$ represents the number of solvent atoms per unit volume, with Avogadro's number $N_A = 6.0232 \times 10^{23}$ atoms/mol and the molar volume of the host lattice V_M.

The trapped hydrogen population C_T can be expressed accordingly,

$$C_T = \theta_T \alpha N_T(\bar{\varepsilon}^p) \tag{3}$$

with the number of hydrogen atoms per trap α and $N_T(\bar{\varepsilon}^p)$ representing the number of traps per unit volume as a function of the local equivalent plastic strain $\bar{\varepsilon}^p$.

The governing equation for the transient hydrogen distribution considering hydrostatic stress dependency and trapping reads [13]

$$\frac{D}{D_{eff}}\frac{dC_L}{dt} + \underbrace{\alpha\theta_T \frac{\partial N_T}{\partial \bar{\varepsilon}^p}\frac{\partial \bar{\varepsilon}^p}{dt}}_{\textit{trap generation}} - \underbrace{D\,\nabla^2 C_L}_{\textit{diffusion}} + \underbrace{\nabla\left(\frac{DV_H}{RT} C_L \nabla p\right)}_{\textit{advection}} = 0. \tag{4}$$

The trap generation term depends on the equivalent plastic strain $\bar{\varepsilon}_p$. This term is zero for vanishing equivalent plastic strain rates. The term marked by diffusion accounts for the normal interstitial lattice side diffusion of the hydrogen concentration. The advective part includes the gradient of the hydrostatic stress $p = \sigma_{kk}/3$, which can be interpreted as the advective flow direction for the hydrogen concentration. Further details on the definition is given in the following.

The notation $\frac{\partial \star}{\partial t}$ and $\frac{\partial \star}{\partial \bar{\varepsilon}^p}$ represent the derivatives in time and plastic strain, respectively. Furthermore, $V_H = 2\,cm^3$/mol represents the partial molar volume of hydrogen. The parameter D is the hydrogen diffusion coefficient within the interstitial lattice sites whereas the effective diffusion coefficient D_{eff} is taking into account trapping effects. The coefficient $\frac{D}{D_{eff}}$ represents the reduction of the hydrogen concentration in lattice sites C_L while the traps are not filled and can be expressed like [13]

$$\frac{D}{D_{eff}} = \left(1 + \frac{\partial C_T}{\partial C_L}\right) = \left(1 + \frac{K_T \alpha \beta N_L N_T}{[\beta N_L + (K_T - 1)C_L]^2}\right). \tag{5}$$

According to Oriani's theorem, the trap filling kinetics are very rapid. Thus, this coefficient alters the lattice hydrogen concentration by being less than one, in case the hydrogen concentrations in lattice and trapping sites are not in equilibrium or in case of the creation of new traps during plastic processes [13].

With Eqs. 1–3, the concentration of trapped hydrogen is derived,

$$C_T = \frac{K_T\, \alpha\, N_T(\bar{\varepsilon}_p)\, C_L}{\beta N_L + (K_T - 1)C_L}. \tag{6}$$

Further details can be found in [13, 15].

The trapped hydrogen concentration is strongly dependent on the available trapping sites, represented by N_T, which is a function of the equivalent plastic strain $\bar{\varepsilon}_p$. This relation was derived from empirical investigations by Kumnick and Johnson for α−iron [17] as

$$\log(N_T) = 23.3 - 2.33\, e^{-5.5\,\bar{\varepsilon}_p}. \tag{7}$$

Since no equivalent studies are available for the low carbon steel investigated in this work, the empiric study for α−iron for the trap density N_T is utilized herein.

2.2 Mechanical Model

An elasto-plastic continuum material model under plane strain conditions considering the von Mises yield criterion is used within this work. The balance equation for a static problem reads [18]

$$\nabla \underset{\sim}{\sigma} + \rho\, \underset{\sim}{b} = 0, \tag{8}$$

with the gradient of the stress tensor $\underset{\sim}{\sigma}$, the mass density ϱ and the applied loads $\underset{\sim}{b}$. The stresses can be determined by the expression [18]

$$\underset{\sim}{\sigma} = C^{ep}\, (\underset{\sim}{\varepsilon} - \underset{\sim}{\varepsilon}^p) \tag{9}$$

with the elasto plastic tangent modulus C^{ep}, the plastic strain $\underset{\sim}{\varepsilon}^p$ and the total strain $\underset{\sim}{\varepsilon} = sym\, \nabla\, \underset{\sim}{u}$. The plastic deformation is defined by the von Mises yield criterion

$$f = ||\underset{\sim}{\tilde{\sigma}}|| - \underset{\sim}{\sigma_0}(c) \tag{10}$$

where $\underset{\sim}{\tilde{\sigma}}$ is the deviatoric part of the stress tensor.

The influence of hydrogen is assumed to be dominated by local softening (HELP mechanism) and can be expressed according to [15] in the linear form

$$\sigma_0(c) = [(\xi - 1)c + 1]\sigma_y \tag{11}$$

with the material dependent softening parameter $\xi \leq 1$, the yield stress σ_y and the relative hydrogen concentration is defined as

$$c = (C_L + C_T)/N_L. \tag{12}$$

which is measured in atoms per solvent atom. It can be noticed that the hydrogen concentration within the lattice C_L as well as in the traps C_T is assumed to alter the yield stress. The hydrogen concentration is related to the theoretically available number of solvent atoms per unit volume N_L. No isotropic hardening is considered in Eq. 11 but can be extended in a straightforward manner.

The material dependent softening parameter ξ represents the susceptibility of the material to hydrogen. By choosing $\xi = 0$, the material is strongly affected by hydrogen. In the opposite case, materials which are not sensitive to hydrogen can be numerically modeled with $\xi = 1$.

3 Numerical Solution Scheme

The numerical realization of the previously discussed coupled elasto-plastic hydrogen model is described in the following. To solve the mechanical elasto-plastic problem, an institute-internal finite element code with von Mises yield condition is used. The flow rule is integrated using the radial return mapping scheme. The transient stress driven hydrogen diffusion problem is solved using finite elements for spatial discretization and the time-discontinuous Galerkin method for the transient part. This leads to stable results even in the presence of steep gradients resulting from the hydrostatic stress field.

In order to solve the coupled system of equations, first of all the mechanical part is calculated considering a zero hydrogen concentration. After the global equilibrium of the external and internal forces is achieved with the help of the Newton Raphson scheme, the required data for the hydrogen distribution calculation, namely the hydrostatic stress p and the equivalent plastic strain $\bar{\varepsilon}^p$, are provided.

For the advective part of Eq. 4 the derivative of the hydrostatic stress ∇p is needed. Therefore, the gauss point solutions are projected onto the nodes to achieve a smooth stress field. The gradient of the hydrostatic stress field is attained by using the derivative of the shape functions. This gradient represents the advective flow direction of the hydrogen. In other words, the hydrogen accumulates in regions where the lattice is volumetrically strained to a maximum. Here, on atomic length scale, more free space between the molecules exists, where in turn the hydrogen can fit in more easily.

The second parameter for coupling the elasto-plastic model with the hydrogen distribution is the equivalent plastic strain. The trap generation term of Eq. 4 represents the trap generation and thus the corresponding reduction of the lattice hydrogen concentration. Only if new plastic strain was accumulated in the previous mechanical step, this term is nonzero.

When the lattice hydrogen concentration C_L is determined by the transient advection diffusion relation given in Eq. 4, the trapped hydrogen concentration can be calculated by Eq. 6. Here, also the equivalent plastic strain $\bar{\varepsilon}^p$ is used to determine the trap density N_T. Thus, the trapped hydrogen concentration strongly depends on the level of the equivalent plastic strain. Using the results of the lattice and trapped hydrogen concentration, the value for the reduction of the yield stress c can be determined by Eq. 12. This value is then forwarded to the mechanical calculation.

This procedure is repeated until the mechanical load and the hydrogen concentration are in equilibrium.

4 Numerical Investigations

The effect of hydrogen on the material properties considering the numerical investigation of the HELP mechanism is shown in the following. First of all, the theoretical effect of hydrogen on the yield strength is considered. On the basis of these theoretical and experimental observations, the numerical model of a cracked pipeline is chosen and the results are discussed.

4.1 Model Setup

Firstly, the influence of hydrogen on a theoretically perfect pipeline tube is investigated. The hydrogen concentration at the surface can be determined by Sivert's law [11]

$$C_{L,0} = 6.331 \cdot 10^{28} \sqrt{P}\, e^{-\frac{\overline{\Delta H_S}}{RT}}, \tag{13}$$

with the hydrogen gas pressure within the tube P in Pa and the heat of solution $\overline{\Delta H_S} = 28.6\,\mathrm{kJ/mol}$. In the case of a perfect tube the maximal hydrostatic stress is located at the inner surface. Consequently, the maximal lattice hydrogen concentration will also be accumulated at the surface.

In order to study the hydrogen embrittlement effect on the perfect tube, in a first step the lattice hydrogen concentration at the surface $C_{L,0}$ is calculated by using

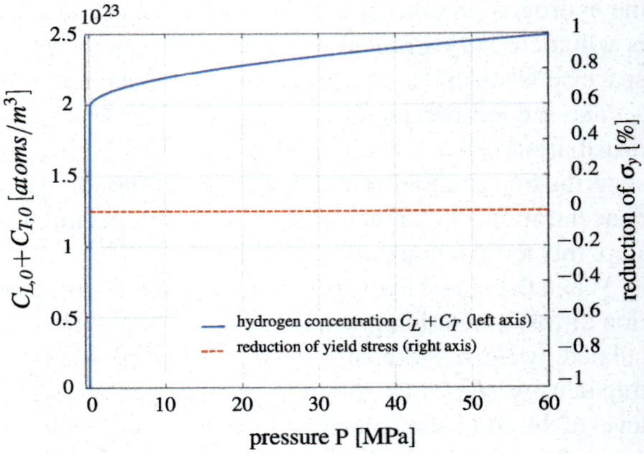

Fig. 1 Hydrogen concentration at the surface of a perfect pipeline according to Sievert's law $C_{L,0}$ plus the trapped hydrogen concentration $C_{T,0}$ for $\bar{\varepsilon}_p = 0$ (left axis); Reduction of the yield stress σ_y according to the induced hydrogen concentration (right axis)

Sivert's law. This result is then used to calculate the trapped hydrogen concentration $C_{T,0}$ for the equivalent plastic strain being zero with Eq. 3.

The total hydrogen concentration $C_{L,0} + C_{T,0}$ is depicted in Fig. 1 with respect to the pressure. The corresponding reduction of the yield stress σ_y is calculated according to the relation in Eq. 12 in percent and depicted as a dashed line, being equal to zero for all investigated concentrations. Since the usual pressure is up to 16 MPa in a natural gas pipelines system [19], an undamaged tube is considered not to be affected by hydrogen embrittlement under static loading conditions.

This theoretical consideration is in agreement with the experimental findings of Hoover et al. [20]. The authors investigated different pipeline steels by tensile testing in a high pressure cell by 6.9 MPa and observed that the embrittlement of the specimens coexists with the formation of surface cracks. According to these findings, it is unlikely that the accumulated hydrogen at defects, like dislocations and grain boundaries, causes a pronounced embrittlement effect.

As stated above, the basic idea is to use the existing natural gas pipeline system to transport hydrogen gas. Here, it has to be assumed that these pipelines can be damaged during their lifetime by corrosion or small cracks. Consequently, in contrast to the investigations on a perfect pipeline, in this work the effect of hydrogen on a damaged pipeline with a crack is investigated.

Figure 2 schematically depicts a cracked pipeline. This pipeline can be simplified into a two dimensional plane strain problem using this symmetry planes and considering a notch in one of the symmetry planes. Since this full field model is computationally costly, it is feasible to choose a smaller domain size.

Dadfarnia et al. [13] showed that the two dimensional full field model provides the same results in stress and strain states as the surrogate numerical model of a blunted crack tip. Only the steady state time for the hydrogen diffusion differs, which in the context of the discussion on the safety of the pipeline in general is not considered to be a primary parameter. Therefore, the surrogate numerical model of a blunted crack tip is used in the following analyses.

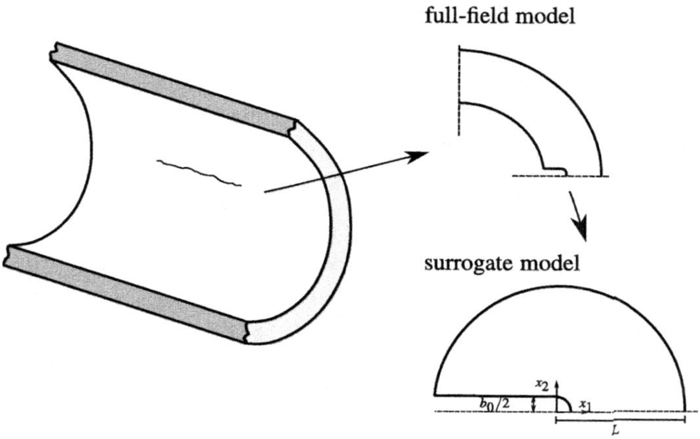

Fig. 2 Schematic representation of a pipeline damaged by a radial crack and the numerical plane strain model of the pipeline wall (full-field model). A further reduction of the investigated domain leads to a practical surrogate model (adapted from [13])

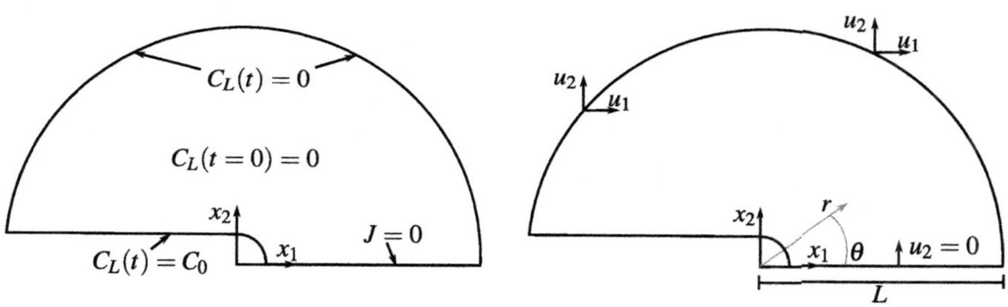

Fig. 3 Representation of the applied boundary conditions on the surrogate model for the hydrogen (left) and the mechanical model (right) (adapted from [13])

The surrogate model with the applied mechanical and hydrogen boundary conditions is depicted in Fig. 3. For the mechanical model, the displacement in x_2 direction is set to zero in the symmetry plane. Mode one opening conditions are applied according to [21] on the surface with the relation

$$\begin{bmatrix} u_1 \\ u_2 \end{bmatrix} = \frac{K_I}{2G}\sqrt{\frac{r}{2\pi}} \begin{bmatrix} \cos\left(\frac{\theta}{2}\right)\left(\psi - 1 + 2\sin^2\left(\frac{\theta}{2}\right)\right) \\ \sin\left(\frac{\theta}{2}\right)\left(\psi + 1 - 2\cos^2\left(\frac{\theta}{2}\right)\right) \end{bmatrix}. \tag{14}$$

The polar coordinates, the radius r and the angle θ, are marked in Fig. 3. By assuming plane strain conditions, ψ is equal to $3 - 4\nu$. The parameter G denotes the shear modulus and K_I the stress intensity factor for mode one opening conditions.

The boundary conditions for the transient hydrogen model are given in Fig. 3, left. In the symmetry plane, a zero flux boundary condition is applied and for time step

$t = 0$ s it is assumed that no hydrogen is within the domain. At the crack surface, a Dirichlet boundary condition of the concentration $C_{L,0}$ is applied.

In the numerical calculation, firstly the load is applied on the blunted crack, simulating the present stress state in the pipeline while natural gas is ducted through. Here, it is assumed that no hydrogen is within the material. After reaching the final load level the Dirichlet boundary condition $C_{L,0}$ is applied on the surface of the crack. Since, according to the assumed HELP mechanism, the hydrogen alters the yield strength of the material and the stress state alters the hydrogen concentration, the mechanical and hydrogen model have to be iterated until convergence is reached.

The applied lattice concentration $C_{L,0}$ at the crack surface is varied, while the stress intensity factor $K_I = 19\,\text{MPa}\sqrt{\text{m}}$ is fixed. The aim of this parametric study is to investigate the influence of the coupling of the mechanical and hydrogen model.

4.2 Results

The results of the numerical simulation for the coupled elasto-plastic and hydrogen model are discussed in the following. The material properties for a X70/X80 pipeline steel are given in Table 1.

Using the surrogate model presented above, the local reduction of the yield stress can be determined for different surface hydrogen concentrations $C_{L,0}$. In Fig. 4 the reduction of the yield strength σ_y is depicted in the symmetry plane ($\theta = 0$). The higher the hydrogen concentration is at the surface, the higher is also the effect on the reduction of the yield strength. The effect of a hydrogen concentration in the

Table 1 Material properties of a X70/X80 pipeline steel [13]

Properties	Symbol	Value
Young's modulus	E	201.88 GPa
Poisson's ratio	ν	0.3 [−]
Yield stress	σ_y	595 MPa
Material softening parameter	ξ	0 [−]
Trap binding energy	W_B	60 kJ/mol
Diffusion coefficient	D	$2\,10^{-8}\,\text{m}^2/\text{s}$
Partial molar volume of H	V_H	$2\,\text{cm}^3/\text{mol}$
Molar volume host material	V_M	$7.116\,\text{cm}^3/\text{mol}$
No. of H-atoms per trap	α	1 [−]
No. of interstitial lattice sides per host atom	β	1 [−]
Universal gas constant	R	8.314 J/mol K
Temperature	T	300 K
Avogadro's number	N_A	6.0232×10^{23} atoms/mol

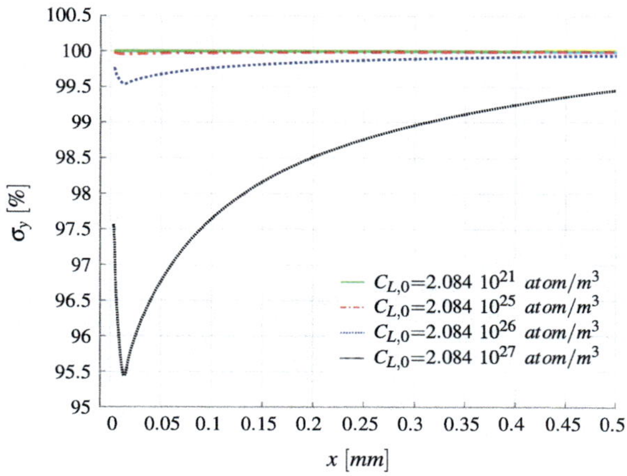

Fig. 4 Reduction of the yield stress for varying hydrogen surface concentrations $C_{L,0}$ with a constant stress intensity factor $K_I = 19\,\mathrm{MPa}\sqrt{\mathrm{m}}$ in the symmetry plane ($\theta = 0$)

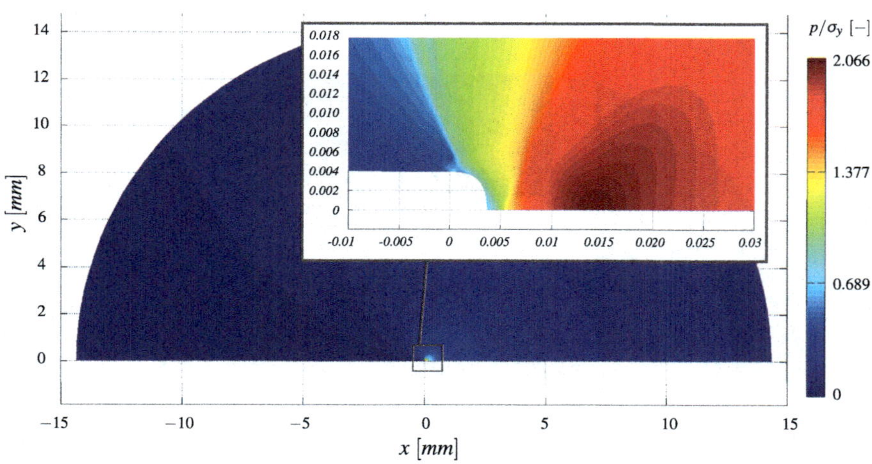

Fig. 5 Hydrostatic pressure p normalized by the yield strength σ_y in the vicinity of the blunting crack tip with a stress intensity factor $K_I = 19\,\mathrm{MPa}\sqrt{\mathrm{m}}$

range between 10^{21} and 10^{25} is considered to be minor. Only when applying higher hydrogen concentrations the reduction of the yield stress gets more pronounced. The maximum reduction of the yield stress spatially coexists with the maximum of the lattice hydrogen concentration. This is shown in more detail for a surface hydrogen concentration of $C_{L,0} = 2.084 \times 10^{27}\,\mathrm{atoms/m^3}$ with a constant stress intensity factor $K_I = 19\,\mathrm{MPa}\sqrt{\mathrm{m}}$.

The corresponding hydrostatic pressure for mode one opening displacements normalized by the yield stress σ_y is depicted in Fig. 5. It is noticed that the maximal

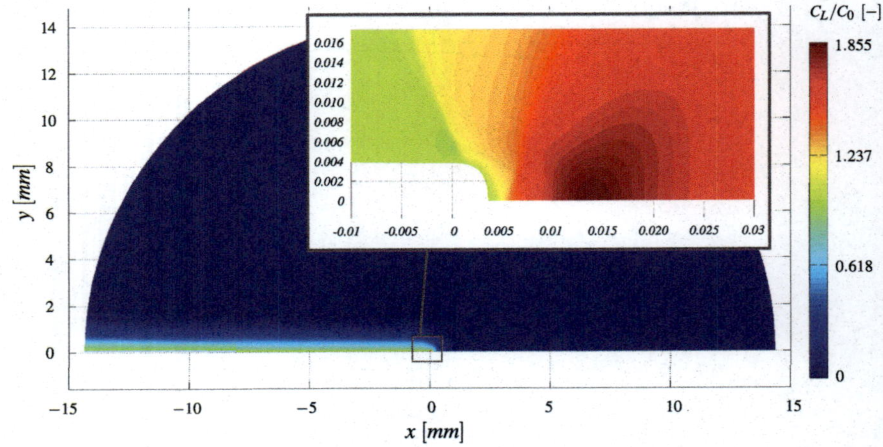

Fig. 6 Lattice hydrogen concentration C_L normalized by C_0 the at the blunting crack tip with a stress intensity factor $K_I = 19\,\text{MPa}\sqrt{\text{m}}$ and material dependent softening parameter $\xi = 0$ at the time t = 10s

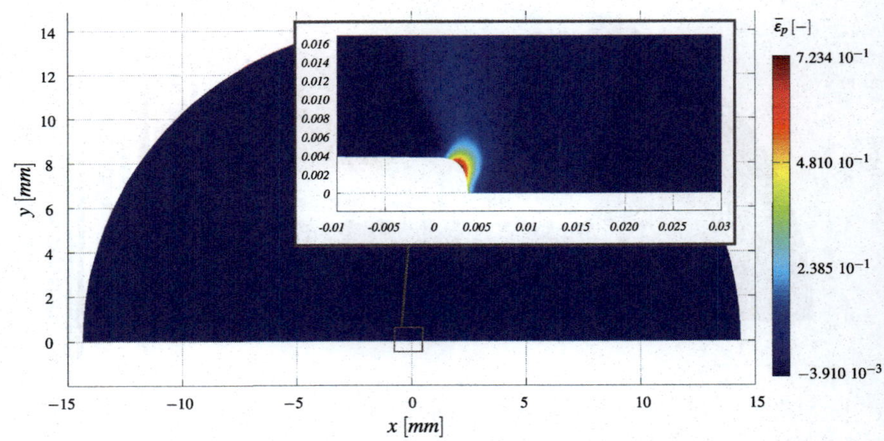

Fig. 7 Equivalent plastic strain $\bar{\varepsilon}_p$ with a stress intensity factor $K_I = 19\,\text{MPa}\sqrt{\text{m}}$ and material dependent softening parameter $\xi = 0$ at the time t = 10s

hydrostatic stress is about 2.5 times the original crack opening displacement away from the blunted crack tip. The gradient of the hydrostatic stress p gives the equivalent to an advective direction in the transient hydrogen Eq. 4. Therefore, at the maximal hydrostatic stress also the maximal lattice hydrogen concentration C_L can be found (see Fig. 6). In other words, the hydrogen accumulates in the region where the lattice is volumetrically strained to a maximum. Here, on atomic length scale, more free space between the molecules exists, where in turn the hydrogen can fit in more easily.

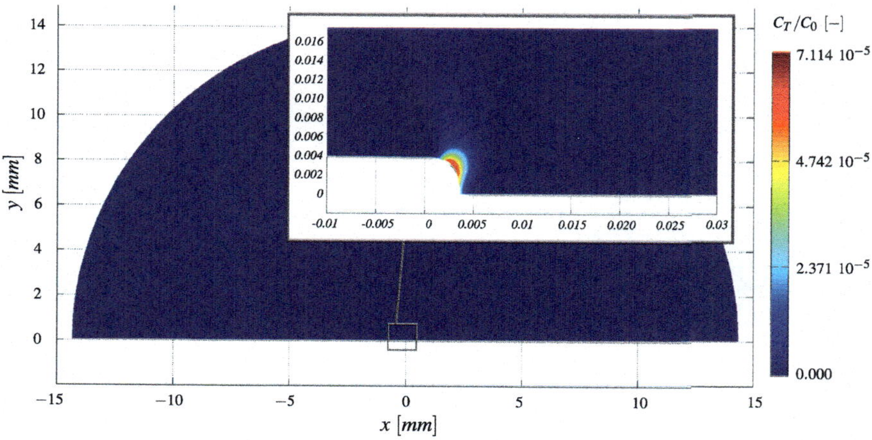

Fig. 8 Trapped hydrogen concentration C_T normalized by C_0 with a stress intensity factor $K_I = 19\,\text{MPa}\sqrt{\text{m}}$ and material dependent softening parameter $\xi = 0$ at the time t = 10s

The equivalent plastic strain $\bar{\varepsilon}_p$ is maximal at the surface of the crack tip, as shown in Fig. 7. According to Eq. 6, the maximal trapped hydrogen concentration C_T can also be found in this region (compare Fig. 8). This is due to the dependence of N_T on the equivalent plastic strain. In other words, the more traps are available, the more hydrogen can generally be accumulated within that region in form of trapped hydrogen.

The hydrogen concentration related to the number of solvent atoms per unit volume c is the parameter influencing the yield stress in Eq. 11. For the chosen material parameter $\xi = 0$ this factor represents the reduction of the yield stress. In Fig. 9, the parameter c is depicted in the vicinity of the crack tip and the maximal reduction of the yield stress can be straightforwardly determined to 4.6%. This is in agreement with the maximal reduced yield strength shown in Fig. 4. The trend of the lattice hydrogen concentration can be clearly seen in Fig. 9.

Thus, by applying a stress intensity factor $K_I = 19\,\text{MPa}\sqrt{\text{m}}$ and the discussed numerical realization of the HELP mechanism the lattice hydrogen concentration clearly leads to a slight softening of the material in front of the crack tip.

Since the von Mises yield criterion is used the reduction of the yield stress can be implicitly seen in the distribution of the von Mises stress σ_{vM}. A small area of reduced yield strength can be found located in the vicinity of the crack tip in Fig. 10. This agrees well with the location of maximal c, where, according to the mechanical model, the detrimental effect of the local hydrogen concentration on the local yield strength should be the highest. Thus, the coupled elasto-plastic model with the transient hydrogen model can describe in general the theoretical model of the HELP mechanism.

A hydrogen gas pressure of up to 16 MPa can be found for natural gas pipelines in industrial [19] and this leads according to Siverts law and Eq. 3 to a surface

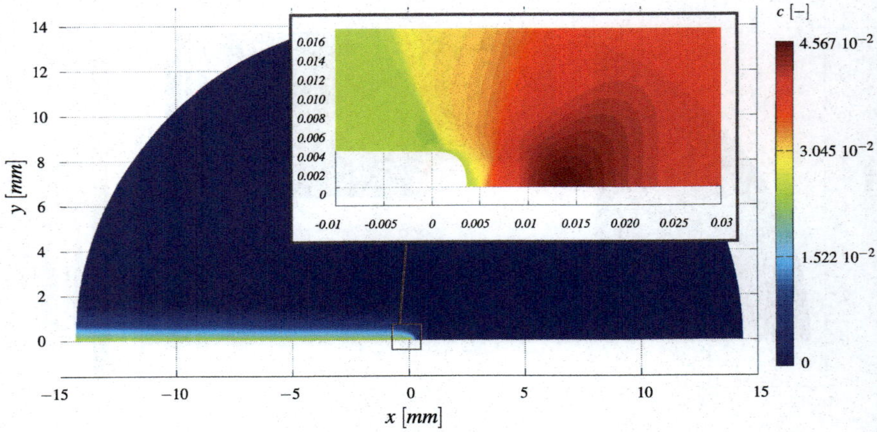

Fig. 9 The total hydrogen concentration $c = (C_L + C_T)/N_L$ with a stress intensity factor $K_I = 19\,\mathrm{MPa}\sqrt{\mathrm{m}}$ and the material dependent softening parameter $\xi = 0$ at the time t = 10s

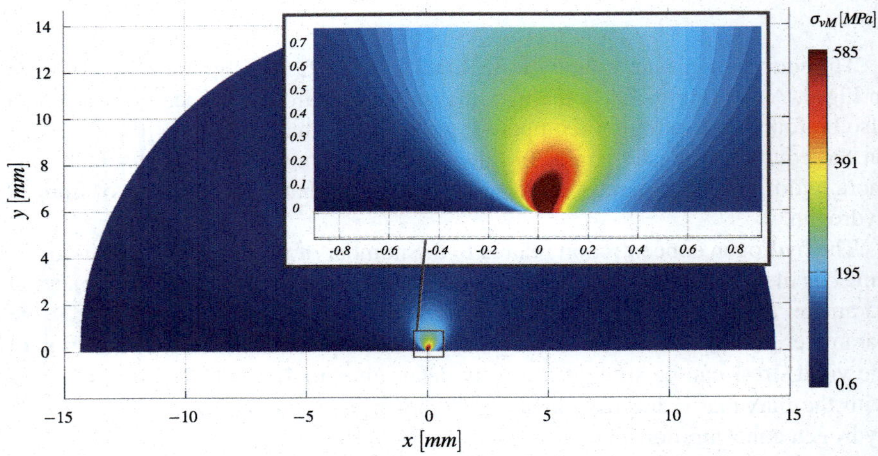

Fig. 10 Von Mises stress σ_{vM} with a stress intensity factor $K_I = 19\,\mathrm{MPa}\sqrt{\mathrm{m}}$ at the time t = 10s

hydrogen concentration $C_{L,0} + C_{T,0}$ of about $2.62 \times 10^{22}\mathrm{atoms/m^3}$. By applying this boundary condition on the investigated model the embrittlement effect can be described as negligible (compare Fig. 4).

Since the link of the internal hydrogen gas pressure and the applied mechanical boundary conditions of the surrogate model are not straightforward, the numerical effect of hydrogen needs to be investigated under realistic boundary conditions. It is possible that higher hydrogen gas pressures within the pipeline result in a more pronounced effect on the yield stress.

Therefore, the results of Takayama et al. [14] are used exemplary. The authors investigated a model similar to the presented full-field model depicted in Fig. 2 under varying gas pressure within the pipeline. Thus the mechanical loading can be directly linked to the surface hydrogen concentration. For a hydrogen pressure of 60 MPa within the tube, a total hydrogen concentration $C_L + C_T$ of about 105 C_0 and a surface lattice concentration of about 4.658×10^{22} $H\text{atoms}/\text{m}^3$ results in $c = 4.3 \times 10^{-5}$ with Eq. 12. This reduction of the yield stress of $4.3 \times 10^{-3}\%$ implies that the numerical approximation of the HELP mechanism does not have an effect in the numerical calculation for higher gas pressures.

5 Conclusion

To study the effect of hydrogen on the material of a pipeline a transient hydrogen model considering trapping was coupled with an elasto-plastic material model considering von Mises plasticity. The HELP mechanism was assumed to be active in the investigated pipeline and was numerically realized by defining a hydrogen dependent reduction of the yield stress. A numerical iterative solution scheme was applied to achieve equilibrium for the mechanical and hydrogen properties.

In a first step, a perfect pipeline with hydrogen being ducted through was investigated. The effect on the yield stress was shown to be negligible considering the numerical realization of the HELP mechanism. A pre-damaged pipeline showed to be more susceptible to hydrogen embrittlement by accumulating hydrogen in front of the crack tip, leading to higher hydrogen concentrations than in the case of a perfect pipeline. The accumulation of hydrogen at the maximal hydrostatic stresses as well as the enhancement of local plasticity could be observed for high hydrogen concentrations in front of the blunted crack tip, which is in agreement with the HELP mechanism.

Nevertheless, for surface hydrogen concentrations under realistic boundary conditions considering Sivert's law no pronounced embrittlement could be observed. One reason could be that the applied stress state in the calculation was not high enough to reach a critical value where hydrogen affects the material properties. Thus, the results of the hydrogen distribution calculations of Takayama et al. [14] were discussed exemplarily. However, even when assuming a comparatively high gas pressure of 60 MPa in the pipeline no pronounced effect of hydrogen on the yield stress could be observed.

It is therefore concluded that the numerical realization of the HELP mechanism did not cause a pronounced effect on the material properties for a blunted crack in a pipeline under static mode one opening conditions when applying realistic hydrogen concentrations. This result however does not allow for a judgment whether or not the HELP mechanism in general could be the prevalent mechanism for failure.

In a further study, another model of hydrogen altering the mechanical properties by a pure dilatation straining dependent on the hydrogen concentration could be investigated (compare Dadfarnia et al. [13]). Since Dadfarnia et al. focused on the steady state conditions for the hydrogen concentration and not on how the material properties are influenced by hydrogen, it could provide further insight to investigate the magnitude of this effect in the future.

References

1. Johnson, W.H.: On some remarkable changes produced in iron and steel by the action of hydrogen and acids. In: Proceedings of the Royal Society of London, vol.23, pp. 168–179 (1874)
2. Gerberich, W.: Modeling hydrogen induced damage mechanisms in metals. In: Gangloff, R.P., Somerday, B.P. (eds.) Gaseous Hydrogen Embrittlement of Materials in Energy Technologies, Mechanisms, Modelling and Future Developments, vol. 2, pp. 209–246. Woodhead Publishing Limited (2012)
3. Robertson, I.M., Sofronis, P., Nagao, A., Martin, M.L., Wang, S., Gross, D.W., Nygren, K.E.: Hydrogen embrittlement understood. Metall. Mater. Trans. B **46**(3), 1085–1103 (2015)
4. Zapffe, C.A., Sims, C.E.: Hydrogen embrittlement, internal stress and defects in steel. Trans. Amer. Inst. Mining Met. Engrs. Iron Steel Div. **145** 47 (1941)
5. Haeseldonckx, D., D'haeseleer, W.: The use of the natural-gas infrastructure for hydrogen transport in a changing market structure. Int. J. Hydrogen Energy. **32**, 1381–1386 (2007)
6. Barnoush, A.: Hydrogen embrittlement, revisited by in situ electrochemical nanoindentation. Ph.D. thesis, Universität des Saarlandes (2007)
7. Van Leeuwen, H.P.: The kinetics of hydrogen embrittlement: a quantitative diffusion model. Engineering Fracture Mechanics (1974)
8. Lynch, S.P.: Hydrogen embrittlement phenomena and mechanisms. Corros. Rev. **30**, 105–123 (2012)
9. Robertson, I.M., Teter, D.: Controlled environment transmission electron microscopy. J. Microsc. Res. Tech. **42**, 260 (1998)
10. Teter, D.F., Robertson, I.M., Birnbaum, H.K.: The effects of hydrogen on the deformation and fracture of beta titanium. Acta Mater. **49**, 4313–4323 (2001)
11. Sofronis, P., McMeeking, R.M.: Numerical analysis of hydrogen transport near a blunting crack tip. J. Mech. Phys. Solids **37**(3), 317–350 (1989)
12. Gobbi, G., Colombo, C., Miccoli, S., Vergani, L.: A numerical model to study the hydrogen embrittlement effect. Procedia Engineering. XVII International Colloquium on Mechanical Fatigue of Metals (ICMFM17), vol. 74, pp. 460 – 463 (2014)
13. Dadfarnia, M., Somerday, B.P., Sofronis, P., Robertson, I.M., Stalheim D.: Interaction of hydrogen transport and material elastoplasticity in pipeline steels. J. Pressure Vessel Technol. **131**(4), 041404–1 -041404–13 (2009)
14. Takayama, K., Matsumoto, R., Taketomi, S., Miyazaki, N.: Hydrogen diffusion analyses of a cracked steel pipe under internal pressure. Int. J. Hydrogen Energy **36**(1), 1037–1045. In: 11th International Conference: Hydrogen Materials Science; Chemistry of Carbon Nanomaterials (2011)
15. Liang, Y., Sofronis, P.: Micromechanics and numerical modelling of the hydrogen particle matrix interactions in nickel-base alloys. Modell. Simul. Mater. Sci. Eng. **11**(4), 523 (2003)
16. Oriani, R.A.: The diffusion and trapping of hydrogen in steel. Acta Metall. **18**(1), 147–157 (1970)
17. Krom, A.H.M.: Numerical modelling of hydrogen transport in steel. Ph.D. thesis, TU Delft (1998)
18. Peri, D., de Souza Neto, E.A., Owen, D.R.J.: Computational Methods for Plasticity. Wiley (2008)
19. Schumacher, T.: Vertikale Integration im Erdgasmarkt: eine industrieökonomische Betrachtung. Phd thesis, Universität Trier (2011)

20. Hoover, W.R., Iannucci, J.J., Robinson, S.L., Spingarn, J.R., Stoltz, R.E.: Hydrogen compatibility of structural materials for energy storage and transmission. NASA STI/Recon Technical Report N, 81 (1981)
21. Rice, J.R.: Mathematical analysis in the mechanics of fracture. In: Liebowitz, H. (ed.) Fracture: an Advanced Treatise. Mathematical Fundamentals, vol. 2, pp. 191–311. Academic Press, NY (1968)

Printed by Printforce, the Netherlands